INTEGRALS

I1 $\int u^n\, du = \dfrac{u^{n+1}}{n+1} + C \qquad (n \neq -1)$ p. 134

I2 $\int (u \pm v)dx = \int u\, dx \pm \int v\, dx + C$ p. 135

I3 $\int kf(u)\, du = k \int f(u)\, du + C$ p. 135

I4 $\int \dfrac{du}{u} = \ln|u| + C \qquad (u(x) \neq 0)$ p. 198

I5 $\int \ln u\, du = u \ln u - u + C$ p. 199

I6 $\int e^u\, du = e^u + C$ p. 201

I7 $\int \cos u\, du = \sin u + C$ p. 248

I8 $\int \sin u\, du = -\int \cos u + C$ p. 248

I9 $\int \sec^2 u\, du = \tan u + C$ p. 248

I10 $\int \csc^2 u\, du = -\cot u + C$ p. 248

I11 $\int \sec u \tan u\, du = \sec u + C$ p. 248

I12 $\int \csc u \cot u\, du = -\csc u + C$ p. 248

I13 $\int \dfrac{du}{1+u^2} = \mathrm{Tan}^{-1}\, u + C$ p. 253

I14 $\int \dfrac{du}{\sqrt{1-u^2}} = \mathrm{Sin}^{-1}\, u + C$ p. 253

I15 $\int u\, dv = uv - \int v\, du + C$ *(Integration by parts)* p. 258

THE BRIEF CALCULUS

THE BRIEF
WITH APPLICATIONS

CALCULUS
IN THE SOCIAL SCIENCES

SECOND EDITION

JAMES E. SHOCKLEY
Virginia Polytechnic Institute

HOLT, RINEHART AND WINSTON
New York Chicago San Francisco Atlanta Dallas Montreal Toronto London Sydney

Library of Congress Cataloging in Publication Data

Shockley, James E
The brief calculus, with applications in the social sciences.

1. Calculus. I. Title.
QA303.S5538 1976 515 75-45135
ISBN 0-03-089397-6

Printed in the United States of America

6789 032 123456789

To Steven, David, and Susanne

PREFACE

This text is designed for a one-year course in calculus for students in business and the social sciences. Traditionally, calculus has been the language of science. Now that it has become part of the language of social science, students in these fields need early training in its fundamental concepts. They need to learn to use the basic tools of the calculus (the derivative and the integral) in their own fields and in related courses in probability and statistics. They should be exposed to enough of the theory to be able to understand mathematical derivations based on it, and they need to study applications that are relevant to their later work.

In an attempt to meet these needs I have tried to present a readable treatment of the essential topics of calculus with a good balance between theory, technique, and relevant applications. The theory is needed to help the students to understand mathematical derivations in future courses, the technique to help them master the basic skills, and the applications to show them that mathematical analysis is an incisive tool in the social, as well as the physical sciences.

Changes in the Second Edition Most of the text has been rewritten for the second edition in an attempt to simplify and clarify the concepts. In particular, the proofs of a number of theorems have been simplified and others have been omitted. In addition, the exercises have been completely rewritten and the figures have been redrawn.

The following structural changes have been made in the text: (1) The remedial work on algebra has been placed in the Appendix, allowing the well-prepared student to begin the study of the calculus earlier in the course. (2) The chapter on integration (Chapter 4) has been reorganized so that the antiderivative is introduced before the definite integral. (3) Sections on probability density functions and the normal density function have been added to Chapters 4 and 5. (4) A number of applications to economics, social science, wildlife management, and so on, have been added. As with the first edition, applications have been chosen that amplify and illuminate the development of the calculus, rather than distract the reader's attention from it.

Exercises and Examples The exercises are the most important part of the course. It is while working them that the student masters the subject. For this reason I have included a large number of graded exercises. Most are designed to illuminate the theory and imprint the basic techniques on the mind of the student—others involve elementary applications. A great many exercises are included to allow the instructor to vary his assignment for several years without duplication. The techniques needed to work the exercises are illustrated by worked-out examples, as are all new topics at the time they are introduced. As mentioned above, the exercises have been completely rewritten for the second edition.

Appendix on Algebra Because the backgrounds of the students vary widely, many remedial topics have been included. Most of these are developed in the Appendix on algebra, which contains a much more complete review of the subject than is found in most books of this type. (Essentially, the Appendix is a mini-text on algebra, complete with examples and exercises.) The Appendix is designed for use in three ways: (1) It can be taught as an integral part of the course if the background level of the class is low. (2) Selected topics can be taught in class or assigned to be read as needed. (For example, an instructor may wish to review systems of linear equations before covering Lagrange multipliers in Chapter 8.) (3) It can be used as a text for private study concurrent with the study of the calculus.

Organization Chapter 1 (*Techniques of Graphing*) contains the basic topics of analytic geometry needed to understand elementary geometrical arguments in the remainder of the text.

Chapters 2 (*The Derivative*), 3 (*Applications of the Derivative*), and 4 (*The Integral*) form the heart of the book. In these chapters the principles of the calculus are developed for algebraic functions. They can be used as the basis of a one-semester course in "polynomial calculus."

Chapters 5 (*Exponential and Logarithmic Functions*) and 6 (*Trigonometric Functions*) develop the calculus of transcendental functions. Chapter 5 contains a number of applications of interest to social science students. Chapter 6 is optional. It can be omitted without loss of continuity.

Chapter 7 (*Techniques of Integration*) introduces the student to two of the most basic of the special techniques of integration—integration-by-parts and the use of partial fractions. These topics are followed by a discussion of the use of integral tables and two sections on approximate integration.

Chapter 8 (*Calculus of Higher Dimensions*) opens with an introduction to the analytic geometry of three-dimensional space. This is followed by a study of partial derivatives. Several topics of importance to economics majors are discussed near the end of the chapter, including Lagrange multipliers.

Chapter 9 (*Introduction to Differential Equations*) presents the rudimentary concepts of the theory of differential equations, illustrated mainly with first-order linear differential equations. A number of applications to the social sciences round out the chapter and the book.

Acknowledgements I am indebted to a number of individuals for their assistance with either the first or second edition. Valuable help was given by Professors Klaus Witz, Emory Pace, and B. C. Horne, who offered many excellent suggestions during the development of the book. John Lane and Roderick Coleman wrote the first drafts of the exercises for the first and second editions, respectively. Jane Ross and Dorothy Crane of Holt, Rinehart and Winston worked closely with me on the second edition. A. E. Montague typed the second edition and Evelyn Rubin helped with the preparation of the manuscript for the editors and compositors.

January 1976 JAMES E. SHOCKLEY

CONTENTS

CHAPTER 3

CHAPTER 4

CHAPTER 5

CHAPTER 6

THE TRIGONOMETRIC FUNCTIONS 221

CHAPTER 7

TECHNIQUES OF INTEGRATION 255

CHAPTER 8

CALCULUS IN HIGHER DIMENSIONS 275

CHAPTER 9

DIFFERENTIAL EQUATIONS 327

---

* Optional section.

1

TECHNIQUES OF GRAPHING

1.1 INTRODUCTION

Analytic geometry is the study of the relationships between equations and their graphs. In this chapter we develop a few topics from analytic geometry. We make no pretense of a thorough study. Some of the topics are needed for our study of the calculus; others enable us to make accurate graphs of a large number of functions and equations with a minimum of difficulty.

This branch of geometry was invented independently by two of the great seventeenth-century mathematicians: *René Descartes* (1596–1650) and *Pierre de Fermat* (1601–1665). To some extent, the early developers of the subject were guided by practical considerations. The technology of the day demanded a quantitative method of studying curves. Descartes realized that algebraic methods are most efficient for solving certain types of problems. He showed that he could translate a problem involving curves into a new problem involving equations, thereby obtaining specific information that could be used to solve the original problem and related problems.

Consider the following example. It is known that a cannonball fired from a cannon will traverse a parabolic arc (Figure 1.1).[1] If we know the initial velocity of the cannonball and the angle of elevation, α, we can calculate the equation of

[1] The trajectory diverges slightly from a parabola owing to air resistance, wind velocity, and other factors. It is very close to a parabola if the cannonball does not go too high or too far and the wind is not blowing too hard.

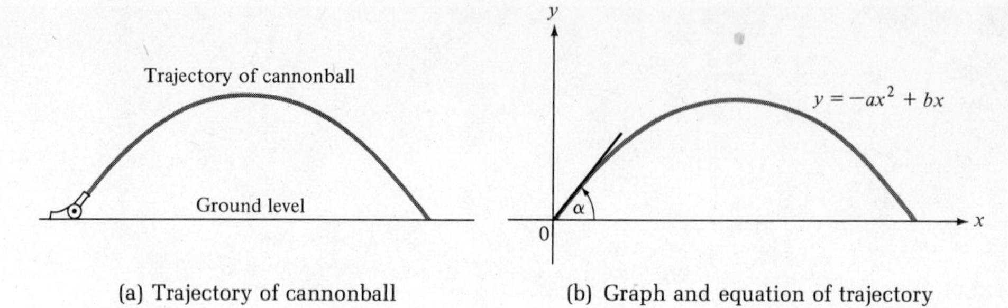

(a) Trajectory of cannonball

(b) Graph and equation of trajectory

FIGURE 1.1

the parabola. This equation is of form

$$y = -ax^2 + bx,$$

where a and b are positive constants that depend on α and the initial velocity.

Once we have derived the general equation of the parabola, we can use algebraic methods to compute the exact angle α needed to make the cannonball strike any point that lies on a possible path. For example, to make the cannonball strike the point A shown on the mountainside in Figure 1.2 we calculate the coordinates of $A(x_0, y_0)$ by surveying methods and substitute these into the equation

$$y = -ax^2 + bx.$$

This gives us the equation

$$y_0 = -ax_0{}^2 + bx_0.$$

We can obtain another equation relating a and b in terms of α and the initial velocity. These two equations can then be solved to determine the exact value of α that will cause the cannonball to strike point A.

In most of our work we will take a different approach. We will start with an equation that relates certain quantities of interest and obtain pertinent information by studying the graph. Consider the following example:

Statisticians of the Zeus Company have discovered that the company's profit is determined by output according to the equation

$$P = -0.1x^2 + 2.4x - 6.3,$$

where the output, x, is measured in hundreds of thousands of units and the

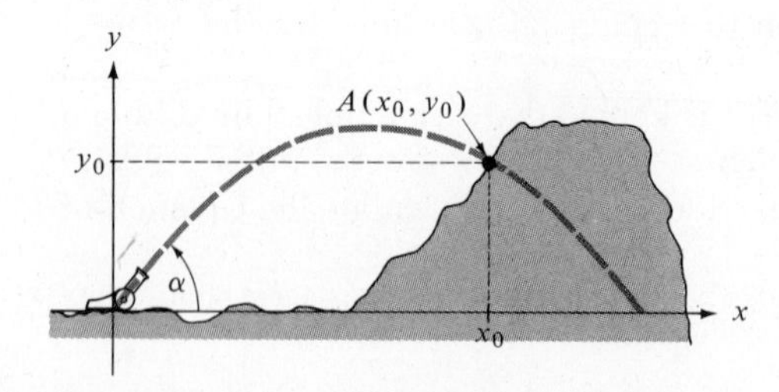

FIGURE 1.2

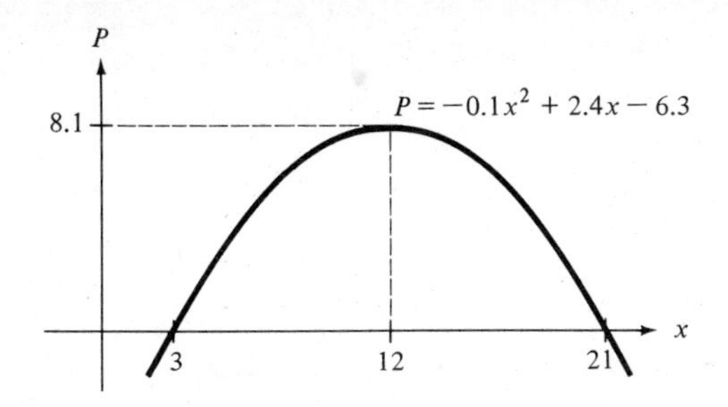

FIGURE 1.3 Profit function for the Zeus Company.

profit is measured in millions of dollars. What level of output yields the maximum profit? At what levels will the company operate at a loss?

To solve these problems we graph the equation. (See Figure 1.3.) Observe that this graph also is a parabola that opens downward. The parabola crosses the x-axis at the points (3, 0) and (21, 0). The highest point on the curve is (12, 8.1). The profit is *positive* if $3 < x < 21$. It is *negative*, indicating a loss, if $0 < x < 3$ or $x > 21$. Thus, the company operates at a loss if it produces less than three hundred thousand or more than 21 hundred thousand units. The maximum profit is 8.1 million dollars, obtained at an output of 12 hundred thousand units.

1.2 THE DISTANCE FORMULA. CIRCLES

Let $|P_1P_2|$ denote the distance between the points $P_1(x_1, y_1)$ and $P_2(x_2, y_2)$ in the coordinate plane. Before considering the general problem of calculating $|P_1P_2|$, we examine some special cases.

Observe that if P_1 and P_2 are both on the x-axis, then $|P_1P_2| = |x_2 - x_1|$. Similar results hold if P_1 and P_2 are two points on a line parallel to either coordinate axis (Figure 1.4). That is:

1. If $P_1(x_1, b)$ and $P_2(x_2, b)$ are on the same horizontal line, then

$$|P_1P_2| = |x_2 - x_1|.$$

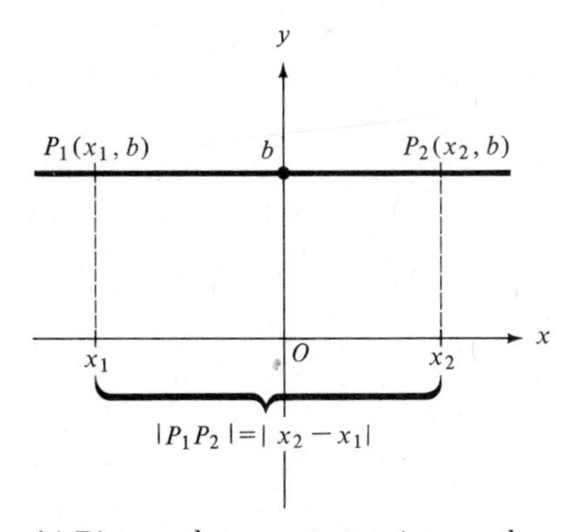

(a) Distance between two points on the same horizontal line

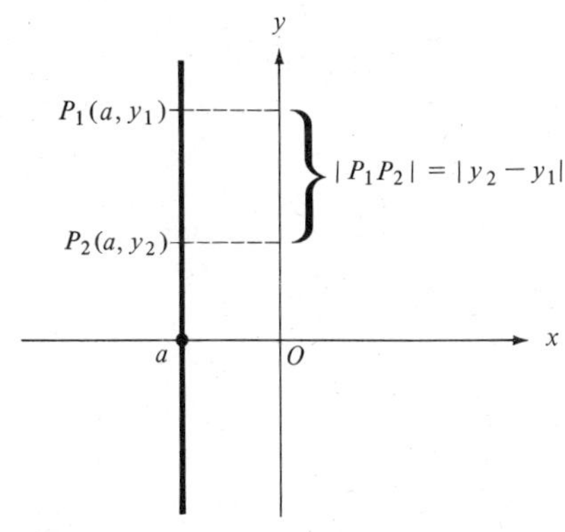

(b) Distance between two points on the same vertical line.

FIGURE 1.4

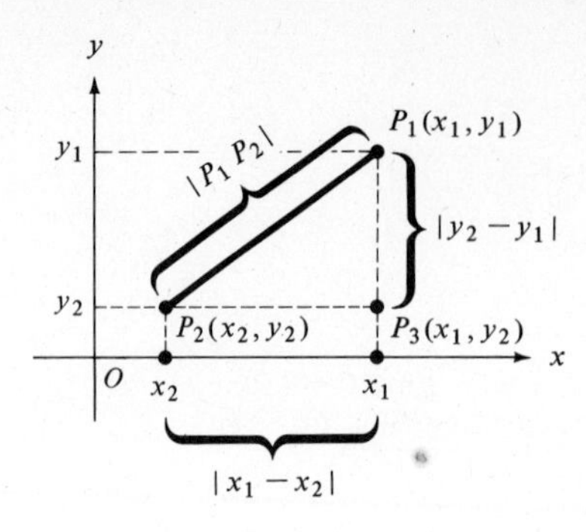

FIGURE 1.5 The distance formula: $|P_1P_2| = \sqrt{(x_2 - x_1)^2 + (y_2 - y_1)^2}$.

2. If $P_1(a, y_1)$ and $P_2(a, y_2)$ are on the same vertical line, then

$$|P_1P_2| = |y_2 - y_1|.$$

We now consider the general distance problem: Determine $|P_1P_2|$ if $P_1(x_1, y_1)$ and $P_2(x_2, y_2)$ are points in the xy-plane.

Let $P_3(x_1, y_2)$ be the point obtained by intersecting the vertical line through P_1 with the horizontal line through P_2 (Figure 1.5). Since P_2 and P_3 are on the same horizontal line, then

$$|P_2P_3| = |x_1 - x_2|.$$

Similarly,

$$|P_1P_3| = |y_2 - y_1|.$$

Because the triangle $P_2P_3P_1$ is a right triangle,[2] we can apply the Pythagorean Theorem.[3] It follows that

$$\begin{aligned} |P_1P_2|^2 &= |P_2P_3|^2 + |P_1P_3|^2 \\ &= |x_1 - x_2|^2 + |y_2 - y_1|^2 \\ &= (x_2 - x_1)^2 + (y_2 - y_1)^2. \end{aligned}$$

Therefore,

The distance formula

$$\boxed{|P_1P_2| = \sqrt{(x_2 - x_1)^2 + (y_2 - y_1)^2}}$$

where the symbol $\sqrt{\ }$ indicates the positive square root.

[2] The careful reader will note that if P_1 and P_2 lie on the same horizontal or vertical line, then we do not actually have a triangle. The formula we are deriving will hold, however, even in these degenerate cases.

[3] The Pythagorean Theorem: If a and b are the lengths of the sides of a right triangle and c is the length of the hypotenuse, then $a^2 + b^2 = c^2$.

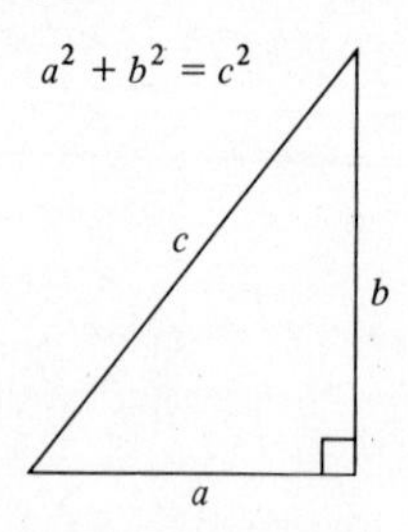

Example 1

(a) The distance between $P_1(2, 3)$ and $P_2(-1, 7)$ is

$$|P_1P_2| = \sqrt{(-1-2)^2 + (7-3)^2} = \sqrt{(-3)^2 + 4^2}$$
$$= \sqrt{9 + 16} = \sqrt{25} = 5.$$

(b) The distance between $A(5, -1)$ and $B(2, 2)$ is

$$|AB| = \sqrt{(2-5)^2 + (2+1)^2} = \sqrt{3^2 + 3^2} = \sqrt{18} = 3\sqrt{2}.$$

The Circle The graph of the equation

$$x^2 + y^2 = r^2 \qquad (r > 0)$$

is a circle of radius r with center at the origin. To show that this is the case we must establish two facts:

1. Every point on the graph of the equation is r units from the origin.
2. Every point that is r units from the origin is on the graph of the equation.

PROOF of 1 Let $P(x, y)$ be a point on the graph. (See Figure 1.6.) Then

$$x^2 + y^2 = r^2.$$

If we take the positive square root of each side, we have

$$\sqrt{x^2 + y^2} = r,$$
$$\sqrt{(x-0)^2 + (y-0)^2} = r,$$
$$|PO| = r.$$

Thus, the distance from P to O is r units.

PROOF of 2 Let $Q(X, Y)$ be a point on the circle. Then the distance from Q to O is equal to r. Therefore

$$|QO| = r,$$
$$\sqrt{(X-0)^2 + (Y-0)^2} = r,$$
$$\sqrt{X^2 + Y^2} = r,$$
$$X^2 + Y^2 = r^2.$$

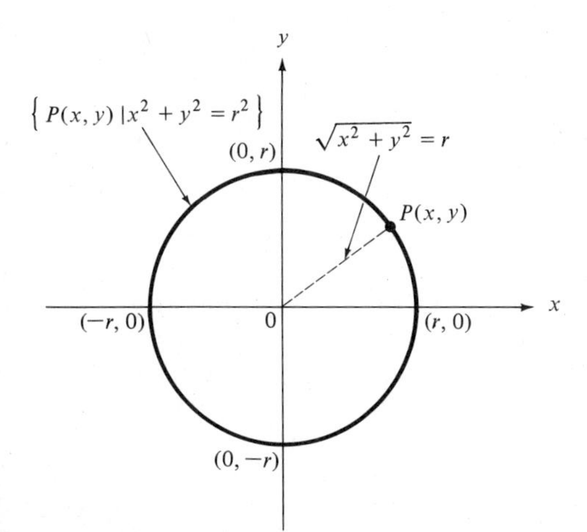

FIGURE 1.6 The graph of

$$x^2 + y^2 = r^2$$

is the circle with center at the origin and radius r.

Therefore, the coordinates of Q satisfy the equation

$$x^2 + y^2 = r^2,$$

and so Q is on the graph. It follows that every point on the graph is on the circle and every point on the circle is on the graph. Consequently, the graph of $x^2 + y^2 = r^2$ is the circle.

Essentially the same argument establishes the following theorem. The proof is omitted.

Equation of circle

Theorem 1.1 The graph of the equation

$$(x - a)^2 + (y - b)^2 = r^2 \qquad (r > 0)$$

is the circle with radius r and center at (a, b).

Example 2 Sketch the graph of

$$x^2 - 2x + 1 + y^2 + 4y + 4 = 9.$$

Solution By inspection we see that the equation can be rewritten as

$$(x - 1)^2 + (y + 2)^2 = 3^2$$
$$(x - 1)^2 + (y - (-2))^2 = 3^2.$$

The graph is the circle with radius 3 and center (1, −2) (Figure 1.7).

The following example shows that certain equations that appear to be of the form

$$(x - a)^2 + (y - b)^2 = r^2$$

may not have circles for graphs because the number r either is zero or is not real.

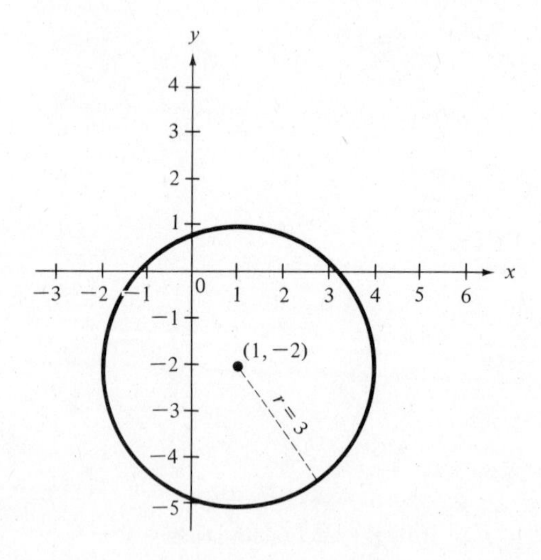

FIGURE 1.7 Example 2. The graph of $x^2 - 2x + 1 + y^2 + 4y + 4 = 9$ is the circle with center (1, −2) and radius 3. (The equation reduces to

$$(x - 1)^2 + (y - [-2])^2 = 3^2.)$$

Example 3 Describe the graphs of the equations

(a) $$(x - 2)^2 + (y - 3)^2 = 0,$$
(b) $$(x - 2)^2 + (y - 3)^2 = -1.$$

Solution

(a) If x and y are any fixed real numbers, then

$$(x - 2)^2 \geqslant 0 \quad \text{and} \quad (y - 3)^2 \geqslant 0$$

with the equality sign holding only if $x = 2$ and $y = 3$. Therefore, if

$$(x - 2)^2 + (y - 3)^2 = 0,$$

we must have $x = 2$, $y = 3$. Consequently, the graph of $(x - 2)^2 + (y - 3)^2 = 0$ consists of the single point $(2, 3)$.

(b) If x and y are any fixed real numbers, then

$$(x - 2)^2 + (y - 3)^2 \geqslant 0 > -1.$$

Thus, no real numbers x, y satisfy the equation

$$(x - 2)^2 + (y - 3)^2 = -1.$$

The graph is the empty set $\varnothing$.

Any equation of form

$$x^2 + y^2 + Ax + By + C = 0$$

can be rewritten in the form

$$(x - a)^2 + (y - b)^2 = R$$

by "completing the square" on the terms involving x and on the terms involving y.

1. If $R > 0$, the graph of the equation is the circle with center (a, b) and radius $r = \sqrt{R}$.
2. If $R = 0$, the graph is the single point (a, b).
3. If $R < 0$, there is no graph.

Since, in the only case of interest, the graph of the equation is a circle, the equation

General equation of circle

$$\boxed{x^2 + y^2 + Ax + By + C = 0}$$

is called the *general equation of a circle.*

The following example shows how the general equation of a circle can be reduced to the form

$$(x - a)^2 + (y - b)^2 = R$$

by completing the square on the x- and on the y-terms. (Recall that to complete the square on

$$x^2 + Ax$$

we add $(A/2)^2$ to the expression.)

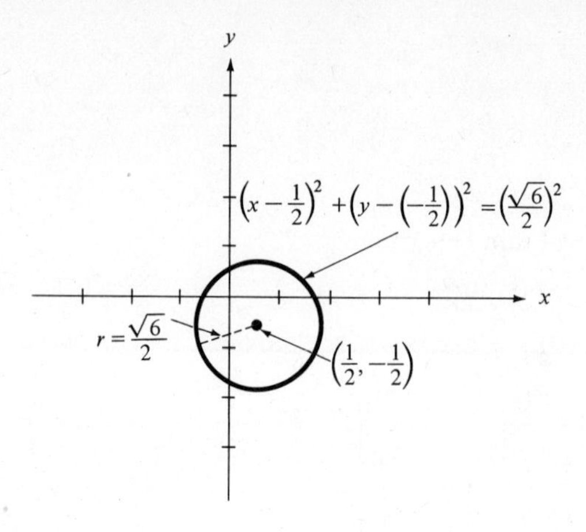

FIGURE 1.8 Example 4. The graph of $x^2 - x + y^2 + y = 1$.

Example 4 Show that the graph of

$$x^2 - x + y^2 + y = 1$$

is a circle. Determine the center and the radius.

Solution We complete the square on the terms involving x and on the terms involving y separately.

$$x^2 - x + y^2 + y = 1,$$

$$(x^2 - x \qquad\qquad) + (y^2 + y \qquad\qquad) = 1,$$

$$\left(x^2 - x + \left(\frac{1}{2}\right)^2\right) + \left(y^2 + y + \left(\frac{1}{2}\right)^2\right) = 1 + \left(\frac{1}{2}\right)^2 + \left(\frac{1}{2}\right)^2,$$

$$\left(x - \frac{1}{2}\right)^2 + \left(y + \frac{1}{2}\right)^2 = 1 + \frac{1}{4} + \frac{1}{4} = \frac{6}{4},$$

$$\left(x - \frac{1}{2}\right)^2 + \left(y - \left(-\frac{1}{2}\right)\right)^2 = \left(\frac{\sqrt{6}}{2}\right)^2.$$

On comparing this equation with Theorem 1.1, we see that the graph is the circle with radius $\sqrt{6}/2$ and center $(\frac{1}{2}, -\frac{1}{2})$ (Figure 1.8).

EXERCISES 1.2

1. Calculate the distances between the following pairs of points:

   (a) (7, 3), (−2, 3)
   • (b) (−2, −3), (1, 1)
   (c) (0, 4), (2, 1)
   • (d) (−1, 6), (2, 2)
   (e) (−2, −1), (1, 2)
   (f) (4, 2), (4, −8)

2. The following points are vertices of a triangle. Which are isosceles triangles? Which are right triangles?

   (a) (2, 3), (1, 4), (1, 2)
   • (b) (−3, −1), (1, 2), (−6, 3)
   (c) (−1, −1), (1, 5), (3, −3)

3. It can be proved that the midpoint of the line segment with endpoints (x_1, y_1)

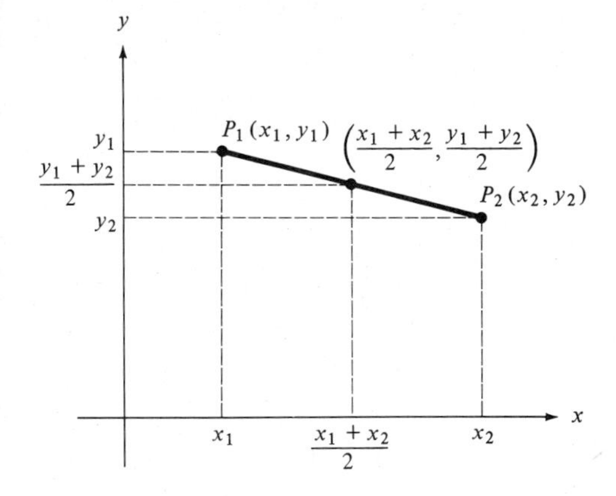

FIGURE 1.9 Exercise 3. The point $\left(\frac{x_1+x_2}{2}, \frac{y_1+y_2}{2}\right)$ is the midpoint of the line segment connecting $P_1(x_1, y_1)$ and $P_2(x_2, y_2)$.

The midpoint formula

and (x_2, y_2) is $((x_1+x_2)/2, (y_1+y_2)/2)$. (See Figure 1.9.) Thus, the x-coordinate of the midpoint is the midpoint of the x-coordinates and the y-coordinate of the midpoint is the midpoint of the y-coordinates. Use this result to find the midpoint of the line segment with the given endpoints.

(a) $(2, 1), (6, 3)$
• (b) $(4, 0), (0, -1)$
• (c) $(2, 4), (-4, -2)$
(d) $(2, 2), (-3, -5)$

4. Write the equation in x and y whose graph is a circle with the given point as center and the given number as radius.

(a) $(0, 0)$, 2
• (b) $(1, -1)$, 2
(c) $(2, -1)$, 2
(d) $(1, 1)$, 3
(e) $(-1, 1)$, 2.
• (f) $(3, -4)$, 4
(g) $(-2, -3)$, 3
• (h) $(4, -1)$, 2

5. Sketch the graphs of the following equations. If the graph is the empty set, $\varnothing$, state this.

• (a) $x^2 + y^2 = 9$
(b) $x^2 + y^2 = 36$
(c) $(x+1)^2 + (y+2)^2 = 1$
(d) $(x-2)^2 + (y-1)^2 = 0$
• (e) $x^2 + (y+2)^2 = 0$
(f) $x^2 + y^2 = -2$
(g) $x^2 + (y-3)^2 = 4$
• (h) $(x-3)^2 + (y+3)^2 - 4 = 0$
(i) $x^2 + y^2 = 0$
(j) $(x+1)^2 + y^2 = 2$
• (k) $(x+1)^2 + (y-1)^2 = -1$
(l) $(x-3)^2 + (y+1)^2 = 4$

6. Show that each of the following is the equation of a (possibly degenerate) circle and calculate the center and radius. If the graph reduces to a single point or the empty set, $\varnothing$, state that fact.

(a) $x^2 + 2x + y^2 + 2y = 1$
• (b) $x^2 - x + y^2 + y = 0$
(c) $x^2 + y^2 + 4x - 8y + 3 = 0$
• (d) $x^2 - 4x + y^2 - 6y + 15 = 0$
(e) $x^2 + 5x + y^2 - y + 1 = 0$
• (f) $x^2 + x + y^2 - 3y + 2 = 0$
(g) $x^2 - 3x + y^2 + 4y + \frac{25}{4} = 0$
(h) $x^2 - 2x + y^2 - 3y - 2 = 0$

7. Prove that the graph of $(x-a)^2 + (y-b)^2 = r^2$, $r > 0$, is a circle with center at (a, b) and radius equal to r.

•8. Find an equation satisfied by every point $P(x, y)$ that is equidistant from $A(-1, 2)$ and $B(4, 7)$. (*Hint:* First draw a figure.)

1.3 THE LINE $y = mx + b$

Increments Δx and Δy

Let $\mathscr{L}$ be a line on the xy-plane. Let $P_1(x_1, y_1)$ and $P_2(x_2, y_2)$ be two distinct points on $\mathscr{L}$. Denote the quantities $x_2 - x_1$ and $y_2 - y_1$ by Δx and Δy, respectively.[4] These quantities are called *increments*. Note that the increment Δx is the horizontal change as we go from P_1 to P_2 and Δy is the corresponding vertical change. These increments may be positive, negative, or zero.

Assume that the line $\mathscr{L}$ is not vertical. (Vertical lines will be considered separately.) Then $\Delta x \neq 0$ and we can form the ratio $\Delta y/\Delta x$. It can be shown that this ratio is always the same, no matter which distinct points on the line are used to define Δy and Δx. In other words, $\Delta y/\Delta x$ is a number that depends only on the line $\mathscr{L}$ and is completely independent of the points used to define Δx and Δy. The quantities Δx and Δy, on the other hand, depend both on the line $\mathscr{L}$ and on the points P_1 and P_2. (See Figure 1.10.)

Slope

The ratio $\Delta y/\Delta x$ is called the *slope* of the line $\mathscr{L}$ and measures its steepness. It is customary to denote the slope by the letter m. Thus

$$m = \frac{\Delta y}{\Delta x} = \frac{y_2 - y_1}{x_2 - x_1}.$$

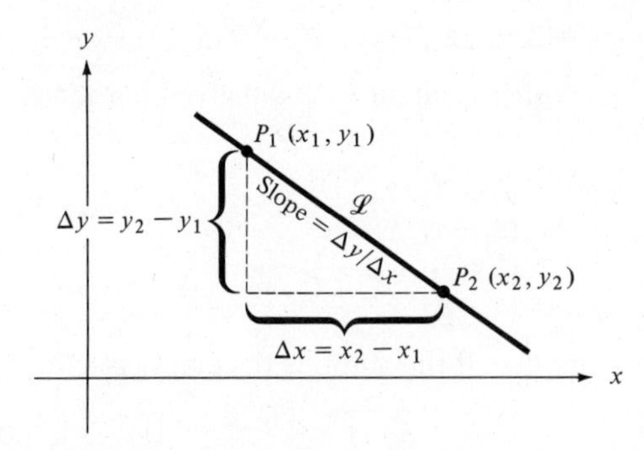

FIGURE 1.10 The slope of $\mathscr{L}$ is $m = \dfrac{\Delta y}{\Delta x} = \dfrac{y_2 - y_1}{x_2 - x_1}$, where

Δx = change in x from P_1 to P_2 $= x_2 - x_1$

Δy = change in y from P_1 to P_2 $= y_2 - y_1$

Example 1 Calculate the slope of the line that contains the points $P_1(1, -2)$ and $P_2(3, 5)$. (See Figure 1.11.)

Solution

$$\Delta x = x_2 - x_1 = 3 - 1 = 2,$$
$$\Delta y = y_2 - y_1 = 5 - (-2) = 7,$$
$$m = \frac{\Delta y}{\Delta x} = \frac{7}{2}.$$

Properties of slope

A line rises to the right if $m > 0$, rises to the left if $m < 0$, and is horizontal if $m = 0$. The greater the magnitude of m, the steeper is the rise. Two nonvertical lines are parallel if and only if they have the same slope. A *vertical line has no slope* in the sense that its slope is undefined. (See Figure 1.12.)

---

[4] The symbol "Δx" stands for a single quantity (read "delta-x") and does not indicate a product.

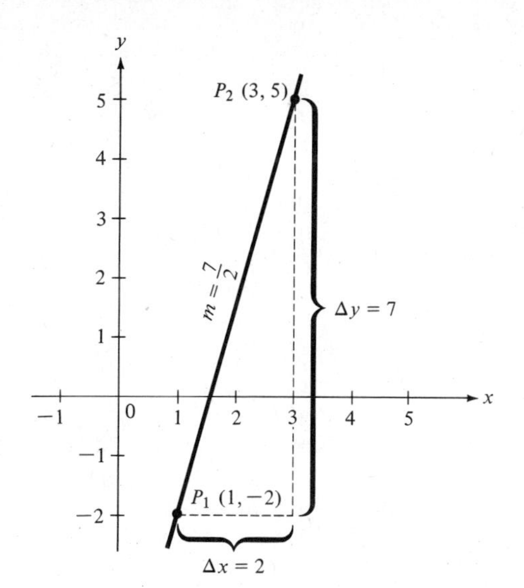

FIGURE 1.11 Example 1. The slope of the line through $P_1(1, -2)$ and $P_2(3, 5)$ is $m = \Delta y/\Delta x = 7/2$.

If we have one point $P_1(x_1, y_1)$ on the line, then we can use the slope $m = \Delta y/\Delta x$ to locate a different point also on the line. Observe that if $P_2(x_2, y_2)$ is on the line and $\Delta x = x_2 - x_1$, then $x_2 = x_1 + \Delta x$ and $y_2 = y_1 + \Delta y$. Thus, the point is $P_2(x_1 + \Delta x, y_1 + \Delta y)$. (See Figure 1.13.)

Example 2 Sketch the line through the point (4, 1):
(a) with slope $m = -3$;
(b) with slope $m = \frac{3}{7}$.

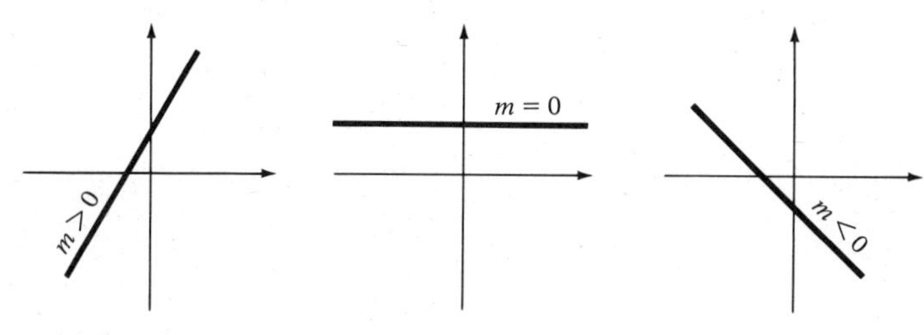

(a) Positive slope (b) Horizontal line, slope = 0 (c) Negative slope

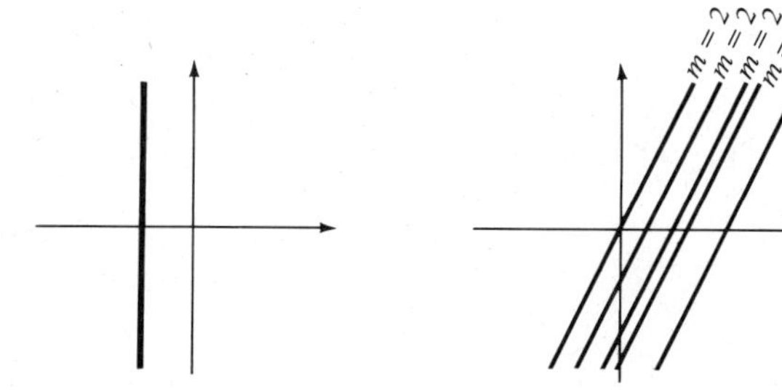

(d) Vertical line, no slope (e) Parallel lines have same slope

FIGURE 1.12

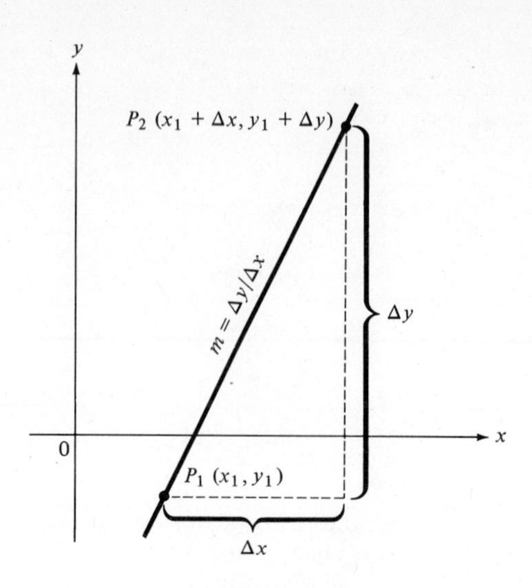

FIGURE 1.13 If $P_1(x_1, y_1)$ is on the line with slope $m = \Delta y/\Delta x$, then the point $P_2(x_1 + \Delta x, y_1 + \Delta y)$ is also on the line.

Solution

(a) One point on the line is (4, 1). The slope is $m = -3 = -3/1$. Let $\Delta x = 1$, $\Delta y = -3$. Then the point $(4 + \Delta x, 1 + \Delta y)$—that is, $(5, -2)$—is also on the line. We draw the line through these points (Figure 1.14a).

(b) Again, the point (4, 1) is on the line. In this case the slope is

$$m = \frac{\Delta y}{\Delta x} = \frac{3}{7}.$$

Let $\Delta x = 7$, $\Delta y = 3$. Since Δx represents the change in x and Δy the change in y, then the point (11, 4) is also on the line. We draw the line through these two points (Figure 1.14b).

Vertical Lines As we mentioned before a vertical line has no slope. It is easy to find the equation of such a line. Suppose it passes through the point x_0 on the

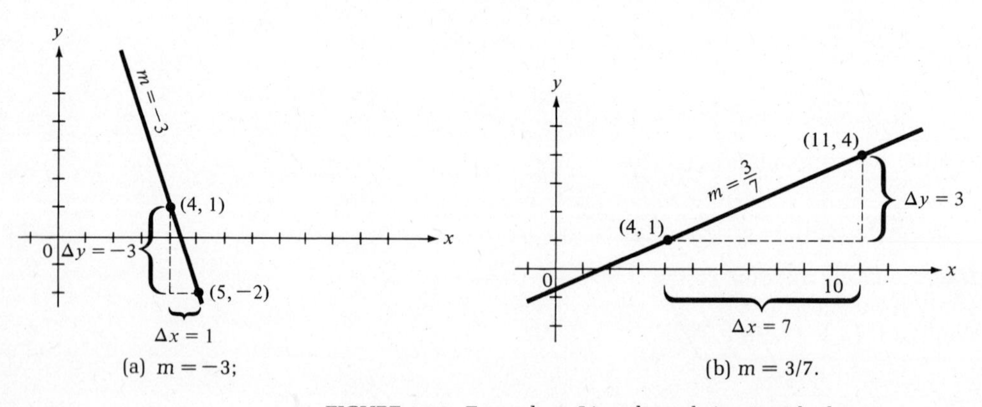

(a) $m = -3$; (b) $m = 3/7$.

FIGURE 1.14 Example 2. Line through (4, 1) with slope m.

x-axis. Then every point on the line has its x-coordinate equal to x_0. Thus, the line is

$$\{P(x, y)|x = x_0\}.$$

The equation of the line (see Figure 1.15) is

$$x = x_0.$$

Equation of vertical line

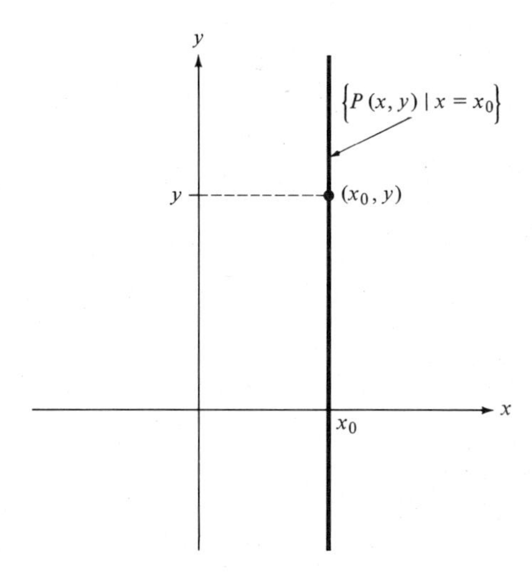

FIGURE 1.15 Every point on the vertical line through the point $(x_0, 0)$ has its x-coordinate equal to x_0. The equation of the line is $x = x_0$.

For example, the graph of the equation

$$x = 3$$

is

$$\{P(x, y)|x = 3\},$$

the vertical line three units to the right of the y-axis.

The Point-Slope Form The following theorem can be used to obtain the equation of a line if we know its slope and one point.

Theorem 1.2 The line through $P_0(x_0, y_0)$ with slope m is the graph of the equation

$$y - y_0 = m(x - x_0).$$

PROOF Let $\mathscr{L}$ be the line through $P_0(x_0, y_0)$ with slope m. We must show that

$$\mathscr{L} = \{(x, y)|y - y_0 = m(x - x_0)\}.$$

Let $P(x, y)$ be a point in the plane different from P_0. (See Figure 1.16.) Then P is on the line $\mathscr{L}$ if and only if the slope of the line containing P and P_0 is equal to m. That is, $P(x, y)$ is on the line $\mathscr{L}$ if and only if

$$\frac{y - y_0}{x - x_0} = m.$$

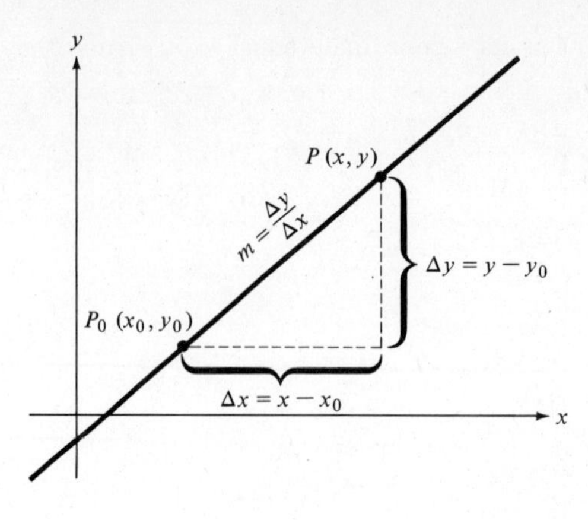

FIGURE 1.16 The point-slope form of the equation of a line:
$y - y_0 = m(x - x_0)$

If we multiply this equation by $x - x_0$, we obtain the equivalent equation

$$y - y_0 = m(x - x_0).$$

Thus, a point different from P_0 is on the line if and only if its coordinates satisfy the equation

$$y - y_0 = m(x - x_0).$$

We can verify by direct substitution that the coordinates of P_0 satisfy this equation also. Consequently, the equation

$$y - y_0 = m(x - x_0)$$

is satisfied by the coordinates of the points on $\mathscr{L}$ and by no other points. Therefore,

$$\mathscr{L} = \{(x, y) | y - y_0 = m(x - x_0)\}.$$

The equation

Point-slope form of equation of line

$$\boxed{y - y_0 = m(x - x_0)}$$

is called the *point-slope* form of the equation of the line.

Example 3 Find the equation of the line through $(-3, 0)$ with slope $m = \frac{3}{2}$. Sketch the line.

Solution We use the point-slope form:

$$\begin{aligned} y - y_0 &= m(x - x_0), \\ y - 0 &= \tfrac{3}{2}(x - (-3)), \\ 2y &= 3x + 9, \\ 3x - 2y &= -9. \end{aligned}$$

To sketch the graph we first plot the point $(-3, 0)$. Since the slope is $\frac{3}{2}$, we make a change of two units along the x-axis and one of three units along

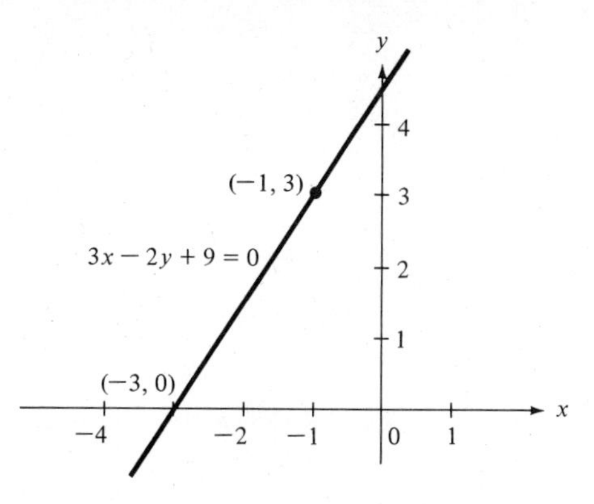

FIGURE 1.17 Example 3. The graph of $3x - 2y + 9 = 0$.

the y-axis, obtaining the point $(-1, 3)$, which must also be on the line. Given the two points, we can easily complete the sketch. (See Figure 1.17.)

The Slope-Intercept Form The point-slope form of the equation of a line is primarily used to find the equation of a line if we know its slope and a point on the line. If, on the other hand, we start with the equation and wish to find the slope an alternate form is more useful.

Consider the special case of the point-slope form where the point is $(0, b)$, the point where the line crosses the y-axis. (See Figure 1.18.) The equation is

$$y - b = m(x - 0),$$

which we can write as

$$\boxed{y = mx + b.}$$

This form of the equation is called the *slope-intercept form*.

Slope-intercept form of equation of line.

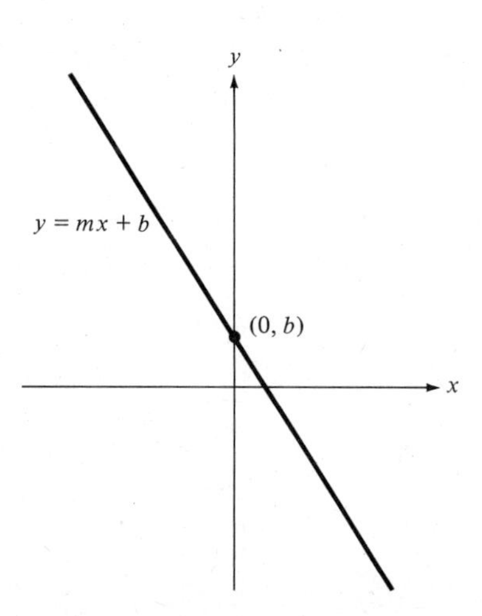

FIGURE 1.18 The slope-intercept form of the equation of a line:

$$y = mx + b$$

Example 4 The equation of a line is

$$3x - 2y = 8.$$

Find the slope and a point on the line.

Solution We solve the equation for y:

$$\begin{aligned} 3x - 2y &= 8, \\ 2y &= 3x - 8, \\ y &= \tfrac{3}{2}x - 4. \end{aligned}$$

On comparison with the slope-intercept form we see that the slope is $m = \frac{3}{2}$. The y-intercept is the point $(0, -4)$.

The General Form Observe that the graph of any equation of form $ax + by = c$, with a and b not both equal to zero, is a straight line. If $b \neq 0$, we reduce it to the slope-intercept form as in Example 4. If $b = 0$, the graph is

$$\left\{(x, y) | x = \frac{c}{a}\right\},$$

which is a vertical line c/a units from the y-axis. The equation

General form of equation of line

$$\boxed{ax + by = c}$$

is called the *general form* of the equation of the line. It is the usual form in which the equations are written.

Example 5 Calculate the equation of the line through $P_1(1, 5)$ and $P_2(-1, 8)$.

Solution *Method I* The slope of the line is

$$m = \frac{y_2 - y_1}{x_2 - x_1} = \frac{8 - 5}{-1 - 1} = -\frac{3}{2}.$$

Since P_1 is on the line, we use the point-slope form and obtain

$$y - 5 = -\frac{3}{2}(x - 1).$$

Simplifying, we obtain the general form

$$3x + 2y = 13.$$

Method II Let the equation be $ax + by = c$ (where a, b, c are to be determined). Since the point $P_1(1, 5)$ is on the line, then

$$a \cdot 1 + b \cdot 5 = c.$$

Similarly, since $P_2(-1, 8)$ is on the line, then

$$-a + 8b = c.$$

Thus, the numbers a, b, c must satisfy the system of equations

$$\begin{cases} a + 5b = c, \\ -a + 8b = c. \end{cases}$$

The solution is $a = 3c/13$, $b = 2c/13$. Thus, the equation of the line is

$$\frac{3c}{13}x + \frac{2c}{13}y = c.$$

Canceling c and multiplying by 13, we obtain

$$3x + 2y = 13.$$

EXERCISES 1.3

1. Calculate the slope of the line through the two given points. Sketch the line.

   (a) $(0, 0)$, $(6, 3)$
   (b) $(2, 1)$, $(2, 3)$
   • (c) $(6, -2)$, $(3, 7)$
   (d) $(5, -2)$, $(1, -3)$
   • (e) $(-4, 1)$, $(4, 2)$
   (f) $(1, -2)$, $(1, -1)$
   • (g) $(1, 3)$, $(-4, -5)$
   (h) $(2, 0)$, $(0, 1)$

2. Calculate the equation of the line with the given slope that passes through the given point. Write the equation in point-slope form and general form. Sketch the line.

   (a) $(4, 1)$, $m = -7$
   • (b) $(2, 4)$, $m = -4$
   (c) $(1, 3)$, $m = 5$
   (d) $(7, 3)$, $m = 0$
   • (e) $(2, 1)$, $m = 2$
   (f) $(3, -2)$, $m = -2$
   (g) $(-5, 2)$, $m = \frac{5}{3}$
   (h) $(-4, -6)$, $m = \frac{1}{2}$

3. Write the general equation of the line through each of the pairs of points.

   • (a) $(1, 1)$, $(0, 6)$
   • (b) $(1, 2)$, $(3, 1)$
   (c) $(1, 3)$, $(-3, 4)$
   (d) $(-3, 5)$, $(-3, 3)$
   (e) $(1, 1)$, $(-3, 2)$
   (f) $(3, 5)$, $(-3, 2)$
   (g) $(4, 3)$, $(2, 1)$
   • (h) $(0, 5)$, $(5, 0)$

4. Write the following equations in slope-intercept form. Find a point on each line, the slope of the line (provided it exists), and sketch the graph.

   (a) $2x - 4y = 13$
   • (b) $4x + 2y = 5$
   (c) $3x - 8y = 9$
   (d) $2x - y = 0$
   • (e) $x + y = 1$
   (f) $7x + 3y = 21$
   (g) $4x + 2y = 3$
   (h) $6x - 3 = 0$
   • (i) $2x + 3y = 0$
   (j) $5x - 3y = 1$

1.4 SYMMETRY. THE PARABOLA $y = ax^2$

A curve $\mathscr{C}$ in the xy-plane is said to be *symmetric about the line* $\mathscr{L}$ if the following condition is met:

Symmetry about the line $\mathscr{L}$

For each point P on $\mathscr{C}$ there is a corresponding point Q such that $\mathscr{L}$ is the perpendicular bisector of the line segment connecting P and Q. (See Figure 1.19.)

Symmetry can be most useful in sketching curves. If a curve $\mathscr{C}$ is symmetric about $\mathscr{L}$, then the part of $\mathscr{C}$ on one side of the line is the mirror image of the part on the other side. If we graph the part on one side of $\mathscr{L}$ we can use the symmetry to graph the part on the other side.

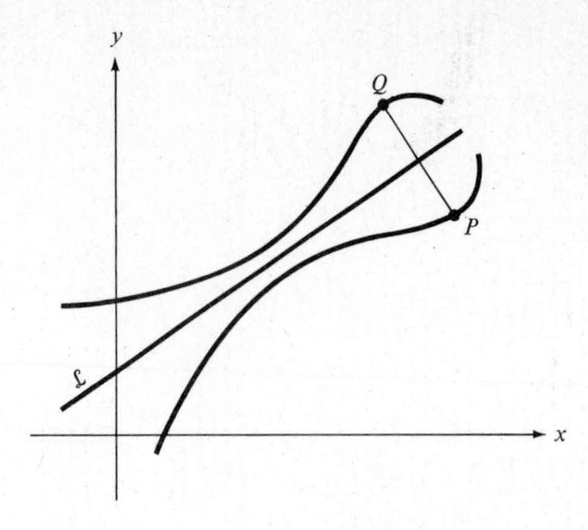

FIGURE 1.19 Symmetry about the line $\mathscr{L}$.

Symmetry about the y-axis

Symmetry about the y-axis Let the curve $\mathscr{C}$ be symmetric about the y-axis. If $P(x, y)$ is a point on $\mathscr{C}$, then the corresponding point $Q(-x, y)$ must also be on the graph. (See Figure 1.20a.) This leads us to a simple test for this symmetry when the curve is defined by an equation. If the graph is symmetric about the y-axis, then $(-x, y)$ must be on it whenever (x, y) is on it. Thus, $(-x, y)$ must satisfy the defining equation whenever (x, y) satisfies it. To test the equation we substitute $-x$ for x. If the new equation is equivalent to the original equation, then the graph is symmetric about the y-axis.

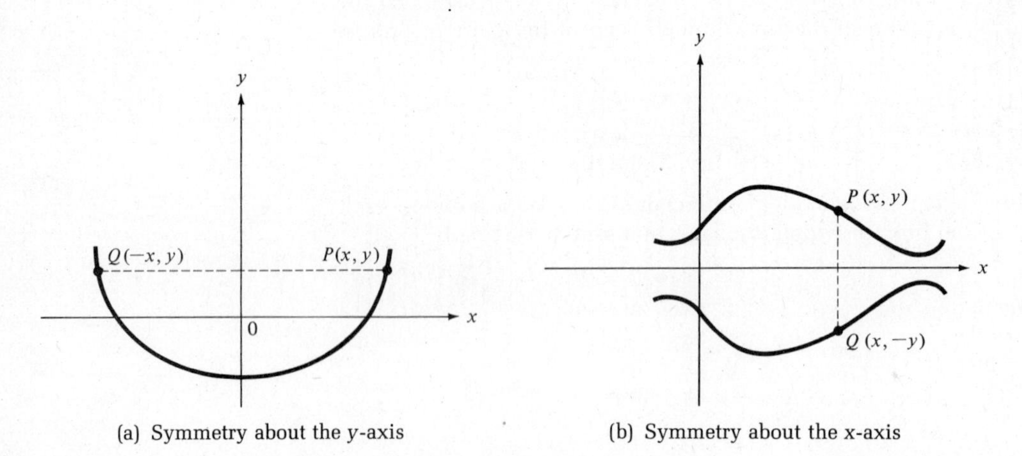

(a) Symmetry about the y-axis

(b) Symmetry about the x-axis

FIGURE 1.20

Example 1 Show that the graph of $y = x^2$ is symmetric about the y-axis.

Solution To test the equation we substitute $-x$ for x. The new equation is

$$y = (-x)^2 \qquad \text{(new equation)}.$$

Since $(-x)^2 = x^2$, then this last equation reduces to

$$y = x^2,$$

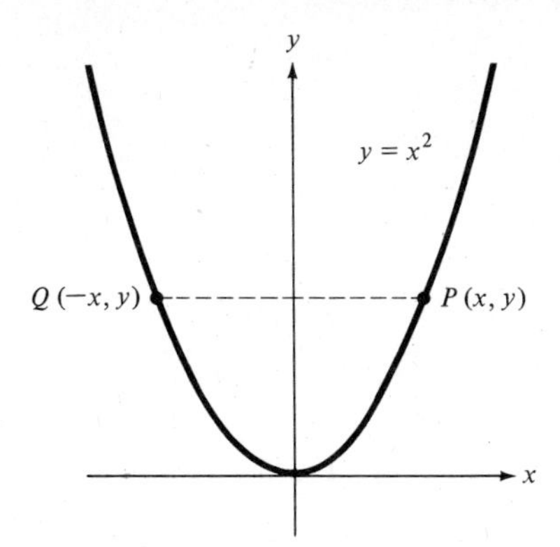

FIGURE 1.21 Example 1. The graph of $y = x^2$ is symmetric about the y-axis.

which is the original equation. Since the new equation reduces to the original equation, the graph is symmetric about the y-axis. (See Figure 1.21.)

Symmetry about the x-axis

Symmetry about the x-axis The graph of an equation is symmetric about the x-axis provided $Q(x, -y)$ is on the graph whenever $P(x, y)$ is on it. (See Figure 1.20b.) To test an equation we substitute $-y$ for y. If the new equation reduces to the original one, then the graph is symmetric about the x-axis.

Example 2 Show that the graph of

$$x^2 + 2y^2 = 4$$

is symmetric about both axes.

Solution *Symmetry about the x-axis* Substitute $-y$ for y. Then

$$x^2 + 2(-y)^2 = 4 \qquad \text{(new equation).}$$
$$x^2 + 2y^2 = 4 \qquad \text{(since } (-y)^2 = y^2\text{).}$$

Since the new equation reduces to the original one, the graph is symmetric about the x-axis.

Symmetry about the y-axis Substitute $-x$ for x. Then

$$(-x)^2 + 2y^2 = 4 \qquad \text{(new equation).}$$
$$x^2 + 2y^2 = 4 \qquad \text{(since } (-x)^2 = x^2\text{).}$$

Since the new equation reduces to the original one, the graph is symmetric about the x-axis.

Symmetry about the origin

Symmetry about the Origin We consider one additional type of symmetry. If the point $Q(-x, -y)$ is on the curve $\mathscr{C}$ whenever $P(x, y)$ is on it, then the curve $\mathscr{C}$ is *symmetric about the origin.* (See Figure 1.22.) To test an equation we substitute $-x$ for x and $-y$ for y. If the new equation reduces to the original one, then the graph is symmetric about the origin.

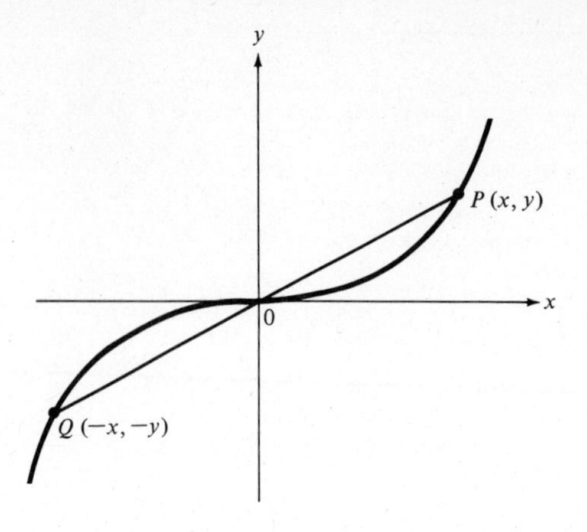

FIGURE 1.22 Symmetry about the origin.

Example 3 Show that the graph of $xy = 1$ is symmetric about the origin.

Solution We substitute $-x$ for x and $-y$ for y obtaining

$$(-x)(-y) = 1 \qquad \text{(new equation)},$$

which reduces to

$$xy = 1 \qquad \text{(original equation)}.$$

The graph is symmetric about the origin. (See Figure 1.23.)

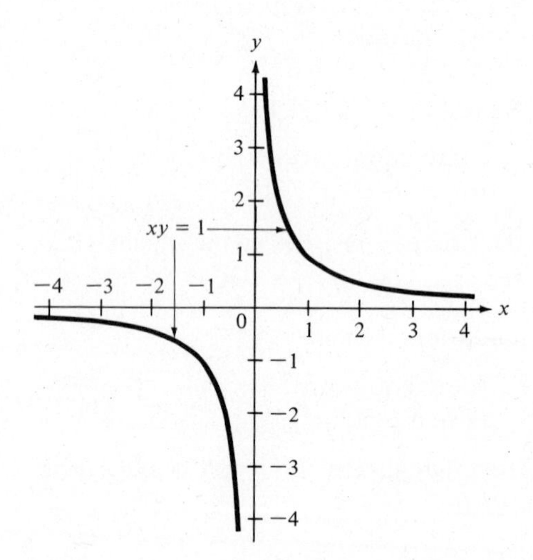

FIGURE 1.23 Example 3. The graph of $xy = 1$ is symmetric about the origin.

Remark If a curve is symmetric about any two of the following, then it is symmetric about all three: the x-axis, the y-axis, and the origin.

The three tests for symmetry are summarized below:

Tests for symmetry

Type of Symmetry	*Test*
x-axis	Substitute $-y$ for y
y-axis	Substitute $-x$ for x
Origin	Substitute $-x$ for x and $-y$ for y

The parabola $y = ax^2$

The Parabola $y = ax^2$ The graph of $y = x^2$, shown in Figure 1.21, is an example of a *parabola*. In general, the graph of $y = ax^2$ $(a \neq 0)$ is a parabola. It is a cup-shaped curve that passes through the origin and is symmetric about the y-axis. If $a > 0$, the parabola opens upward. If $a < 0$, it opens downward. (See Figure 1.24.)

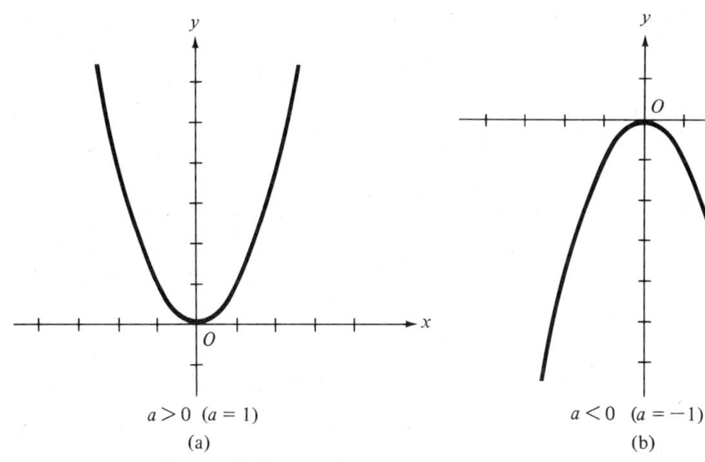

FIGURE 1.24 The parabola $y = ax^2$.

The point where a parabola crosses its line of symmetry is known as its *vertex*. This is the point where the parabola has its maximum curvature. Observe that the parabola $y = ax^2$ has its vertex at the origin.

Several parabolas are shown in Figure 1.25. Observe that the larger the numerical value of a the "steeper" the sides of the "cup." Thus, the graph of $y = 3x^2$ is "steeper" than the graph of $y = x^2$, which, in turn, is "steeper" than the graph of $y = x^2/4$.

In the next section we show that the graph of $y = ax^2 + bx + c$ $(a \neq 0)$ is identical to the graph of $y = ax^2$ in size and orientation, but is shifted to a new location in the plane. Thus, the graph of $y = ax^2 + bx + c$ is a parabola.

EXERCISES 1.4

1. Examine each equation for symmetry about the x-axis, y-axis, and the origin.

   • (a) $xy^2 = 2$
   (b) $3x^2y + xy^2 - xy = 10$
   • (c) $x^3 + y^3 = 1 - xy$
   (d) $x^4y^2 - xy = -y^2 - x^2$

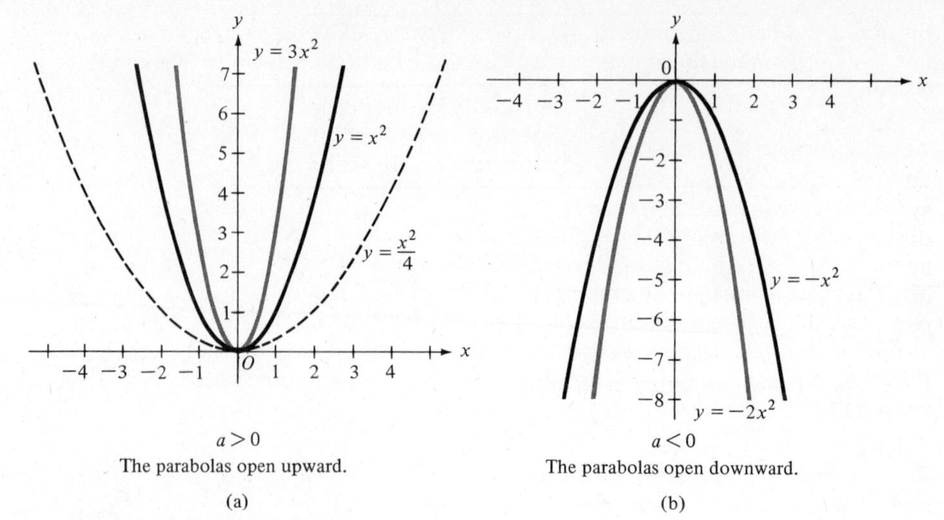

FIGURE 1.25 Parabolas $y = ax^2$.

(e) $xy^3 - x^3y + 6 = 0$
(f) $x^2 + y^2 = 9x^2y^2$
• (g) $9x^2 + 4y^2 = 36$
(h) $xy = -7$
(i) $x^2 - y^2 = 2x^4$
(j) $x^4y + x^2y^3 + x^3y^2 + xy^4 = 1$
• (k) $(3x - y)(3x + y) = 0$
(l) $3x^2 + 2y^2 = x^2y^2 + 6y^6$

2. Sketch the graphs of the following parabolas:

(a) $y = x^2$
(b) $y = 2x^2$
(c) $y = -x^2/4$
(d) $3y + x^2 = 0$

3. Prove: If a curve is symmetric about both of the axes, then it is also symmetric about the origin. (*Hint:* Let $P(x, y)$ be a point on the curve. Explain why $Q(-x, y)$ and $R(-x, -y)$ must also be on the curve.)

4. Sketch a smooth curve with the following properties: (1) the curve is symmetric about the origin and the line $y = x$; (2) in quadrant II the curve increases from $(-1, 0)$ to $(0, 1)$; (3) in quadrant I the curve increases from $(0, 1)$ to $(3, 3)$.

5. Sketch a smooth curve with the following properties: (1) in quadrant I the curve decreases from $(0, 2)$ to $(1, 1)$, increases from $(1, 1)$ to $(3, 4)$, decreases from $(3, 4)$ to $(4, 0)$; (2) the curve is symmetric with respect to the origin.

6. The graph of a certain equation has the following properties: (1) The graph is a smooth curve symmetric about both of the axes; (2) in quadrant I the curve decreases from $(0, 1)$ to $(3, 0)$; (3) the area of the figure bounded by the portion of the curve in quadrant I and the coordinate axes is $\frac{5}{2}$ square units. What is the total area enclosed by the curve?

1.5 TRANSLATION. THE PARABOLA $y = ax^2 + bx + c$

The equation of a curve depends as much on its location in the xy-plane as on its shape. There is a simple technique to minimize the importance of the location of the curve and emphasize the properties of its shape. We construct a set of auxiliary axes and reduce the equation to a standard form with respect to the new axis system.

Auxiliary Axes Let (a, b) be a point in the plane. Construct a new axis system, the $x'y'$-system, with its origin at (a, b) and with the same orientation as the xy-system (as in Figure 1.26a). Each point P in the plane now has two sets of coordinates. Relative to the xy-system it has coordinates $P(x, y)$. Relative to the $x'y'$-system it has coordinates $P[x', y']$. (In this section we use brackets to indicate coordinates in the new system.) For example, the point (a, b) in the original system has coordinates $[0, 0]$ in the new system.

Coordinates in the $x'y'$-system

As we see in Figure 1.26b there is a simple relationship between the coordinates in the two systems. In every case we have

Equations of translation

$$x' = x - a, \qquad y' = y - b.$$

These equations can be used to convert equations in x and y into equations in x' and y' which have the same graph. The process is known as a *translation of axis.*

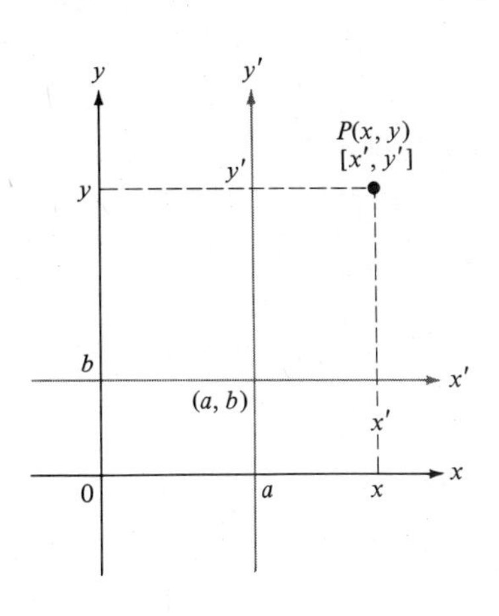

(a) The $x'y'$–system

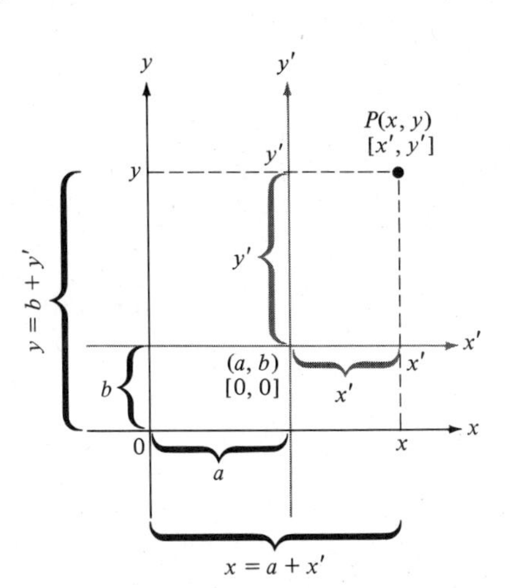

(b) The $x'y'$–system $\begin{cases} x' = x - a \\ y' = y - b \end{cases}$

FIGURE 1.26 Auxiliary axes.

Example 1 By a translation of axis change the equation

$$2(x - 1) + 3(y + 2) = 0$$

into the equation of a line that passes through the new origin.

Solution If we let

$$x' = x - 1, \qquad y' = y + 2,$$

then the new equation is

$$2x' + 3y' = 0,$$

the equation of a line through the origin of the $x'y'$-system.

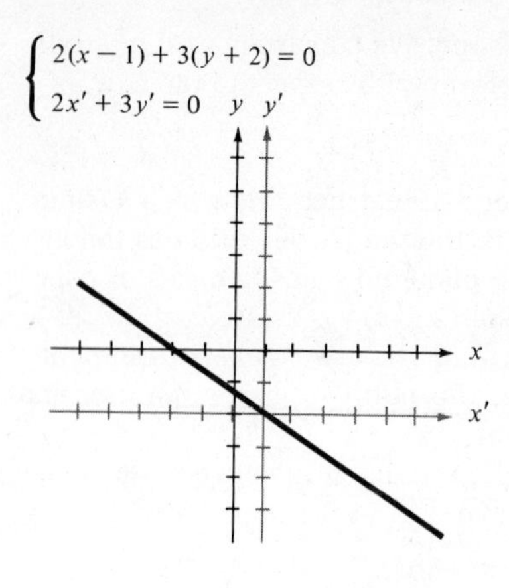

FIGURE 1.27 Example 1. Graph of $2(x-1)+3(y+2)=0$.

The equations of the translation are

$$x' = x - a = x - 1$$

and

$$y' = y - b = y - (-2)$$

so that $a = 1$, $b = -2$. The new origin is the point $(1, -2)$ in the xy-system. (See Figure 1.27.)

Example 2 Simplify the equation

$$x^2 + 6x + y^2 - 10y + 30 = 0$$

by a translation of axis. Sketch the graph.

Solution We complete the square on the terms involving x and on the terms involving y. (See Section A.14 of the Appendix on Algebra for the method of completing the square.)

$$\begin{aligned} x^2 + 6x + y^2 - 10y + 30 &= 0 \\ (x^2 + 6x \qquad) + (y^2 - 10y \qquad) &= -30 \\ (x^2 + 6x + 9) + (y^2 - 10y + 25) &= -30 + 9 + 25 = 4 \\ (x+3)^2 + (y-5)^2 &= 2^2. \end{aligned}$$

$$\begin{cases} x^2 + 6x + y^2 - 10y + 30 = 0 \\ x'^2 + y'^2 = 2^2 \end{cases}$$

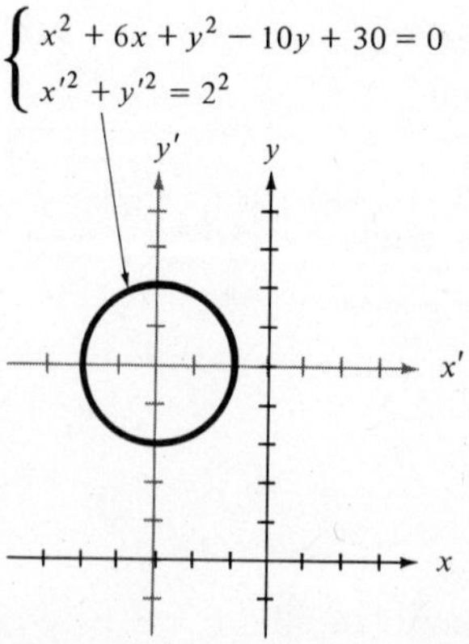

FIGURE 1.28 Example 2. Graph of $x^2 + 6x + y^2 - 10y + 30 = 0$.

Let $x' = x + 3$, $y' = y - 5$. The equation becomes

$$x'^2 + y'^2 = 2^2.$$

The graph is a circle of radius 2 with center at the origin of the $x'y'$-system (the point $(-3, 5)$ of the xy-system). The graph is shown in Figure 1.28.

The parabola $y = ax^2 + bx + c$

The Parabola $y = ax^2 + bx + c$ As we saw in Section 1.4 the graph of $y = ax^2$ is a parabola with vertex at the origin. The y-axis is a line of symmetry. If $a > 0$, it opens upward, if $a < 0$, it opens downward. The larger the value of a the "steeper" the sides of the parabola.

We will show that the graph of

$$y = ax^2 + bx + c$$

is a translate of the graph of $y = ax^2$. Thus, it is a parabola identical to the parabola $y = ax^2$, but is shifted in location.

We rewrite the equation

$$y = ax^2 + bx + c$$

and complete the square:

$$y - c = a\left(x^2 + \frac{b}{a}x \qquad\right),$$

$$y - c + a \cdot \frac{b^2}{4a^2} = a\left(x^2 + \frac{b}{a}x + \frac{b^2}{4a^2}\right) \qquad \left(\text{adding } a \cdot \frac{b^2}{4a^2} \text{ to both sides}\right),$$

$$y - \frac{4ac - b^2}{4a} = a\left(x + \frac{b}{2a}\right)^2,$$

$$y' = ax'^2,$$

where

$$x' = x + \frac{b}{2a}, \qquad y' = y - \frac{4ac - b^2}{4a}.$$

The graph is a parabola identical in size, shape, and orientation to the graph of $y = ax^2$, with vertex at the point $(-b/2a, (4ac - b^2)/4a)$ (Figure 1.29).

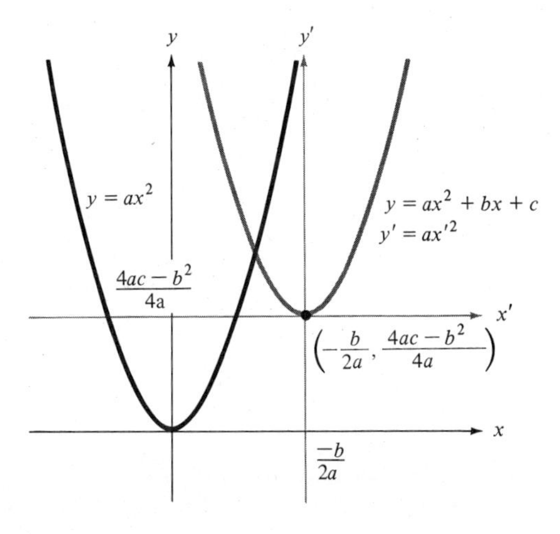

FIGURE 1.29 The graph of $y = ax^2 + bx + c$ is a translate of the graph of $y = ax^2$.

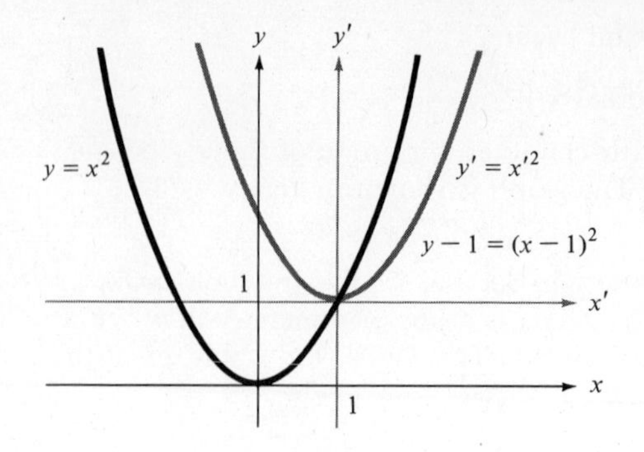

FIGURE 1.30 The graph of
$$y - 1 = x^2 - 2x + 1$$
is a translate of the graph of $y = x^2$.

The parabola opens upward if $a > 0$, downward if $a < 0$. The vertical line through the vertex is the line of symmetry.

Example 3 Simplify the equation

$$y = x^2 - 2x + 2$$

by a translation of axis. Sketch the curve.

Solution We rewrite the equation as

$$y - 2 = x^2 - 2x$$

and complete the square on the right-hand side:

$$\begin{aligned} y - 2 + 1 &= x^2 - 2x + 1 \\ y - 1 &= (x - 1)^2 \\ y' &= x'^2 \qquad \text{(where } x' = x - 1,\ y' = y - 1\text{)} \end{aligned}$$

The graph is a parabola that opens upward. The vertex is at the origin of the $x'y'$-system (the point (1, 1) of the xy-system). The parabola is shown in Figure 1.30.

EXERCISES 1.5

1. Simplify the following equations by a translation of axes. Show that all of the graphs are parabolas. Sketch the graphs, showing both sets of axes.

   • (a) $y = x^2 + 4x + 4$
   (b) $y = 2x^2 - 4x + 2$
   • (c) $y = x^2 - 2x - 3$
   (d) $y = 2x^2 - 4x - 3$
   • (e) $y = 6x^2 + 12x - 7$
   (f) $y = -3x^2 - 6x + 24$

2. Simplify the equations by a translation of axes. Identify and sketch the graphs, showing both sets of axes.

   (a) $2(x - 1) = 3(y + 2)$
   • (b) $5(x + 3) + (y + 1) = 0$
   (c) $(x - 2)^2 + (y + 3)^2 = 9$
   (d) $x^2 + y^2 + 6x - 4y + 13 = 0$
   • (e) $4x^2 + 4y^2 - 8x + 16y + 11 = 0$
   (f) $2(x - 3)^2 + y + 2 = 0$
   (g) $(x + 3)^2 = 4(y - 3)$
   • (h) $x^2 + 2x + y^2 - 4y + 7 = 0$

1.6 THE HYPERBOLA $xy = c$. THE ELLIPSE $\frac{x^2}{a^2} + \frac{y^2}{b^2} = 1$

In Section 1.4 we showed that the graph of $xy = 1$ is symmetric with respect to the origin (Example 3). It is easy to show that the graph is not symmetric about either axis. If we substitute $-x$ for x the new equation is $-xy = 1$, which is not equivalent to the original equation. Thus, the graph is not symmetric with respect to the y-axis. A similar argument shows that it is not symmetric with respect to the x-axis.

To study the graph we write the equation in the form

$$y = \frac{1}{x}.$$

If $0 < x_1 < x_2$, then

$$y_1 = \frac{1}{x_1} > \frac{1}{x_2} = y_2 > 0.$$

Therefore, as we choose larger values of x, the points on the graph decrease towards the x-axis. Obviously we can make (x, y) as close to the x-axis as we wish by choosing x large enough. We calculate a few points for reference and connect them with a smooth *decreasing* curve.

$$\begin{array}{ll}
x = \frac{1}{4},\ y = 4: & (\frac{1}{4}, 4). \\
x = \frac{1}{3},\ y = 3: & (\frac{1}{3}, 3). \\
x = \frac{1}{2},\ y = 2: & (\frac{1}{2}, 2). \\
x = 1,\ y = 1: & (1, 1). \\
x = 2,\ y = \frac{1}{2}: & (2, \frac{1}{2}). \\
x = 3,\ y = \frac{1}{3}: & (3, \frac{1}{3}). \\
x = 4,\ y = \frac{1}{4}: & (4, \frac{1}{4}).
\end{array}$$

This curve is the portion of the graph in quadrant I. We now use the fact that the graph is symmetric about the origin and sketch the portion in quadrant III. There is no portion in quadrant II or quadrant IV. (See Figure 1.31.)

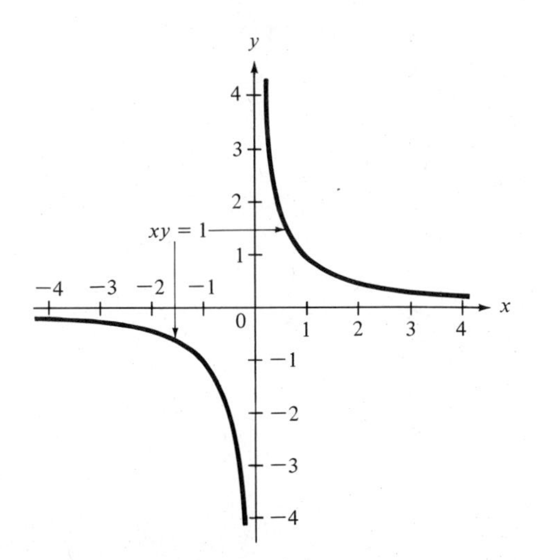

FIGURE 1.31 The hyperbola $xy = 1$.

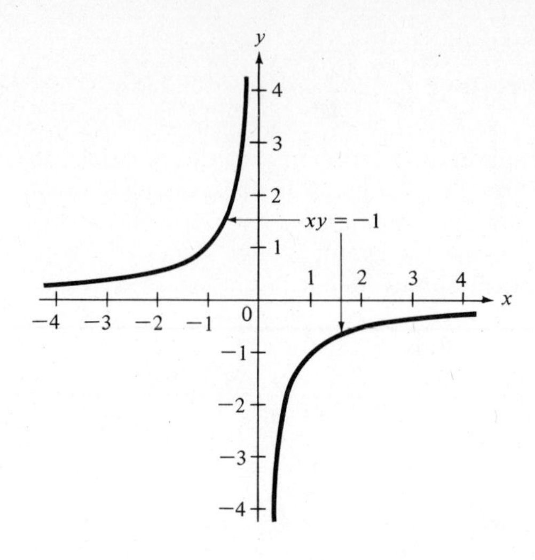

FIGURE 1.32 The hyperbola $xy = -1$.

Asymptotes

The coordinate axes are known as *asymptotes* to the graph. The points on the graph get closer and closer to these asymptotes as we go out along the branches of the graph.

The graph of

$$xy = -1$$

is similar in form to the graph of $xy = 1$. It is located in quadrants II and IV (Figure 1.32).

The graphs of $xy = 1$ and $xy = -1$ are examples of a *hyperbola*. Some typical hyperbolas are shown in Figure 1.33. A hyperbola has the following properties:

Properties of hyperbolas

1. It consists of two distinct branches.
2. It has two asymptotes which intersect at a point.
3. It has two lines of symmetry. These lines are perpendicular to each other and bisect the angles formed by the asymptotes.

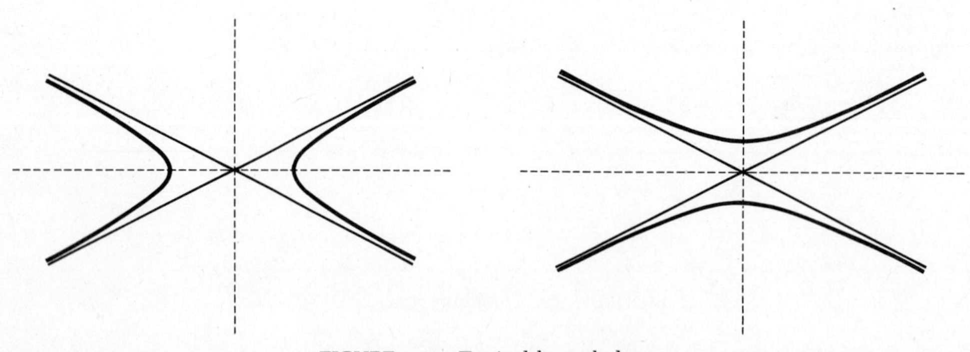

FIGURE 1.33 Typical hyperbolas.

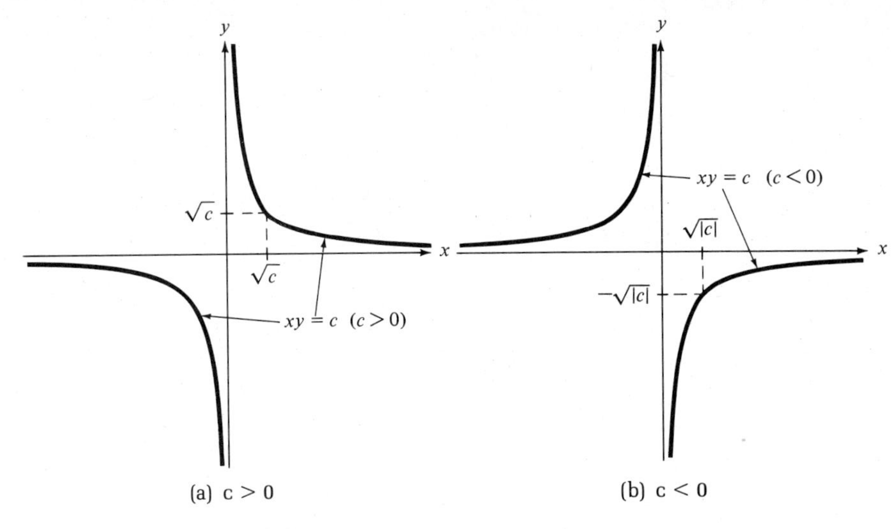

FIGURE 1.34 The hyperbola $xy = c$.

The defining equation for a hyperbola can be rather complicated if the curve is not in some standard position with respect to the origin. We only consider the case where the asymptotes are parallel to the coordinate axes.

The hyperbola

$$xy = c \qquad (c > 0)$$

is similar to the hyperbola $xy = 1$. It consists of two branches, one in quadrant I, the other in quadrant III. It is symmetric about the origin (Figure 1.34a).

The hyperbola

$$xy = c \qquad (c < 0)$$

is similar to the hyperbola $xy = -1$. It consists of two branches, one in quadrant II, the other in quadrant IV (Figure 1.34b).

The equation

$$(x - a)(y - b) = c$$

can be reduced to the form

$$x'y' = c$$

by the substitution $x' = x - a$, $y' = y - b$. Its graph is a translate of the hyperbola $xy = c$.

Example 1 The equation

$$xy - 2x + y = 5$$

can be rewritten as

$$xy - 2x + y - 2 = 3$$

and factored as

$$(x + 1)(y - 2) = 3.$$

If we let $x' = x + 1$, $y' = y - 2$ the equation becomes

$$x'y' = 3,$$

the equation of a hyperbola (Figure 1.35).

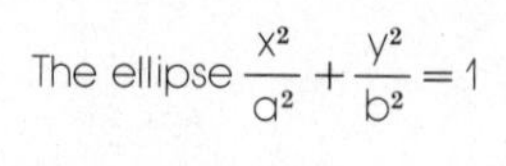

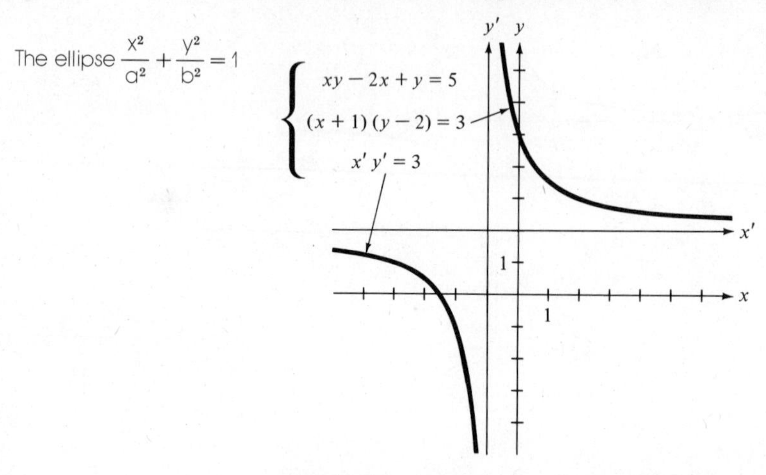

FIGURE 1.35 Example 1. The hyperbola $xy - 2x + y = 5$.

The Ellipse $x^2/a^2 + y^2/b^2 = 1$ The graph of $x^2/a^2 + y^2/b^2 = 1$, where a and b are positive constants, $a \neq b$, is known as an *ellipse*.

To sketch the graph we observe that it is symmetric about both axes and the origin. We first graph the portion in quadrant I. The rest can be obtained from the symmetry. The portion of the graph in quadrant I is defined by

$$\frac{x^2}{a^2} + \frac{y^2}{b^2} = 1, \qquad x \geqslant 0, y \geqslant 0,$$

$$\frac{y^2}{b^2} = 1 - \frac{x^2}{a^2} = \frac{a^2 - x^2}{a^2}, \qquad x \geqslant 0, y \geqslant 0,$$

$$\frac{y}{b} = \frac{\sqrt{a^2 - x^2}}{a}, \qquad x \geqslant 0,$$

$$y = \frac{b}{a}\sqrt{a^2 - x^2}, \qquad x \geqslant 0.$$

We make two observations from this last equation: (1) x must be less than or equal to a in order to have y defined and real; (2) y *decreases* in quadrant I *as* x *increases*. Thus, the points on the curve in quadrant I decrease from the point $(0, b)$ on the y-axis down to the point $(a, 0)$ on the x-axis. The graph does not extend to the right of the point $(a, 0)$. We sketch the portion of the graph in quadrant I. The rest of the graph can be obtained by using the symmetry (Figure 1.36).

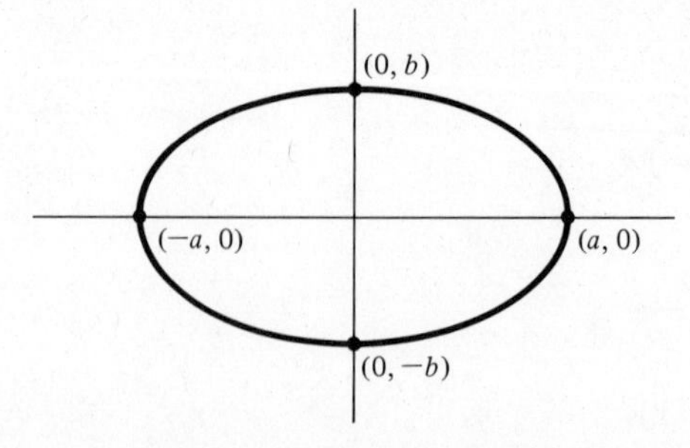

FIGURE 1.36 The ellipse $\frac{x^2}{a^2} + \frac{y^2}{b^2} = 1, a > 0, b > 0$.

Example 2 Sketch the graphs of

(a) $\frac{x^2}{3^2} + \frac{y^2}{5^2} = 1$

(b) $4x^2 + 9y^2 - 36 = 0$

Solution

(a) The equation $x^2/3^2 + y^2/5^2 = 1$ is in the standard form with $a = 3$, $b = 5$. The graph is shown in Figure 1.37a.

(b) The equation $4x^2 + 9y^2 - 36 = 0$ can be reduced to standard form as follows:

$$4x^2 + 9y^2 = 36$$

$$\frac{4x^2}{36} + \frac{9y^2}{36} = 1$$

$$\frac{x^2}{9} + \frac{y^2}{4} = 1$$

$$\frac{x^2}{3^2} + \frac{y^2}{2^2} = 1.$$

The equation is now in standard form. The graph is shown in Figure 1.37b.

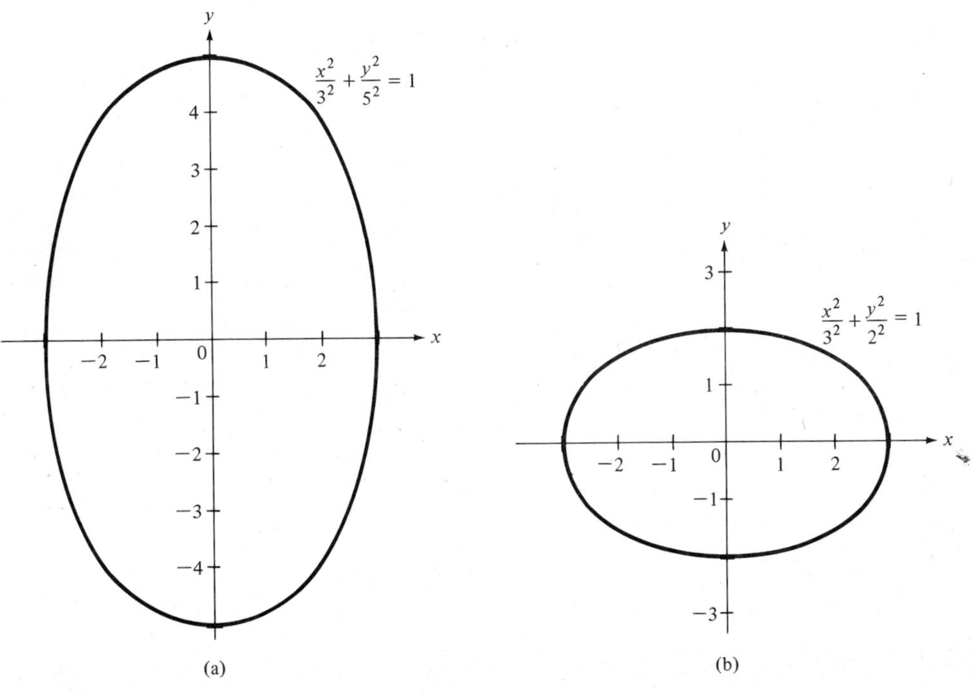

FIGURE 1.37 Example 2.

The ellipse has the simple equation

$$\frac{x^2}{a^2} + \frac{y^2}{b^2} = 1$$

only when it is symmetric about the two coordinate axes. If it is a translate of the graph of $x^2/a^2 + y^2/b^2 = 1$, then it can be reduced to the form

$$\frac{x'^2}{a^2} + \frac{y'^2}{b^2} = 1$$

by a substitution and a translation of axes.

Example 3 Sketch the graph of

$$25x^2 - 100x + 9y^2 - 18y = 116.$$

Solution We "complete the square" on the terms involving x and y:

$$\begin{aligned}
25x^2 - 100x \quad + 9y^2 - 18y \quad &= 116, \\
25(x^2 - 4x \quad) + 9(y^2 - 2y \quad) &= 116, \\
25(x^2 - 4x + 4) + 9(x^2 - 2y + 1) &= 116 + 25 \cdot 4 + 9 \cdot 1, \\
25(x - 2)^2 \quad + 9(y - 1)^2 \quad &= 116 + 100 + 9 = 225 = 25 \cdot 9. \\
\frac{(x - 2)^2}{9} + \frac{(y - 1)^2}{25} &= 1, \\
\frac{(x - 2)^2}{3^2} + \frac{(y - 1)^2}{5^2} &= 1, \\
\frac{x'^2}{3^2} + \frac{y'^2}{5^2} &= 1,
\end{aligned}$$

where $x' = x - 2$, $y' = y - 1$.

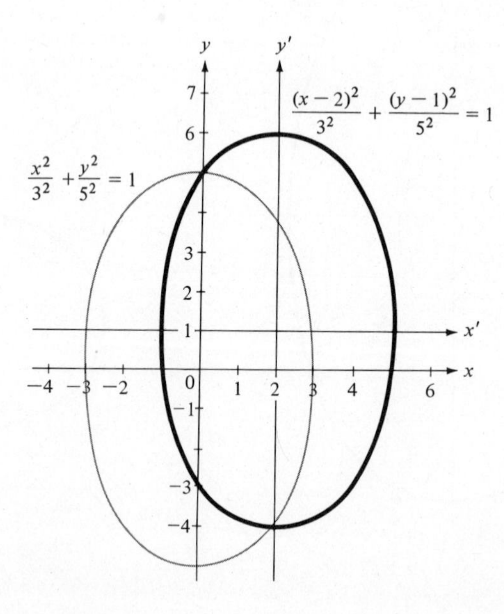

FIGURE 1.38 Example 3. The graph of $25x^2 - 100x + 9y^2 - 18y = 116$.

$$\left(\frac{(x - 2)^2}{3^2} + \frac{(y - 1)^2}{5^2} = 1\right)$$

The graph is identical to the graph of

$$\frac{x^2}{3^2} + \frac{y^2}{5^2} = 1$$

except that it is shifted 2 units to the right and 1 unit up. The graph is shown in Figure 1.38.

EXERCISES 1.6

1. Sketch the graphs of the following hyperbolas:

(a) $xy = -3$
(b) $xy = 3$
(c) $xy = -\frac{1}{3}$
(d) $xy = \frac{1}{4}$
(e) $(-x)y = -4$
(f) $-xy = 4$
(g) $x(-y) = -\frac{1}{4}$
(h) $xy = -\frac{1}{4}$

2. Sketch the graph of $x^2y^2 = 1$. (*Hint:* By factoring the equation show that the graph is the union of the graphs of $xy = 1$ and $xy = -1$.)

3. Sketch the graphs of the following ellipses:

(a) $\frac{x^2}{16} + \frac{y^2}{3^2} = 1$
(b) $\frac{x^2}{4^2} + \frac{y^2}{25} = 1$
(c) $\frac{x^2}{4^2} + \frac{y^2}{6^2} = 1$
(d) $\frac{x^2}{5^2} + \frac{y^2}{2^2} = 1$
(e) $\frac{x^2}{16} + \frac{y^2}{4} = 1$
(f) $16x^2 + 9y^2 = 144$
(g) $4x^2 + y^2 = 16$
(h) $16x^2 + 25y^2 = 400$
(i) $36x^2 + 16y^2 = 576$
(j) $4x^2 + 25y^2 = 100$

4. Simplify the following equations by a translation of axes. Identify and sketch the graphs, showing both sets of axes.

(a) $x^2 + 4x + 16y^2 - 32y = -16$
• (b) $xy - 5 = 2x - 2y$
(c) $2x - 3y + xy = -3$
• (d) $3y - 4x = 2 - xy$
(e) $xy - 2x + y = 3$
• (f) $y = 2x^2 + 8x + 8$
(g) $y + 3x^2 - 18x = 0$
• (h) $9x^2 + 18x + 4y^2 + 8y = 23$
(i) $4y^2 + 8y + 5x^2 + 50x = 71$
• (j) $2x^2 + 3y^2 + 4x + 30y = -71$

2 THE DERIVATIVE

2.1 INTRODUCTION

What is the calculus? As we shall see, this question has no simple answer. (Some students are not even sure of the answer after completing a course in the subject.) Essentially, the calculus is the body of mathematical knowledge concerned with rates of change and with the cumulative effect of small changes. The following example illustrates a problem that can be at least partially solved by its use.

Example An economist studies the profit of a large automobile manufacturing company. He has determined the sales that result when various discounts are given from the list prices. This information will be used to find the rate at which the profit changes with respect to change in price. The ultimate goal is a formula relating price and profit which can be used to determine the pricing policy that will yield the maximum profit to the company.

Problems involving rates of change were studied piecemeal for hundreds of years. Finally, in the seventeenth century several important breakthroughs occurred. In the early part of that century Descartes published his monumental work *La géométrie,* which laid the foundation for analytic geometry. Shortly thereafter mathematicians discovered that rates of change can be studied geometrically by the use of tangent lines to curves. Finally, at the end of the seventeenth century *Isaac Newton* (English, 1642–1727) and *Gottfried Wilhelm Leibniz* (German, 1646–1716) independently developed the calculus.

The work of Newton and Leibniz can be best interpreted in geometrical terms. As we mentioned above rates of change can be studied by using tangent lines to curves. Similarly, the cumulative effects of small changes can be studied by using areas under curves. Newton and Leibniz were able to establish a fundamental relationship between tangent lines and areas of certain curves.

Figure 2.1 illustrates these concepts. The graph is the velocity graph of a boy riding a bicycle. He starts from rest at a time $t = 0$, accelerates to 40 ft/sec at 25 seconds, then slows to a stop at 60 seconds. It can be shown that the acceleration (the rate of change of the velocity) is numerically equal to the slope of the tangent line at any given time. He speeds up when the tangent line has positive slope and slows down when it has negative slope. The maximum velocity occurs at $t = 25$ when the tangent line has zero slope. The area under the curve is numerically equal to the distance traveled. This holds for any part of the ride. For example, the area between $t = 0$ and $t = 25$ is equal to the distance traveled over the first 25 seconds.

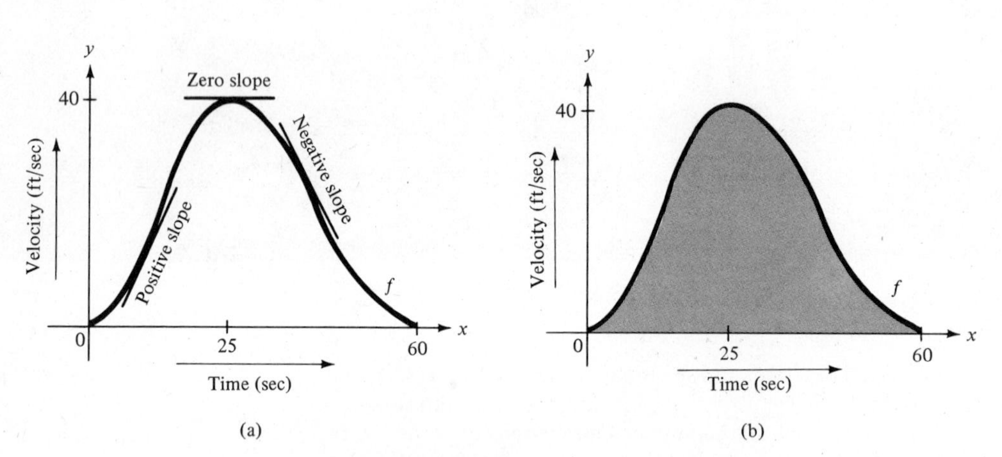

FIGURE 2.1 (a) The slope of the tangent line is numerically equal to the acceleration; (b) the area under the curve is numerically equal to the total distance traveled.

We note in passing that the independent discovery of the calculus by Newton and Leibniz caused a rift in the mathematical world that lasted for centuries. Traditionally, the English gave full credit to Newton and ignored Leibniz, and the Germans did the opposite. Most impartial scholars now give the men equal credit. In a sense Leibniz has triumphed over Newton, even in England. His notation was so superior that it has been universally adopted.

The beginning student may have difficulty recognizing that the calculus is the study of rates of change. Part of the difficulty arises with the geometrical nature of the development. Most of the basic concepts can be explained most efficiently by recourse to graphs, points in the plane, tangent lines, and so on. Thus, the development of a topic may be based on tangent lines to curves even though many of the meaningful applications involve rates of change.

2.2 LIMITS. TANGENT LINES

The study of the calculus involves a concept that is new to most students—the concept of the limit of a function. This concept will be considered in more detail in Section 2.5. In this section we discuss limits from an intuitive standpoint and use them to calculate slopes of tangent lines.

Limits Let $f(x)$ be a function of x; let a be a number. To find the limit of $f(x)$ as x approaches a we ask, "What number is $f(x)$ approaching when x approaches a?" In other words, we must find what number $f(x)$ is "near" when x is "near" a.

Example 1 Let $f(x) = 2x - 1$. If x is near 3, then $2x$ is near 6 and $2x - 1$ is near 5. Thus,

$$\lim_{x\to 3} f(x) = \lim_{x\to 3} (2x - 1) = 5.$$

Example 2 Find the following limits:

(a) $\lim_{x\to 1} (x + 5 - 2x^2)$

(b) $\lim_{y\to 3} \left(\frac{2y - 1}{y + 7}\right)$

Solution

(a) When x is close to 1, then $2x^2$ is close to 2 and $x + 5 - 2x^2$ is close to $1 + 5 - 2 = 4$. Thus,

$$\lim_{x\to 1} (x + 5 - 2x^2) = 4.$$

(b) As y approaches 3 the quantity $2y - 1$ approaches $2 \cdot 3 - 1 = 5$ and $y + 7$ approaches $3 + 7 = 10$. Thus, the quotient $(2y - 1)/(y + 7)$ approaches $\frac{5}{10}$.

$$\lim_{y\to 3} \frac{2y - 1}{y + 7} = \frac{5}{10} = \frac{1}{2}.$$

The indeterminant form $\frac{0}{0}$

The Indeterminant Form 0/0 Some of the most interesting limits involve fractions in which both numerator and denominator approach zero. For example,

$$\lim_{x\to 0} \frac{3x^2 - 2x}{x} \quad \text{and} \quad \lim_{x\to 2} \frac{x^2 - 4}{x - 2}$$

are limits of this type. We say that such a limit *has the indeterminant form 0/0*. (This does not mean that we are considering 0/0 to be a number. The symbol only indicates that both numerator and denominator approach zero as a limit.)

Example 3 Evaluate the limits

(a) $\lim_{x\to 0} \frac{3x^2 - 2x}{x}$

(b) $\lim_{x\to 2} \frac{x^2 - 4}{x - 2}$

Solution

(a) We are not concerned with evaluating $(3x^2 - 2x)/x$ when $x = 0$. We are only interested in finding the number that $(3x^2 - 2x)/x$ approaches as x approaches zero. Thus, we may assume that $x \neq 0$ and cancel x from both the numerator and the denominator:

$$\lim_{x\to 0} \frac{3x^2 - 2x}{x} = \lim_{x\to 0} (3x - 2) = 0 - 2 = -2.$$

(b) We are only interested in values of x that are near 2 but different from 2. Therefore, we may cancel $x - 2$ from the numerator and the denominator:

$$\lim_{x\to 2} \frac{x^2 - 4}{x - 2} = \lim_{x\to 2} \frac{(x - 2)(x + 2)}{x - 2} = \lim_{x\to 2} (x + 2) = 2 + 2 = 4.$$

An indeterminant form 0/0 may have a finite limit as in Example 3, or may have no limit at all. We shall have examples of many such limits in the remainder of this book.

Tangent lines

Tangent Lines What do we mean by a tangent line to a curve? Most people have good intuitive ideas about what constitutes a tangent line, but would be hard pressed to actually define the term. Basically, the tangent line to a curve at a point P is the line that best approximates the curve in the vicinity of P. If we restrict ourselves to a very small portion of the curve near P, then the points on the curve are better approximated by the points on the tangent line than by the points on any other line. (See Figure 2.2.) In the discussion that follows we assume that the line in Figure 2.2 actually is a tangent line.

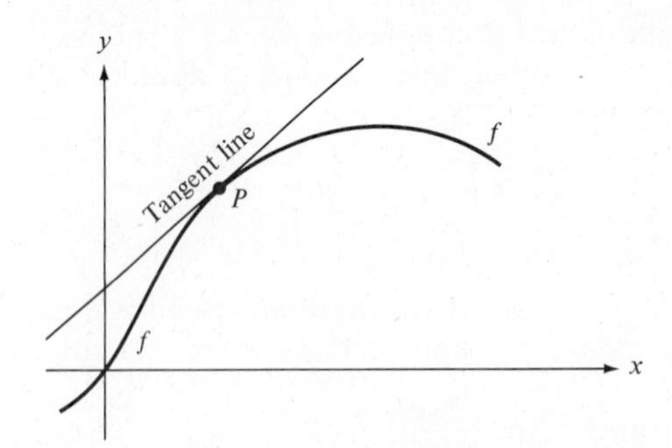

FIGURE 2.2 Tangent line to the graph of f. The tangent line is the line that best approximates the curve within the immediate vicinity of P.

The point P where the tangent line touches the curve is called the *point of tangency*. Let Q be a point on the curve different from P, and construct a line through P and Q. This line is called a *secant line*. While this secant line is not a tangent line, it should be a good approximation to the tangent line, provided P and Q are close together. Figure 2.3 shows several such lines.

If the secant line is a good approximation to the tangent line, then the slope of the secant line should be a good approximation to the slope of the tangent line. Thus, we first calculate the slope of the secant line. To do this we assume that the curve is the graph of a function f, and that P is the point $P(x_0, f(x_0))$. Let Δx be the difference between the x-coordinates of P and Q. Then

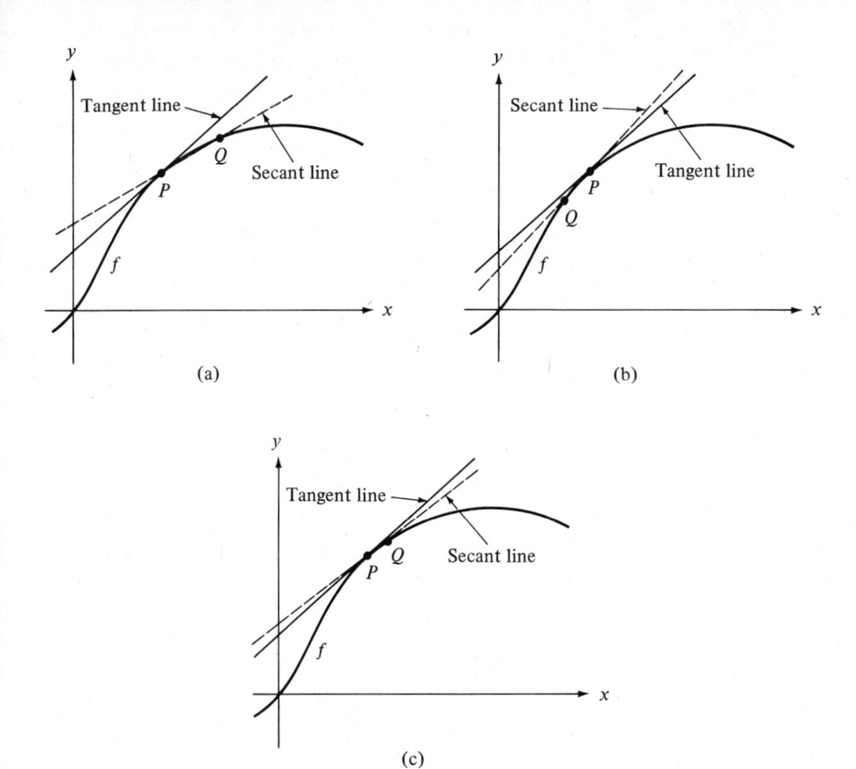

FIGURE 2.3 The secant line through P and Q approximates the tangent line at P. The closer Q is to P the better the approximation.

Q is the point $Q(x_0 + \Delta x, f(x_0 + \Delta x))$. Observe that the slope of the line through these two points is

$$\frac{\Delta y}{\Delta x} = \frac{f(x_0 + \Delta x) - f(x_0)}{\Delta x}.$$

For each value of $\Delta x \neq 0$ the expression

$$\frac{\Delta y}{\Delta x} = \frac{f(x_0 + \Delta x) - f(x_0)}{\Delta x}$$

is an approximation to the slope of the tangent line. Observe in Figure 2.3 that, in general, the closer Q is to P the better the secant line approximates the tangent line. Thus, the smaller the value of Δx, the closer Q is to P and the better the approximation of $\Delta y/\Delta x$ to the slope of the tangent line. In the limiting case, as Δx approaches zero, we would expect $\Delta y/\Delta x$ to approach the slope of the tangent line:

Slope of tangent line

$$m = \textit{slope of tangent line} = \lim_{\Delta x \to 0} \frac{\Delta y}{\Delta x}$$

$$= \lim_{\Delta x \to 0} \frac{f(x_0 + \Delta x) - f(x_0)}{\Delta x}.$$

For this reason it is customary to *define* the tangent line to be the line through P with slope equal to

$$m = \lim_{\Delta x \to 0} \frac{f(x_0 + \Delta x) - f(x_0)}{\Delta x}.$$

(See Figure 2.4.) Once we have calculated the slope of the tangent line we can use the point-slope form to calculate the equation of the line.

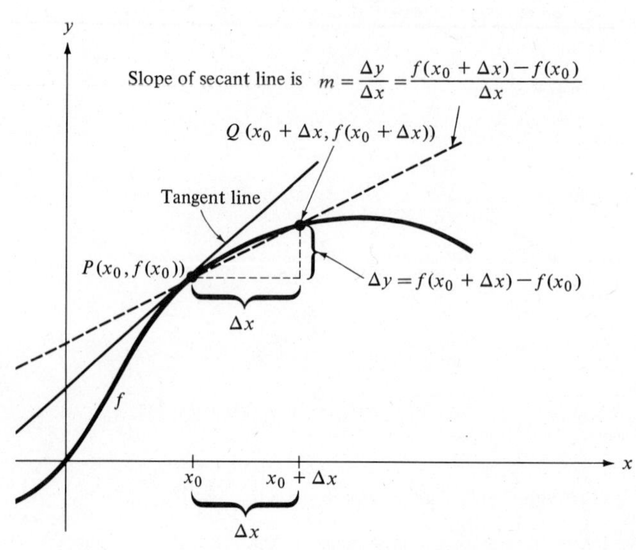

FIGURE 2.4 The slope of the secant line through $P(x_0, f(x_0))$ and $Q(x_0 + \Delta x, f(x_0 + \Delta x))$ is

$$m = \frac{f(x_0 + \Delta x) - f(x_0)}{\Delta x}$$

The slope of the tangent line at P is

$$m = \lim_{\Delta x \to 0} \frac{f(x_0 + \Delta x) - f(x_0)}{\Delta x}$$

Example 4 Let $f(x) = x^2 - 1$.

(a) Find the slope of the tangent line at the point $P_0(2, 3)$.

(b) Find the equation of the tangent line at the point $P_0(2, 3)$.

Solution

(a) The slope is

$$m = \lim_{\Delta x \to 0} \frac{f(x_0 + \Delta x) - f(x_0)}{\Delta x}, \quad \text{where } x_0 = 2.$$

We break the problem of calculating this limit into several parts:

1. We calculate $f(x_0)$ and $f(x_0 + \Delta x)$:

$$f(x_0) = f(2) = 2^2 - 1 = 3,$$
$$f(x_0 + \Delta x) = f(2 + \Delta x) = (2 + \Delta x)^2 - 1 = 4 + 4\Delta x + \Delta x^2 - 1$$
$$= 3 + 4\Delta x + \Delta x^2.$$

2. We calculate and simplify

$$\frac{f(x_0+\Delta x)-f(x_0)}{\Delta x}$$

$$\frac{f(x_0+\Delta x)-f(x_0)}{\Delta x}=\frac{(3+4\Delta x+\Delta x^2)-3}{\Delta x}=\frac{4\Delta x+\Delta x^2}{\Delta x}$$

$$=\frac{(4+\Delta x)\cdot\Delta x}{\Delta x}=4+\Delta x.$$

3. We take the limit as $\Delta x \to 0$:

$$m=\lim_{\Delta x\to 0}\frac{f(x_0+\Delta x)-f(x_0)}{\Delta x}=\lim_{\Delta x\to 0}(4+\Delta x)=4+0=4.$$

The slope of the tangent line is $m = 4$.

(b) To find the equation of the tangent line we use the point-slope form

$$y-y_0=m(x-x_0)$$

with $x_0 = 2$, $y_0 = 3$, and $m = 4$:

$$\begin{aligned} y-3&=4(x-2),\\ y-3&=4x-8,\\ y&=4x-5. \end{aligned}$$

The equation of the tangent line is $y = 4x - 5$ (Figure 2.5).

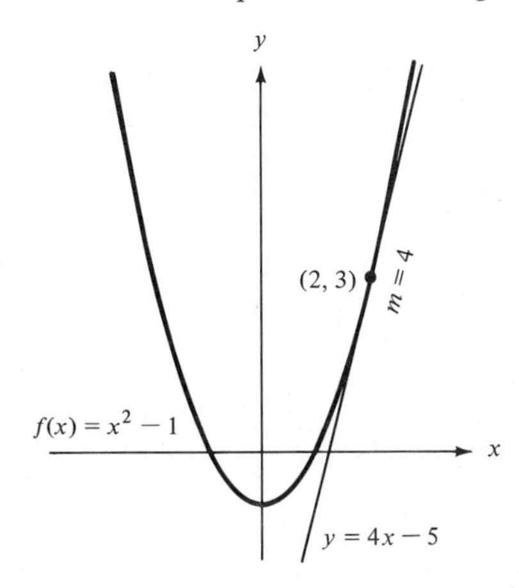

FIGURE 2.5 Example 4. The tangent line to the graph of $f(x) = x^2 - 1$ at the point (2,3) is $y = 4x - 5$.

EXERCISES 2.2

1. Use your intuitive notions to calculate the following limits:

(a) $\lim_{x\to 2}(4x^2+1)$

• (b) $\lim_{x\to 1}(x^2-5x)$

(c) $\lim_{x\to 3}(x^2+x-5)$

• (d) $\lim_{x\to 1/4}(2x^2-3x+1)$

(e) $\lim_{x\to 4}(x^3-3x^2-7)$

(f) $\lim_{x\to 1}(x^3-4x^2-3x+1)$

2. Find the limits of the indeterminate forms 0/0:

• (a) $\lim_{x\to 0}\frac{x}{x^2+2x}$

(b) $\lim_{x\to -1}\frac{x^2-1}{x+1}$

(c) $\lim_{y\to 2} \frac{5(y-2)^2 + 3(y-2)}{(y-2)(y+2)}$ • (e) $\lim_{\Delta x\to 0} \frac{5x\Delta x - 3\Delta x}{\Delta x}$

• (d) $\lim_{\Delta x\to 0} \frac{3\Delta x^2 - 2\Delta x}{\Delta x}$ (f) $\lim_{\Delta x\to 0} \frac{3x^2\Delta x^2 + 5x\Delta x - 3\Delta x}{\Delta x}$

3. Find the slope of the tangent line at the given point on the graph of f.

(a) $f(x) = 3x + 2$, (1, 5)
• (b) $f(x) = x^2$, (2, 4)
• (c) $f(x) = 5x - 7$, (2, 3)
(d) $f(x) = x^2 + 1$, (1, 2)
(e) $f(x) = -x + 7$, (4, 3)
(f) $f(x) = 2x^2 + 1$, (1, 3)
• (g) $f(x) = x^2 + 2x - 1$, (1, 2)
(h) $f(x) = x^2 + x - 1$, (1, 1)

4. Find the equation of the tangent line at the given point for each of the functions in Exercise 3.

2.3 THE DERIVATIVE

Many problems in mathematics can be reduced to a study of the limit

$$\lim_{\Delta x\to 0} \frac{f(x_0 + \Delta x) - f(x_0)}{\Delta x}$$

As we saw in the preceding section this limit is equal to the slope of the tangent line at the point $P_0(x_0, y_0)$ on the graph of f. This is only one of many applications of the limit.

The above limit, if it exists, is called the *derivative* of the function f at x_0. It is denoted by the symbol $f'(x_0)$.* That is,

The derivative

$$f'(x_0) = \lim_{\Delta x\to 0} \frac{f(x_0 + \Delta x) - f(x_0)}{\Delta x},$$

provided the limit exists.

If $f'(x_0)$ exists, then the above limit must be an indeterminate form 0/0. (Since the denominator Δx approaches zero, then the numerator $f(x_0 + \Delta x) - f(x_0)$ must also approach zero or there is no limit.) Thus some type of simplification is always required before the limit can be calculated.

Example 1 Let $f(x) = x^2 + 2$. Calculate $f'(1)$.

Solution

$$\begin{aligned} f'(1) &= \lim_{\Delta x\to 0} \frac{f(1+\Delta x) - f(1)}{\Delta x} \\ &= \lim_{\Delta x\to 0} \frac{[(1+\Delta x)^2 + 2] - [1^2 + 2]}{\Delta x} \\ &= \lim_{\Delta x\to 0} \frac{1 + 2\Delta x + \Delta x^2 + 2 - 1 - 2}{\Delta x} \\ &= \lim_{\Delta x\to 0} \frac{2\,\Delta x + \Delta x^2}{\Delta x} = \lim_{\Delta x\to 0} (2 + \Delta x) = 2. \end{aligned}$$

* Read "f-prime at x_0."

If we calculate the derivative of f at each point where it exists, we have a new (or derived) function f', defined where the limit exists by

$$f'(x) = \lim_{\Delta x \to 0} \frac{f(x + \Delta x) - f(x)}{\Delta x}$$

(In calculating $f'(x)$ we treat x as a constant and consider the expression

$$\frac{f(x + \Delta x) - f(x)}{\Delta x}$$

as if it were a function of Δx.)

Example 2 Let $f(x) = 2x + 7$. Calculate $f'(x)$.

$$f'(x) = \lim_{\Delta x \to 0} \frac{f(x + \Delta x) - f(x)}{\Delta x}$$

$$= \lim_{\Delta x \to 0} \frac{[2(x + \Delta x) + 7] - [2x + 7]}{\Delta x}$$

$$= \lim_{\Delta x \to 0} \frac{2x + 2\Delta x + 7 - 2x - 7}{\Delta x}$$

$$= \lim_{\Delta x \to 0} \frac{2\Delta x}{\Delta x} = \lim_{\Delta x \to 0} 2 = 2.$$

There is an alternate way of writing the limit that is the derivative of f at x_0. If we let x represent the number $x_0 + \Delta x$, then

$$x \to x_0 \qquad \text{as } \Delta x \to 0.$$

Furthermore, $\Delta x = x - x_0$, so the relationship

$$f'(x_0) = \lim_{\Delta x \to 0} \frac{f(x_0 + \Delta x) - f(x_0)}{\Delta x}$$

can be written in the form

$$f'(x_0) = \lim_{x \to x_0} \frac{f(x) - f(x_0)}{x - x_0}$$

For example, if $f(x) = 3x^2 - 1$, then

$$f'(2) = \lim_{x \to 2} \frac{f(x) - f(2)}{x - 2} = \lim_{x \to 2} \frac{(3x^2 - 1) - (3 \cdot 2^2 - 1)}{x - 2}$$

$$= \lim_{x \to 2} \frac{3x^2 - 1 - 3 \cdot 4 + 1}{x - 2} = \lim_{x \to 2} \frac{3(x^2 - 4)}{x - 2} = \lim_{x \to 2} 3(x + 2)$$

$$= 12.$$

There are places where it is more convenient to write the derivative in the alternate form

$$f'(x_0) = \lim_{x \to x_0} \frac{f(x) - f(x_0)}{x - x_0}$$

Alternate way of writing the derivative

than in the form given in the definition. In the remainder of this chapter we will use whichever form is most convenient.

Notations for the derivative

Notations for the Derivative There are a number of different notations for the derivative. Each of these is the most natural to use in at least one setting, so all of them have survived as more-or-less standard notations. If the original function is written as $y = f(x)$ then the derivative can be denoted by any of the expressions $f'(x)$, y', $D_x y$, dy/dx, Dy, $Df(x)$ or $d(f(x))/dy$. In the elementary study of the calculus the notations $f'(x)$, y' and dy/dx are the most convenient to use. Thus, for the most part we will restrict ourselves to these three notations. Each of these means the same thing:

$$\frac{dy}{dx} = f'(x) = y' = \lim_{\Delta x \to 0} \frac{f(x + \Delta x) - f(x)}{\Delta x}.$$

Example 3 Let $y = 2x^2 - 5x$. Calculate dy/dx.

Solution Let $f(x) = 2x^2 - 5x$. Then

$$\begin{aligned} \frac{dy}{dx} = y' &= \lim_{\Delta x \to 0} \frac{f(x + \Delta x) - f(x)}{\Delta x} \\ &= \lim_{\Delta x \to 0} \frac{[2(x + \Delta x)^2 - 5(x + \Delta x)] - [2x^2 - 5x]}{\Delta x} \\ &= \lim_{\Delta x \to 0} \frac{2x^2 + 4x \cdot \Delta x + 2\Delta x^2 - 5x - 5\Delta x - 2x^2 + 5x}{\Delta x} \\ &= \lim_{\Delta x \to 0} \frac{4x \cdot \Delta x + 2\Delta x^2 - 5\Delta x}{\Delta x} = \lim_{\Delta x \to 0} (4x + 2\Delta x - 5) \\ &= 4x - 5. \end{aligned}$$

Remark The expression dy/dx represents the derivative with respect to x. This expression is not a fraction. The fractional-like notation was chosen by Leibniz, one of the founders of the calculus, because it simplified many of the formulas.

If we use variables other than x and y, the symbols for the derivative must be modified accordingly.

Example 4 Let $v = g(u) = 3u + 1$. Calculate dv/du.

Solution

$$\begin{aligned} \frac{dv}{du} = g'(u) &= \lim_{\Delta u \to 0} \frac{g(u + \Delta u) - g(u)}{\Delta u} \\ &= \lim_{\Delta u \to 0} \frac{(3(u + \Delta u) + 1) - (3u + 1)}{\Delta u} \\ &= \lim_{\Delta u \to 0} \frac{3u + 3\Delta u + 1 - 3u - 1}{\Delta u} = \lim_{\Delta u \to 0} \frac{3\Delta u}{\Delta u} \\ &= \lim_{\Delta u \to 0} 3 = 3. \end{aligned}$$

Example 5 Let $y = f(x) = 1/x$. Calculate dy/dx.

Solution

$$\frac{dy}{dx} = f'(x) = \lim_{\Delta x \to 0} \frac{f(x+\Delta x) - f(x)}{\Delta x}$$

$$= \lim_{\Delta x \to 0} \frac{\dfrac{1}{x+\Delta x} - \dfrac{1}{x}}{\Delta x} = \lim_{\Delta x \to 0} \frac{\dfrac{x-(x+\Delta x)}{(x+\Delta x)\cdot x}}{\dfrac{\Delta x}{1}}$$

$$= \lim_{\Delta x \to 0} \frac{x - x - \Delta x}{(x+\Delta x)\cdot x} \cdot \frac{1}{\Delta x} = \lim_{\Delta x \to 0} \frac{-\Delta x}{(x+\Delta x)\cdot x} \cdot \frac{1}{\Delta x}$$

$$= \lim_{\Delta x \to 0} \frac{-1}{(x+\Delta x)\cdot x} = -\frac{1}{x^2}$$

EXERCISES 2.3

1. Calculate the derivative at the indicated value of x. Use $f'(x_0) = \lim_{\Delta x \to 0} \frac{f(x_0+\Delta x) - f(x_0)}{\Delta x}$ for (a) through (e). Use $f'(x_0) = \lim_{x \to x_0} \frac{f(x) - f(x_0)}{x - x_0}$ for (f) through (j).

   (a) $f(x) = -3x + 2, x = 1$
   (b) $f(x) = x^2, x = 3$
   • (c) $f(x) = x^2 - 2x - 3, x = 2$
   (d) $f(x) = 3x^2 - 4x + 1, x = 1$
   (e) $f(x) = -x^2 + 2x + 1, x = 2$
   • (f) $f(x) = x^3, x = -1$
   (g) $f(x) = x^3 - x^2 + x - 1, x = -2$
   • (h) $f(x) = 2x^3 - 3x^2 + x - 3, x = -1$
   (i) $f(x) = 4x^3 - x^2 + 4, x = 0$
   • (j) $f(x) = 3x^3 - 2x + 1, x = -1$

2. Calculate the derivative:

   (a) $y = 2x + 5$
   • (b) $y = 3x^3 + 2x^2 + 6x - 1$
   (c) $y = 2x^2 - 5x + 2$
   (d) $y = 2x^3 + 2x - 1$
   • (e) $y = 3x^2 + 5x - 2$
   (f) $v = 4u^3 + 3u^2 - 6u + 7$
   • (g) $v = u^2$
   (h) $s = 1/t$
   • (i) $s = t^2 - t$
   (j) $x = s$

3. Calculate the equation of the tangent line at the given point.

   (a) $f(x) = x^2 - 1$, (1, 0)
   • (b) $f(x) = x^3$, (1, 1)
   (c) $f(x) = 2x^2 + 3x - 1$, (−1, −2)
   (d) $f(x) = 2x^3 - 6x - 1$, (1, −5)
   • (e) $f(x) = 3x^3 + 4x + 1$, (−2, 5)
   (f) $f(x) = x^2 - 2x + 1$, (0, 1)
   (g) $f(x) = -3x^3 + 4x^2 - 2x + 1$, (1, 0)
   • (h) $f(x) = x$, (3, 3)
   • (i) $f(x) = 1/x$, (3, 1/3) (see Example 5)
   (j) $f(x) = 5x^2 - x - 4$, (1, 0)

4. For each of the following functions find the points, if any, where the tangent line has slope equal to zero.

   (a) $f(x) = x^2 + 2x - 3$
   (b) $f(x) = 2x^3 + 6x^2 + 6x - 1$
   • (c) $f(x) = 2x^2 - 2x - 1$
   (d) $f(x) = 2x + 1$
   • (e) $f(x) = x^3 + x^2 - x - 2$
   (f) $f(x) = 3x^2 - 4x + 2$
   (g) $f(x) = 3x^3 - x + 1$
   (h) $f(x) = 2x^3 + 6x^2 + 6x + 1$
   • (i) $f(x) = x$
   (j) $f(x) = -1/x$

2.4 RATES OF CHANGE

The derivative of a function can be considered as a rate of change. Suppose that a particle moves along a line. We put a coordinate system on the line and let $s(t)$ be the coordinate of the particle at time t. Then $s(t)$ measures the distance from the origin O to the particle at time t. This distance is considered to be positive on one side of t and negative on the other side.

The *average velocity* of the particle over a period of time is defined to be the change in distance over the period of time divided by the length of the period of time. For example, if the particle travels 100 miles in the positive direction in two hours, the average velocity is

$$\frac{100 \text{ miles}}{2 \text{ hours}} = 50 \text{ mph}.$$

More precisely, if the period of time is measured from t to $t + \Delta t$, the *average velocity* is

Average velocity

$$v_{\text{av}} = \frac{s(t + \Delta t) - s(t)}{\Delta t}.$$

(If the particle moves 12 feet in the positive direction over a two-minute period, the average velocity is $12/2 = 6$ ft/min. If it moves 12 feet in the negative direction, the average velocity is $-12/2 = -6$ ft/min.)

The *velocity*, or *rate of change of distance*, at time t is defined to be the limit of the average velocity as $\Delta t \to 0$:

$$v(t) = \textit{velocity at time } t = \lim_{\Delta t \to 0} \frac{s(t + \Delta t) - s(t)}{\Delta t}.$$

This limit is equal to the derivative of s with respect to t. Therefore,

Velocity

$$v(t) = \textit{velocity at time } t = s'(t).$$

The *speed* of the particle is defined as the absolute value of the velocity:

Speed

$$\textit{speed at time } t = |v(t)| = |s'(t)|.$$

The rate of change of the velocity is called the *acceleration* of the particle:

Acceleration

$$a(t) = \textit{acceleration at time } t = \lim_{\Delta t \to 0} \frac{v(t + \Delta t) - v(t)}{\Delta t} = v'(t).$$

The acceleration is measured in units of *feet per second per second, miles per hour per hour,* and so on. (If the acceleration is 10 feet per second per second, then during each second the velocity changes at the rate of 10 ft/sec.) These units are abbreviated as ft/sec^2, miles/hour^2, and so on.

We will show in Chapter 4 that if an object is thrown upward from an initial height of s_0 feet above the earth with an initial velocity of v_0 feet per second, then the height of the object above the earth after t seconds is given by the formula[1]

$$s(t) = -16t^2 + v_0 t + s_0.$$

---

[1] This formula is obtained after several simplifying assumptions. In actual practice it is a very good approximation to the distance at time t, provided the object has little air resistance and does not go too high.

Example 1 A rock is thrown upward from a height of 96 feet above the ground with an initial velocity of 80 ft/sec.

(a) Derive formulas for the velocity and acceleration at time t. Calculate the speed when $t = 4$.

(b) For which values of t are the formulas in (a) valid? (In other words, when does the rock strike the ground?)

(c) What is the maximum height of the rock above the earth?

Solution The formula for distance is

$$s(t) = -16t^2 + 80t + 96.$$

(a) At the time t the velocity is

$$\begin{aligned} v(t) = s'(t) &= \lim_{\Delta t \to 0} \frac{s(t + \Delta t) - s(t)}{\Delta t} \\ &= \lim_{\Delta t \to 0} \frac{(-16(t + \Delta t)^2 + 80(t + \Delta t) + 96) - (-16t^2 + 80t + 96)}{\Delta t} \\ &= \lim_{\Delta t \to 0} \frac{-32t \cdot \Delta t - 16\Delta t^2 + 80\,\Delta t}{\Delta t} = \lim_{\Delta t \to 0} (-32t - 16\Delta t + 80) \\ &= -32t + 80 \text{ ft/sec.} \end{aligned}$$

The acceleration at time t is

$$\begin{aligned} a(t) = v'(t) &= \lim_{\Delta t \to 0} \frac{v(t + \Delta t) - v(t)}{\Delta t} \\ &= \lim_{\Delta t \to 0} \frac{(-32(t + \Delta t) + 80) - (-32t + 80)}{\Delta t} \\ &= \lim_{\Delta t \to 0} \frac{-32\Delta t}{\Delta t} = -32 \text{ ft/sec}^2. \end{aligned}$$

When $t = 4$, the velocity is $v(4) = -48$ ft/sec. (The negative sign indicates that the object is falling.) The speed is $|v(4)| = 48$ ft/sec.

(b) The object will strike the earth when $s(t) = 0$. Setting $s(t) = 0$ and solving for t, we obtain

$$\begin{aligned} -16t^2 + 80t + 96 &= 0, \\ -16(t + 1)(t - 6) &= 0, \\ t = -1 \quad \text{or} \quad t &= 6. \end{aligned}$$

Since we are interested only in the values of t that are nonnegative, then $t = 6$. Thus, the object strikes the ground at the end of 6 seconds. The formulas for $v(t)$, $s(t)$, and the acceleration are valid only for $0 \leq t \leq 6$.

(c) $v(t) = -32t + 80$. Thus, $v(t) > 0$ if $0 \leq t < 2\frac{1}{2}$ and $v(t) < 0$ if $2\frac{1}{2} < t \leq 6$. This means that the object is *rising* until $t = 2\frac{1}{2}$ and is *falling* from $2\frac{1}{2}$ seconds until 6 seconds. Thus, the maximum height is obtained when $t = 2\frac{1}{2}$. This height is

$$s(2\tfrac{1}{2}) = -16(2\tfrac{1}{2})^2 + 80(2\tfrac{1}{2}) + 96 = 196 \text{ feet.}$$

Interpretations of the derivative

We have several interpretations of the derivative:

1. If we graph $y = f(x)$, then $f'(x_0)$ is the slope of the tangent line at the point $(x_0, f(x_0))$.

2. If $s(t)$ is the directed distance from a fixed point on a line to a particle that moves on the line, then $s'(t_0)$ is the *velocity* of the particle at time t_0.
3. If $v(t)$ is the velocity of the particle on the line at time t, then $v'(t_0)$ is the acceleration of the particle at time t_0.

Even when $f(x)$ does not measure velocity or distance the quotient

$$\frac{\Delta y}{\Delta x} = \frac{f(x_0 + \Delta x) - f(x_0)}{\Delta x}$$

measures the ratio of the change in y to the change in x. Thus,

$$\frac{dy}{dx} = \lim_{\Delta x \to 0} \frac{f(x_0 + \Delta x) - f(x_0)}{\Delta x}$$

is a measure of the rate at which y changes with respect to x at $x = x_0$.

Example 2 The Knight Chess Company produces a deluxe chess set sold to serious chess players. If the company produces x sets in a month, it finds its total cost in dollars to be

$$C(x) = 500 + 4x + \frac{1}{10{,}000}\, x^2.$$

Calculate the rate of change of the cost function C with respect to x. Compare the rates when $x = 1000$ and $x = 5000$.

Solution $C'(x)$ is the rate of change of C with respect to x. Therefore,

$$C'(x) = \lim_{\Delta x \to 0} \frac{C(x+\Delta x) - C(x)}{\Delta x}$$

$$= \lim_{\Delta x \to 0} \frac{\left[500 + 4(x+\Delta x) + \frac{1}{10{,}000}(x+\Delta x)^2\right] - \left[500 + 4x + \frac{1}{10{,}000}x^2\right]}{\Delta x}$$

$$= \lim_{\Delta x \to 0} \frac{500 + 4x + 4\,\Delta x + \frac{x^2 + 2x \cdot \Delta x + \Delta x^2}{10{,}000} - 500 - 4x - \frac{x^2}{10{,}000}}{\Delta x}$$

$$= \lim_{\Delta x \to 0} \frac{4\,\Delta x + \frac{2x \cdot \Delta x + \Delta x^2}{10{,}000}}{\Delta x} = \lim_{\Delta x \to 0} \left(4 + \frac{2x + \Delta x}{10{,}000}\right)$$

$$= 4 + \frac{2x + 0}{10{,}000} = 4 + \frac{x}{5000}.$$

When $x = 1000$, $C'(x) = C'(1000) = 4 + \frac{1000}{5000} = 4\frac{1}{5}$.
When $x = 5000$, $C'(x) = C'(5000) = 4 + \frac{5000}{5000} = 5$.

Remark The number $f'(x_0)$, the rate of change of f with respect to x, is approximately equal to the change in f that corresponds to a unit change in x from x_0 to $x_0 + 1$. This derivative is given special names in many applications. For example, the derivative $C'(n)$ in Example 2 is called the *marginal cost* of the nth item. It is approximately equal to the cost of producing the nth item after the first

Marginal cost

$n - 1$ items have been produced. (It costs the company an additional \$4.20 to produce the 1000th set after the first 999 sets have been produced.) Marginal cost will be discussed in more detail in the next chapter.

EXERCISES 2.4

1. (a) Let a, b, c be constants. Use the definition of the derivative to show that

$$\frac{d}{dx}(ax^2 + bx + c) = 2ax + b.$$

Derivative of a quadratic polynomial

This formula can be used for the direct calculation of the derivatives of first and second degree polynomials.

Use the result of (a) to calculate the derivatives of the following functions:

(b) $f(x) = x^2 - 3x + 2$
• (c) $f(x) = 5x^2 + 2$
• (d) $f(x) = 7x^2 - 32x + \sqrt{7}$
(e) $f(x) = 8x + 3$

2. A particle moves along a straight line with its position at each instant given by $s(t)$. For each of the following distance functions obtain functions for the velocity, acceleration, and speed. Calculate the values of these functions at $t = 1$ and $t = 2$.

(a) $s(t) = 6t^2 - 5t + 8$
• (b) $s(t) = -8t^2 + 16t + 1$
(c) $s(t) = 7t^2 - 30t + 1$
• (d) $s(t) = 0.5t^2 - 6t + 2$
(e) $s(t) = 30t^2 - t - 1$
(f) $s(t) = t + 2$
(g) $s(t) = 2t^2 - 7t + 4$
• (h) $s(t) = 2t + 1/t$
(i) $s(t) = 3t^2 + 1$
• (j) $s(t) = 3t^2 + 2t + 1$

3. A ball is thrown upward from a 128-foot tower with a velocity of 32 ft/sec. Determine the maximum height of the ball above the ground, the length of time the ball is in the air, and its velocity on impact with the earth.

•4. A ball is thrown upward from the ground with a velocity of 96 ft/sec. Show that its acceleration is constant. For how many seconds is the ball rising? What is its maximum height above the ground? For how many seconds is the ball falling? At what speed does it strike the ground?

5. From a bridge 48 feet above the water a boy throws a rock downward with a velocity of 32 ft/sec. At what velocity does the rock strike the water? How far above the water is the stone when its velocity is -48 ft/sec?

•6. The Bimbo Bottle Company finds that the cost of producing x cases of bottles is given by $C(x) = 80 + \frac{9x}{100} + \frac{70x^2}{10{,}000}$ dollars. Determine the marginal cost function and compare its values for $x = 10$, 50, and 100.

7. The Luckee Seven Dice Company accountants have determined that if x cartons of dice, $0 \leq x \leq 10{,}000$, are produced in a day, the costs involved for the day are: (1) a fixed cost of \$200; (2) a production cost of \$20 per carton; and (3) an allowance for machine depreciation, repairs, and so on, of $\frac{3x^2}{1{,}000{,}000}$ dollars. Determine the cost function $C(x)$ and the marginal cost function $C'(x)$. Compare the marginal costs for $x = 10$, 100, 1000, and 10,000.

2.5 LIMITS

Thus far we have considered limits from an intuitive standpoint. We have noted, for example, that if

$$f(x) = 2 + 3x - x^2,$$

then $3x$ and x^2 are close to zero whenever x is close to zero, so that $2 + 3x - x^2$ is close to 2. We indicate this fact by writing

$$\lim_{x \to 0} f(x) = \lim_{x \to 0} (2 + 3x - x^2) = 2 + 0 - 0 = 2.$$

From the time of Newton and Leibniz until the early part of the nineteenth century the development of the calculus depended on arguments similar to the one above. Finally, in the 1830s the French mathematician *A. L. Cauchy* (1789–1857) developed a proper definition of the term "limit." Cauchy reasoned that the idea of "closeness" could not be used in the definition, because the word "close" is, by its very nature, imprecise. (Two objects may be "close" on a global scale and very far apart on a molecular scale.) Thus, he developed a different approach to the concept.

The mathematical definition of the word "limit," as devised by Cauchy, is somewhat similar to the idea of "tolerance" used by engineers and architects. When an architect plans a structure he specifies the dimensions to within certain tolerances. For example, his plans may call for a concrete beam 100 feet long, 2 feet high, and 1 foot thick. To insure the proper strength he may require that the volume be 200 cubic feet with a maximum tolerance of 1 cubic foot. This means that the actual volume must be no less than 199 cubic feet and no more than 201 cubic feet. He can then set up tolerances for the length, height, and width to insure that the volume is in the proper range. For example, he may decide that each dimension must be within $\frac{1}{32}$ inch of the stated value. Thus, the length must be between 99 feet $11\frac{31}{32}$ inches and 100 feet $\frac{1}{32}$ inch, the height must be between $23\frac{31}{32}$ inches and $24\frac{1}{32}$ inches, and so on. If this is done, the volume will be within the required range.

Observe that any predetermined tolerance range for the volume of the beam results in a corresponding set of tolerance ranges for the dimensions. The smaller the tolerance range for the volume, the smaller the tolerance ranges for the dimensions.

The concept of the limit of a function is similar to the idea of tolerance. We say that

$$\lim_{x \to a} f(x) = L$$

if we can keep the values $f(x)$ within any given tolerance range about L by requiring x to be within a corresponding tolerance range about a.

We first assign a tolerance range about L. To do this we choose a small positive number ϵ (epsilon). The tolerance range is the set of numbers between $L - \epsilon$ and $L + \epsilon$. We then find a small positive number δ (delta), which depends on ϵ, to set the tolerance range about a. This tolerance range is the set of numbers between $a - \delta$ and $a + \delta$, *excluding a itself*. If we can always find this δ such that for x between $a - \delta$ and $a + \delta$ ($x \neq a$) the values of $f(x)$ are between $L - \epsilon$ and $L + \epsilon$, then we say that L is the limit of $f(x)$ as x approaches a.

We exclude the value a from the tolerance range because we are not actually concerned with the value of f at a. We only care about the values near a. In fact the function f may not be defined at a. This occurs, for example, whenever we calculate a limit that involves an indeterminate form 0/0.

Observe that the inequality $a - \delta < x < a + \delta$, $x \neq a$, is equivalent to the inequality

$$0 < |x - a| < \delta$$

and the inequality $L - \epsilon < f(x) < L + \epsilon$ is equivalent to

$$|f(x) - L| < \epsilon.$$

We are now ready for the formal definition.

Definition of limit

DEFINITION Let f be defined for all values of x near a except possibly at $x = a$. We say that L *is the limit of* $f(x)$ *as* x *approaches* a provided:

If $\epsilon > 0$ there exists a corresponding number $\delta > 0$ (which depends on ϵ) such that

$$\text{if } 0 < |x - a| < \delta, \text{ then } |f(x) - L| < \epsilon.$$

When we apply the definition we first choose $\epsilon > 0$ (which sets the tolerance range about L). We then must find a corresponding number $\delta > 0$ (which sets the tolerance range about a) such that if x is within the δ-range about a (and $x \neq a$) then $f(x)$ is within the ϵ-range about L. If this can be done for every choice of $\epsilon > 0$, then L is the limit as x approaches a.

Example 1 Use the definition to show that

$$\lim_{x \to 3} (2x - 5) = 1.$$

Solution Let $\epsilon > 0$. We must find $\delta > 0$ (where δ is in terms of ϵ) such that

$$\text{if} \quad 0 < |x - 3| < \delta, \qquad \text{then} \quad |(2x - 5) - 1| < \epsilon.$$

Now

$$\begin{aligned} |(2x - 5) - 1| &= |2x - 6| = |2(x - 3)| \\ &= 2|x - 3|. \end{aligned}$$

If $0 < |x - 3| < \epsilon/2$, then

$$|(2x - 5) - 1| = 2|x - 3| < 2 \cdot \frac{\epsilon}{2} = \epsilon.$$

Thus, δ may be any positive number $\leqslant \epsilon/2$.

Limits can be interpreted geometrically as follows: Draw two horizontal lines at a distance of ϵ above and below the point L on the y-axis. If $\lim_{x \to a} f(x) = L$, there must exist a pair of vertical lines at a distance of δ to the left and right of a such that, with the exception of the point $(a, f(a))$, the portion of the graph of f between the vertical lines is between the horizontal lines (Figure 2.6a). This must be the case for every such pair of horizontal lines.

Frequently, this geometric interpretation can be used to show that a certain number L is not the limit. To do this we choose a fixed number $\epsilon > 0$ and

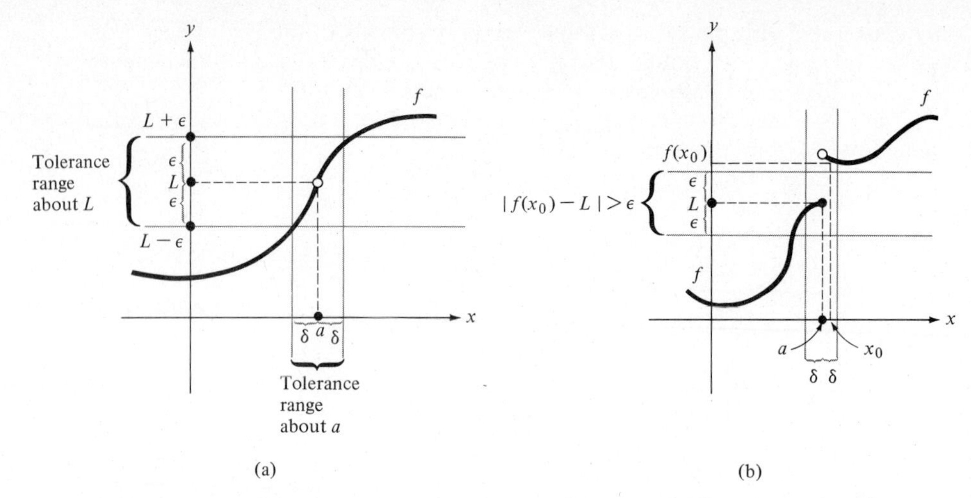

(a) $\lim_{x \to a} f(x) = L$. When x is within the tolerance range about a, then $f(x)$ is within the tolerance range about L.

(b) L is not the limit as x approaches a. If δ is any positive number, there exists a number x_0 such that $0 < |x_0 - a| < \delta$ and $|f(x_0 - L| > \epsilon$.

FIGURE 2.6

draw the horizontal lines about L. If it is impossible to find a corresponding pair of vertical lines about L such that the portion of the graph between the vertical lines (excepting the point $(a, f(a))$) is also between the horizontal lines, then L is not the limit as x approaches a.

This method is illustrated in Figure 2.6b. No matter how close we put the vertical lines about a, a portion of the graph between the vertical lines will not be between the horizontal lines. Thus, L is not the limit as x approaches a.

The greatest integer function

The Greatest Integer Function The function $f(x) = [x]$ is defined for each real number x to be the greatest integer that is no larger than x. Thus,

$$\begin{aligned} f(1.9) &= [1.9] = 1 \\ f(2.0001) &= [2.0001] = 2, \\ f(-1.3) &= [-1.3] = -2, \\ f(7) &= [7] = 7, \end{aligned}$$

and so on.

Observe that

$$\begin{aligned} f(x) &= 0 \quad \text{if} \quad 0 \leqslant x < 1, \\ f(x) &= 1 \quad \text{if} \quad 1 \leqslant x < 2, \\ f(x) &= 2 \quad \text{if} \quad 2 \leqslant x < 3, \end{aligned}$$

and so on. Thus, the graph consists of line segments parallel to the x-axis with "breaks" at the integers. (See Figure 2.7.)

Example 2 Show that the function $f(x) = [x]$ has no limit as $x \to 1$.

Solution Recall that $f(x) = 0$ if $0 \leqslant x < 1$ and $f(x) = 1$ if $1 \leqslant x < 2$. Thus, 0 and 1 are the only likely candidates for the limit as $x \to 1$. We shall show that 1 is not the limit.

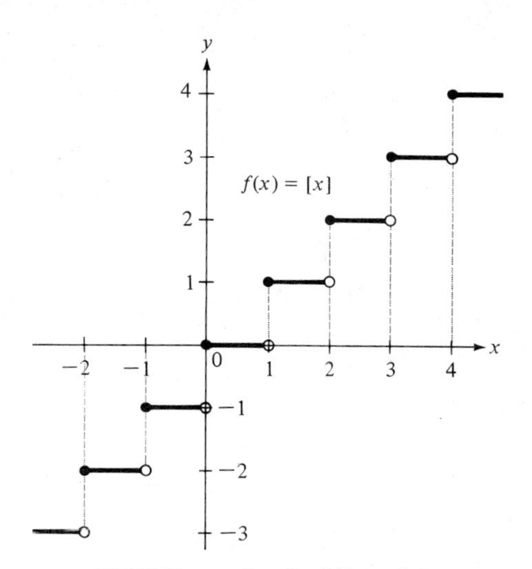

FIGURE 2.7 Graph of $f(x) = [x]$. The greatest integer function.

FIGURE 2.8 Example 2. The function $f(x) = [x]$ has no limit as $x \to 1$.

We choose a value of ϵ that is fairly small, say $\epsilon = \frac{1}{4}$, and draw horizontal lines at a distance of ϵ above and below 1 (located on the y-axis). We see from Figure 2.8 that it is not possible to find a pair of vertical lines with 1 (located on the x-axis) midway between them such that the portion of the graph between the vertical lines also is between the horizontal lines. Any such pair of vertical lines must contain a point x_0 between them where $0 < x_0 < 1$. Then $f(x_0) = 0$ and so the point $(x_0, f(x_0))$ is not between the two horizontal lines.

Since there is at least one number ϵ such that the conditions of the definition cannot be satisfied, then 1 is not the limit of $f(x)$ as $x \to 1$. A similar argument shows that 0 is not the limit as $x \to 1$. The argument can be modified slightly to prove that no number L is the limit as $x \to 1$.

When we find the limit of $f(x)$ as x approaches a we are not concerned with the value of $f(x)$ at $x = a$. We are only interested in what happens to $f(x)$ when x is near a, not equal to a. The functional value of $f(a)$ may be different

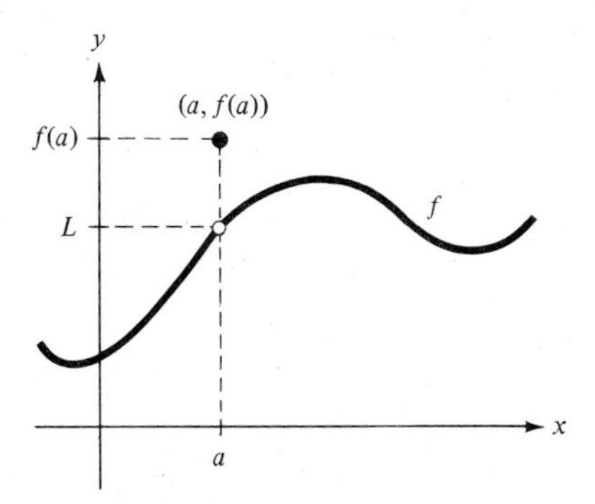

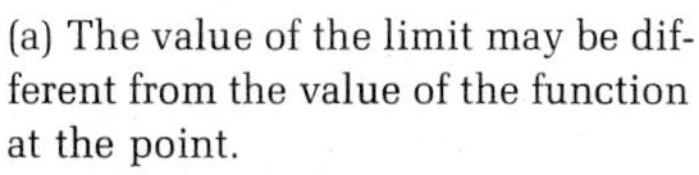
(a) The value of the limit may be different from the value of the function at the point.

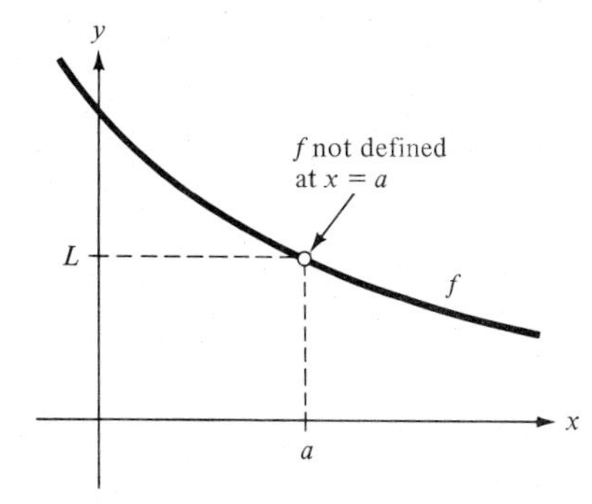

(b) The function may have a limit as x approaches a even though it is not defined at $x = a$.

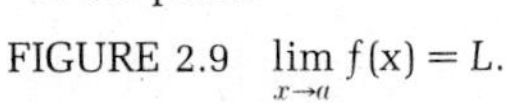
FIGURE 2.9 $\lim_{x \to a} f(x) = L$.

from the limit or the function may not even be defined at $x = a$. These possibilities are pictured in Figure 2.9. In (a) the value of $f(a)$ is different from the limit L. In (b) the function is not defined at a (indicated by the open circle on the graph).

EXERCISES 2.5

1. Use the definition of limit to restate the following limits in terms of inequalities involving ϵ and δ.

   • (a) $\lim_{x \to 2} (3x - 1) = 5$

   (b) $\lim_{x \to -1} (x^2 + 2) = 3$

   (c) $\lim_{x \to 0} (5x^3 + 2x - 1) = -1$

   (d) $\lim_{x \to 20} (x^2 + 7) = 407$

2. Let $f(x) = [x]$. Examine the graph of f to determine if $f(x)$ has a limit as

   • (a) $x \to 2$

   (b) $x \to \frac{1}{3}$

   (c) $x \to 5$

   • (d) $x \to -\frac{1}{2}$

3. A function f has the graph shown in Figure 2.10. Find each of the following limits or explain why no limit exists:

   (a) $\lim_{x \to 1} f(x)$

   (b) $\lim_{x \to 2} f(x)$

   • (c) $\lim_{x \to 3} f(x)$

   (d) $\lim_{x \to 4} f(x)$

   (e) $\lim_{x \to 5} f(x)$

   (f) $\lim_{x \to 6} f(x)$

   • (g) $\lim_{x \to 7} f(x)$

   (h) $\lim_{x \to 8} f(x)$

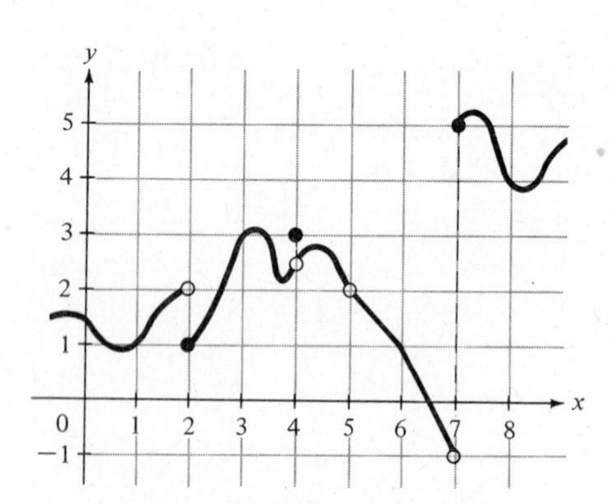

FIGURE 2.10 Exercise 3.

4. Let $\epsilon = 0.1$. Find a value of $\delta > 0$ such that if $0 < |x - 1| < \delta$, then $|f(x) - L| < \epsilon = 0.1$.

   (a) $f(x) = 2x, L = 2$;

   (b) $f(x) = x + 2, L = 3$;

   • (c) $f(x) = -x + 1, L = 0$;

   (d) $f(x) = 5x - 3, L = 2$.

5. Use the definition to establish the following limits:

   (a) $\lim_{x \to 2} (3x - 1) = 5$

   (b) $\lim_{x \to -1} (2x + 1) = -1$

   (c) $\lim_{x \to 1} (5x + 7) = 12$

   (d) $\lim_{x \to 0} (3x + 1) = 1$

2.6 THE LIMIT THEOREM

In actual practice we normally do not use the definition to calculate limits. Rather, we establish useful results about limits and apply them as needed. These results are stated in the following theorem. The proof is omitted.

The limit theorem

Theorem 2.1 (*The Limit Theorem*) Suppose that $\lim_{x\to a} f(x) = L$ and $\lim_{x\to a} g(x) = M$. Then

(a) $$\lim_{x\to a} [f(x) \pm g(x)] = L \pm M = \lim_{x\to a} f(x) \pm \lim_{x\to a} g(x).$$

The limit of the sum (or difference) is the sum (or difference) of the limits.

(b) $$\lim_{x\to a} [f(x) \cdot g(x)] = LM = \lim_{x\to a} f(x) \cdot \lim_{x\to a} g(x).$$

The limit of the product is the product of the limits.

(c) $$\lim_{x\to a} \frac{f(x)}{g(x)} = \frac{L}{M} = \frac{\lim_{x\to a} f(x)}{\lim_{x\to a} g(x)} \quad \text{provided } M \neq 0.$$

The limit of the quotient is the quotient of the limits, provided the limit of the denominator is not zero.

(d) $$\lim_{x\to a} \sqrt[n]{f(x)} = \sqrt[n]{L} = \sqrt[n]{\lim_{x\to a} f(x)}$$

provided $L > 0$ when n is an even integer.

The limit of the nth root is the nth root of the limit provided the limit is positive when n is even.

In order to calculate limits of polynomials we also need the following two special limits. The proof of the first is left for the reader. The proof of the second is in Example 3.

If $f(x) = C$ for all x, where C is a constant, then $\lim_{x\to a} f(x) = C$ for any real number a.

If $f(x) = x$ for all x, then $\lim_{x\to a} f(x) = a$ for any real number a.

These six rules for limits are summarized below in a convenient shorthand notation. The variables u and v are functions of x and C is a constant.

Shorthand notation

L1 $\lim_{x\to a} [u \pm v] = \lim_{x\to a} u \pm \lim_{x\to a} v.$

L2 $\lim_{x\to a} [uv] = \lim_{x\to a} u \cdot \lim_{x\to a} v.$

L3 $\lim_{x\to a} \dfrac{u}{v} = \dfrac{\lim_{x\to a} u}{\lim_{x\to a} v}$ provided $\lim_{x\to a} v \neq 0.$

L4 $\lim_{x\to a} \sqrt[n]{u} = \sqrt[n]{\lim_{x\to a} u}$ provided $\lim_{x\to a} u > 0$ when n is even.

L5 $\lim_{x\to a} C = C.$

L6 $\lim_{x\to a} x = a.$

Example 1 Calculate the following limits:

(a) $\lim_{x \to 7} (3x^2 + 8x + 2)$.

(b) $\lim_{x \to 1} \dfrac{x^2 - 1}{x - 1}$.

Solution

(a)
$$\begin{aligned}
&\lim_{x\to 7} (3x^2 + 8x + 2) \\
&= \lim_{x\to 7} 3x^2 + \lim_{x\to 7} 8x + \lim_{x\to 7} 2 \qquad \text{(by L1)} \\
&= \lim_{x\to 7} 3 \cdot \lim_{x\to 7} x \cdot \lim_{x\to 7} x + \lim_{x\to 7} 8 \cdot \lim_{x\to 7} x + 2 \\
&\qquad\qquad \text{(by L2 and L5)} \\
&= 3 \cdot 7 \cdot 7 + 8 \cdot 7 + 2 = 205 \qquad \text{(by L5 and L6).}
\end{aligned}$$

(b) Observe that $(x^2 - 1)/(x - 1)$ is not defined when $x = 1$, but is equal to $x + 1$ when $x \neq 1$. Since we are not interested in the value of the function when $x = 1$, we may assume that $x \neq 1$. Thus

$$\lim_{x\to 1} \frac{x^2 - 1}{x - 1} = \lim_{x\to 1} (x + 1) = \lim_{x\to 1} x + \lim_{x\to 1} 1 = 1 + 1 = 2.$$

Example 2 Use the properties of limits to show that if $g(x) = Cf(x)$ for each x (where C is a constant), then

$$\lim_{x\to a} g(x) = C \cdot \lim_{x\to a} f(x).$$

The limit of a constant times a function is equal to the constant times the limit of the function.

Solution

$$\begin{aligned}
\lim_{x\to a} g(x) &= \lim_{x\to a} C \cdot f(x) \\
&= \lim_{x\to a} C \cdot \lim_{x\to a} f(x) \qquad \text{(by L2)} \\
&= C \cdot \lim_{x\to a} f(x) \qquad \text{(by L5).}
\end{aligned}$$

This useful fact is written in shorthand notation as

L7	$\lim_{x\to a} Cu = C \cdot \lim_{x\to a} u.$

Example 3 Use the definition of a limit to prove **L6**. If $f(x) = x$ for all x then $\lim_{x\to a} f(x) = a$.

Solution Let ϵ be a positive number, let $L = a$. We choose $\delta = \epsilon$. If $0 < |x - a| < \delta$, then

$$|f(x) - L| = |x - a| < \delta = \epsilon.$$

Since $|f(x) - L| < \epsilon$ whenever $0 < |x - a| < \delta$, then

$$\lim_{x\to a} f(x) = L,$$

$$\lim_{x\to a} x = a.$$

EXERCISES 2.6

1. Use the basic properties of limits to calculate the following limits. Be sure to state explicitly which formula you are using at each step. Do not omit any steps. (For (q) use Exercise 2.)

(a) $\lim_{x\to 2} (2x^4 - 5x^3 + 6x - 10)$

• (b) $\lim_{x\to 0} \frac{x^2 - 2x - 1}{x + 1}$

(c) $\lim_{x\to 1} \frac{x^2 - 2x + 1}{x - 1}$

• (d) $\lim_{x\to 2} \frac{x^4 - 16}{x^2 + 4}$

(e) $\lim_{x\to -3} \frac{x^2 + 2x - 3}{x + 3}$

(f) $\lim_{x\to -2} \frac{x^3 + 8}{x^2 - 4}$

(g) $\lim_{x\to 4} \frac{x - 2}{x - 4}$

(h) $\lim_{x\to 1} \sqrt[3]{x^2 - x - 8}$

(i) $\lim_{x\to 2} \sqrt[3]{\frac{x^2 - 4}{x - 2}}$

(j) $\lim_{x\to 1} \frac{x^2 + 2x - 3}{x^2 - 1}$

(k) $\lim_{x\to 2} \frac{x^3 + 2x^2 + x + 5}{x^2 - 7x + 13}$

(l) $\lim_{x\to -1} \frac{x^3 + x^2}{x^2 - x - 2}$

(m) $\lim_{x\to 1} \frac{x^4 - 1}{x^2 + 2x + 1}$

(n) $\lim_{x\to 1} \frac{x^4 - 2x^2 + 1}{(x - 1)(x^2 - 1)}$

(o) $\lim_{x\to 2} \frac{(x - 1)(1 - x^2)}{x^4 + 2x^2 + 1}$

• (p) $\lim_{x\to -1} \frac{2x^2 + x - 1}{x^4 - 1}$

(q) $\lim_{x\to 2} x - 2|x|$

(r) $\lim_{x\to 2} \frac{\sqrt[3]{x^2 + 4}}{x}$

(s) $\lim_{x\to 2} \frac{x^2 + 1 - x}{1 + x}$

(t) $\lim_{x\to 1} \frac{x^2 + 3x - 1}{x + 1}$

2. Draw the graph of $f(x) = |x|$. Show that $\lim_{x\to a} |x| = |a|$. (*Hint:* Consider the three cases $a > 0$, $a = 0$, $a < 0$.)

2.7 CONTINUOUS FUNCTIONS

Most of the functions considered thus far have had the property that the limit as x approaches a is equal to the value of the function at a. Such functions are said to be continuous at a. As we shall see later, it is usually necessary to have a function continuous in order to apply the principles of calculus.

Continuity at a point

DEFINITION We say that the function f is *continuous at the point a* provided

$$\lim_{x\to a} f(x) = f(a).$$

Thus, if f is continuous at a, then f must be defined at a, the limit of $f(x)$ as x approaches a must exist, and, furthermore, this limit must be equal to the number $f(a)$ (Figure 2.11). Figure 2.12 illustrates several of the ways that a function can be *discontinuous* (that is, *not continuous*) at a point. Most of the functions considered thus far are continuous at all points of their domains.

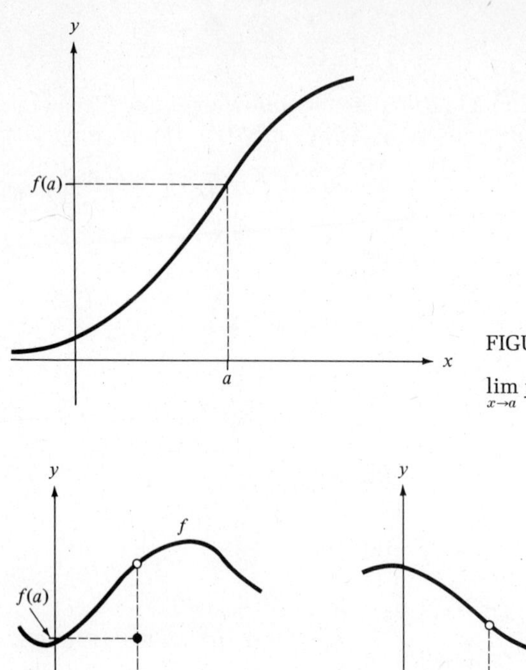

FIGURE 2.11 f is continuous at a: $\lim_{x \to a} f(x) = f(a)$.

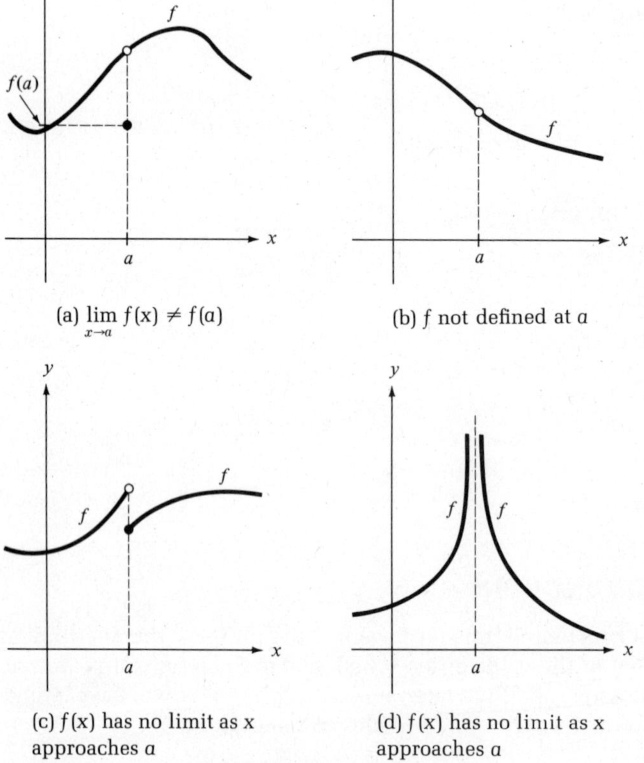

(a) $\lim_{x \to a} f(x) \neq f(a)$

(b) f not defined at a

(c) $f(x)$ has no limit as x approaches a

(d) $f(x)$ has no limit as x approaches a

FIGURE 2.12 Examples of discontinuous functions.

Example 1 Let $f(x) = (x^2 - 1)/(x + 2)$. Discuss the continuity of f.

Solution f is defined for all values of x except $x = -2$. If a is any point of the domain of f, then by the limit theorem:

$$\lim_{x\to a} f(a) = \lim_{x\to a} \frac{x^2 - 1}{x + 2} = \frac{\lim_{x\to a} (x^2 - 1)}{\lim_{x\to a} (x + 2)}$$

$$= \frac{\lim_{x\to a} x \cdot \lim_{x\to a} x - 1}{\lim_{x\to a} x + 2} = \frac{a^2 - 1}{a + 2} = f(a).$$

Thus, f is continuous at each point a that is not equal to -2. Since f is not defined for $x = -2$, then f is discontinuous there.

It is easy to use the limit theorem to establish the following results.

Theorem 2.2 Let f and g be continuous at $x = a$. Then

(a) $[f(x) \pm g(x)]$ is continuous at $x = a$,
(b) $[f(x) \cdot g(x)]$ is continuous at $x = a$,
(c) $[f(x)/g(x)]$ is continuous at $x = a$ provided $g(a) \neq 0$,
(d) $\sqrt[n]{f(x)}$ is continuous at $x = a$ provided $f(a) > 0$ when n is even.

PROOF OF (c)

$$\lim_{x\to a} \left[\frac{f(x)}{g(x)}\right] = \frac{\lim_{x\to a} f(x)}{\lim_{x\to a} g(x)} = \frac{f(a)}{g(a)} \qquad \text{provided } g(a) \neq 0.$$

The proofs of (a), (b), and (d) are similar.

Properties L5 and L6 can be restated for continuous functions as follows:

Theorem 2.3 (a) The constant function $f(x) = C$ is continuous everywhere. (b) The function $f(x) = x$ is continuous everywhere.

COROLLARY If $f(x)$ is a polynomial, then f is continuous everywhere.

PROOF Observe first that, by Theorems 2.2b and 2.3b, the power functions $x, x^2, x^3, \ldots$ are continuous everywhere. If we write

$$f(x) = b_n x^n + b_{n-1} x^{n-1} + \cdots + b_1 x + b_0$$

we see that f is a sum of products of continuous functions (the constant functions $b_n, b_{n-1}, \ldots, b_1, b_0$ and the functions $x, x^2, x^3, \ldots, x^n$). Thus, it follows from (a) and (b) of Theorem 2.2 that f is continuous everywhere.

Example 2 Use Theorem 2.2 and the Corollary to discuss the continuity of

$$f(x) = \frac{7x^2 - 3x + 7}{x^2 - 3x + 2}.$$

Solution $f(x)$ is a quotient of two polynomials, both of which are continuous at each real number a. The quotient is continuous at each number for which the denominator is not zero. It is easy to see that the denominator is zero only when $x = 1$ or $x = 2$. Therefore, $f(x)$ is continuous for all real numbers except 1 and 2.

Example 3 Calculate $\lim_{x\to 3} \sqrt{4x^2 - 7}$.

Solution The function $4x^2 - 7$ has a positive limit as $x \to 3$. Therefore, by Theorem 2.2d,

$$\lim_{x\to 3} \sqrt{4x^2 - 7} = \sqrt{\lim_{x\to 3} (4x^2 - 7)} = \sqrt{4 \cdot 3^2 - 7} = \sqrt{29}.$$

Continuity on an interval

In most of our work we shall deal with functions that are continuous at all points of an interval on the x-axis. (An interval is the set of all points between two fixed points a and b. The points a and b may or may not be included.) The graph of such a function is unbroken on the interval. Hence, it can be drawn without lifting the pencil from the paper. (See Figure 2.13.)

In our earlier work on graphing we assumed that the graphs of the power functions are continuous; the reader should verify that this is true. (See Exercise 4.)

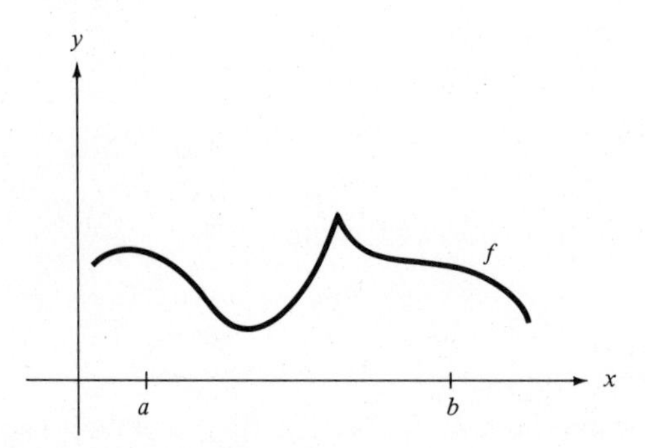

FIGURE 2.13 f is continuous on an interval containing a and b. The graph of f can be drawn between a and b without lifting the pencil from the paper.

Example 4 The function $f(x) = [x]$ (the greatest integer function) is continuous everywhere except at the integers. (See Figure 2.14.) If a and b are between consecutive integers on the x-axis, then the graph of f is unbroken on the interval $[a, b]$. If, on the other hand, one or more integers are between a and b then the graph is broken at each of these integers.

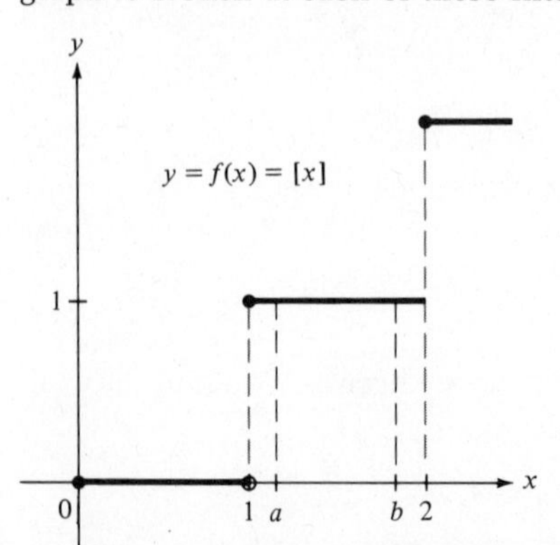

(a) The graph is unbroken between a and b if a and b are between consecutive integers.

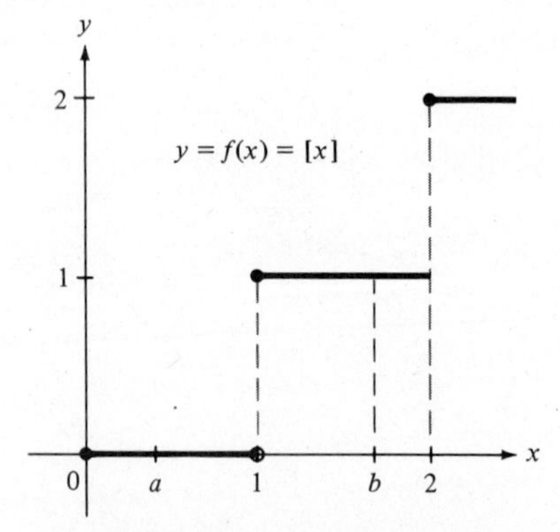

(b) The graph is broken between a and b if an integer is between a and b.

FIGURE 2.14 Example 4. The greatest integer function is discontinuous at the integers and is continuous everywhere else.

Differentiability and Continuity Most of the functions that we consider in this book have derivatives almost everywhere in their domains. The following theorem, which is needed in Section 2.9, shows that such functions are continuous everywhere that they are differentiable. In other words, if the function g has a derivative at the point $x = a$, then

$$\lim_{x\to a} g(x) = g(a).$$

Theorem 2.4 If the function g is differentiable at $x = a$, then

Differentiability implies continuity

$$\lim_{x\to a} g(x) = g(a).$$

PROOF The function g must be defined at a in order for $g'(a)$ to exist. Write the identity

$$g(x) = g(a) + \frac{g(x) - g(a)}{x - a} \cdot (x - a)$$

and take the limit as $x \to a$:

$$\lim_{x\to a} g(x) = \lim_{x\to a} g(a) + \lim_{x\to a} \frac{g(x) - g(a)}{x - a} \cdot \lim_{x\to a} (x - a) \qquad \text{(by L1 and L2)}$$

$$= g(a) + g'(a) \cdot 0 = g(a).$$

Theorem 2.4 shows that it is impossible for a function to be discontinuous at a point where it has a derivative. Thus, such a function cannot have a break in its graph at a point where the derivative exists.

Many of the functions that we study in this book have derivatives everywhere. Such functions are continuous everywhere.

Example 5 Show that the function $f(x) = 2x^2 - 5x$ is continuous everywhere.

Solution We established in Example 3 of Section 2.3 that $f'(x) = 4x - 5$ for every value of x. Since f has a derivative everywhere it is continuous everywhere.

EXERCISES 2.7

1. The graph of a function f is pictured in Figure 2.15. Decide if it is continuous or discontinuous at each of the following points:

- • (a) $x = a$
- • (b) $x = b$
- (c) $x = c$
- (d) $x = d$
- (e) $x = e$
- (f) $x = f$

2. Discuss the continuity of the following functions. You may use Exercise 5 if it is needed.

- (a) $f(x) = 5x + 7$
- (b) $f(x) = x^2 + 2x + 1$
- • (c) $f(x) = \dfrac{2}{x + 1}$
- (d) $f(x) = \dfrac{-x}{x - 1}$
- • (e) $f(x) = \dfrac{x - 2}{x + 3}$
- • (f) $f(x) = \dfrac{x^2 - 2x + 1}{x^2 + 2x - 3}$

(g) $f(x) = \sqrt{x^2 + 2}$

(h) $f(x) = \dfrac{2x^2 - 3x + 1}{x^2 - 2x + 1}$

(i) $f(x) = \dfrac{4x^2 - 12x + 9}{2x^2 + x - 6}$

• (j) $f(x) = \dfrac{x^2 - 2x + 6}{x^2 + x + 1}$

(k) $f(x) = \dfrac{x^2 + x + 1}{2x^2 - 3x + 2}$

(l) $f(x) = \dfrac{x^2}{(x + 1)(x^2 - 1)}$

• (m) $f(x) = x - |x|$

(n) $f(x) = \dfrac{x}{|x|}$

3. Determine all points of discontinuity, if any, of the following functions. You may use Exercise 5 if it is needed.

(a) $f(x) = \sqrt{x^2 + 1}$

(b) $f(x) = \dfrac{|x|}{x + 1}$

(c) $f(x) = \dfrac{5}{\sqrt{x} - 2}$

(d) $f(x) = \dfrac{x^2 + 1}{x^2 - 9}$

(e) $f(x) = \dfrac{x^3 - x^2 - x - 1}{(x - 1)^2(x - 2)^2x^4}$

(f) $f(x) = \sqrt[3]{\dfrac{x^2 - 1}{x^2 + 2}}$

(g) $f(x) = \dfrac{\sqrt{|x|}}{|x|}$

(h) $f(x) = 1/x^2 + 1$

Continuity of power functions and the absolute value function

4. Show that the power functions are continuous everywhere.

5. Show that the function $f(x) = |x|$ is continuous everywhere.

6. A continuous function f is known to have the values $f(0) = -1$ and $f(1) = 1$. Explain why the graph must cross the x-axis between 0 and 1. Would this conclusion necessarily hold if the function were not continuous?

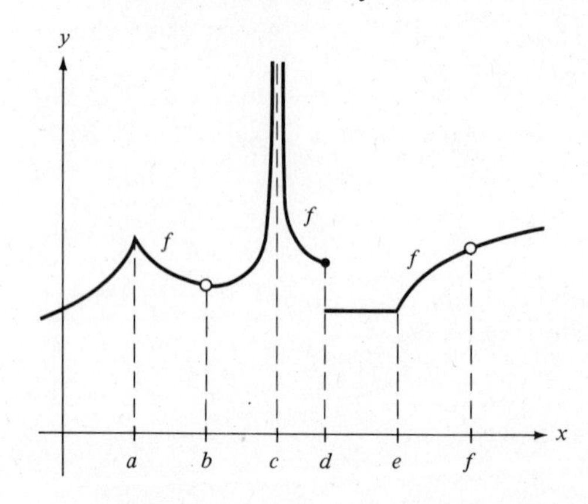

FIGURE 2.15 Exercise 1.

2.8 BASIC DERIVATIVE FORMULAS

As we have seen, the computation of a derivative from the definition is quite laborious. The work can be greatly simplified by establishing a few basic formulas. These formulas enable us to calculate most derivatives with comparative ease.

Two special derivatives are easy to compute:

D1 *The derivative of a constant function is zero.*

If $\quad f(x) = C \quad$ for all x,

then

$$f'(x) = 0.$$

D2 *The derivative of the function*

$$f(x) = x \quad \text{is} \quad f'(x) = 1.$$

These formulas are usually written in shorthand notation as

D1 $$\frac{d}{dx}(C) = 0$$

D2 $$\frac{d}{dx}(x) = 1.$$

The following two formulas are basic. They are used in almost all of the derivative calculations in this book. (The symbols u and v represent functions of x; C is a constant.)

D3 $$\frac{d}{dx}(u \pm v) = \frac{du}{dx} \pm \frac{dv}{dx}.$$

The derivative of a sum (or difference) is the sum (or difference) of the derivatives.

D4 $$\frac{d}{dx}(Cu) = C\frac{du}{dx}.$$

The derivative of a constant times a function of x is the same constant times the derivative of the function.

Example 1

$$\frac{d}{dx}(5x + 7) = \frac{d}{dx}(5x) + \frac{d}{dx}(7) \qquad \text{(by D3)}$$
$$= 5 \cdot \frac{d(x)}{dx} + 0 \qquad \text{(by D4, D1)}$$
$$= 5 \cdot 1 = 5 \qquad \text{(by D2).}$$

In Section 2.9 we will establish that

The power rule

D5 $$\frac{d}{dx}(x^n) = nx^{n-1} \qquad \text{(the power rule)}$$

for every positive integer n. This useful formula enables us to calculate a very large number of derivatives almost automatically.

Example 2

(a) $$\frac{d}{dx}(x^5) = 5x^4 \qquad \text{(by D5).}$$

(b) $$\frac{d}{dt}(7t^3) = 7 \cdot \frac{d}{dt}(t^3) = 7 \cdot 3t^2 = 21t^2.$$

(c) $$\frac{d}{dx}(4x^2 - 6x + 2) = 4 \cdot \frac{d}{dx}(x^2) - 6 \cdot \frac{d}{dx}(x) + \frac{d}{dx}(2)$$
$$= 4 \cdot 2x - 6 \cdot 1 + 0 = 8x - 6.$$

The derivative of any polynomial can be calculated by the process illustrated in Example 2c. If the polynomial is

$$f(x) = a_n x^n + a_{n-1}x^{n-1} + \cdots a_2x^2 + a_1x + a_0,$$

then its derivative is

$$f'(x) = na_n x^{n-1} + (n-1)a_{n-1}x^{n-2} + \cdots 2a_2x + a_1.$$

Example 3

(a) $$\frac{d}{dx}(5x^3 + 4x^2 - x + 2) = 5 \cdot 3x^2 + 4 \cdot 2x - 1$$
$$= 15x^2 + 8x - 1.$$

(b) $$\frac{d}{dx}(3x^{100} + x^{50} - 3x + \sqrt{17}) = 300x^{99} + 50x^{49} - 3.$$

The statements of the derivative formulas involve functions of x. If a different variable is used, then the statements can be modified accordingly as in Example 2b.

In the remainder of this section we give proper statements and proofs of derivative formulas D1 to D4. The proofs of the formulas are simplified by use of the expression

$$f'(a) = \lim_{x \to a} \frac{f(x) - f(a)}{x - a}$$

for the derivative.

Theorem 2.5 **(D1)** If $f(x) = C$ for all x, then $f'(x) = 0$.

PROOF Let a be a real number. Then

$$f'(a) = \lim_{x \to a} \frac{f(x) - f(a)}{x - a} = \lim_{x \to a} \frac{C - C}{x - a}$$
$$= \lim_{x \to a} 0 = 0 \qquad \text{(by L5)}.$$

Since a can be any real number, then

$$f'(x) = 0 \qquad \text{for all } x.$$

Theorem 2.6 **(D2)** If $f(x) = x$ for all x, then

$$f'(x) = 1.$$

The proof, which is almost identical to the proof of Theorem 2.5, is left for the reader. (Exercise 7.)

Theorem 2.7 **(D3)** Let f and g be differentiable functions. If $h(x) = f(x) \pm g(x)$, then

$$h'(x) = f'(x) \pm g'(x).$$

PROOF Let a be a real number. Then

$$h'(a) = \lim_{x \to a} \frac{h(x) - h(a)}{x - a} = \lim_{x \to a} \frac{[f(x) \pm g(x)] - [f(a) \pm g(a)]}{x - a}$$

$$= \lim_{x \to a} \frac{[f(x) - f(a)] \pm [g(x) - g(a)]}{x - a}$$

$$= \lim_{x \to a} \frac{f(x) - f(a)}{x - a} \pm \lim_{x \to a} \frac{g(x) - g(a)}{x - a} \qquad \text{(by L1)}$$

$$= f'(a) \pm g'(a).$$

Since a can be any real number, then

$$h'(x) = f'(x) \pm g'(x) \quad \text{for all } x.$$

Theorem 2.8 (D4) Let f be a differentiable function. Let $h(x) = Cf(x)$, where C is a constant. Then

$$h'(x) = Cf'(x).$$

The proof is left for the reader. (Exercise 8.)

EXERCISES 2.8

1. Calculate the derivative. (Use the results of this section.)

   (a) $f(x) = 5$
   (b) $f(x) = 5x$
   (c) $f(x) = 2x + 1$
   • (d) $f(x) = 1 + \sqrt{2}$
   (e) $f(x) = 4x - 1$
   • (f) $f(x) = 5x^2$
   (g) $f(x) = 3x^2 + 6x$
   (h) $f(x) = 4x^2 - 3x + 1$
   (i) $f(x) = -2x^2 - 4x + 2$
   (j) $f(x) = 6x^3$
   • (k) $f(x) = 2x^3 + 3x^2 - 5x + 8$
   (l) $f(x) = x^5 + 7x^4 - 4x^2 + 3$
   (m) $f(x) = -x^4 + 4x^3 + 2x^2 - 3x + 1$
   (n) $f(x) = 4x^4 + 3x^3 + 2x^2 + x$
   • (o) $f(x) = x^9 + 6x^2 - x + 1$
   (p) $f(x) = x^7 - 6x^5 + 12x^2 + 3$
   (q) $f(x) = 10x^3 - 3x + 7$
   • (r) $f(x) = 2x^{12} + x^7 + 2x^5 + 4x - 1$
   (s) $f(x) = 8x^7 + 2x^3 - 16x^2 + 7x$
   (t) $f(x) = 4x^{10} - 8x^8 + 7x^5 - 124$

2. Determine all points on the graph of f where the tangent line has slope equal to zero.

   (a) $f(x) = 2x^3$
   (b) $f(x) = -2x^2 + 1$
   • (c) $f(x) = 3x^2 + 2x + 3$
   (d) $f(x) = 3x^2 - x + 1$
   (e) $f(x) = 2x^3 + 2.5x^2 - 4x - 7$
   • (f) $f(x) = x^3 + 1.5x^2 + 1$
   (g) $f(x) = 6x^2 - 3x + 1$
   (h) $f(x) = x^3 + 3x^2 - 2$
   • (i) $f(x) = 3x^3 + 3x^2 + x - 2$
   (j) $f(x) = x^3 + x^2 - x - 1$

•3. Let $f(x) = x^3 + 4x^2 - 5x - 2$. Find where the line tangent to the graph at the point $(1, -2)$ crosses the x-axis and where it crosses the y-axis.

4. Let $f(x) = 5x^2 + 6x - 7$. Determine the point on the graph of f where the tangent line has slope equal to 16.

5. Find the rate of change of the volume of a cube with respect to an edge.

•6. Find the rate of change of the area of a circle with respect to the radius.

7. Prove Theorem 2.6.

8. Prove Theorem 2.8.

2.9 THE PRODUCT, QUOTIENT, AND POWER RULES

It would be natural to conjecture that the derivative of the product of two functions should be the product of their derivatives. This is not the case, as we can see by choosing almost any two functions. For example, if $f(x) = 5x$ and $g(x) = x^2$, then the derivative of the product is $15x^2$, while the product of the derivatives is $10x$.

The correct formula for the derivative of a product is given in shorthand notation as

The product rule

D6 $$\frac{d}{dx}(uv) = u \cdot \frac{dv}{dx} + v \cdot \frac{du}{dx}.$$

The derivative of the product of two functions is equal to the first function times the derivative of the second function plus the second function times the derivative of the first.

Example 1 Calculate the derivative of $(5x^2 - 2)(2x + 3)$.

Solution Letting $u = 5x^2 - 2$ and $v = 2x + 3$ we have

$$\begin{aligned}\frac{d}{dx}(uv) &= u \cdot \frac{dv}{dx} + v \cdot \frac{du}{dx}\\ &= (5x^2 - 2) \cdot \frac{d}{dx}(2x + 3) + (2x + 3) \cdot \frac{d}{dx}(5x^2 - 2)\\ &= (5x^2 - 2) \cdot 2 + (2x + 3) \cdot 10x\\ &= 30x^2 + 30x - 4.\end{aligned}$$

The formal statement of the product rule is given in the following theorem:

Theorem 2.9 (D6) Let f and g be differentiable functions of x. Let $h(x) = f(x) \cdot g(x)$. Then

$$h'(x) = f(x)g'(x) + g(x)f'(x).$$

PROOF Let a be a fixed number. Then

$$h'(a) = \lim_{x\to a} \frac{h(x) - h(a)}{x - a}.$$

Before calculating the limit we modify the form of the difference quotient $(h(x) - h(a))/(x - a)$. In the second step we add and subtract $f(x)g(a)$ to the numerator.

$$\begin{aligned}\frac{h(x) - h(a)}{x - a} &= \frac{f(x)g(x) - f(a)g(a)}{x - a}\\ &= \frac{f(x)g(x) - f(x)g(a) + f(x)g(a) - f(a)g(a)}{x - a}\\ &= f(x)\frac{g(x) - g(a)}{x - a} + \frac{f(x) - f(a)}{x - a}\,g(a).\end{aligned}$$

The limit of the first term (by Theorem 2.4) is

$$\lim_{x \to a} f(x) = f(a)$$

while the limits of the difference quotients are the derivatives $g'(a)$ and $f'(a)$. Thus,

$$\begin{aligned} h'(a) &= \lim_{x \to a} \frac{h(x) - h(a)}{x - a} \\ &= \lim_{x \to a} \left[f(x) \frac{g(x) - g(a)}{x - a} + \frac{f(x) - f(a)}{x - a} g(a) \right] \\ &= f(a)g'(a) + f'(a)g(a). \end{aligned}$$

Since a could be any number, then

$$h'(x) = f(x)g'(x) + g(x)f'(x)$$

for every x.

The Power Rule For the special case where $v = u$, the product rule gives us

$$\frac{d}{dx}(u^2) = \frac{d}{dx}(u \cdot u) = u\frac{du}{dx} + u\frac{du}{dx} = 2u\frac{du}{dx}.$$

For the special case $v = u^2$, we obtain

$$\begin{aligned} \frac{d}{dx}(u^3) &= \frac{d}{dx}(u \cdot u^2) = u\frac{d}{dx}(u^2) + u^2\frac{du}{dx} \\ &= u \cdot 2u\frac{du}{dx} + u^2\frac{du}{dx} = 3u^2\frac{du}{dx}. \end{aligned}$$

If we continue on in this fashion, we obtain

$$\frac{d}{dx}(u^4) = \frac{d}{dx}(u \cdot u^3) = 4u^3\frac{du}{dx},$$

$$\frac{d}{dx}(u^5) = \frac{d}{dx}(u \cdot u^4) = 5u^4\frac{du}{dx},$$

and, in general, if n is a positive integer,

$$\frac{d}{dx}(u^n) = nu^{n-1}\frac{du}{dx}.$$

For the special case where $u(x) = x$ we obtain

$$\frac{d}{dx}(x^n) = nx^{n-1},$$

the formula used in Section 2.8 to differentiate polynomials.

We thus have two very important formulas obtained from D6:

D5 (The power rule)

$$\frac{d}{dx}(x^n) = nx^{n-1},$$

$$\frac{d}{dx}(u^n) = nu^{n-1}\frac{du}{dx}.$$

Remark The general power rule D5 can be used when the exponent n is any real number. The exponent does not have to be an integer.

Example 2 Calculate $\frac{d}{dx}\sqrt{2x^2-3}=\frac{d}{dx}(2x^2-3)^{1/2}$.

Solution Let $u = 2x^2 - 3$. Then

$$\frac{d}{dx}(2x^2-3)^{1/2}=\frac{d}{dx}(u^{1/2})=\frac{1}{2}u^{-1/2}\frac{du}{dx}$$
$$=\frac{1}{2}(2x^2-3)^{-1/2}(4x)=\frac{2x}{\sqrt{2x^2-3}}.$$

The Quotient Rule The formula for the derivative of a quotient is given in shorthand notation as

The quotient rule

D7	$\frac{d}{dx}\left(\frac{u}{v}\right)=\frac{v\frac{du}{dx}-u\frac{dv}{dx}}{v^2}$	(provided $v \neq 0$).

The derivative of a quotient is equal to the denominator times the derivative of the numerator minus the numerator times the derivative of the denominator, divided by the square of the denominator.

Example 3 Calculate

$$\frac{d}{dx}\left(\frac{3x^2-1}{x^2+5}\right).$$

Solution Let $u = 3x^2 - 1$, $v = x^2 + 5$. Then

$$\frac{d}{dx}\left(\frac{u}{v}\right)=\frac{v\cdot\frac{du}{dx}-u\frac{dv}{dx}}{v^2}$$
$$=\frac{(x^2+5)\cdot\frac{d}{dx}(3x^2-1)-(3x^2-1)\cdot\frac{d}{dx}(x^2+5)}{(x^2+5)^2}$$
$$=\frac{(x^2+5)6x-(3x^2-1)2x}{(x^2+5)^2}=\frac{32x}{(x^2+5)^2}.$$

Theorem 2.10 Let f and g be differentiable functions of x. Let $h(x)=f(x)/g(x)$. Then

$$h'(x)=\frac{g(x)f'(x)-f(x)g'(x)}{[g(x)]^2}\qquad\text{(provided } g(x)\neq 0).$$

PROOF Let a be a fixed number. Then

$$h'(a)=\lim_{x\to a}\frac{h(x)-h(a)}{x-a}.$$

As in the previous proof we work with the difference quotient before calculating the limit:

$$\frac{h(x)-h(a)}{x-a}=\frac{\dfrac{f(x)}{g(x)}-\dfrac{f(a)}{g(a)}}{x-a}=\frac{\dfrac{g(a)f(x)-f(a)g(x)}{g(x)g(a)}}{\dfrac{x-a}{1}}$$

$$=\frac{1}{g(x)g(a)}\cdot\frac{g(a)f(x)-f(a)g(x)}{x-a}$$

$$=\frac{1}{g(x)g(a)}\frac{g(a)f(x)-g(a)f(a)+f(a)g(a)-f(a)g(x)}{x-a}$$

$$=\frac{1}{g(x)g(a)}\left\{g(a)\frac{f(x)-f(a)}{x-a}+f(a)\frac{g(a)-g(x)}{x-a}\right\}.$$

We now let x approach a. Then (by Theorem 2.4)

$$g(x)\to g(a),$$

while

$$\frac{f(x)-f(a)}{x-a}\to f'(a)$$

and

$$\frac{g(a)-g(x)}{x-a}\to -g'(a).$$

Therefore,

$$h'(a)=\lim_{x\to a}\frac{h(x)-h(a)}{x-a}$$

$$=\lim_{x\to a}\frac{1}{g(x)g(a)}\left\{g(a)\frac{f(x)-f(a)}{x-a}+f(a)\frac{g(a)-g(x)}{x-a}\right\}$$

$$=\frac{1}{g(a)\cdot g(a)}\{g(a)f'(a)+f(a)[-g'(a)]\}$$

$$=\frac{g(a)f'(a)-f(a)g'(a)}{[g(a)]^2}.$$

Since a can be any number, then

$$h'(x)=\frac{g(x)f'(x)-f(x)g'(x)}{[g(x)]^2}.$$

EXERCISES 2.9

1. Use the product rule D6 to calculate the derivative.
   - (a) $(x^3+1)(x^2-1)$
   - • (b) $(2x^2-1)(x^2+x+1)$
   - (c) $(3x^2+x)(x^2+5x-1)$
   - (d) $(x^3+x^2+x)(x^2-2x+1)$
   - • (e) $(5x^4+1)(x^5+1)$
   - (f) $(x^7+x^5+x^3+x)(1+x^2-x^4+x^6)$
2. Use the quotient rule D7 to calculate the derivative.

(a) $\frac{x}{x-1}$

• (b) $\frac{2}{x}$

(c) $\frac{x^2}{x+1}$

• (d) $\frac{x^2+1}{x^2-1}$

(e) $\frac{x^2-x+1}{2x^2+x+1}$

• (f) $\frac{x^2+3x+1}{x^2-x}$

(g) $\frac{x^4+3x^2-1}{(x^2+1)(x^2-1)}$

(h) $\frac{x^3-x^2+x-2}{x^2+2x+1}$

• (i) $\frac{2x^3-x^2}{2x^2+4x}$

(j) $\frac{x^3+6x^2-3x+1}{7x^3+3x^2-3x+2}$

3. Use the general power rule D5 to calculate the derivative.

(a) $(2x+5)^3$
• (b) $(3-2x)^4$
(c) $(2x^4+3x^2+4x)^7$
• (d) $(x^2-2x+3)^9$
(e) $(7x^5-12x^3+5x)^5$
(f) $(3x^6+4x^5-7x^3+8)^4$
(g) $(4x^7+8x^5-4x^3)^2$
(h) $(x^3+8x^2-4)^9$
• (i) $(x^2+x+17)^{1/5}$
(j) $\sqrt{3x+1}$
• (k) $\sqrt{8+4x}$
(l) $\sqrt[3]{x^3-13}$
(m) $(7x^2+17x+13)^{1/4}$
• (n) $(x^2+4)^{5/2}$
(o) $(x^3+3x^2-8x+1)^{-1/2}$
(p) $(3x^2-4x-2)^{-4}$
(q) $(2x^3+x^2)^{-3}$
• (r) $(5x^2+10x-7)^{5/4}$
(s) $(6x^2-1)^{1/2}$
(t) $(7x^2+3x+8)^{-3/4}$

4. Calculate the derivative.

(a) $(x^2+1)^5(x^2-1)^3$

• (b) $\frac{(x^2-1)^5}{4x}$

(c) $(x+\sqrt{x})^{1/2}$

• (d) $\frac{3x^2-x+2}{\sqrt{x-1}}$

(e) $\left(\frac{x^2-1}{x^2+1}\right)^{1/3}$

(f) $(5x^2+1)^{3/2}$

• (g) $(x^2-1)\sqrt{x^2+1}$

(h) $\frac{3x^2-1}{x^2+1}$

5. Write the equation of the tangent line to the graph of f at the given point.

• (a) $f(x)=\frac{x+1}{x-1}$, (2, 3)

(b) $f(x)=(x^2-1)^2(x^2+1)^3$, (0, 1)

(c) $f(x)=\frac{1}{\sqrt{x-1}}$, (2, 1)

• (d) $f(x)=\sqrt{x}$, (9, 3)

2.10 COMPOSITION OF FUNCTIONS. THE CHAIN RULE

Let $y=f(u)$ be a function of u where $u=g(x)$ is a function of x. Then y is also a function of x, say $y=h(x)$, where

$$y=h(x)=f(u)=f(g(x))$$

for each x in the domain of h.

For example, if $y=3u-1$ and $u=2x+7$, then

$$y=3u-1=3(2x+7)-1=6x+20.$$

Thus, $h(x)=6x+20$.

This new function h is called the *composition of f with g*. It is denoted by writing

$$h(x) = f(g(x)).^*$$

The domain of $f(g(x))$ consists of all x in the domain of g with the property that g(x) is in the domain of f. In order to calculate $f(g(x))$ we substitute the functional value of g(x) into the functional value of $f(x)$ as in the above example. Composition is represented schematically in Figure 2.16.

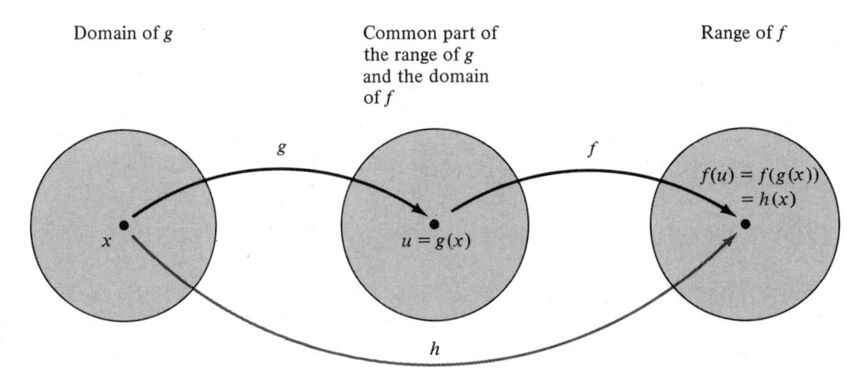

FIGURE 2.16 Schematic representation of composition of functions. The function $h(x) = f(g(x))$ is calculated in two stages:
(1) Calculate $u = g(x)$;
(2) Calculate $f(u) = f(g(x))$.
This process defines $h(x) = f(g(x))$ as a function of x.

Composition of functions

Composition of functions occurs naturally in many practical applications. For example, the monthly production (p) of a factory may be considered to be a function of the number (m) of man-hours worked,

$$p = f(m).$$

The number (m) of man-hours, however, is a function of the money (w) paid for wages,

$$m = g(w).$$

Thus

$$p = f(m) = f(g(w)),$$

so that p is a composite function of the wages. Consequently, it is as valid to consider production to be a function of wages as of man-hours.

In many of the derivative problems that occur later in this book it will be almost essential to consider complicated functions as compositions of simpler ones. Essentially, this involves making a substitution as an intermediate step. The following example illustrates the process.

Example 1 Let $y = (3/x)^2 - 4 \cdot (3/x) + 7$. Simplify the expression for y by a substitution $u = g(x)$.

---

* Some books use $f \circ g$ for the composition of f with g.

Solution For all practical purposes, y is a function of $3/x$. If we make the substitution

$$u = g(x) = \frac{3}{x}$$

then

$$y = u^2 - 4u + 7 = f(u) \qquad \text{where } u = \frac{3}{x}.$$

Observe that

$$y = f(u) = f(g(x)),$$

so that y is the composition of f with g.

The Derivative of a Composite Function We have already considered the derivative of one type of composite function. If we analyze the general power rule we can see that if $y = u^n$, then

$$\frac{dy}{du} = nu^{n-1}$$

so that

$$\frac{dy}{dx} = nu^{n-1}\frac{du}{dx} = \frac{dy}{du} \cdot \frac{du}{dx}.$$

It can be shown that the formula

$$\frac{dy}{dx} = \frac{dy}{du} \cdot \frac{du}{dx}$$

holds for an arbitrary composite function, not just for a power of u. This formula, known as *the Chain Rule*, can be stated alternately as

$$\text{If } y = f(u), \text{ then } \frac{dy}{dx} = f'(u)\frac{du}{dx}.$$

Example 2 Use the Chain Rule to calculate dy/dx if $y = u^2 - 4u + 7$ and $u = 3/x$.

Solution

$$\frac{dy}{du} = 2u - 4, \qquad \frac{du}{dx} = \frac{-3}{x^2}.$$

Thus,

$$\frac{dy}{dx} = \frac{dy}{du} \cdot \frac{du}{dx} = (2u - 4)\left(-\frac{3}{x^2}\right).$$

If we wish, we can express dy/dx as a function of x by substituting $u = 3/x$:

$$\frac{dy}{dx} = (2u - 4)\left(-\frac{3}{x^2}\right) = \left(2 \cdot \frac{3}{x} - 4\right)\left(-\frac{3}{x^2}\right) = -\frac{18}{x^3} + \frac{12}{x^2}.$$

The general proof of the Chain Rule is rather involved. The comparatively simple proof that we give requires that the function $\Delta u = g(x + \Delta x) - g(x)$ never be zero when Δx is "small." This restriction can be removed, but the corresponding proof is much more difficult.

The proof will be simplified if we recall that

$$\frac{dy}{dx} = \lim_{\Delta x \to 0} \frac{\Delta y}{\Delta x}.$$

Similarly,

$$\frac{dy}{du} = \lim_{\Delta u \to 0} \frac{\Delta y}{\Delta u}, \qquad \frac{du}{dx} = \lim_{\Delta x \to 0} \frac{\Delta u}{\Delta x},$$

and so on.

Theorem 2.11 (*The Chain Rule*) If $y = f(u)$ and $u = g(x)$ where f and g are differentiable functions, then dy/dx exists and is equal to

D8 $$\frac{dy}{dx} = \frac{dy}{du} \cdot \frac{du}{dx} = f'(u)\frac{du}{dx} \qquad \text{(the Chain Rule).}$$

The chain rule

PROOF Let $\Delta u = g(x + \Delta x) - g(x)$ for each Δx. Let $\Delta y = f(u + \Delta u) - f(u)$. Since the function g is differentiable, then it is continuous so that

$$\lim_{\Delta x \to 0} \Delta u = \lim_{\Delta x \to 0} g(x + \Delta x) - g(x) = g(x) - g(x) = 0.$$

Thus, $\Delta u \to 0$ as $\Delta x \to 0$. If we assume that Δu is never equal to zero unless Δx is equal to zero, then

$$\frac{\Delta y}{\Delta x} = \frac{\Delta y}{\Delta u} \cdot \frac{\Delta u}{\Delta x}.$$

If we take the limit as $\Delta x \to 0$ we obtain

$$\begin{aligned}
\frac{dy}{dx} = \lim_{\Delta x \to 0} \frac{\Delta y}{\Delta x} &= \lim_{\Delta x \to 0} \frac{\Delta y}{\Delta u} \cdot \frac{\Delta u}{\Delta x} \\
&= \lim_{\Delta x \to 0} \frac{\Delta y}{\Delta u} \cdot \lim_{\Delta x \to 0} \frac{\Delta u}{\Delta x} \qquad \text{(by L2)} \\
&= \lim_{\Delta u \to 0} \frac{\Delta y}{\Delta u} \cdot \lim_{\Delta x \to 0} \frac{\Delta u}{\Delta x} \qquad \text{(since } \Delta u \to 0 \text{ as } \Delta x \to 0) \\
&= \frac{dy}{du} \cdot \frac{du}{dx}.
\end{aligned}$$

Example 3 Let $y = \left(\frac{5}{x-1}\right)^2 + 2\left(\frac{5}{x-1}\right) + 4.$

(a) Simplify the functional expression by a substitution $u = g(x)$.
(b) Use the Chain Rule to calculate dy/dx.

Solution

(a) If we let $u = g(x) = 5/(x - 1)$, then $y = u^2 + 2u + 4$.

(b) $$\begin{aligned}
\frac{dy}{dx} = \frac{dy}{du} \cdot \frac{du}{dx} &= \frac{d}{du}(u^2 + 2u + 4) \cdot \frac{d}{dx}\left(\frac{5}{x-1}\right) \\
&= (2u + 2) \cdot \left(\frac{-5}{(x-1)^2}\right) \\
&= \left(2 \cdot \frac{5}{x-1} + 2\right)\left(-\frac{5}{(x-1)^2}\right).
\end{aligned}$$

The Chain Rule may be applied to composite functions $f(g(x))$ even if the original functions f and g are not differentiable everywhere. If $y=f(u)$ and $u=g(x)$, then the Chain Rule can be used to calculate dy/dx at the point $x=x_0$ provided g is differentiable at x_0 and f is differentiable at $g(x_0)$. In other words, if $h(x)=f(g(x))$, then

$$h'(x_0)=f'(g(x_0))g'(x_0)$$

provided both of the derivatives on the right exist.

Remark on notation

Remark 1 The beginning student has a tendency to believe that the Chain Rule is a consequence of the cancellation law of fractions. When we write

$$\frac{dy}{dx}=\frac{dy}{du}\cdot\frac{du}{dx}$$

it is natural to think that the du-terms cancel each other, leaving dy/dx. The reader must remember that dy/dx is a symbol for the derivative, not a fraction. One of the main reasons for using this particular symbol is that it makes the Chain Rule and certain other formulas resemble corresponding rules for fractions. In other words, we choose the notation because of the Chain Rule. The Chain Rule is not a consequence of the notation.

Change of variable

Remark 2 The Chain Rule is involved whenever we have a derivative problem that involves a change of variables. As the careful reader will have noticed, we do not need it to calculate derivatives of composite functions $f(g(x))$ if both f and g are rational functions. We could first carry out the substitution and then take the derivative. Later in the book, however, we will study exponential and logarithmic functions. With these functions even the simplest derivative calculations require the Chain Rule.

Our immediate needs for the Chain Rule will become apparent in Chapter 4. In that chapter we shall be concerned with finding *antiderivatives*—functions that have given functions for their derivatives. In most cases we must change the variables by substitution. An effective use of this process requires the use of the Chain Rule as well as a thorough knowledge of all the basic derivative formulas.

EXERCISES 2.10

1. Find the law of correspondence for the composite function $h(x)=f(g(x))$.
   - • (a) $f(u)=3u-1,\ u=g(x)=2x^2+7$
   - (b) $f(u)=5u^2+2,\ u=g(x)=\dfrac{1}{x-1}$
   - (c) $f(x)=\dfrac{7}{x-1},\ g(x)=\dfrac{2}{x+2}$
   - • (d) $f(x)=15x+2,\ g(x)=9x^2-1$
2. In each of the following y is given as a function of u, where $u=\sqrt{x-1}$. Calculate dy/dx, first in terms of du/dx, then in terms of x.
   - (a) $y=\sqrt{u^2-1}$
   - (c) $y=\dfrac{u-1}{2u}$
   - • (b) $y=(u^3+1)^3$
   - • (d) $y=(u+3)^{12}$

(e) $y = \dfrac{\sqrt{u}}{u^2 + 1}$ • (f) $y = (3u^2 - 2u + 5)^2$

3. Simplify the function by a substitution $u = g(x)$. Then use the Chain Rule to calculate dy/dx. (For example, in (a) let $u = g(x) = x^2 + 1$.)

• (a) $y = (x^2 + 1)^2 + 5(x^2 + 1) - 1$ (d) $y = 2\left(\dfrac{1}{x}\right)^3 + 3\left(\dfrac{1}{x}\right)^2 + \dfrac{5}{x} + 2$

(b) $y = (5x)^3 + 3(5x)^2 + 20x - 2$ • (e) $y = \dfrac{(\sqrt{x})^2 - 3\sqrt{x} + 1}{\sqrt{x} + 2}$

• (c) $y = \left(\dfrac{1}{x} + x\right)^{1/2} + \left(\dfrac{1}{x} + x\right)^{-1/2}$ (f) $y = (\sqrt{x} + 2)^2 + 3(\sqrt{x} + 2) + 7$

4. In each of the following, y is a function of x and x is a function of t. Find dy/dt for the given value of t.

(a) $y = 3x^2$, $x = t^2 - 1$, $t = 2$ • (d) $y = \dfrac{x}{x - 1}$, $x = \sqrt{t}$, $t = 9$

• (b) $y = \dfrac{x^2}{x + 1}$, $x = 2t^2 - 2$, $t = 1$ (e) $y = x^2 + 1$, $x = \sqrt{t}$, $t = 4$

(c) $y = \dfrac{x^2}{x^2 + 1}$, $x = t^2 + 1$, $t = 0$ (f) $y = \dfrac{x^2 + 1}{x^2}$, $x = \dfrac{t}{t + 1}$, $t = 1$

APPLICATIONS OF THE DERIVATIVE

3.1 INTRODUCTION TO MAXIMUM AND MINIMUM PROBLEMS

Intervals In the work of this chapter we will use certain types of intervals on the x-axis. An *interval* is the set of all points on the axis between two fixed points a and b. The numbers a and b, called the *endpoints* of the interval, may or may not be included as part of the interval. The points between the endpoints are called *interior points* of the interval.

Open and closed intervals

An interval is said to be *closed* if it contains both of its endpoints. It is *open* if it contains neither endpoint. Observe that each point of an open interval is an interior point.

Let a and b be fixed real numbers with $a < b$. The closed interval with endpoints a and b is denoted by $[a, b]$. That is,

$$[a, b] = \{x \mid a \leqslant x \leqslant b\}.$$

The open interval with endpoints a and b is denoted by (a, b). That is,

$$(a, b) = \{x \mid a < x < b\}.$$

These intervals are pictured in Figure 3.1.

It is unfortunate that the symbol (a, b) is used to denote both points in the plane and open intervals. We will specify which we mean wherever there is any

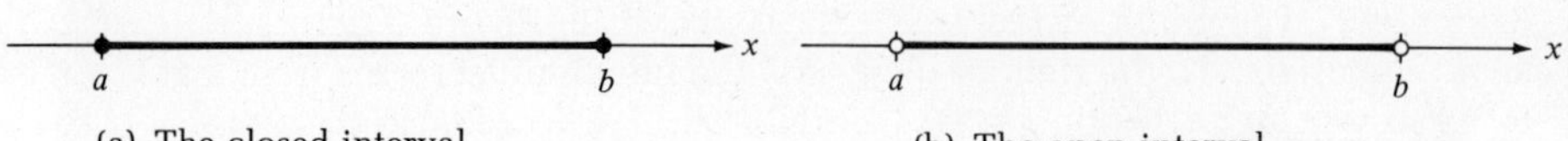

(a) The closed interval $[a, b] = \{x | a \leq x \leq b\}$

(b) The open interval $(a, b) = \{x | a < x < b\}$

FIGURE 3.1

chance of confusion. Thus, we will write "the open interval (a, b)" or "the point (a, b)."

Maximum and Minimum Problems Practical problems frequently involve extreme values of functions. We want the *largest profit*, the *smallest cost*, the *least effort*. Many problems of this type can be solved by the use of the derivative. We consider a few examples of extreme values of functions.

Example 1 The graph of $h(x) = x^2 - 2x - 1$ is a parabola with vertex at the point $(1, -2)$ that opens upward. Thus, the minimum value of $h(x)$ is -2 (taken when $x = 1$). There is no maximum value. (See Figure 3.2.)

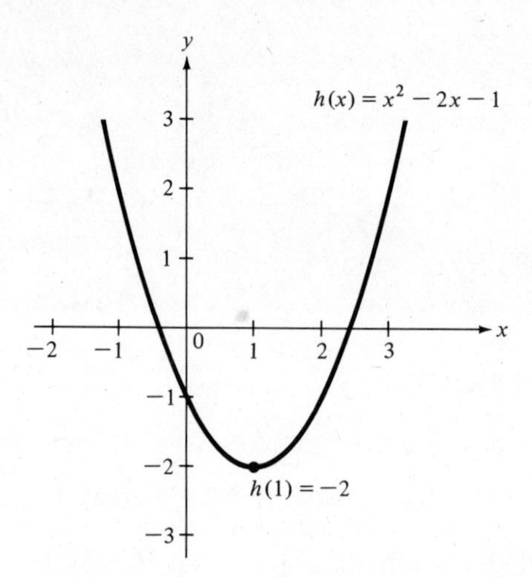

FIGURE 3.2 Example 1. The minimum value of $h(x) = x^2 - 2x - 1$ is $h(1) = -2$. The function has no maximum value.

Example 2 The graph of $g(x) = 4 - x$, $1 \leq x \leq 2$, is a line segment. The minimum value of $g(x)$ is 2 (at $x = 2$) and the maximum value is 3 (at $x = 1$). (See Figure 3.3.)

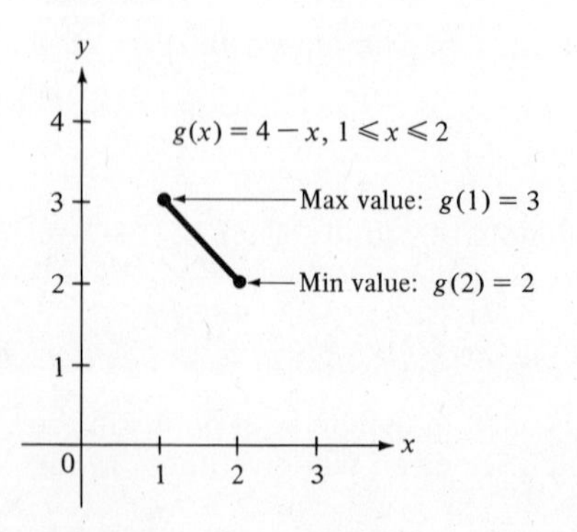

FIGURE 3.3 Example 2. The function $g(x) = 4 - x$, $1 \leq x \leq 2$. The maximum value of g is $g(1) = 3$, the minimum is $g(2) = 2$.

Example 3 The graph of $f(x) = x^3 - 3x$ is shown in Figure 3.4. It is symmetric about the origin. To the right of the y-axis the function decreases from 0 at $x = 0$ to -2 at $x = 1$ and then increases without bound. Thus, f has no maximum. The corresponding portion to the left of the y-axis decreases without bound. Thus, f has no maximum and no minimum.

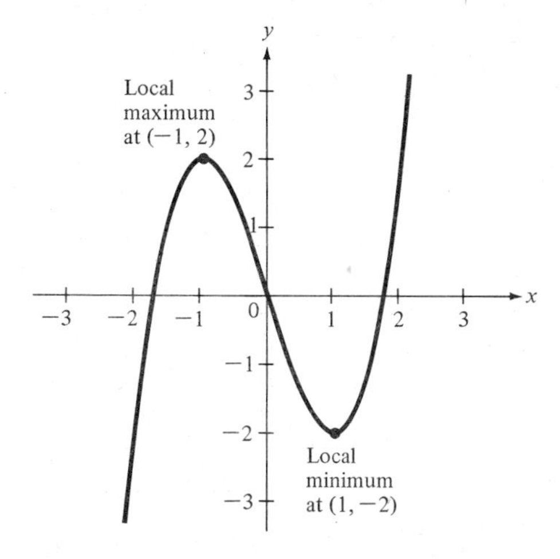

FIGURE 3.4 Example 3. The function $f(x) = x^3 - 3x$ has no maximum and no minimum. It has a local maximum at $(-1, 2)$ and a local minimum at $(1, -2)$.

This last example is worth considering in more detail. Consider the behavior of $f(x)$ when x is close to 1. Although $f(1) = -2$ is not the minimum of f on the entire domain, it is the minimum when f is restricted to a small interval about 1—the open interval $(0, 2)$, for example. We say that -2 is a *local minimum* of f. Similarly, $f(-1) = 2$ is a *local maximum*. ($f(-1)$ is the maximum value of f on the open interval $(-2, 0)$.)

Local minimum

DEFINITION We say that the function f has a *local minimum* at the point a if there exists an open interval I that contains a as an interior point, on which f is defined, such that $f(a)$ is the minimum value of $f(x)$ when $x \in I$. (That is, $f(x) \geq f(a)$ for all $x \in I$.)

Local maximum

A similar definition holds for a *local maximum*. There must exist an open interval I containing a such that f is defined on I and $f(a)$ is the maximum value of $f(x)$ when $x \in I$. A local maximum or local minimum may be called a *local extremum*. This open interval I is not unique. Any smaller open interval containing a also has the above property. (Other books may use the terms *relative minimum* and *relative maximum* rather than *local minimum* and *local maximum*. The meanings are the same.)

The function h in Example 1 (Figure 3.2) has a local minimum at $x = 1$. ($h(1)$ is the minimum value of $h(x)$ on the open interval $(0, 2)$, for example.) This function has no local maximum.

The requirement that there must exist about a an open interval I on which f is defined is not as arbitrary as it may seem. We will soon apply the principles of calculus to the problem of finding local extrema. In order to calculate the derivative at a we must at least have f defined at all points on an open interval containing a. Because of this restriction, a function may fail to have a local extremum at an extreme value. The function $g(x) = 4 - x$, $1 \leq x \leq 2$, of Example 2 (Figure 3.3), is a case in point. Since the maximum and minimum occur at the endpoints of the interval of definition, neither is a local extremum.

The graph in Figure 3.5 pictures a function with several local extrema. There are local maxima at x_1, x_3, and at each point in the range $x_6 \leqslant x \leqslant x_7$. There are local minima at x_2, x_5, and x_8. Several tangent lines are also shown in Figure 3.5. Observe in this example that if there is a tangent line at a local extremum, then it is horizontal. Of course, a tangent line may be horizontal at other points as well—x_4, for example. In the following section we will show that this property holds in general, and we will use it to determine local extrema.

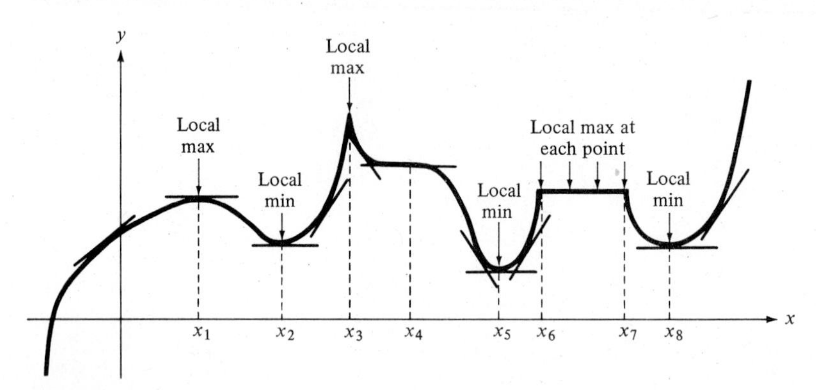

FIGURE 3.5

EXERCISES 3.1

1. Give the endpoints and interior points for each of the following intervals. State whether the interval is open or closed or neither.

   (a) $\{x|1 < x \leqslant 2\}$
   • (b) $\{x|-2 \leqslant x < 1\}$
   (c) $\{x|0 \leqslant x \leqslant 2\}$
   (d) $\{x|3 < x < 6\}$
   (e) $\{x|-1 \leqslant x \leqslant 2\}$
   (f) $\{x|0 < x < 3\}$
   • (g) $\{x||x| \leqslant 1\}$
   (h) $\{x||x| < 3\}$

2. Graph the following functions. Use the graphs to find the maximum and minimum values when they exist. Find all local extrema.

   (a) $f(x) = x^2 - 4x + 4$
   • (b) $f(x) = -x^2 - 2x - 1$
   (c) $f(x) = 3x - 1,\ 0 \leqslant x \leqslant 1$
   • (d) $f(x) = -3x - 1,\ 0 \leqslant x \leqslant 1$
   (e) $f(x) = -x^2 + 2x - 1$
   (f) $f(x) = x^3,\ |x| \leqslant 2$
   • (g) $f(x) = |x|,\ |x| \leqslant 2$
   (h) $f(x) = -2x^2 - 2x + 1$

3. Show that the constant function defined by $f(x) = C$ has a local extremum at each point.

4. Explain why the function pictured in Figure 3.5 has a local maximum at each point of the closed interval $[x_6, x_7]$.

5. Sketch the graph of a continuous function that has one local maximum and two local minima, but no maximum.

6. Sketch the graph of a differentiable function that has one local maximum and one local minimum. Sketch the tangent lines at these two points. Do these lines appear to have zero slope?

7. Sketch the graph of a continuous function with one local minimum where the tangent line is not defined.

8. Sketch the graph of a continuous function f with $f'(2) = 0$ but no local extremum at $x = 2$.

9. Sketch the graph of a continuous function f, defined on the closed interval $[-1, 2]$, that has a local maximum at each point of the closed interval $[0, 1]$.

3.2 MAXIMUM AND MINIMUM PROBLEMS

The basic concepts of maxima and minima were introduced in the preceding section. We observed in the discussion of Figure 3.5 that the tangent line appeared to have slope equal to zero at a local extremum, provided the tangent line existed at the point. We now show that this is indeed the case. (The reader should not jump to the conclusion that a local extremum always exists when the derivative is zero. The graph may just level off at the point.)

Theorem 3.1 If f has a local extremum at the point a and $f'(a)$ exists, then $f'(a) = 0$.

PROOF Recall that $f'(a)$ is equal to

$$\lim_{x \to a} \frac{f(x) - f(a)}{x - a}.$$

This limit is the same if x approaches a through values greater than or less than a.

Suppose that f has a local minimum at a. Then there exists an open interval $I = (c, d)$ that contains a as an interior point such that f is defined on I and $f(a)$ is the minimum of $f(x)$ when $x \in I$. We now restrict ourselves to values of x between c and d. We can do this because we are interested only in values of x near a when we let x approach a. Since $f(a)$ is the minimum value of $f(x)$ when $c < x < d$, then for every such x

$$f(x) \geqslant f(a).$$

Suppose $x > a$. Then

$$f(x) - f(a) \geqslant 0 \quad \text{and} \quad x - a > 0, \quad \text{so} \quad \frac{f(x) - f(a)}{x - a} \geqslant 0.$$

(See Figure 3.6a.) If we now let x approach a through values greater than a, we have

$$f'(a) = \lim_{x \to a} \frac{f(x) - f(a)}{x - a} \geqslant 0.$$

(Because the function $(f(x) - f(a))/(x - a)$ is nonnegative, it cannot have a negative limit.) Similarly, if $x < a$, then

$$f(x) - f(a) \geqslant 0 \quad \text{and} \quad x - a < 0, \quad \text{so} \quad \frac{f(x) - f(a)}{x - a} \leqslant 0$$

and

$$f'(a) = \lim_{x \to a} \frac{f(x) - f(a)}{x - a} \leqslant 0.$$

(See Figure 3.6b.) Since $f'(a)$ is both nonnegative and nonpositive, it must be equal to zero.

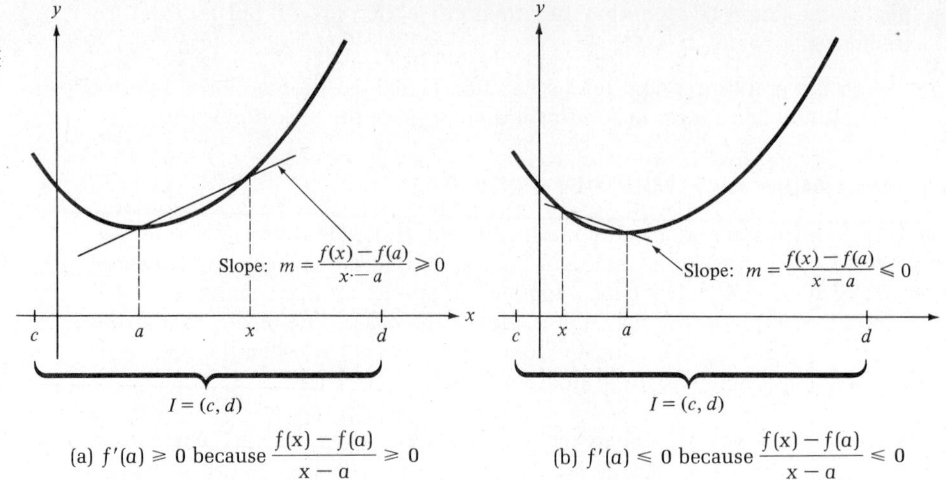

(a) $f'(a) \geq 0$ because $\frac{f(x) - f(a)}{x - a} \geq 0$ when $x > a$

(b) $f'(a) \leq 0$ because $\frac{f(x) - f(a)}{x - a} \leq 0$ when $x < a$

FIGURE 3.6 The derivative is zero at a local extremum if it exists.

A similar proof holds if f has a local maximum at a. The details are left for the reader.

As we saw in Section 3.1, a function does not necessarily have extreme values. The members of one important class of functions, however, always have maximum and minimum values. The following theorem is proved in advanced courses:

Theorem 3.2 If f is continuous on a closed interval, then f has a maximum and a minimum on the interval.

Obviously, if f has an extreme value on a closed interval, it must occur either at an endpoint or at an interior point. If it occurs at an interior point, then it is a local extremum and the derivative is zero if it exists. This leads us to the following procedure for calculating the maximum and minimum of a function that is continuous on a closed interval:

Steps for finding extrema on a closed interval

Let f be continuous on a closed interval. To find the maximum (minimum) of f on the interval we:

1. Calculate $f'(x)$.
2. Find all x on the interval such that $f'(x) = 0$.
3. Find all x on the interval where the derivative does not exist.
4. Calculate the values of f at these points and at the endpoints of the interval. The maximum (minimum) of f on the interval is the largest (smallest) of the calculated values.

The points where $f'(x) = 0$ or where f has no derivative are called the *critical points* for the function. The extreme values of f must occur at critical points or at the endpoints of the interval under consideration.

Example 1 Calculate the maximum and minimum of $f(x) = x^3 - 6x^2 + 9x + 5$ on the interval $[0, 4]$. Locate all local maxima and minima.

Solution

$$f'(x) = 3x^2 - 12x + 9 = 3(x-1)(x-3).$$

Thus, $f'(x)$ exists at all points on the interval and $f'(x) = 0$, if and only if $x = 1$ or $x = 3$. Consequently, an interior local maximum or minimum can occur only at one of these critical points. The maximum (or minimum) must occur at one of the critical points or at one of the endpoints $x = 0$, $x = 4$. We compute

$$f(0) = 5,$$
$$f(1) = 9,$$
$$f(3) = 5,$$
$$f(4) = 9.$$

The minimum of f is 5 and the maximum is 9 (Figure 3.7).

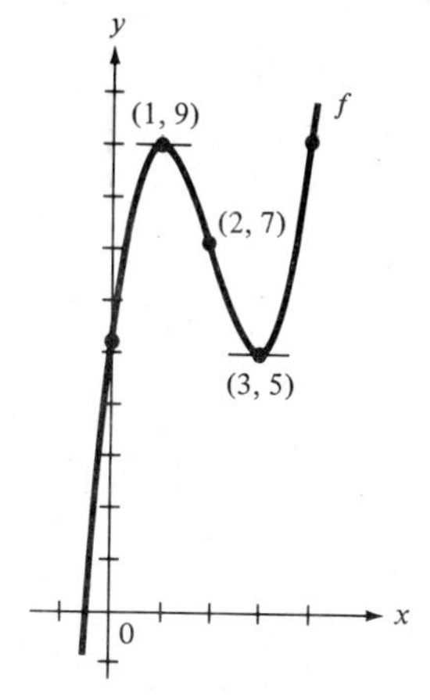

FIGURE 3.7 Example 1. Graph of $f(x) = x^3 - 6x^2 + 9x + 5$.

Example 2 A prominent government official has purchased a summer home near a small city. The telephone company has agreed to furnish additional communication lines to the estate and wishes to minimize the installation costs. The location of the estate is shown in Figure 3.8. Major communication lines exist at point A in the city. The company plans to extend these lines along a highway to point P using existing poles. From P new lines will be constructed across the countryside to the estate (D). The cost is \$500 per mile between A and P and is \$1300 per mile between P and D. Where should P be located in order to minimize costs?

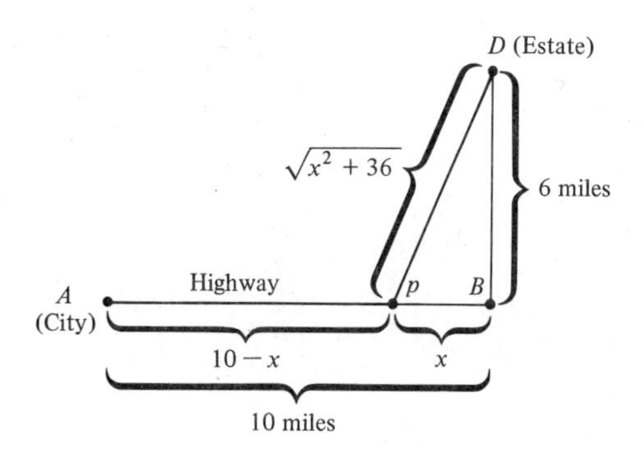

FIGURE 3.8 Example 2.

Solution Let B be the point on the highway nearest the estate. (See Figure 3.8.) If x is the distance from B to P, then the distance from P to the estate is $\sqrt{x^2 + 36}$ (by the Pythagorean Theorem). The distance from A to P is $10 - x$. The total cost (measured in hundreds of dollars) is

$$C(x) = 5(10 - x) + 13\sqrt{x^2 + 36}.$$

Obviously $0 \leq x \leq 10$. (At one extreme the lines run directly from A to D; at the other extreme they run from A to B and then to D.)

To minimize $C(x)$ we calculate $C'(x)$, set it equal to zero, and solve for x:

$$C(x) = 5(10 - x) + 13\sqrt{x^2 + 36},$$

$$C'(x) = -5 + 13 \cdot \tfrac{1}{2}(x^2 + 36)^{-1/2}(2x) = -5 + \frac{13x}{\sqrt{x^2 + 36}}.$$

Therefore, $C'(x) = 0$ if and only if

$$\begin{aligned} 13x &= 5\sqrt{x^2 + 36}, \\ 169x^2 &= 25(x^2 + 36), \\ 144x^2 &= 25 \cdot 36, \\ 12x &= 30, \\ x &= \tfrac{5}{2}. \end{aligned}$$

The only critical point is $x = \frac{5}{2}$. The minimum of C must occur at $x = \frac{5}{2}$ or at one of the endpoints $x = 0$, $x = 10$. Since $C(0) = 128$, $C(\frac{5}{2}) = 122$, and $C(10) = 151.6$ (approximately), then the minimum installation cost is $12,200, which occurs if *P* is located $2\frac{1}{2}$ miles from *B*.

Example 3 The profit (p) of the Wonder Wicket Company can be expressed as a function of the output (x) by the formula

$$p = -\frac{x^3}{30} + \frac{x^2}{20} + 2x - 1,$$

where x is measured in thousands of crates of wickets and p is measured in tens of thousands of dollars. What output yields the largest profit? What is the largest possible profit?

Solution We calculate $p'(x)$, set it equal to zero, and solve for x:

$$p(x) = -\frac{x^3}{30} + \frac{x^2}{20} + 2x - 1$$

$$\begin{aligned} p'(x) &= -\frac{x^2}{10} + \frac{x}{10} + 2 = -\frac{1}{10}(x^2 - x - 20) \\ &= -\frac{1}{10}(x - 5)(x + 4). \end{aligned}$$

Thus, $p'(x) = 0$ if and only if $x = 5$ or $x = -4$.

The graph of the profit function shows that there is a local maximum at $x = 5$ (Figure 3.9). Observe that the function $p(x) = -x^3/30 + x^2/20 + 2x - 1$ has a local minimum at $x = -4$, but that this value has no significance for our problem. We are only concerned with values of x that are greater than or equal to zero.

The largest possible profit is

$$p(5) = -\tfrac{125}{30} + \tfrac{25}{20} + 10 - 1 \approx 6.08.$$

Since p is measured in tens of thousands of dollars the maximum profit is approximately $60,800.

One final word of caution. The fact that a function *f* has its derivative equal to zero at a point $x = a$ does not necessarily mean that it has a local max-

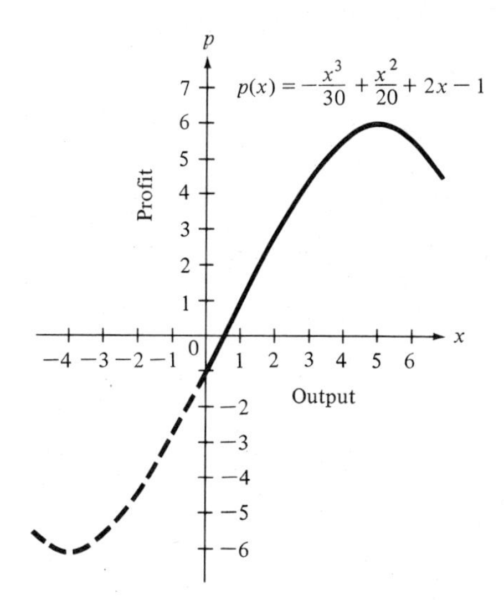

FIGURE 3.9 Example 3

imum or minimum there. The graph may just level off at the point and have a horizontal tangent line there. (See Figure 3.5 and Exercise 2.)

EXERCISES 3.2

1. Calculate the maximum and minimum of f on the indicated interval:

   (a) $f(x) = 2x^4 - x^2$, $[0, 2]$
   • (b) $f(x) = x^4 - 2x^2$, $[-3, 3]$
   (c) $f(x) = x^2 + 1$, $[-2, 1]$
   • (d) $f(x) = x^3 - 3x^2 - 9x + 10$, $[-1, 3]$
   (e) $f(x) = x^4 - 2x^2 + 2$, $[-2, 1]$
   (f) $f(x) = 2x^3 - 8x^2 - 6x + 1$, $[-3, 2]$
   • (g) $f(x) = 4x^3 - 15x^2 + 12x - 1$, $[-1, 2]$
   (h) $f(x) = \frac{4x^5}{5} + 2x^4 - 4x^3$, $[-2, 1]$
   • (i) $f(x) = \frac{x^4}{4} - \frac{x^3}{3} - \frac{x^2}{2} + x + 1$, $[-2, 2]$
   (j) $f(x) = 3x^4 - 4x^3 + 2$, $[-1, 2]$
   (k) $f(x) = x^3 - 6x^2 + 9x - 6$, $[1, 3]$
   • (l) $f(x) = 2x^3 - 3x^2 - 12x$, $[-2, 2]$

2. Sketch the graph of $f(x) = x^3$, defined for x in $[-1, 1]$. Show that $f'(x) > 0$ for $x \neq 0$ and $f'(0) = 0$. Note that a local maximum or a minimum does not occur at $x = 0$. (This is an example where $f'(x_0) = 0$, but $f(x_0)$ is not a local maximum or minimum.)

3. Sketch the graph of $f(x) = 3x^{2/3}$, defined for x in $[-1, 1]$. Show that $f'(x)$ is not defined for $x = 0$. Show that $f(0)$ is a local minimum. (This is an example where $f'(x_0)$ fails to exist and $f(x_0)$ is a minimum.)

4. Sketch the graph of $f(x) = -5x^{2/5}$, defined for x in $[-1, 1]$. Show that $f'(x)$ is not defined for $x = 0$. Show that $f(0)$ is a maximum. (This is an example where $f'(x_0)$ fails to exist and $f(x_0)$ is a maximum.)

•5. The P.U. Fertilizer Company packages its product in a 27-cubic-foot rectangular package with a square base. What dimensions should be selected for such a package in order to have the least amount of surface area?

6. A farmer has 200 feet of fencing with which he wishes to construct a rectangular hog pen. What dimensions for the pen will enclose the maximum area?

7. A Farmer wishes to enclose a rectangular field. He has 1200 feet of fencing. In order to minimize costs he plans to use an existing fence for one of the boundaries of the field. What is the area of the largest field that can be enclosed?

•8. The U. H. F. Television Rental Company can rent 120 color television sets for \$50 per month each. For each dollar that the monthly rental is increased, two sets will go unrented; that is, at \$51 per month, 118 sets will be rented, at \$52 per month, 116 sets will be rented, and so on. What rental charge will produce the greatest gross income?

3.3 TESTS FOR LOCAL EXTREMA

In the preceding section we discussed maxima and minima of functions defined over closed intervals. In many problems we cannot restrict ourselves to such intervals. For example, we may need to find the maximum of a function that is defined on the entire x-axis, or on an open interval, or, perhaps, on the positive x-axis. It is rather easy to make up examples of such functions that do not have maximum values even though they may have local maxima. It is not hard to see, however, that if such a function has a maximum then it must occur at a local maximum. (The reader should be able to justify this remark by recalling the definition of a local maximum.)

In general it may be difficult to decide if a given local maximum is the maximum of the function. There is one special case, however, in which this can be established at once:

Let f be continuous on the entire x-axis. Suppose that c is the only critical point for f. If f has a local maximum (minimum) at c then $f(c)$ is the maximum (minimum) value of the function.

The proof of the above statement is left for the reader (Exercise 8). The proof can be modified for functions which are continuous on open intervals or on the positive x-axis.

As we can see, in certain cases we only need to know if a function has a local maximum at a critical point in order to find its maximum value. In this section we consider two tests which can be used to test critical points for local extrema.

Let c be a critical point in the interior of an interval. Recall that if $f(c)$ is a local maximum, then $f(c) \geq f(x)$ for all x "near" c. Suppose we choose x_1 to the left and x_2 to the right of c. If $f(x_1) \leq f(c)$ and $f(x_2) \leq f(c)$, it would appear that f should have a local maximum at c (Figure 3.10). This line of reasoning is valid *provided c is the only critical point between* x_1 and x_2. A similar proposition is valid for a local minimum at c.

It is not necessary to specify that f is continuous on a closed interval that contains c as an interior point. It is sufficient to require that f be continuous at c. Thus we have the following test:

Test 1 For Interior Local Extrema Let f be continuous on an interval I that contains the critical point c. Let x_1 and x_2 be chosen on the interval so that c is the only critical point in the range $x_1 < c < x_2$.

Test 1 for interior local extrema

1. If $f(c)$ is greater than both $f(x_1)$ and $f(x_2)$, then f has a local maximum at $x = c$. (See Figure 3.10a.)
2. If $f(c)$ is less than both $f(x_1)$ and $f(x_2)$, then f has a local minimum at $x = c$. (See Figure 3.10b.)
3. If $f(c)$ is between $f(x_1)$ and $f(x_2)$, then f has neither a local maximum nor a local minimum at $x = c$. (See Figure 3.10c.)

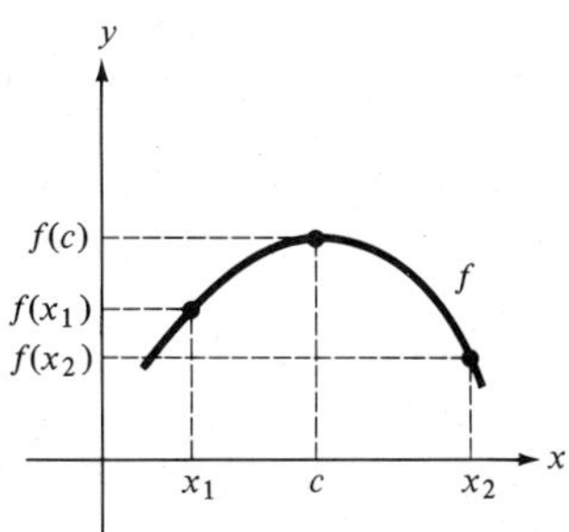

(a) Interior local maximum if $f(c)$ is greater than $f(x_1)$ and $f(x_2)$.

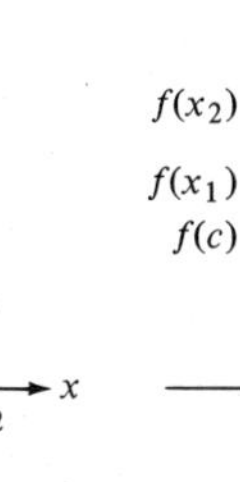

(b) Interior local minimum if $f(c)$ is less than $f(x_1)$ and $f(x_2)$.

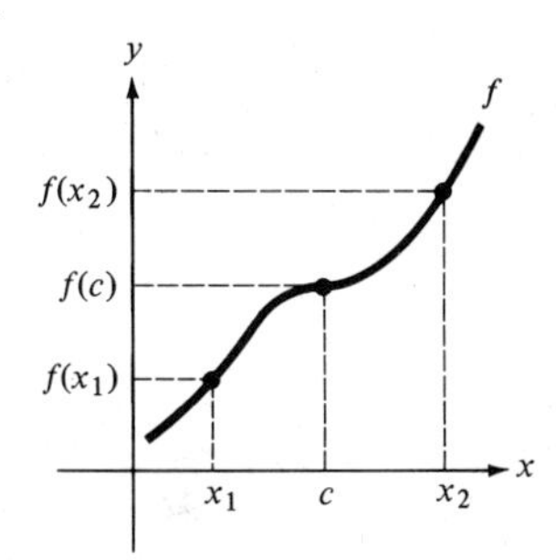

(c) No local extremum at $x = c$ if $f(c)$ is between $f(x_1)$ and $f(x_2)$.

FIGURE 3.10 Test number 1 for interior local extrema.

In applying Test 1 it is essential that c be the only critical point between x_1 and x_2, although x_1 and x_2 may themselves be critical points. Figure 3.11 pictures a situation in which the test fails because another critical point is between x_1 and x_2. The number $f(c)$ is greater than both $f(x_1)$ and $f(x_2)$, but $f(c)$ is neither a local maximum nor a local minimum.

Example 1 Locate all local extrema of $f(x) = x^3 + 12x^2 + 45x + 51$.

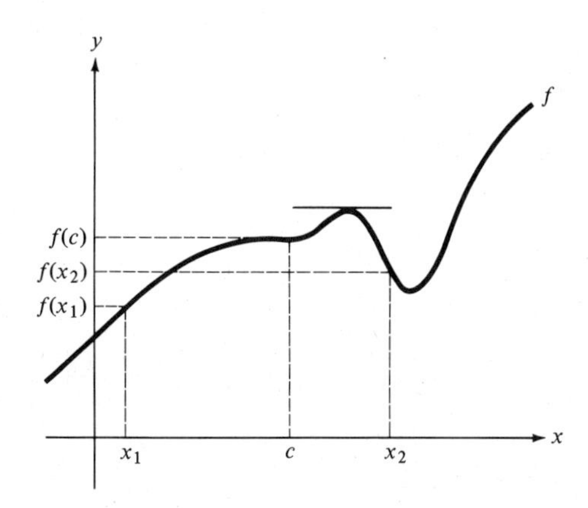

FIGURE 3.11 Illustration showing that Test 1 may fail if c is not the only critical point between x_1 and x_2.

Solution We first calculate the derivative

$$\begin{aligned} f(x) &= x^3 + 12x^2 + 45x + 51, \\ f'(x) &= 3x^2 + 24x + 45 \\ &= 3(x^2 + 8x + 15) \\ &= 3(x+5)(x+3). \end{aligned}$$

Then $f'(x) = 0$ if and only if

$$\begin{aligned} 3(x+5)(x+3) &= 0, \\ x+5 = 0 \quad &\text{or} \quad x+3 = 0, \\ x = -5 \quad &\text{or} \quad x = -3. \end{aligned}$$

The only critical points are $x = -5$ and $x = -3$.

Since f is a polynomial, it is continuous everywhere. Thus, we may apply the test. As we mentioned above it is permissible to let the testing points x_1 and x_2 be critical points. To test the critical point $x = -5$ we choose $x_1 = -6$, $c = -5$, and $x_2 = -3$. Then

$$\begin{aligned} f(x_1) &= f(-6) = (-6)^3 + 12(-6)^2 + 45(-6) + 51 = -3, \\ f(c) &= f(-5) = (-5)^3 + 12(-5)^2 + 45(-5) + 51 = 1, \\ f(x_2) &= f(-3) = (-3)^3 + 12(-3)^2 + 45(-3) + 51 = -3. \end{aligned}$$

Thus, f has a local maximum at $x = -5$.

To test the critical point $x = -3$ we let $x_1 = -5$, $c = -3$, and $x_2 = 0$. Then

$$\begin{aligned} f(x_1) &= f(-5) = 1, \\ f(c) &= f(-3) = -3, \end{aligned}$$

and

$$f(0) = 0^3 + 12 \cdot 0^2 + 45 \cdot 0 + 51 = 51.$$

Thus, f has a local minimum at $x = -3$. (See Figure 3.12.)

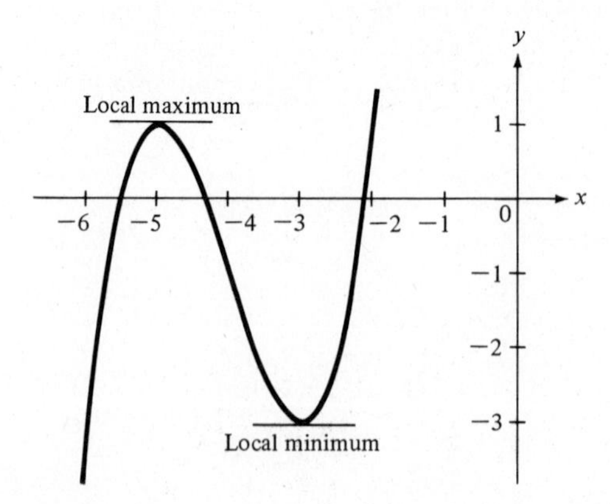

FIGURE 3.12 Example 1. $f(x) = x^3 + 12x^2 + 45x + 51$.

We now show that the second case of Test 1 is valid.

Proof of (2) Suppose that $f(x_1)$ and $f(x_2)$ are both greater than $f(c)$. By the hypothesis it follows that f is continuous on the interval $[x_1, x_2]$ and (by Theorem 4.2) has a minimum on that interval. Since $f(x_1) > f(c)$ and $f(x_2) > f(c)$, the minimum does not occur at either of the endpoints. Therefore, the minimum is a local minimum that occurs at one of the interior points. Since the only interior critical point is c, this is the only possible location for this minimum. Therefore, f has a local minimum at $x = c$.

The Derivative Test We will prove in Theorem 3.6 (Section 3.7) that if $f'(x)$ is positive at each point of an interval, then f increases on the interval. That is, if x_1 and x_2 are on the interval and $x_1 < x_2$, then $f(x_1) < f(x_2)$. Similarly, if $f'(x) < 0$ on an interval, then f decreases on the interval. This information can be used to obtain a new test for local maxima and minima.

Suppose that f is continuous at the critical point c, that $f'(x) < 0$ for all x near c and less than c, and $f'(x) > 0$ for all x near c and greater than c. Then f decreases to the left of c and increases to the right of c, so f has a local minimum at c. Similarly, if f increases to the left of c and decreases to the right, then f has a local maximum at c. This gives us the following test for local extrema (see Figure 3.13):

Derivative test for interior local extrema

Test 2 for Local Extrema (*The Derivative Test*) Let f be continuous at the critical point c.

1. If $f'(x) \geq 0$ for all x near c and less than c and $f'(x) \leq 0$ for all x near c and greater than c, then $f(c)$ is a local maximum.
2. If $f'(x) \leq 0$ for all x near c and less than c and $f'(x) \geq 0$ for all x near c and greater than c, then $f(c)$ is a local minimum.

When we say that x is "near" c, we mean that for some predetermined positive real number δ the number x is between $c - \delta$ and $c + \delta$.

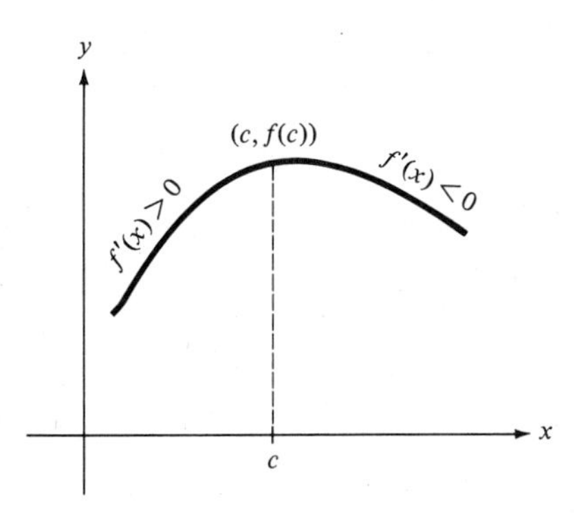

(a) Interior local maximum at c:

$f'(x) > 0$ if $x < c$
$f'(x) < 0$ if $x > c$

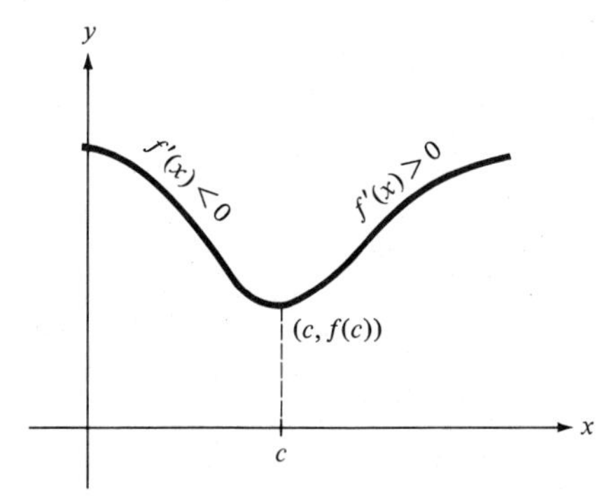

(b) Interior local minimum at c:

$f'(x) < 0$ if $x < c$
$f'(x) > 0$ if $x > c$

FIGURE 3.13 The derivative test.

Example 2 Let $f(x) = x^3 - 3x + 1$. Locate all local extrema of f.

Solution $f'(x) = 3x^2 - 3 = 3(x - 1)(x + 1)$. Thus, $x = -1$ and $x = +1$ are the only critical points. Note:

If $x < -1$ then $f'(x) > 0$ (f *increasing*).
If $-1 < x < 1$ then $f'(x) < 0$ (f *decreasing*).
If $x > 1$ then $f'(x) > 0$ (f *increasing*).

It follows from Test 2 that f has a local maximum at $x = -1$ and a local minimum at $x = +1$ (Figure 3.14).

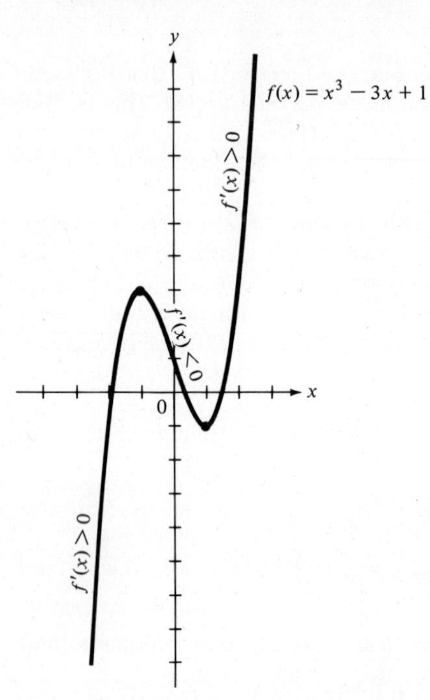

FIGURE 3.14 Example 2. $f(x) = x^3 - 3x + 1$.

A number of geometrical problems have been solved by use of the calculus. The following example establishes that the square is the rectangle of maximum area that can be inscribed in a circle.

Example 3 Determine the rectangle of maximum area that can be inscribed in a circle of radius r.

Solution The location of the circle in the plane obviously makes no difference in this problem. Thus, we may assume that the circle has its center at the origin. The equation of the circle is

$$x^2 + y^2 = r^2.$$

Also, we can rotate the inscribed rectangle until it is in the position shown in Figure 3.15a. Let (X, Y) be the coordinates of the corner in the first quadrant. Since (X, Y) is on the circle, then

$$Y = \sqrt{r^2 - X^2}.$$

Thus the area of the rectangle is

$$A = (2X) \cdot (2Y) = 4X\sqrt{r^2 - X^2}.$$

The problem then is to maximize

$$A(X) = 4X\sqrt{r^2 - X^2}$$

subject to the condition $0 < X < r$.

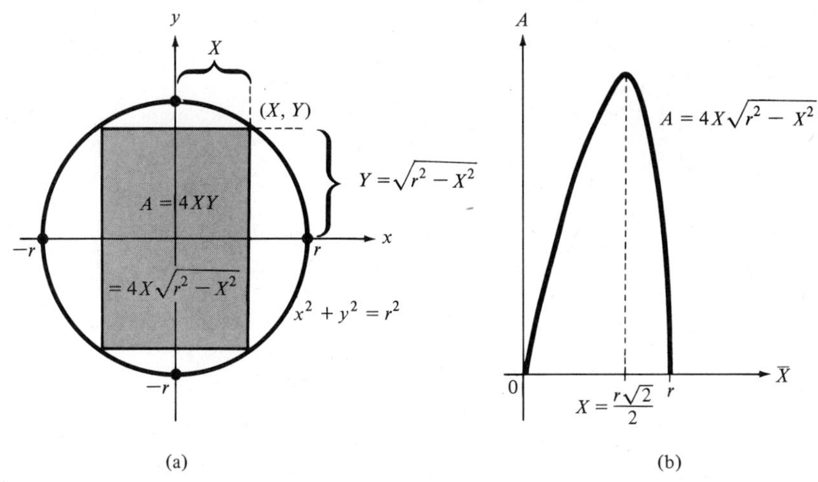

FIGURE 3.15 Example 3. (a) Area $= 4XY = 4X\sqrt{r^2 - X^2}$. (b) Graph of $A = 4X\sqrt{r^2 - X^2}$.

We calculate

$$\begin{aligned} A'(X) &= 4X \cdot \tfrac{1}{2}(r^2 - X^2)^{-1/2}(-2X) + 4\sqrt{r^2 - X^2} \\ &= \frac{-4X^2}{\sqrt{r^2 - X^2}} + 4\sqrt{r^2 - X^2} \\ &= \frac{-4X^2 + 4(r^2 - X^2)}{\sqrt{r^2 - X^2}} = \frac{4(r^2 - 2X^2)}{\sqrt{r^2 - X^2}}. \end{aligned}$$

$A'(X) = 0$ if and only if $X = r\sqrt{2}/2$. Therefore, the only critical point is $X = r\sqrt{2}/2$. If $0 < X < r\sqrt{2}/2$, then $A'(x) > 0$. If $r\sqrt{2}/2 < X < r$, then $A'(X) < 0$. By Test 2, the maximum value of A occurs when $X = r\sqrt{2}/2$. In this case

$$Y = \sqrt{r^2 - \left(\frac{r\sqrt{2}}{2}\right)^2} = \frac{r\sqrt{2}}{2}$$

and so $X = Y$. Thus, the rectangle of maximum area that can be inscribed in a given circle of radius r is a square with the length of each side equal to $2X = r\sqrt{2}$.

Example 4 An open-topped box with a square base is to be made of thin sheet metal. The total volume must be 4 cubic feet. Determine the dimensions so that the total amount of sheet metal used will be a minimum.

Solution Let x be the length of one side and y be the height. (See Figure 3.16.) Then volume $= x^2y = 4$ and so $y = 4/x^2$. The amount of sheet metal needed is

$$\begin{aligned} A &= \text{area of bottom} + \text{area of sides} \\ &= x^2 + 4xy = x^2 + 4x \cdot \frac{4}{x^2} \\ &= x^2 + 16x^{-1}. \end{aligned}$$

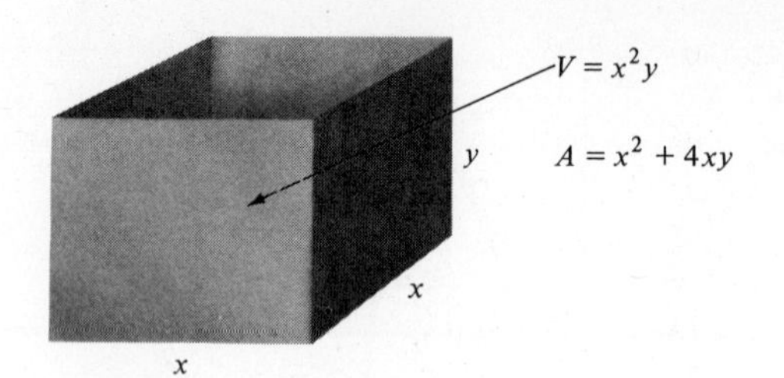

FIGURE 3.16 Example 4

The problem is to determine the minimum value of $A(x) = x^2 + 16x^{-1}$, where $x > 0$. We compute

$$A'(x) = 2x + (-1)16x^{-2} = 2x - 16x^{-2}.$$

The only critical point is $x = 2$.

$$\begin{aligned} \text{If } 0 < x < 2 &\quad \text{then } A'(x) < 0. \\ \text{If } x > 2 &\quad \text{then } A'(x) > 0. \end{aligned}$$

It follows from Test 2 that A has its minimum value when $x = 2$. The dimensions of the box that require the minimum amount of sheet metal are $x = 2$, $y = 1$.

Demand function

Demand Functions Many important applications to business involve *demand functions*. These are functions that describe the number of items that can be sold (the *demand* for the commodity) at a given price. Demand functions can be given by mathematical formulas or by statements that describe the number of items that can be sold.

In the following example (Example 5) a demand function is stated in terms of the number of additional books that will remain unsold for each increase in price of one cent. The same demand function is described later in the example by the mathematical formula

$$y = 40{,}000 - (x - 3.50)(10{,}000).$$

For each value of $x \geq 3.50$ we can use the formula to calculate y, the number of books that can be sold at $x per book.

Example 5 shows how demand functions can be used to establish the selling price that yields the maximum profit to the company.

Example 5 The Macabre Book Company publishes an inexpensive line of murder mysteries. Each book costs the company $10,000 (fixed cost) plus $2.00 per copy that is printed and distributed. The marketing division has found that 40,000 copies of a book can be sold at a wholesale price of $3.50 per book, and that sales decrease by 100 copies for each cent the wholesale price is raised. (For example, at a price of $3.51 only 39,900 copies would be sold. At a price of $4.00 only 35,000 would be sold.)

Find the wholesale price which yields the maximum net profit to the company.

Solution Let $x =$ the wholesale price per copy (in dollars). Then $x - 3.50 =$ the amount that x is greater than 3.50.

For each dollar that x is greater than 3.50 ten thousand fewer books will be sold. (One hundred books per cent is equivalent to 10,000 per dollar.) Thus,

$$\begin{aligned}&\text{The number of books that will be sold}\\&= 40{,}000 - (x - 3.50)(10{,}000)\\&= 40{,}000 - 10{,}000x + 35{,}000 = 75{,}000 - 10{,}000x.\end{aligned}$$

The gross revenue for the company is equal to

$$\begin{aligned}&(\text{selling price per book}) \cdot (\text{number of books sold})\\&= x \cdot (75{,}000 - 10{,}000x) = 75{,}000x - 10{,}000x^2.\end{aligned}$$

The cost of producing the books is equal to

$$\begin{aligned}&(\text{fixed cost}) + (\text{cost per copy}) \cdot (\text{number of copies})\\&= 10{,}000 + 2 \cdot (75{,}000 - 10{,}000x)\\&= 160{,}000 - 20{,}000x.\end{aligned}$$

The net profit is equal to

$$\begin{aligned}p(x) &= (\text{gross revenue}) - (\text{total cost})\\&= (75{,}000x - 10{,}000x^2) - (160{,}000 - 20{,}000x)\\&= -160{,}000 + 95{,}000x - 10{,}000x^2.\end{aligned}$$

To find the selling price that makes the net profit a maximum we calculate $p'(x)$, set it equal to zero, and solve for x:

$$p'(x) = 95{,}000 - 20{,}000x.$$

$p'(x) = 0$ if and only if

$$x = \frac{95{,}000}{20{,}000} = 4.75.$$

Note that $p'(x) > 0$ if $x < 4.75$ and $p'(x) < 0$ if $x > 4.75$. Thus, the maximum net profit is obtained at a wholesale price of \$4.75 per book. (See Figure 3.17.)

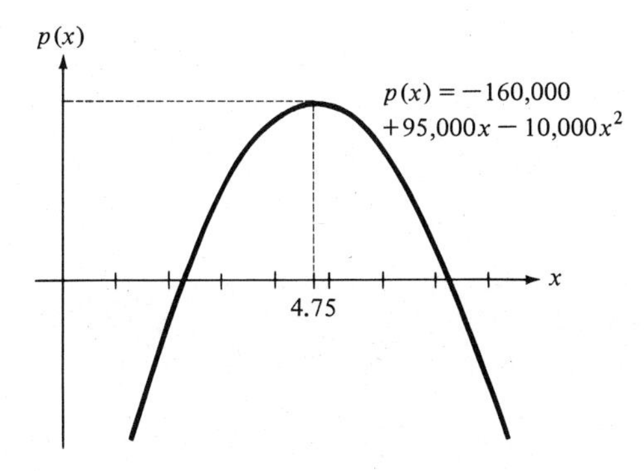

FIGURE 3.17 Example 5. (Different scales are used on the two axes.)

EXERCISES 3.3

1. Find all local extrema. Use Test 1 or 2.

• (a) $f(x) = (x + 1)^4$

(b) $f(x) = 4x^3 - 12x^2 + 12x - 7$

(c) $f(x) = x^4 - x^2/2 + 2$

• (d) $f(x) = x^3 - 6x + 3$

(e) $f(x) = x^3 - 12x$

(f) $f(x) = \dfrac{2x}{x - 1}$

(g) $f(x) = \dfrac{x^3}{3} - \dfrac{3x^2}{2} - 4x$

• (h) $f(x) = \dfrac{x^4}{4} - x^3 - \dfrac{x^2}{2} + 3x$

(i) $f(x) = 2x^3 - 24x + 1$

• (j) $f(x) = 2x - \sqrt{x}$

(k) $f(x) = \dfrac{(x - 1)^2}{x + 1}$

• (l) $f(x) = \dfrac{\sqrt{x}}{x + 1}$

2. Sketch the graph of a continuous function f, defined on the open interval (0, 5), which has a local minimum at $x = 1$ and no minimum value on the interval.

• 3. In a well-managed orchard, if no more than 50 apple trees are planted per acre, each tree will yield 500 apples. The average yield per tree is reduced by 10 apples for each additional tree over 50 trees per acre. (If 51 trees are planted, the average yield is 490, if 52 are planted, the average yield is 480, and so on.) Find the number of trees per acre that will maximize the total yield per acre.

4. The demand function for the "Cutie" doll is

$$d(x) = 500 + 20x - \frac{x^2}{3},$$

where x is the wholesale price per box of dolls and $d(x)$ is measured in hundreds of boxes. Find the value of x that yields the maximum revenue.

5. Find the dimensions of a tin can (right circular cylinder) with a capacity of 16 cubic inches that uses a minimum amount of material in its construction.

• 6. Find the dimensions of an open-topped can (right circular cylinder) with a capacity of V_0 cubic feet that uses a minimum amount of material in its construction.

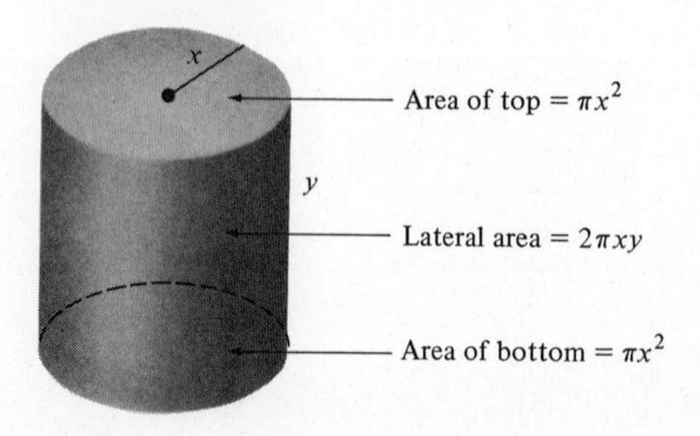

Figure for Exercises 5 and 6.

7. A cereal manufacturer plans to introduce a new style of rectangular cereal box which has a square front (length equals height). In order to stack the

boxes properly on the grocery shelves the depth of each box must be at least $1\frac{1}{2}$ inches and no more than 3 inches. Find the dimensions of the box that has the smallest total surface area for the given volume:
• (a) $V = 27$ cubic inches.
(b) $V = 300$ cubic inches.

8. Prove the following theorem: *Let f be continuous on the entire x-axis. Suppose that c is the only critical point for f. If f has a local maximum at c then f(c) is the maximum value of the function.* (Hint: Suppose that $f(x_1) > f(c)$. Show that there must exist a critical point x_2 between x_1 and c at which f has a local minimum. Since no such critical point x_2 exists, then no such x_1 can exist.)

3.4 SECOND DERIVATIVES. CONCAVITY

Let $y = f(x)$. If $f'(x)$, the derivative of $f(x)$, is a differentiable function, then we can calculate its derivative, obtaining a new derived function

$$\begin{aligned} y'' = f''(x) &= \text{the derivative of } f'(x) \\ &= \frac{d}{dx}(f'(x)) = \frac{d}{dx}\left(\frac{d}{dx}(f(x))\right). \end{aligned}$$

For example, if

$$y = f(x) = 5x^3 - 7x^2 + 3x + 2,$$

then

$$y' = f'(x) = 15x^2 - 14x + 3 \qquad \textit{(first derivative)},$$

and

$$y'' = f''(x) = 30x - 14 \qquad \textit{(second derivative)}.$$

Frequently this process can be continued, yielding derivatives of still higher orders. In our example we obtain

$$y''' = f'''(x) = \frac{d}{dx}(30x - 14) = 30 \qquad \textit{(third derivative)},$$

$$y^{IV} = f^{IV}(x) = \frac{d}{dx}(30) = 0 \qquad \textit{(fourth derivative)},$$

$$y^{V} = f^{V}(x) = \frac{d}{dx}(0) = 0 \qquad \textit{(fifth derivative)},$$

and so on.

In general, the higher derivatives are defined by

$$y' = f'(x) = \frac{dy}{dx} = \frac{d}{dx}(f(x)) \qquad \text{(first derivative)},$$

$$y'' = f''(x) = \frac{d}{dx}(y') = \frac{d}{dx}(f'(x)) \qquad \text{(second derivative)},$$

$$y''' = f'''(x) = \frac{d}{dx}(y'') = \frac{d}{dx}(f''(x)) \qquad \text{(third derivative)},$$

and so on, the $n + 1$st derivative being the derivative of the nth derivative.

Several notations are used for higher order derivatives. The notations

d^2y/dx^2 and $d^2(f(x))/dx^2$ are used for the second derivative because of the similarity of the expressions

$$\frac{d}{dx}\left(\frac{dy}{dx}\right) = \frac{d}{dx}\left(\frac{d}{dx}(f(x))\right)$$

to products. These are formal symbols and do not represent "squares of the derivative." Using the first of these notations we can represent the higher order derivatives by

Notations for higher derivatives

$$\frac{d^2y}{dx^2} = \frac{d}{dx}\left(\frac{dy}{dx}\right) = y'' \qquad \text{(second derivative)},$$

$$\frac{d^3y}{dx^3} = \frac{d}{dx}\left(\frac{d^2y}{dx^2}\right) = y''' \qquad \text{(third derivative)},$$

$$\frac{d^4y}{dx^4} = \frac{d}{dx}\left(\frac{d^3y}{dx^3}\right) = y^{IV} \qquad \text{(fourth derivative)},$$

and so on. Some books also use the notation $y^{(n)} = f^{(n)}(x)$ for the nth derivative.

For example, if

$$y = f(x) = x^4 - 3x^3 + 7x^2 + x - 1,$$

then

$$y' = \frac{dy}{dx} = 4x^3 - 9x^2 + 14x + 1,$$

$$y'' = \frac{d^2y}{dx^2} = 12x^2 - 18x + 14,$$

$$y''' = \frac{d^3y}{dx^3} = 24x - 18,$$

$$y^{IV} = \frac{d^4y}{dx^4} = 24,$$

and

$$y^{(n)} = f^{(n)}(x) = \frac{d^ny}{dx^n} = 0 \quad \text{if} \quad n \geq 5.$$

We will restrict our study to the second derivative. In the next section we will use the second derivative to develop a new test for local maxima and minima. In the remainder of this section we consider the concept of *concavity*.

Concavity

Consider the graph of $y = f(x)$ in Figure 3.18. Between a and b it bends in a clockwise direction; between b and c it bends in a counterclockwise direction. At $x = b$ the direction of bending changes. We say that the graph is *concave downward* between a and b and is *concave upward* between b and c.

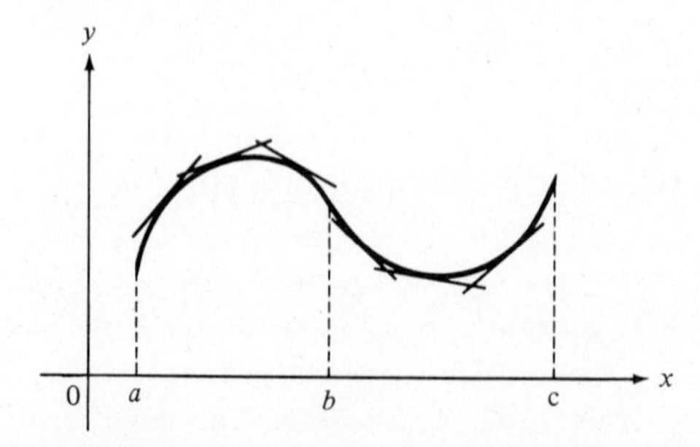

FIGURE 3.18 Clockwise bending between a and b, counterclockwise bending between b and c.

If we examine the slopes of the tangent lines in Figure 3.18, we see that the slopes decrease when the graph bends in a clockwise direction and increase when it bends in a counterclockwise direction. It can be shown that this is true in general. That is, if f is a differentiable function, then the graph of f is concave downward over an interval if and only if the derivative f' is a decreasing function on the interval. It is concave upward if and only if f' is an increasing function.

Recall that a function is known to be decreasing over an interval if its derivative is negative there. Suppose then that $f''(x)$ is negative on an interval. Since f'' is the derivative of f', then f' is a decreasing function over the interval; consequently the graph of f is concave downward there. Similarly, if $f''(x)$ is positive on an interval, then f' is an increasing function and the graph of f is concave upward. (See Figure 3.19.)

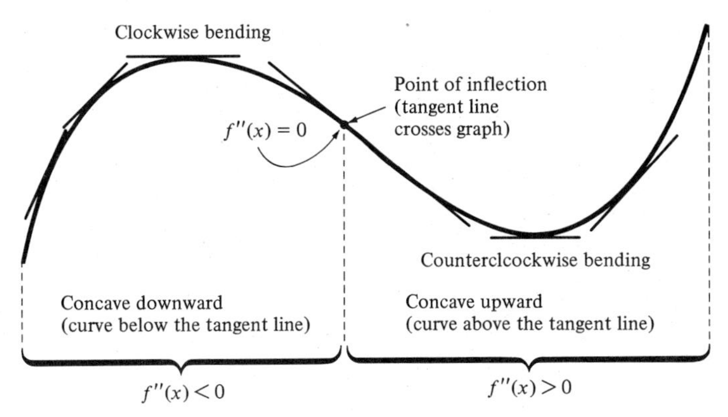

FIGURE 3.19 Geometrical properties of concavity.

In summary: If $f''(x) < 0$ on the interval I, the graph of f is concave downward on I. If $f''(x) > 0$ on the interval I, the graph of f is concave upward on I.

Example 1 Let $f(x) = x^3 - 3x + 1$. Where is the graph of f concave downward, concave upward?

Solution

$$f(x) = x^3 - 3x + 1,$$
$$f'(x) = 3x^2 - 3,$$
$$f''(x) = 6x.$$

Then $f''(x) < 0$ if $x < 0$ and $f''(x) > 0$ if $x > 0$. Therefore, the graph is concave downward if $x < 0$ and is concave upward if $x > 0$. (See Figure 3.20.)

Point of inflection

DEFINITION A point $(a, f(a))$ at which the graph of f has a tangent line and that separates a portion of the graph that is concave upward from a portion that is concave downward is called a *point of inflection.*

In Example 1 the point (0, 1) is the only point of inflection. At that point the second derivative is zero. In general the second derivative is zero at a point of inflection if it exists at the point.

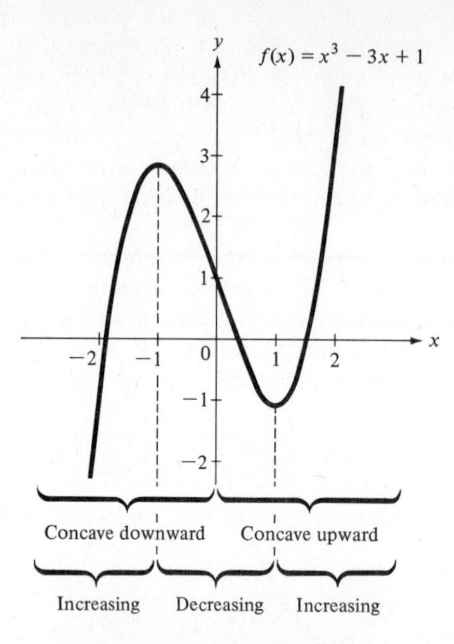

FIGURE 3.20 Example 1.

The expressions "concave downward" and "concave upward" agree with the usual notions of concavity. If we draw the tangent line to a curve at a point, then the tangent line is above the curve if it is concave downward and is below the curve if it is concave upward. At a point of inflection the tangent line crosses the curve. (See Figure 3.19.)

Application to curve sketching

Application to Curve Sketching One can make quick and accurate sketches of graphs by applying the principles of this section. We outline the steps for the graph of a function f:

1. Determine the symmetries. (Section 1.4.)
2. Locate all points of discontinuity.
3. Find all intervals on which f is increasing ($f'(x) > 0$) and all intervals on which f is decreasing ($f'(x) < 0$).
4. Find all local maxima and minima.
5. Find all intervals on which the graph is concave upward ($f''(x) > 0$) and all intervals on which it is concave downward ($f''(x) < 0$).
6. Locate all points of inflection.
7. Plot a few reference points; sketch the graph.

Example 2 Sketch the graph of

$$f(x) = x^4 - 6x^2 - 8x + 15.$$

Solution Since f is a polynomial function, there are no discontinuities. There are no obvious symmetries. We calculate the derivative:

$$\begin{aligned} f'(x) &= 4x^3 - 12x - 8 \\ &= 4(x^3 - 3x - 2) \\ &= 4(x - 2)(x + 1)^2. \end{aligned}$$

Then

$$\begin{array}{lll} f'(x) = 0 & \text{if } x = -1 \text{ or } x = 2 & \text{(critical points)}, \\ f'(x) < 0 & \text{if } x < 2 \text{ and } x \neq -1 & \text{(decreasing)}, \\ f'(x) > 0 & \text{if } x > 2 & \text{(increasing)}. \end{array}$$

Thus, the only local extremum is the local minimum at $x = 2$. The curve has a horizontal tangent line at $x = -1$ and at $x = 2$.

The second derivative is

$$\begin{aligned} f''(x) &= 12x^2 - 12 \\ &= 12(x^2 - 1) \\ &= 12(x - 1)(x + 1). \end{aligned}$$

Then

$$\begin{array}{lll} f''(x) = 0 & \text{if } x = -1 \text{ or } x = 1, & \\ f''(x) > 0 & \text{if } x < -1 & \text{(concave upward)}, \\ f''(x) < 0 & \text{if } -1 < x < 1 & \text{(concave downward)}, \\ f''(x) > 0 & \text{if } x > 1 & \text{(concave upward)}. \end{array}$$

Thus, the points of inflection occur when $x = -1$ and $x = 1$.

We now calculate the values of f at the points determined above and a few additional reference points:

$$\begin{array}{lll} \text{if } x = -2 & \text{then } f(x) = 23: & (-2, 23), \\ \text{if } x = -1 & \text{then } f(x) = 18: & (-1, 18), \\ \text{if } x = 0 & \text{then } f(x) = 15: & (0, 15), \\ \text{if } x = 1 & \text{then } f(x) = 3: & (1, 3), \\ \text{if } x = 2 & \text{then } f(x) = -9: & (2, -9), \\ \text{if } x = 3 & \text{then } f(x) = 18: & (3, 18). \end{array}$$

The graph is shown in Figure 3.21.

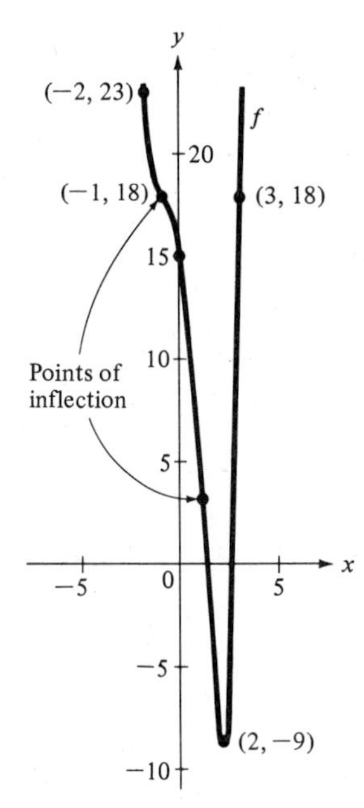

FIGURE 3.21 Example 2. Graph of $f(x) = x^4 - 6x^2 - 8x + 15$.

Application to Population Growth Concavity and points of inflection can be considered in a slightly different light. The graph of $f(x)$ is concave upward if $f''(x) > 0$, which occurs when $f'(x)$, the rate of change of f, is an increasing function. Similarly, the graph is concave downward when the rate of change of f is a decreasing function. A point of inflection occurs when the rate of change stops increasing and starts to decrease (or *vice versa*). Figure 3.22 is the graph of a typical animal population. The maximum possible population that can be supported

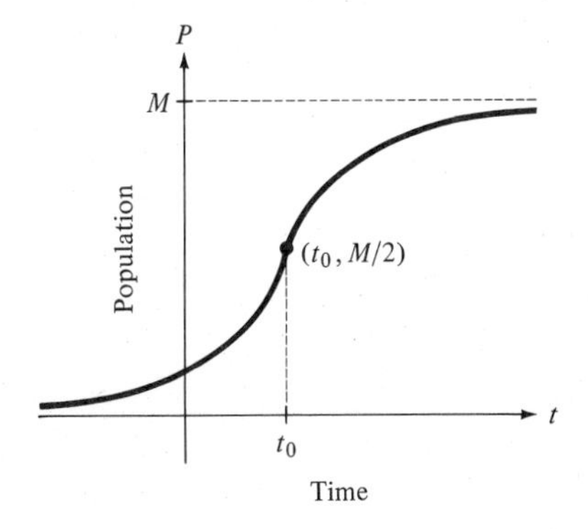

FIGURE 3.22 Graph of an animal population. The rate of change increases until one half the maximum population is attained. The rate then decreases until it is almost zero.

is M. Observe that the rate of increase of the population is an increasing function (the curve is concave upward) for $t < t_0$ and is a decreasing function (concave downward) for $t > t_0$. Thus, the rate of increase becomes greater and greater until time t_0, when it starts to decrease. At $t = t_0$, where the rate of change stops increasing and starts to decrease, there is a point of inflection. Observe that this occurs when the population is $M/2$, one half of the maximum population.

EXERCISES 3.4

1. Find all local maxima and minima and points of inflection.

(a) $f(x) = x^3 - x$
(b) $f(x) = x^2 + 10x + 7$
• (c) $f(x) = x^3 - 2x - 7$
(d) $f(x) = x^5 + 5x^4$
• (e) $f(x) = 6x^5 - 20x^4 + 20x^3$
(f) $f(x) = x^3 + 6x^2 + 9x - 18$
(g) $f(x) = x^3 + 2x$
• (h) $f(x) = (x + 1)^2/x$
(i) $f(x) = 9x - 2x^3$
• (j) $f(x) = (x - 3)^2(x + 5)^2$

2. Determine where the graph is concave upward and where it is concave downward:

(a) $f(x) = x^4 - 4x^3 + 2$
• (b) $f(x) = x^4 - 6x^2 - 24x - 2$
(c) $f(x) = x^4 + 4x^3 - 1$
• (d) $f(x) = (x + 1)^2(x^2 - 1)$
(e) $f(x) = x^3 - 3x^2 + 2x - 2$
(f) $f(x) = x^4 - 4x^3 - 18x^2 + x - 1$
• (g) $f(x) = x^2 - x^{-1}$
(h) $f(x) = x^4 - 12x^3 + 48x^2 - 2$
(i) $f(x) = 3x^4 - 4x^3 - 6x^2 + 4$
• (j) $f(x) = (x^2 + 1)^{-1}$

3. Sketch the graph:

(a) $f(x) = x^3 - 4x$
(b) $f(x) = (x^2 - x)(x - 1)^2$
(c) $f(x) = x^3 - 6x^2 + 15x + 12$
(d) $f(x) = 4x^2 - x^3$
(e) $f(x) = x^3 - 6x^2 - 20$
(f) $f(x) = x^3/3 - x^2/2 - 2x + 1$
(g) $f(x) = x^4 - 32x$
(h) $f(x) = x^4 - 8x^3 + 18x^2 - 24$
(i) $f(x) = 2x^3 - 3x^2 + 3x - 2$
(j) $f(x) = -x^5 + 10x^3 + 20x^2 + 15x + 1$
(k) $f(x) = (5 - x^2)(x^2 - 1)$
(l) $f(x) = 3x^5 - 10x^3 + 15x$

3.5 THE SECOND DERIVATIVE TEST

In this section we establish a third test for local maxima and minima. In many cases this is the easiest test to apply because we only need to consider the function f and its derivatives at the critical point. Before we derive the test we need one additional result:

If the function g has a positive limit as x approaches c, then g(x) must be positive for all values of x sufficiently close to c and different from c. (See Figure 3.23.)

This fact, which seems quite reasonable, follows from the definition of limit. The reader should convince himself of the truth of the statement by drawing a few pictures of functions with positive limits. For reference purposes we give a formal statement of the result in the following theorem.

Theorem 3.3 Let L be a positive number. If $\lim_{x\to c} g(x) = L$, then there exists a positive number δ such that if $0 < |x - c| < \delta$, then $g(x) > 0$.

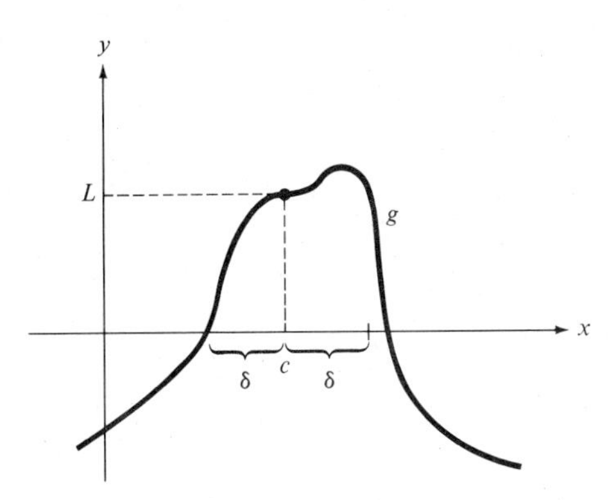

FIGURE 3.23 If $\lim_{x \to c} g(x) = L > 0$ there exists an open interval $(c - \delta, c + \delta)$ on which g is positive.

The proof of the following test will be simplified if we write the derivative in the form

$$f'(c) = \lim_{x \to c} \frac{f(x) - f(c)}{x - c}.$$

Using this form, we can write the second derivative as

$$f''(c) = \lim_{x \to c} \frac{f'(x) - f'(c)}{x - c}.$$

We are now ready for the theorem.

Theorem 3.4 Let f have a second derivative at the interior critical point c.
(a) If $f''(c) > 0$, then f has a local minimum at c.
(b) If $f''(c) < 0$, then f has a local maximum at c.

PROOF Because $f''(c)$ exists, then f and f' must exist and be continuous at c (Theorem 2.4). Since c is an interior critical point, then $f'(c)$ must be zero. Suppose that $f''(c) > 0$. Then

$$\lim_{x \to c} \frac{f'(x) - f'(c)}{x - c} = f''(c) > 0.$$

If we substitute the value $f'(c) = 0$, we obtain

$$\lim_{x \to c} \frac{f'(x) - 0}{x - c} = \lim_{x \to c} \frac{f'(x)}{x - c} = f''(c) > 0.$$

We now apply Theorem 3.3 (with $g(x) = f'(x)/(x - c)$, $L = f''(c)$). There exists a positive number δ such that if

$$0 < |x - c| < \delta, \qquad \text{then} \quad \frac{f'(x)}{x - c} > 0.$$

We now restrict ourselves to values of x such that $c - \delta < x < c$. Then $x - c$ is negative and $f'(x)/(x - c)$ is positive, so the numerator $f'(x)$ must be negative. It follows that f is a decreasing function for x between $c - \delta$ and c.

Similarly, if $c < x < c + \delta$, then $x - c$ is positive and $f'(x)/(x - c)$ is positive, so that $f'(x)$ must be positive. Therefore, f is an increasing function for x between c and $c + \delta$. Therefore, f decreases to the left of c, increases to the right of c, and is continuous at c. It follows from Test 2 that f has a local minimum at c. (See Figure 3.24.)

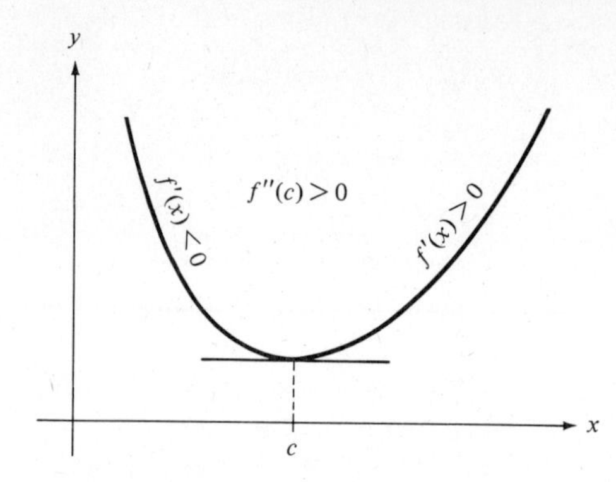

FIGURE 3.24 If $f''(c) > 0$ at the critical point c then $f'(x) < 0$ if $x < c$ and $f'(x) > 0$ if $x > c$ in the vicinity of $x = c$.

A similar proof shows that f has a local maximum at c if $f''(c) < 0$. We omit the details.

No conclusion can be drawn if $f''(c) = 0$. That is, the function may have either a local maximum on a local minimum or neither. (See Exercises 2, 3, and 4.)

The second derivative test for interior local extrema

Test 3 (*The Second-Derivative Test*) Let f have a second derivative at the interior critical point c. Then (see Figure 3.25):

1. If $f''(c) > 0$, then f has a local minimum at c.
2. If $f''(x) < 0$, then f has a local maximum at c.
3. If $f''(c) = 0$, a different test must be used.

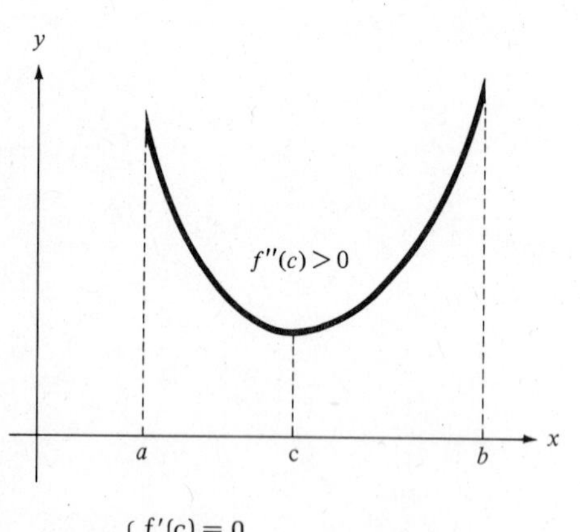

(a) $\begin{cases} f'(c) = 0 \\ f''(c) > 0 \end{cases}$ local minimum

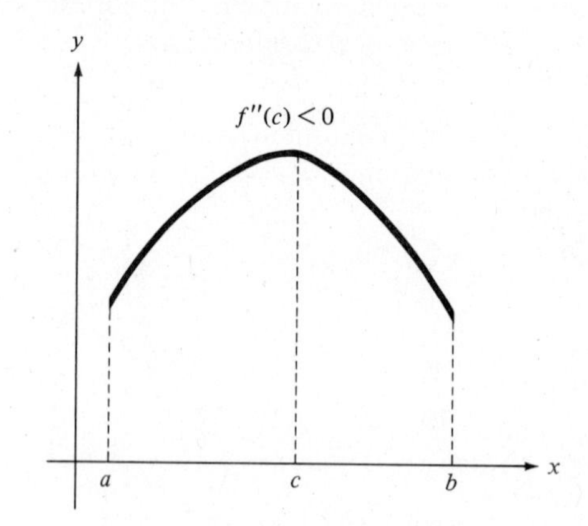

(b) $\begin{cases} f'(c) = 0 \\ f''(c) < 0 \end{cases}$ local maximum

FIGURE 3.25 The second derivative test for interior local extrema.

Example 1 Show that $f(x) = x^2$ has a local minimum at $x = 0$.

Solution

$$f(x) = x^2,$$
$$f'(x) = 2x,$$
$$f''(x) = 2.$$

The only critical point is $x = 0$. At this point we have $f''(0) = 2 > 0$. Thus, f has a local minimum at $x = 0$.

Example 2 Find all local extrema of $f(x) = 2x^3 + 9x^2 - 24x + 5$.

Solution

$$\begin{aligned} f(x) &= 2x^3 + 9x^2 - 24x + 5, \\ f'(x) &= 6x^2 + 18x - 24 \\ &= 6(x^2 + 3x - 4) \\ &= 6(x + 4)(x - 1). \end{aligned}$$

The critical points are $x = -4$ and $x = 1$. The second derivative is

$$f''(x) = 12x + 18.$$

At the critical points we have

$$\begin{aligned} f''(-4) &= 12(-4) + 18 = -30 < 0, \\ f''(1) &= 12 \cdot 1 + 18 = 30 > 0. \end{aligned}$$

Thus, f has a local maximum at $x = -4$ and a local minimum at $x = 1$.

EXERCISES 3.5

1. Use the second derivative test to find all local maxima, minima. Find all points of inflection. Sketch the graphs.

   (a) $y = x^3 - 12x + 1$
   • (b) $y = x^3/3 + 3x^2 - 1$
   (c) $y = 3x^5 - 20x^3 + 4$
   • (d) $y = x(1 - x)^3$
   (e) $y = (x - 1)^2(x - 2)$
   (f) $y = 3x^5 + 10x^3 + 15x$
   • (g) $y = x^2(x + 1)^2$
   (h) $y = x^2(x - 1)^4$
   • (i) $y = x^5 - 10x^3 - 20x^2 - 15x + 4$
   (j) $y = (x^2 - 1)(x^2 - 5)$

   The following three exercises show that the second derivative test fails if $f'(x) = f''(x) = 0$ at a critical point. The curve may have a local maximum, a local minimum, or a point of inflection at the critical point.

2. If $f(x) = x^3$, show that $f'(0) = f''(0) = 0$ and that f has no local extremum at $x = 0$.

3. If $f(x) = -(x - 2)^4$, show that $f'(2) = f''(2) = 0$ and that f has a local maximum at $x = 2$.

4. If $f(x) = (x + 1)^4$, show that $f'(-1) = f''(-1) = 0$ and that f has a local minimum at $x = -1$.

5. Does the curve $y = x^3 + 3x^2 + 6x - 1$ have a point where the slope of the tangent line to the curve has a minimum value? If so, find it.

• 6. On the curve $y = x^3$ find two points whose x-coordinates are one unit apart and such that the line that passes through them has minimum slope. What are the slope and the equation of this line?

3.6 APPLICATION TO BUSINESS ANALYSIS

Optimal Lot Size A company uses N units of a certain item each year. The consumption is distributed evenly over the year, the cost does not vary, the units may be ordered from a distributor and be received almost immediately, and there is no obsolescence. In short, there is no reason for the company to stockpile the item and no reason to order only small quantities. The company wishes to determine the proper number of units to order in each shipment to keep the total cost of stocking the item to a minimum.

The total cost of stocking the item for a year is the sum of the three costs involved: (1) a fixed *unit cost* (price of each unit, shipping cost, and so on); a *reorder cost* that is the same for each shipment, regardless of size; (3) a *storage cost* (depreciation of warehouse facilities, and so on).

Let x be the number of units ordered in each shipment. The number of shipments in a year is N/x. If a is the unit cost and b is the reorder cost, then the cost of each shipment is $ax + b$ and the total cost for all shipments in a year is

$$(ax+b)\frac{N}{x}.$$

If we assume that each shipment arrives as soon as the inventory is zero, then the average size of the inventory is $x/2$. If k is the cost of storing one unit one year, then the storage cost for the year is $kx/2$. The total cost of stocking the item is

$$C(x) = (ax+b)\frac{N}{x} + \frac{kx}{2} = aN + \frac{bN}{x} + \frac{kx}{2}.$$

To find the value of x that minimizes C, we calculate the derivative:

$$C'(x) = -\frac{bN}{x^2} + \frac{k}{2}.$$

$$C'(x) = 0 \quad \text{if and only if} \quad -\frac{bN}{x^2} + \frac{k}{2} = 0, \qquad x = \sqrt{\frac{2bN}{k}}.$$

(We can neglect the negative square root, because x must be positive.) Since $C''(x) = 2bN/x^3$, then C'' is positive at the critical point. Thus C has a local minimum when $x = \sqrt{2bN/k}$. This is the only possible value of x for which C could have a minimum value. Thus, C is small as possible when $x = \sqrt{2bN/k}$. (Figure 3.26.)

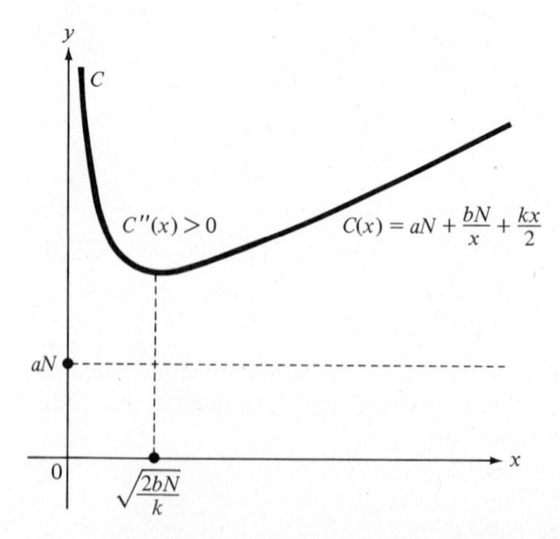

FIGURE 3.26 The optimal lot size problem. Graph of $C(x) = aN + bN/x + kx/2$. C has its minimum value at $x = \sqrt{2bN/k}$.

The result $x = \sqrt{2bN/k}$ is known as the "optimal-lot-size formula." It will always yield the minimum total cost of stocking the item, given the conditions stated at the beginning of this section.[1] Observe that the optimal lot size is

[1] In many cases this value of x will not be an integer. Then we must pick an integer near $\sqrt{2bN/k}$.

proportional to $\sqrt{N}$. Thus if the scale of operations quadruples the optimal lot size doubles. In that case the company should double the size of its order and cut the time between orders in half.

As an example, suppose the company uses 1000 units each year, the reorder cost is $5 per shipment, the storage cost is $0.25 per unit per year. The minimum cost of stocking the item occurs when the number of units in each shipment is x_0, where

$$x_0 = \sqrt{\frac{2 \cdot 5 \cdot 1000}{\frac{1}{4}}} = \sqrt{40{,}000} = 200.$$

(Note that the unit cost does not enter into the calculation. Is this to be expected?)

Marginal Profit and Loss One of the major decisions facing management is that of scale. How many units should be produced? How much land should be bought? How many workmen should be hired? How much money should be spent on advertising?

Before it attempts to answer these questions, the company must know what to expect in the way of costs and revenues from various levels of activity. This information is given in the form of functions which approximate the cost of producing x items, the revenue obtained from selling x items, and so on.

These functions are obtained experimentally. The analyst will determine the actual cost for several different levels of activity and then find a function that closely approximates these values. These formulas are not valid outside of a limited range. (If a cost function were determined by considering levels at 10,000-unit increments between 50,000 and 100,000 units, we would not expect the formula to give a realistic estimate of the cost of producing ten billion units.)

Such an approximating function is almost essential, since the true cost function would be a "step" function. It would probably be described by a table and would be difficult to use. The approximating function, on the other hand, is usually continuous and differentiable. Thus the principles of calculus can be applied to analyze the various levels of activity. For example, one may wish to find the level of production at which the cost would increase the least for a fixed increase in production.

After an approximating function is obtained, much information can be derived from it. We shall restrict ourselves to the consideration of the *marginal* activity of the product. The *true marginal cost* of the nth unit is the cost of producing the nth unit after the $(n-1)$st unit is produced. The *true marginal revenue* obtained from the nth unit is the revenue obtained by selling the nth unit after the $(n-1)$st unit is sold.

True marginal cost

If $C(x)$ is the cost of producing x units, the true marginal cost of the nth unit is $C(n) - C(n-1)$. If $R(x)$ is the revenue obtained from selling x units, the true marginal revenue obtained from the nth unit is $R(n) - R(n-1)$.

For example, if the cost function is $C(x) = 50 + 5x + \frac{1}{100}x^2$ and the revenue function is $R(x) = -25 + 8x + \frac{1}{200}x^2 - \frac{1}{3000}x^2$, then the true marginal cost of the nth item is

$$C(n) - C(n-1) = 5 + \frac{2n-1}{100}$$

and the true marginal revenue brought in by the nth item is

$$R(n) - R(n-1) = 8 + \frac{2n-1}{200} - \frac{3n^2 - 3n + 1}{3000}.$$

The computation involved in calculating the true marginal cost may be prohibitive. What happens if we *approximate* the marginal cost by replacing the cost function between $n-1$ and n with a line? The best linear approximation to a curve near a point is the tangent line at the point. Thus, we use the line tangent to the graph of the cost function at the point $(n, C(n))$ to approximate the marginal cost.

The situation is pictured in Figure 3.27. The true marginal cost is $C(n) - C(n-1)$, the difference in the y-coordinates for two points on the graph.

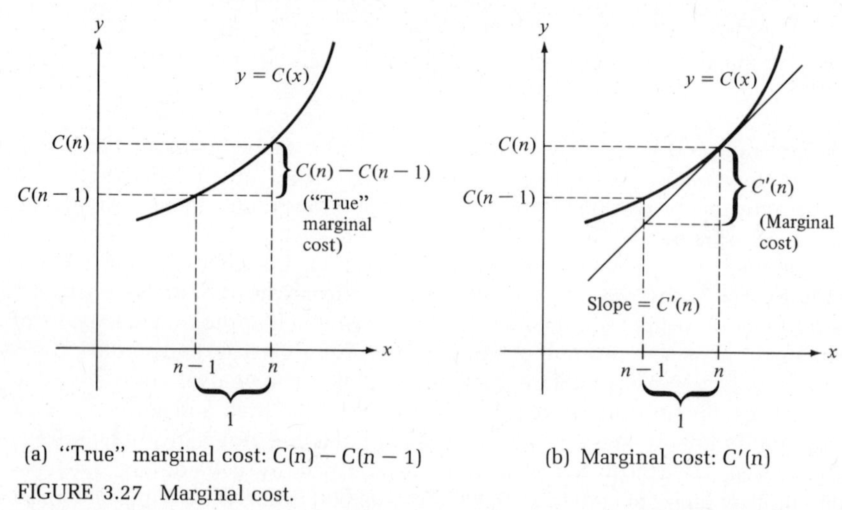

(a) "True" marginal cost: $C(n) - C(n-1)$ (b) Marginal cost: $C'(n)$

FIGURE 3.27 Marginal cost.

If we use the tangent line as an approximation to the marginal cost we see that, since the difference in the x-coordinates is 1, the difference in the y-coordinates is the slope of the line, which equals $C'(n)$. Thus $C'(n)$ is approximately equal to the true marginal cost of the nth unit.

Recall that the true marginal cost is based on a function that approximates the cost function. Thus, it is likely that $C'(n)$ is as good an approximation to the actual cost of producing the nth unit as is the true marginal cost $C(n) - C(n-1)$. Thus, *economists usually define the marginal cost of the nth unit to be* $C'(n)$ *rather than* $C(n) - C(n-1)$. In subsequent discussion we shall follow this convention.

Marginal cost
Marginal revenue
Marginal profit

Similarly, the *marginal revenue* produced by the nth unit is defined to be $R'(n)$, where R is the revenue function. If $P(x)$ is the profit (or loss) obtained by producing and selling x units, then $P'(n)$ is the *marginal profit* on the nth unit.

Observe that $P = R - C$ (*profit is revenue minus cost*). Thus, $P' = R' - C'$ (*marginal profit is marginal revenue minus marginal cost*).

When $P' > 0$, the profit increases as x increases. When $P' < 0$, the profit decreases as x increases. To obtain the maximum profit a company should plan to increase production (provided P' is positive) until $P' = 0$. This occurs when $R' = C'$. In other words, *the maximum profit is obtained when marginal revenue equals marginal cost.*

Example 1 The XYZ Block Company has determined that its cost and revenue functions are

$$C(x) = 50 + 5x + \tfrac{1}{100}x^2$$

and

$$R(x) = -25 + 8x + \tfrac{1}{200}x^2 - \tfrac{1}{3000}x^3$$

where x is the number of production runs per week. The marginal cost of the nth run is

$$C'(n) = 5 + \tfrac{1}{50}n.$$

The marginal revenue produced by the nth run is

$$R'(n) = 8 + \tfrac{1}{100}n - \tfrac{1}{1000}n^2.$$

The marginal profit obtained from the nth run is

$$P'(n) = R'(n) - C'(n) = 3 - \frac{n}{100} - \frac{n^2}{1000}.$$

The maximum profit is obtained when

$$P'(n) = 0$$

$$3 - \frac{n}{100} - \frac{n^2}{1000} = 0$$

$$n^2 + 10n - 3000 = 0$$

$$n = \frac{-10 + \sqrt{100 + 12000}}{2} = 50.$$

Thus the maximum profit is obtained when there are 50 production runs each week.

A typical profit function is pictured in Figure 3.28. The profit is negative when $0 < x < a$ (indicating a loss) and is positive when $a < x < d$. The maximum profit occurs when $x = c$ (note that $P'(c) = 0$). The profit is increasing when $0 < x < c$ $(P'(x) > 0)$ and is decreasing when $x > c$ $(P'(x) < 0)$.

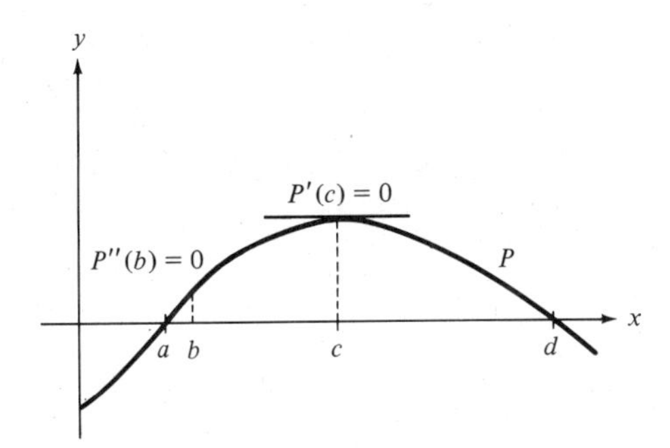

FIGURE 3.28 Typical profit function.

How do we interpret the point of inflection at $x = b$? To the left of this point the curve is concave upward $(P'' > 0)$ and to the right it is concave downward $(P'' < 0)$. Therefore, P' is an increasing function when $0 < x < b$ and is a decreasing function when $x > b$. Consequently, *the maximum marginal profit is obtained at the inflection point* $x = b$.

Marginal Productivity The Chain Rule

$$\frac{dy}{dx} = \frac{dy}{du} \cdot \frac{du}{dx}$$

can be used as part of a theoretical analysis to break quantities into component parts.

As an example, consider a munfacturing plant in which productivity (p) has increased over a period of time. The rate of change of the productivity with respect to time is dp/dt. If x is the size of the labor force, then by the Chain Rule

$$\frac{dp}{dt} = \frac{dp}{dx} \cdot \frac{dx}{dt}.$$

Marginal productivity

In this equation dp/dx is the *marginal productivity of labor*, the rate at which the productivity changes with respect to the size of the labor force, and dx/dt is the rate of change of the labor force with respect to time. It follows that the rate of change of productivity with respect to time is equal to the product of the marginal productivity of labor and the rate of change of the labor force.

As one important application suppose that the marginal productivity of labor is a constant, say $dp/dx = C$. Then

$$\frac{dp}{dt} = \frac{dp}{dx} \cdot \frac{dx}{dt} = C \cdot \frac{dx}{dt},$$

so that the rate of change of productivity is directly proportional to the rate at which the labor force changes.

EXERCISES 3.6

1. Let $C(x)$ be the cost function for the production of x units, $R(x)$ the revenue function for the sale of x units, and $P(x)$ the profit function for the production and sale of x units. For each of the following determine: (1) the marginal cost function, (2) the marginal revenue function, (3) $P(x)$, (4) the marginal profit function, (5) the value of x that maximizes $P(x)$.

   (a) $C(x) = x^2/2 - 50$, $R(x) = 6x + 7$

   • (b) $C(x) = \frac{4}{5}x^2 + 8x + 6$, $R(x) = -\frac{1}{30}x^3 + \frac{6}{5}x^2 + 10x - 4$

   (c) $C(x) = 9x^2 + 40x + 500$, $R(x) = -2x^3 + 30x^2 + 400x - 700$

2. The Jeff-Ri Manufacturing Company uses 100,000 fittings of a certain type each year. The cost is \$20 per thousand with a \$10 reorder cost. The storage cost is \$5 per thousand per year. Let x be the number of fittings in a shipment. Determine a function $C(x)$ that expresses the total cost of stocking the fittings for a year and calculate the value of x that yields the minimum value of $C(x)$.

• 3. The G-C Production Company uses 100,000 magnets each year. The cost of storing a single magnet for one year is five cents. The reorder cost is \$10. What is the optimal number of magnets to be ordered at a single time?

4. Explain why the expression for the optimal lot size does not involve the unit cost.

• 5. The H-2-OH Company manufactures electrical appliances in 300-lot production runs. It can produce x lots of garbage disposals for $300x^2 + 7000x + 400$ dollars. The maximum wholesale price at which x lots can be sold is $(150 - x)/3$ dollars per disposal. Determine x so that the total profit is a maximum.

6. A car-leasing company makes a profit of \$750 per car if it leases no more than 500 cars. For each car over 500 leased, the profit per car decreases by \$5. What is the marginal profit function? How many cars should the company lease to receive maximum profit?

• 7. Each month the Kill-A-Cycle Radio Company can sell x shipments of 100 FM tuners at $\$d$ per tuner, where $d = (375 - 5x)/3$. The cost of producing x shipments is $5000 + 1500x$ dollars.

   (a) How many tuners should the company produce each month in order to make the maximum profit?
   (b) The Federal Government plans to impose an emergency excise tax of \$2 on each FM tuner sold. How will this tax affect production, price, and profit?

• 8. The L and S Book Company finds that its cost function is $C(x) = x^2/10 + x + 10$ and its revenue function $R(x) = 4(x - 3.6)/x$, where x is measured in units of 10,000 books and $C(x)$ and $R(x)$ are in hundreds of dollars. Determine the value of x for which the profit will be maximized; that is, for what x does marginal revenue equal marginal cost?

9. Production at the Kold Kup factory over the past six months is given by the function

$$P(t) = (100 + t^2)10{,}000 \text{ items},$$

where t is measured in months. (Six months ago $(t = 0)$ production was 100(10,000) items, one month later $(t = 1)$ it was 101(10,000) items, and so on.) During the same period the size of the labor force is given by the function

$$x(t) = 75 + 5t.$$

Find dp/dx, the marginal productivity of labor, at the following times:

   • (a) $t = 1$
   (b) $t = 5$

3.7 THE MEAN VALUE THEOREM

Many mathematical propositions that seem quite reasonable are difficult to prove analytically. As an example, consider the following proposition, which was stated earlier without proof:

> *If f has a positive derivative at each point of an interval, then f increases on the interval. That is, if x_1 and x_2 are on the interval with $x_1 < x_2$, then $f(x_1) < f(x_2)$.*

A simple direct proof of this proposition is not easy to find. We can use facts about a function to obtain information about its derivative, but in many cases it is difficult to use facts about the derivative to obtain information about the original function. This is because the derivative at a point x_0 is a number that depends on the value of the function at x_0 and on the values of the function at all points near x_0. In a certain sense the derivative at x_0 is an "average" bit of information about the properties of the function near x_0. It may be difficult to use this "average" to reconstruct properties of the original function.

In this section we discuss a basic theorem that can be used to prove many propositions similar to the one above. This theorem, called the *Mean Value Theorem*, states a proposition that also seems quite reasonable. In all probability, the reader will appreciate it more after he has seen it used as the key step in several of the proofs that follow.

In geometric terms the *Mean Value Theorem* asserts the following proposition: Let f be continuous on the closed interval $[a, b]$ and differentiable on the open interval (a, b). There is at least one number X, $a < X < b$, where the line tangent to the graph of f is parallel to the line connecting the points $(a, f(a))$ and $(b, f(b))$ (Figure 3.29).

Geometrical interpretation of the Mean Value Theorem

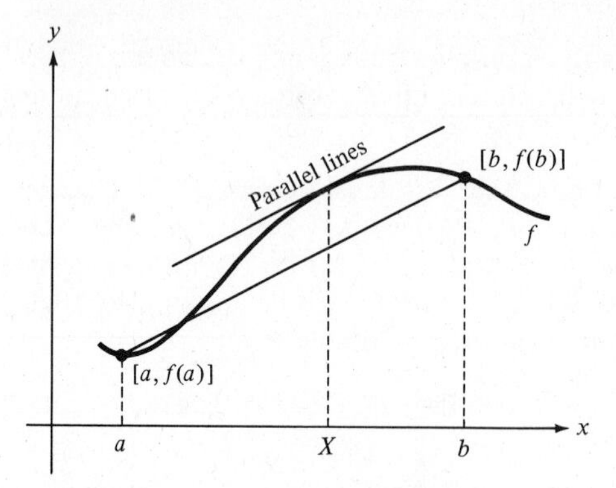

FIGURE 3.29 Geometrical interpretation of the Mean Value Theorem.

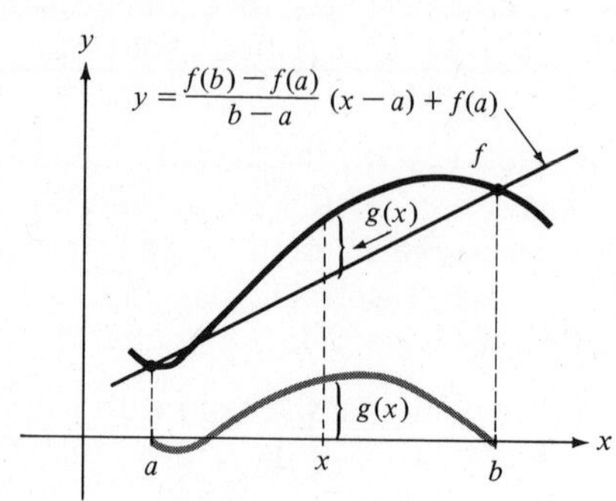

FIGURE 3.30

$$g(x) = f(x) - \left[\frac{f(b) - f(a)}{b - a}(x - a) + f(a)\right]$$

= the difference in the y-coordinates of points on the graph of f and points on the line.

In analytic terms this is equivalent to saying that $f'(X)$, the slope of the tangent line at $(X, f(X))$, is equal to

$$f'(X) = \frac{f(b) - f(a)}{b - a},$$

the slope of the line connecting $(a, f(a))$ and $(b, f(b))$.

The Mean Value Theorem

Theorem 3.5 (*The Mean Value Theorem*) Let f be continuous on the closed interval $[a, b]$ and differentiable at each point of the open interval (a, b). There is at least one number X, $a < X < b$, such that

$$f'(X) = \frac{f(b) - f(a)}{b - a}.$$

PROOF We define the auxiliary function g at each x in the domain of f by

$$g(x) = f(x) - \left[\frac{f(b) - f(a)}{b - a}(x - a) + f(a)\right].$$

Geometrically $g(x)$ represents the difference between the graph of f and the line connecting the points $(a, f(a))$ and $(b, f(b))$. (See Fig. 3.30.) The function g is continuous on the closed interval $[a, b]$ and differentiable on the open interval (a, b). We will show that the derivative of g is zero at at least one point X between a and b.

Observe that $g(a) = g(b) = 0$. Thus, either g is constantly equal to zero on $[a, b]$ or g has a local extremum between a and b.

Case I $g(x) = 0$ for $a \leqslant x \leqslant b$. Then $g'(x) = 0$ for every x, $a < x < b$.

Case II g has a local maximum or minimum between a and b. Let $g(X)$ be this extremum. (If there are several such extrema, choose one of them.) Then $a < X < b$ and so $g'(X)$ exists. By Theorem 3.1

$$g'(X) = 0.$$

Thus, in either case there is at least one number X, $a < X < b$, such that $g'(X) = 0$. But

$$g'(x) = f'(x) - \left[\frac{f(b) - f(a)}{b - a}(1 - 0) + 0\right].$$

Therefore,

$$g'(X) = f'(X) - \frac{f(b) - f(a)}{b - a} = 0.$$

and so

$$f'(X) = \frac{f(b) - f(a)}{b - a}.$$

Remark The statement that

$$\frac{f(b) - f(a)}{b - a} = f'(X)$$

shows that the average rate of change of f over the interval $[a, b]$ is equal to the instantaneous rate of change at the point $x = X$. Thus, there must be at least one point X between a and b where the instantaneous rate of change is equal to the average rate.

For example, if $f(x)$ is the revenue received for a production level of x units, then the average change in the revenue over the levels $x = a$ to $x = b$ is equal to the marginal revenue for at least one value of X between a and b.

Similarly, if $f(t)$ is a distance function for straight line motion, then $[f(b) - f(a)]/(b - a)$ is equal to the average velocity over the interval of time $a \leqslant t \leqslant b$. The Mean Value Theorem shows that this average velocity must be equal to the instantaneous velocity for at least one number $t = X$ between a and b. For example, if a motorist travels 60 miles in one hour then he must have been traveling exactly 60 mph at at least one point of time.

Theoretical Applications The Mean Value Theorem is primarily used as a theoretical tool to establish basic results about a function from information given about its derivative. As an application we return to the problem posed at the beginning of this section and show that a function with a positive derivative over an interval must increase over the interval.

Increasing function

Theorem 3.6 Let f be continuous on the closed interval $[a, b]$ and have a positive derivative at each point of the open interval (a, b). If $a \leqslant x_1 < x_2 \leqslant b$, then $f(x_1) < f(x_2)$.

PROOF It follows from the hypothesis that f is continuous on the closed interval $[x_1, x_2]$ and has a positive derivative at each point of the open interval (x_1, x_2). We apply the Mean Value Theorem to that interval. There exists a number X, $x_1 < X < x_2$, such that

$$\frac{f(x_2) - f(x_1)}{x_2 - x_1} = f'(X).$$

Since $x_2 - x_1 > 0$ and $f'(X) > 0$, then

$$f(x_2) - f(x_1) = (x_2 - x_1)f'(X) > 0.$$

Thus, $f(x_1) < f(x_2)$. (See Figure 3.31.)

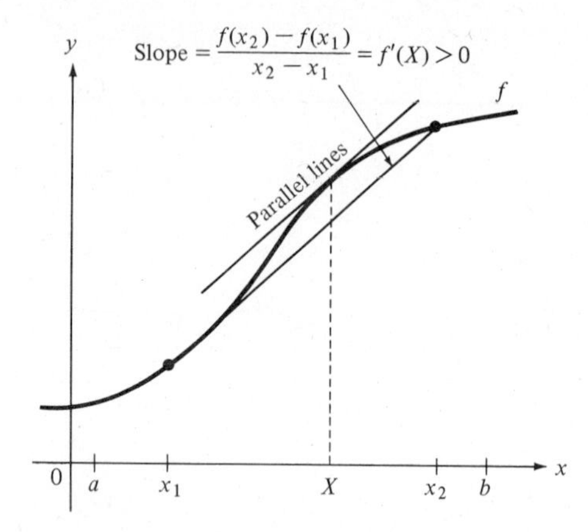

FIGURE 3.31 If $f'(x) > 0$ on (a, b), then f is increasing on the interval.

Since $\dfrac{f(x_2) - f(x_1)}{x_2 - x_1} = f'(X) > 0$, then $f(x_2) > f(x_1)$.

Suppose $f'(x) = 0$ for every x. We would suspect that f must be a constant function. The Mean Value Theorem can be used to show that this is the case.

Theorem 3.7 Let f be continuous on the closed interval $[a, b]$ and differentiable on the open interval (a, b). If $f'(x) = 0$ for every x, $a < x < b$, then there exists a constant C such that

$$f(x) = C$$

for all x, $a \leq x \leq b$.

PROOF Let x be any fixed number in the range $a < x \leq b$. Then f is continuous on $[a, x]$ and differentiable on the interval (a, x). By the Mean Value Theorem there exists a number X, $a < X < x$, such that

$$\frac{f(x) - f(a)}{x - a} = f'(X).$$

But $f'(X) = 0$. Thus,

$$f(x) = f(a).$$

If we let $C = f(a)$, then it follows that

$$f(x) = C$$

for all x, $a \leq x \leq b$.

Antiderivatives

The following theorem, which follows easily from Theorem 3.7 is most important for the work that we do in Chapter 4 and the remainder of the book. It establishes that any two *antiderivatives* of a function can differ by at most a constant. (The function f is an *antiderivative* of the function F if $f'(x) = F(x)$ for every x on some interval.)

Theorem 3.8 Let f and g be continuous on the closed interval $[a, b]$ and differentiable on the open interval (a, b). Suppose that

$$f'(x) = g'(x)$$

for every x, $a < x < b$. Then there exists a constant C such that

$$f(x) = g(x) + C$$

for every x, $a \leq x \leq b$.

PROOF Let $h(x) = f(x) - g(x)$. Then h is continuous on $[a, b]$ and differentiable on the open interval (a, b). Furthermore

$$h'(x) = f'(x) - g'(x) = 0 \qquad \text{for every } x,\ a < x < b.$$

It follows from Theorem 3.7 that there exists a constant C such that

$$\begin{aligned} h(x) &= C && \text{for every } x,\ a \leq x \leq b. \\ f(x) - g(x) &= C && \text{for every } x,\ a \leq x \leq b. \\ f(x) &= g(x) + C && \text{for every } x,\ a \leq x \leq b. \end{aligned}$$

(See Figure 3.32.)

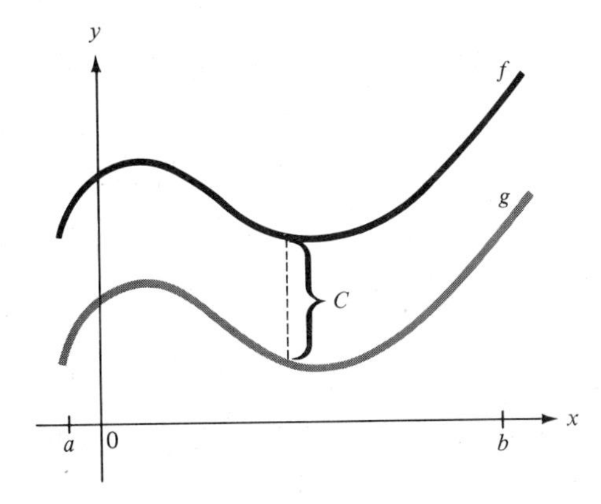

FIGURE 3.32 If $f'(x) = g'(x)$, $a \leq x \leq b$, there exists a constant C such that $f(x) = g(x) + C$, $a \leq x \leq b$.

EXERCISES 3.7

1. Verify the Mean Value Theorem for the following functions defined on the given intervals. (Show that f is continuous on the closed interval and differentiable on the corresponding open interval, and locate a point X that satisfies the conclusion of the Mean Value Theorem.)

   (a) $f(x) = x^2$, $[0, 2]$

   • (b) $f(x) = x^3 - x^2 + 2$, $[0, 1]$

   (c) $f(x) = \dfrac{x}{x - 4}$, $[2, 3]$

   (d) $f(x) = x + \dfrac{1}{x}$, $[2, 3]$

2. Use the Mean Value Theorem to give a direct proof of the following result: If f has a positive derivative at each point of the closed interval $[0, 1]$, then $f(1) > f(0)$.

3. Modify the proof of Theorem 3.6 to prove a comparable result for decreasing functions.

4. Does the Mean Value Theorem apply to the following functions defined on the given intervals? Explain

   (a) $f(x) = 1/(x - 1)$, $[0, 2]$
   • (b) $f(x) = [x]$, $[1, 2]$ (Greatest integer function)
   (c) $f(x) = x^2$, $[-1, 1]$

• 5. Find two functions, f and g, such that $f'(x) = 3x^2 = g'(x)$ and $f(x) \neq g(x)$.

6. Find two functions, f and g, such that $f'(x) = 2x - 1 = g'(x)$ and $f(x) \neq g(x)$.

7. Let $f(x)$ and $g(x)$ be differentiable functions such that $f'(x) = g'(x)$ for all x and $f(0) = g(0)$. Use Theorem 3.8 to show that $f(x) = g(x)$, for all x.

3.8 IMPLICIT DIFFERENTIATION

Most of the functions encountered in this book have been of form $y = f(x)$. Such a relationship defines y *explicitly* as a function of x. If we are given a defining equation of form

Explicit and implicit functions

$$F(x, y) = 0,$$

it may be possible to solve the equation for y in terms of x. Such an equation is said to define y *implicitly* as a function of x.

For example, the equation

$$xy - 3 + 7x = 0$$

can be solved for y as a function of x:

$$y = \frac{3 - 7x}{x}.$$

Thus, the equation above defines y implicitly as a function of x.

For a second example consider the equation

$$x^2 + y^2 - 4 = 0.$$

This equation can be solved for y, yielding

$$y = \pm\sqrt{4 - x^2}.$$

This last equation does not yield y as a function of x, because there are two values of y for each value of x between -2 and 2. However, if we restrict ourselves to either the positive or the negative values of x, we obtain *two* functions of x:

$$y = f_1(x) = \sqrt{4 - x^2} \quad \text{and} \quad y = f_2(x) = -\sqrt{4 - x^2}.$$

Therefore, the equation

$$x^2 + y^2 - 4 = 0$$

implicitly defines two functions of x. Graphically the first of these functions determines the upper semicircle and the second the lower semicircle shown in Figure 3.33.

How do we calculate the derivative of a function defined implicitly by an equation? If we can solve the equation for y explicitly as a function of x, we can

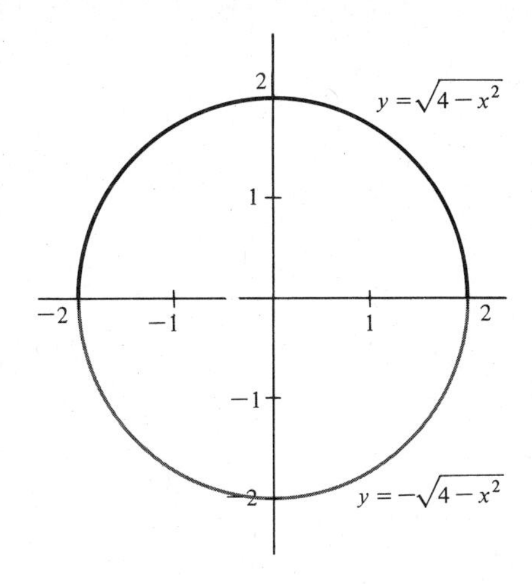

FIGURE 3.33 Graph of $x^2 + y^2 = 4$. The upper simicircle is the graph of $y = \sqrt{4 - x^2}$, the lower semicircle is the graph of $y = -\sqrt{4 - x^2}$.

use the techniques developed in this chapter. In many cases, however, the defining equation is so complicated that it is not practical to solve it for y. (For example, it would be virtually impossible to solve

$$3x^2y^4 + 7x^3y + \frac{3x^9}{y-1} + 5x = 0$$

for y as a function of x.) Even in these cases it may be possible to calculate dy/dx with comparative ease. The process used to calculate dy/dx in these cases is known as *implicit differentiation*.

Steps in implicit differentiation

The process of implicit differentiation is quite simple:

1. Write the defining equation $F(x, y) = 0$
2. Differentiate both sides of the equation, treating y as a differentiable function of x and using the Chain Rule.
3. Solve the resulting equation for dy/dx.

Example 1 Let $x^2 + y^2 - 4 = 0$. Use implicit differentiation to calculate $y' = dy/dx$.

Solution

1. Write the defining equation:

$$x^2 + y^2 - 4 = 0.$$

2. Differentiate both sides of the defining equation using the Chain Rule:

$$\frac{d}{dx}(x^2 + y^2 - 4) = \frac{d}{dx}(0),$$

$$\frac{d}{dx}(x^2) + \frac{d}{dx}(y^2) - \frac{d}{dx}(4) = 0$$

$$2x + 2y \cdot \frac{dy}{dx} - 0 = 0.$$

3. Solve the resulting equation for $\frac{dy}{dx}$:

$$\frac{dy}{dx} = -\frac{x}{y}.$$

How do we interpret the result obtained in Example 1? We know that the equation

$$x^2 + y^2 - 4 = 0$$

defines y as two functions of x. Which one of these functions have we differentiated? Actually, we have differentiated both of them. In order to evaluate $dy/dx = -x/y$ at a point, we must substitute the coordinates of the point for x and y. We can use this result to calculate dy/dx at any point (x, y) with $y \neq 0$ on either of the graphs $y = \sqrt{4 - x^2}$ or $y = -\sqrt{4 - x^2}$. For example, at the point $(\sqrt{3}, 1)$ on the upper semicircle we obtain

$$\frac{dy}{dx} = -\frac{\sqrt{3}}{1} = -\sqrt{3}.$$

At the point $(\sqrt{3}, -1)$ on the lower semicircle we obtain

$$\frac{dy}{dx} = -\frac{\sqrt{3}}{-1} = \sqrt{3}.$$

(See Figure 3.34.)

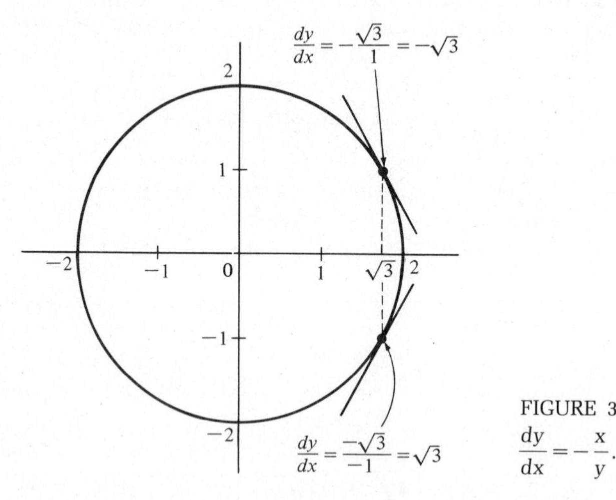

FIGURE 3.34 Graph of $x^2 + y^2 = 4$. $\frac{dy}{dx} = -\frac{x}{y}$.

In most cases the graph of $F(x, y) = 0$ consists of several branches. When we calculate dy/dx by implicit differentiation, we need to know both coordinates of a point in order to evaluate the derivative there. Essentially, this means that we need to know which branch of the graph the point is on.

Example 2 Let $x^2y^2 = 1$. (a) Calculate y' implicitly. (b) Compare the result of (a) with that obtained by first expressing y as a function of x and then differentiating y with respect to x.

Solution

(a)

$$x^2y^2 = 1$$

$$\frac{d}{dx}(x^2y^2) = \frac{d}{dx}(1)$$

$$x^2 \cdot \frac{d}{dx}(y^2) + y^2 \cdot \frac{d}{dx}(x^2) = 0 \qquad \text{(by D6)}$$

$$x^2 \cdot 2y \cdot y' + y^2 \cdot 2x = 0 \qquad \text{(by D5)}$$

$$2x^2yy' + 2xy^2 = 0$$

$$2xy(xy' + y) = 0.$$

From the original equation we see that it is not possible to have either x or y equal to zero. Therefore,

$$xy' + y = 0, \; y' = -\frac{y}{x}.$$

(b) If we solve the equation $x^2y^2 = 1$ for y we obtain two functions

$$y = f_1(x) = \frac{1}{x} \quad \text{and} \quad y = f_2(x) = -\frac{1}{x}.$$

For the first of these functions we have

$$y' = \frac{d}{dx}\left(\frac{1}{x}\right) = \frac{d}{dx}(x^{-1}) = -x^{2} = -\frac{1}{x^2}.$$

To compare this answer with the one we calculated in (a) recall that

$$y' = -\frac{y}{x} \quad \text{and} \quad y = f_1(x) = \frac{1}{x}.$$

Therefore,

$$y' = -\frac{\frac{1}{x}}{x} = -\frac{1}{x^2},$$

so the answers agree.

It is left as an exercise for the reader to show that the two answers are the same for the other function $y = f_2(x)$.

Implicit differentiation enables us to relate derivatives of functions even when the variable of differentiation is not explicitly stated. For example, if $P(x)$, $R(x)$ and $C(x)$ represent the profit, revenue, and cost, respectively, associated with a production level of x units, then each of these functions can be considered to be a function of time. If we differentiate the equation

$$P(x) = R(x) - C(x)$$

implicitly with respect to time we obtain

$$\frac{dP(x)}{dt} = \frac{dR(x)}{dt} - \frac{dC(x)}{dt}.$$

Thus, the rate of change of the profit function with respect to time is equal to the difference of the rate of change of the revenue function and the rate of change of the cost function with respect to time.

Higher order derivatives can be calculated by implicit differentiation, but the method is complicated. In calculating the second derivative we treat both y and y' as functions of x.

Higher derivatives

Example 3 Calculate y' and y'' if

$$x^{1/2} + y^{1/2} = 1.$$

Solution

(a)

$$x^{1/2} + y^{1/2} = 1.$$

$$\frac{d}{dx}(x^{1/2}) + \frac{d}{dx}(y^{1/2}) = \frac{d}{dx}(1).$$

$$\frac{1}{2}x^{-1/2} + \frac{1}{2}y^{-1/2} \cdot y' = 0.$$

$$y' = -y^{1/2}x^{-1/2} = -\sqrt{\frac{y}{x}}.$$

(b) To calculate y'' we recall that

$$y'' = \frac{d}{dx}(y') = \frac{d}{dx}(-y^{1/2} \cdot x^{-1/2})$$

$$= -\left[y^{1/2}\frac{d}{dx}(x^{-1/2}) + x^{-1/2} \cdot \frac{d}{dx}(y^{1/2})\right]$$

$$= -\left[y^{1/2}\left(-\frac{1}{2}\right)x^{-3/2} + x^{-1/2}\left(\frac{1}{2}\right)y^{-1/2}y'\right]$$

$$= \frac{y^{1/2}}{2x^{3/2}} - \frac{1}{2x^{1/2}y^{1/2}} \cdot y'$$

$$= \frac{\sqrt{y}}{2x\sqrt{x}} - \frac{1}{2\sqrt{xy}} \cdot y'.$$

We now substitute the value of y' obtained in (a) and simplify:

$$y'' = \frac{\sqrt{y}}{2x\sqrt{x}} - \frac{1}{2\sqrt{xy}} \cdot \left(-\sqrt{\frac{y}{x}}\right)$$

$$= \frac{\sqrt{y}}{2x\sqrt{x}} + \frac{1}{2x} = \frac{1}{2x}\left(\sqrt{\frac{y}{x}} + 1\right).$$

EXERCISES 3.8

1. Calculate y' in two ways: (1) by implicit differentiation; (2) by expressing y as a function of x. Do the answers agree?

- • (a) $4x^2 + 3y^2 = 12$
- (b) $y^2 - x^2 = 1$
- (c) $x - y^2 + 2y = 10$
- • (d) $x^{1/2} - y^{1/2} = 1$

2. Use implicit differentiation to calculate the slope of the tangent line at the indicated point.

   (a) $x^2 + 3y^3 - xy - y = 0$, $(0, 0)$
   • (b) $2\sqrt{x} - \sqrt{y} + y - 2 = 0$, $(1, 1)$
   (c) $\frac{x}{y} + \frac{y}{x} = 2x^2y$, $(1, 1)$
   • (d) $x^n + y^n = 2$, $(1, 1)$ (n is a positive integer.)

3. Use implicit differentiation to calculate d^2y/dx^2.

   (a) $x^{3/2} + y^{3/2} = 8$
   • (b) $x^2 + 4y^2 = 4$
   (c) $x^2 + y^2 = 1$
   • (d) $xy = 1$

3.9* RELATED RATES

Many problems involve several related variables. If these variables change with respect to time, their rates of change also are related. In this section we show how to calculate one rate if the related rates are all known.

Example 1 Two airplanes in flight cross above a town at 1:00 P.M. One plane travels east at 300 miles per hour, the other north at 400 miles per hour. At what rate does the distance between the planes change at 3:00 P.M.?

Solution Let x be the distance the first plane has traveled from the town at time t and y the distance the second plane has traveled. Let z be the distance between the planes. (See Figure 3.35.) By the Pythagorean theorem we have

$$z^2 = x^2 + y^2.$$

There are two ways we can proceed:

Method I Consider x, y, and z to be functions of t, differentiate both sides of the equation $z^2 = x^2 + y^2$ implicitly with respect to t, and use the resulting equation to calculate dz/dt.

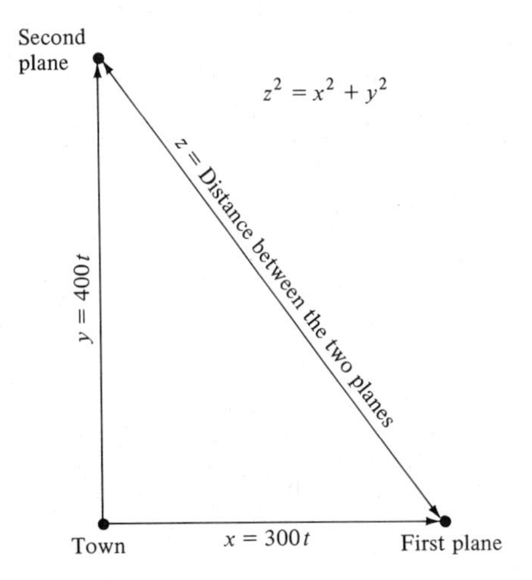

FIGURE 3.35 Example 1

$$z^2 = x^2 + y^2,$$

$$\frac{d}{dt}(z^2) = \frac{d}{dt}(x^2 + y^2),$$

$$2z \cdot \frac{dz}{dt} = 2x \cdot \frac{dx}{dt} + 2y \cdot \frac{dy}{dt},$$

$$\frac{dz}{dt} = \frac{x \cdot \dfrac{dx}{dt} + y \cdot \dfrac{dy}{dt}}{z}.$$

We are given that $dx/dt = 300$ and $dy/dt = 400$. When $t = 2$ (3:00 P.M.), we have $x = 600$, $y = 800$, and

$$z = \sqrt{600^2 + 800^2} = 200\sqrt{3^2 + 4^2} = 200\sqrt{5^2} = 1000.$$

Then

$$\frac{dz}{dt} = \frac{600 \cdot 300 + 800 \cdot 400}{1000} = \frac{500{,}000}{1000} = 500.$$

The planes are traveling apart at the rate of 500 miles per hour when $t = 2$.

Method II Express z as an explicit function of t and calculate dz/dt.

Recall that $x = 300t$ and $y = 400t$ where t is the elapsed time in hours. Then

$$z = \sqrt{x^2 + y^2} = \sqrt{(300t)^2 + (400t)^2} = 100t\sqrt{3^2 + 4^2} = 500t.$$

Therefore $\dfrac{dz}{dt} = \dfrac{d}{dt}(500t) = 500.$

In many cases the first method used in Example 1 is the easier of the two. The solution of the general problem can be obtained as follows:

Problem Calculate the rate of change (with respect to time) of one variable, given the rates of change of one or more related variables.

Steps in the solution

Steps in solution of related rate problems

1. Obtain an equation involving all of the variables.
2. Differentiate the equation implicitly with respect to t, obtaining an equation involving the variables and their derivatives.
3. Calculate the rates of change (derivatives with respect to t) of the related variables.
4. Substitute the values from step 3 into the equation from step 2 and calculate the rate of change of the remaining variable.

These steps are illustrated in the following example:

Example 2 A 10-foot ladder is propped against a wall of a room. The foot of the ladder is moved away from the wall at the rate of 1 ft/sec. At what rate is the top of the ladder sliding down the wall at the instant when the foot of the ladder is 6 feet away from the wall?

Solution Let x be the distance from the foot of the ladder to the wall at time t; let y be the distance from the floor to the top of the ladder. We wish to calculate dy/dt at the instant when $x = 6$.

Step 1 Obtain an equation relating x and y: From the Pythagorean Theorem we have

$$x^2 + y^2 = 10^2.$$

(See Figure 3.36.)

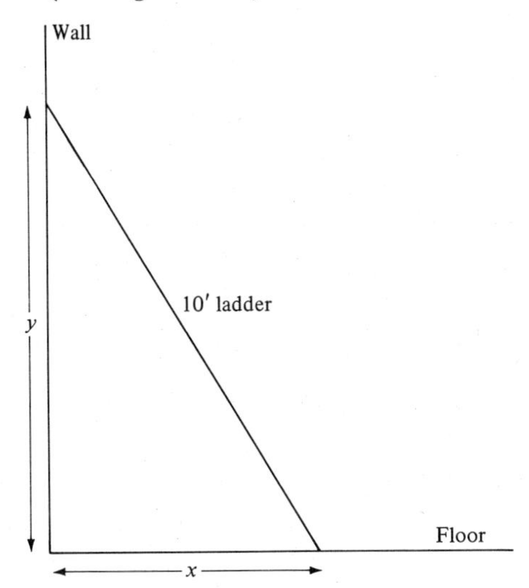

FIGURE 3.36 Example 2

Step 2 Differentiate this equation implicitly with respect to t:

$$x^2 + y^2 = 10^2,$$

$$\frac{d}{dt}(x^2 + y^2) = \frac{d}{dt}(10^2),$$

$$2x \cdot \frac{dx}{dt} + 2y \cdot \frac{dy}{dt} = 0,$$

$$\frac{dy}{dt} = -\frac{x \cdot \frac{dx}{dt}}{y}.$$

Step 3 Calculate dx/dt: We are given that $dx/dt = 1$.

Step 4 Use Steps 2 and 3 to calculate dy/dt:

$$\frac{dy}{dt} = -\frac{x \cdot \frac{dx}{dt}}{y} = -\frac{x \cdot 1}{y} = -\frac{x}{y}.$$

When $x = 6$, $y = 8$ and

$$\frac{dy}{dt} = -\frac{6}{8} = -\frac{3}{4}.$$

Therefore, the top of the ladder is moving towards the floor at $\frac{3}{4}$ ft/sec at the instant when $x = 6$.

Example 3 Hansum Hair, Inc., has a monopoly on a new type of hair conditioner. Demand for the product is steadily increasing and Hansum is increasing production at the rate of three production runs per week. The net profit in dollars for x production runs per week is given by

$$P(x) = -\frac{x^3}{10} + 9x^2 + 1200x - 10{,}000.$$

Find the weekly rate at which the profit is increasing
(a) When $x = 60$.
(b) When $x = 120$.

Solution We differentiate the profit function with respect to time:

$$\frac{dP}{dt} = \frac{dP}{dx} \cdot \frac{dx}{dt} = \left(-\frac{3x^2}{10} + 18x + 1200\right) \cdot \frac{dx}{dt}.$$

$$= \left(-\frac{3x^2}{10} + 18x + 1200\right) \cdot 3$$

since $dx/dt = 3$ runs per week.
(a) When $x = 60$

$$\frac{dP}{dt} = \left(-\frac{3(60)^2}{10} + 18(60) + 1200\right) \cdot 3$$

$$= (-3 \cdot 360 + 18 \cdot 60 + 1200) \cdot 3$$

$$= (-1080 + 1080 + 1200) \cdot 3 = 3600.$$

Profit is increasing at the rate of $3600 per week when $x = 60$.
(b) When $x = 120$

$$\frac{dP}{dt} = \left(\frac{-3(120)^2}{10} + 18(120) + 1200\right) \cdot 3$$

$$= (-3 \cdot 1440 + 18 \cdot 120 + 1200) \cdot 3$$

$$= (-4320 + 2160 + 1200) \cdot 3 = (-960) \cdot 3 = -2880.$$

Profit is decreasing at the rate of $2880 per week when $x = 120$.

EXERCISES 3.9

1. A spherical balloon is inflated at the rate of 10 cubic inches/sec. At what rate is the radius increasing when the radius is 1 inch? 2 inches? ($V = \frac{4}{3}\pi r^3$.)

• 2. Gasoline is pumped into a vertical cylindrical tank at a rate of 32 cubic feet/min. The radius of the tank is 4 feet. How fast is the surface rising? (The formula for the volume of a cylinder of radius r and height h is $V = \pi r^2 h$.)

3. A 20-foot ladder leans against a vertical wall. The lower end of the ladder is pulled along a horizontal floor away from the wall at a rate of 3 ft/sec, causing the top of the ladder to slide down the wall.

   (a) At what rate is the top of the ladder moving when the lower end is 16 feet from the wall?

   • (b) At what rate is the top of the ladder moving when it is 16 feet from the floor?

4. A beam of light shining on a wall outlines a square on the wall. As the light source is moved from the wall the sides of the square increase at the rate of $\frac{1}{2}$ ft/sec. How fast is the area of the square increasing when each side is 10 feet long?

5. The monthly cost and revenue functions of the Coffee Stirrer Corporation are

$$C(x) = x^3 + 20x^2 + 500x + 7500$$

and

$$R(x) = 65x^2 + 3500x,$$

where x is the number of thousands of cases of coffee stirrers produced. Over a five-month period production increases from $x = 40$ to $x = 60$. According to the formula

$$x = 40 + \frac{4t^2}{5}$$

• (a) Calculate dC/dt when $t = 0$ and $t = 5$.
(b) Calculate dR/dt when $t = 0$ and $t = 5$.
• (c) Let $P(x)$ be the profit function. Calculate the monthly rate of change of P with respect to time when $t = 0$ and $t = 5$.

3.10 THE DIFFERENTIAL

Infinitesimals

Thus far in this book we have been careful to emphasize that the dy/dx used for the derivative is not a fraction, but must be considered as a special symbol. Leibniz and the other developers of the calculus, however, actually thought of the derivative as a fraction with dy for the numerator and dx for the denominator. In order to justify the notation they invented the mythical "infinitesimal." Essentially, an infinitesimal, such as dx or dy, was considered to be an object having the same basic properties as the number zero for addition and multiplication. When an infinitesimal was added to a number the sum was the original number. When an infinitesimal was multiplied by a number the product was an infinitesimal. If, however, one divided an infinitesimal into another infinitesimal, the result was usually a number or a function of x (the derivative). The calculus rested on this rather shaky foundation until Cauchy (1789–1857) invented the concept of limit and demolished the infinitesimal.[2]

Actually, it was natural for Leibniz to consider the derivative a fraction. In many ways it acts similarly. For example, the Chain Rule

$$\frac{dy}{dx} = \frac{dy}{du} \cdot \frac{du}{dx}$$

appears to be nothing more than the usual cancellation rule for simplifying fractions.

As a second example consider the problem of implicit functions. Suppose we can solve the equation $y = f(x)$ for x as a function of y: $x = g(y)$. It can be shown (although we have not done it) that under these circumstances, when the two derivatives are evaluated at the point (x_0, y_0), we have

$$f'(x_0) = \frac{1}{g'(y_0)},$$

provided both functions are differentiable and $g'(y_0) \neq 0$.

---

[2] In recent years the infinitesimal has been made respectable in other mathematical settings.

In the Leibniz notation we have

$$\frac{dy}{dx} = f'(x_0) \quad \text{and} \quad \frac{dx}{dy} = g'(y_0)$$

so that

$$\frac{dy}{dx} = \frac{1}{\frac{dx}{dy}},$$

another rule that appears to be a consequence of the manipulation of fractions. These are only two of the many ways in which the derivative acts as if it were a fraction.

Mathematicians did not want to lose the advantages of treating the derivative as a fraction. When Cauchy disposed of the infinitesimal, they invented a new concept to take its place: the *differential*.

The differential

Let f be a function of x. Let dx be another independent variable. (We could just as well use z or t or some other symbol rather than dx, but the notation dx is standard.) The *differential* of f, denoted by $d(f(x))$, is the function of two variables, x and dx, defined by

$$d(f(x)) = f'(x)\, dx.$$

Example 1 Let $f(x) = 3x^2 - 1$.

(a) Calculate $d(f(x))$ as a function of x and dx.

(b) Evaluate $d(f(x))$ at $x = 1$, $dx = 7$; at $x = -3$, $dx = 2$.

Solution

(a) $d(f(x)) = f'(x)\, dx = 6x \cdot dx$.

(b) If $x = 1$ and $dx = 7$, then

$$d(f(x)) = 6 \cdot 1 \cdot 7 = 42.$$

If $x = -3$ and $dx = 2$, then

$$d(f(x)) = 6(-3) \cdot 2 = -36.$$

We also indicate the differential of Example 1 by writing

$$d(3x^2 - 1) = 6x \cdot dx.$$

Observe that $d(x) = 1 \cdot dx = dx$. That is, the function dx is itself the differential of the function defined by $f(x) = x$.

If $y = f(x)$, we also write dy as well as $d(f(x))$. If this is done, then

$$dy = f'(x)\, dx$$

and so

$$\frac{dy}{dx} = f'(x) \qquad (\text{provided } dx \neq 0).$$

Therefore, the derivative can be considered to be a fraction with numerator the differential dy and denominator the differential dx.

At first glance the development above seems to be as mystical as the old approach using infinitesimals. There is a basic difference, however. Infinitesimals were mythical objects invented to explain the limit process, which was not properly understood. Consequently, great gaps were left in the theory when the calculus was based on the infinitesimal. Cauchy's contribution was to give a precise statement of what is meant by a limit and then develop the calculus from that concept. By taking this approach, he put the derivative on a solid mathematical foundation. We have now defined the differential in terms of the derivative (admittedly in a rather strange way), which also puts the differential on a solid foundation.

Geometrical interpretation of the differential

Geometrical Interpretation Let $y = f(x)$ be differentiable at $x = x_0$. For each value of the variable dx the differential of y at $x = x_0$ is

$$dy = f'(x_0)dx.$$

Recall that the tangent line to the graph of f at $x = x_0$ has slope equal to $f'(x_0)$. Thus, if we measure a horizontal distance of dx and a vertical distance of $f'(x_0)dx = dy$ from the point of tangency we obtain another point on the tangent line. (See Figure 3.37a.) Thus, the differential dy (at $x = x_0$) is equal to the difference in the y-coordinates of two points on the tangent line corresponding to a horizontal change of dx from the point of tangency at $x = x_0$.

Earlier in this book we worked with the increments Δx and Δy. Recall that Δx can be any number, positive, negative or zero, and that Δy is defined at $x = x_0$ by

$$\Delta y = f(x_0 + \Delta x) - f(x_0).$$

Thus, Δy is the difference in the y-coordinates of two points on the graph of f corresponding to a change of Δx from the point where $x = x_0$.

Observe that dx and Δx play similar roles in the above discussion. Both are completely arbitrary and both are used to measure horizontal distances from

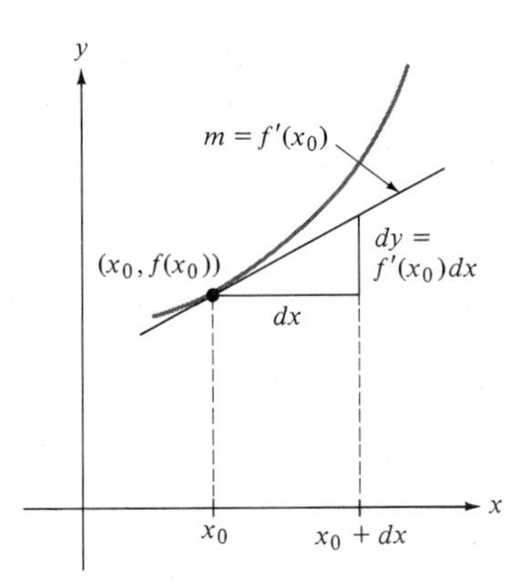

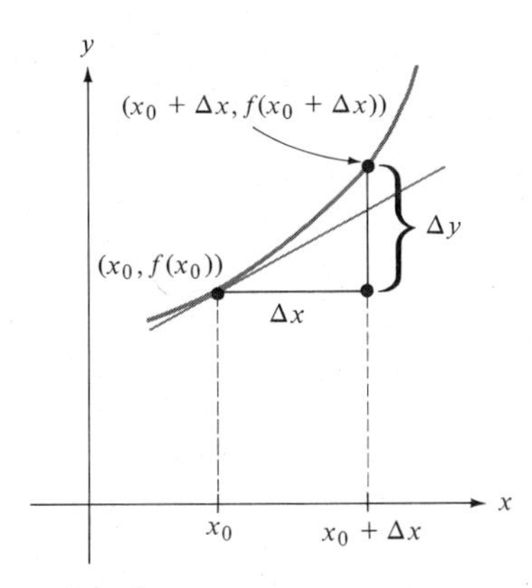

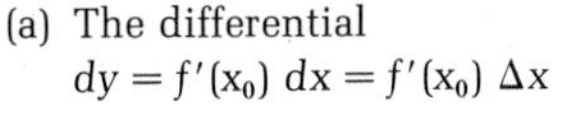

(a) The differential $dy = f'(x_0)\,dx = f'(x_0)\,\Delta x$

(b) The increment $\Delta y = f(x_0 + \Delta x) - f(x_0)$

FIGURE 3.37 When $\Delta x = dx$ is small the increment Δy is approximately equal to the differential dy:

$$\Delta y \approx dy$$
$$f(x_0 + \Delta x) - f(x_0) \approx f'(x_0)\,\Delta x$$
$$f(x_0 + \Delta x) \approx f(x_0) + f'(x_0)\Delta x$$

the point $x = x_0$. For this reason it is customary to choose dx equal to Δx. If this is done, then

$$dy = f'(x_0)dx = f'(x_0) \cdot \Delta x$$

and

$$\Delta y = f(x_0 + \Delta x) - f(x_0) = f(x_0 + dx) - f(x_0).$$

It can be shown that Δy (which measures the difference between the y-coordinates of two points on the curve) and dy (which measures the difference between the y-coordinates of the corresponding two points on the tangent line) are approximately equal when $\Delta x = dx$ is a "small" number. (See Figure 3.37.) Thus

$$\Delta y \approx dy \qquad \text{(if } \Delta x = dx \text{ is small)},$$

so that

Approximation formula

$$f(x_0 + \Delta x) - f(x_0) \approx f'(x_0)dx = f'(x_0)\Delta x \qquad \text{(if } \Delta x = dx \text{ is small)}.$$

We will use this fact in the next section to approximate values of the function f at points where it cannot be calculated conveniently.

Differential Formulas Each of the derivative formulas leads to a corresponding differential formula. As an example we derive the formula for the differential of the product of two differentiable functions of x.

Example 2 Let u and v be differentiable functions of x. Show that

$$d(uv) = u \cdot dv + v \cdot du.$$

Solution

$$\begin{aligned} d(uv) &= \frac{d}{dx}(uv)\,dx \\ &= \left(u \cdot \frac{dv}{dx} + v \cdot \frac{du}{dx}\right) dx \\ &= u \cdot \frac{dv}{dx}\,dx + v \cdot \frac{du}{dx} \cdot dx. \end{aligned}$$

Since the differentials of u and v are

$$du = \frac{du}{dx} \cdot dx \quad \text{and} \quad dv = \frac{dv}{dx} \cdot dx,$$

then

$$\begin{aligned} d(uv) &= u \cdot \frac{dv}{dx} \cdot dx + v \cdot \frac{du}{dx} \cdot dx. \\ &= u \cdot dv + v \cdot du. \end{aligned}$$

The following differential formulas can be established similarly:

Differential formulas

$$\begin{aligned} d(C) &= 0, \\ d(Cu) &= C\,du, \\ d(u^n) &= nu^{n-1}\,du, \\ d(uv) &= u\,dv + v\,du, \\ d\left(\frac{u}{v}\right) &= \frac{v\,du - u\,dv}{v^2} \qquad \text{(provided } v(x) \neq 0). \end{aligned}$$

Differentials and the Chain Rule

The Chain Rule One of the main reasons that we consider differentials is that they greatly simplify the use of the Chain Rule. If we let $y = f(u)$ and $u = g(x)$, where f and g are differentiable functions. Then, considering y to be a function of x,

$$\begin{aligned} dy &= \frac{d}{dx}(f(u))\,dx \\ &= f'(g(x))g'(x)\,dx \\ &= f'(u)\,du. \end{aligned}$$

Thus, $dy = f'(u)\,du$ regardless of whether y is considered to be a function of x or a function of u. Consequently, the use of differentials makes the Chain Rule almost automatic.

Example 3 Let $y = u^{1/2}$, $u = 3x^2 + 1$. Use differentials to calculate dy/dx.

Solution

$$dy = \frac{dy}{du} \cdot du = \frac{1}{2} u^{-1/2}\,du.$$

Since $du = d(3x^2 + 1) = 6x\,dx$, then

$$dy = \frac{1}{2}(3x^2 + 1)^{-1/2}(6x\,dx) = \frac{3x}{\sqrt{3x^2 + 1}}\,dx,$$

and

$$\frac{dy}{dx} = \frac{3x}{\sqrt{3x^2 + 1}}.$$

We can handle derivatives after any number of variable changes by an argument similar to the one in Example 3. When we work with antiderivatives in Chapter 4 we shall find it especially useful to use differentials rather than derivatives. Their use will enable us to work complicated problems involving changes of variables in a completely automatic manner.

EXERCISES 3.10

1. Calculate the differential dy.

- • (a) $y = x^3$
- (b) $y = x^{1/3}$
- (c) $y = 2x^2 + x + 2$
- • (d) $y = (x^3 + 1)^2$
- (e) $y = \dfrac{x}{2x - 1}$
- (f) $y = (x^2 + 1)^2(x^3 + 1)$
- (g) $y = (x + 1)^2\,(x^2 + 1)^3$
- • (h) $y = \dfrac{x^2 + 2x + 1}{x^2 + 1}$
- (i) $y = 2x^3 + 3x^2 - x + 1$
- (j) $y = (x^2 - 1)(x^2 + 1)^{-1}$

2. Evaluate $d(f(x))$ for the given values of x and dx.

- (a) $f(x) = x^3 + 1$, $x = 0$, $dx = \frac{1}{2}$
- • (b) $f(x) = x^2 - 1$, $x = 3$, $dx = 2$
- (c) $f(x) = \dfrac{x}{x + 1}$, $x = 2$, $dx = -1$

• (d) $f(x) = 3x^2 + 5x - 4$, $x = 5$, $dx = \frac{1}{5}$
(e) $f(x) = (x^2 + 1)^3$, $x = 2$, $dx = -\frac{1}{2}$
• (f) $f(x) = (x^3 + 1)(x^2 + 1)^2$, $x = 1$, $dx = \frac{1}{4}$
(g) $f(x) = \dfrac{x}{x - 1}$, $x = 2$, $dx = \frac{1}{5}$
(h) $f(x) = \dfrac{1}{x^2}$, $x = 2$, $dx = \frac{1}{4}$

3. Use differentials to calculate the derivative dy/dx.
   (a) $y = u^2 + 1$, $u = 3x^2 - x + 2$
   • (b) $y = (3u + 1)^2$, $u = \sqrt{x^2 + 1}$
   (c) $y = \dfrac{u + 1}{u^2 + 1}$, $u = x - 1$
   • (d) $y = (u^2 - 1)^2(2u + 1)$, $u = \sqrt{v}$, $v = 3x^2 + 2$
   (e) $y = \dfrac{1}{\sqrt{u^2 + 1}}$, $u = v^2 - 4v + 2$, $v = \sqrt{x}$

4. Let u and v be differentiable functions of x. Show that

$$d\left(\frac{u}{v}\right) = \frac{v\,du - u\,dv}{v^2} \qquad \text{(provided } v \neq 0\text{)}$$

3.11 THE APPROXIMATION OF SMALL CHANGES

We frequently need to find the effect that a small change in one quantity has on a related quantity. As we shall see, differentials can help us in the calculations.

Let $y = f(x)$ be differentiable at $x = x_0$. Let $dx = \Delta x$ be an increment of x, and

$$\Delta y = f(x_0 + \Delta x) - f(x_0)$$

be the corresponding increment of y. Then Δx measures the change in x and Δy measures the corresponding change in y. It can be shown that

Approximation formula

$$\boxed{\Delta y \approx dy = f'(x_0) \cdot \Delta x}$$

if Δx is small. Thus, the differential dy is approximately equal to the change in y that corresponds to a small change in x.

If different variables are used then the approximation formula is modified accordingly.

Example 1 A thin, flat circular metal plate has radius $r = 6$ inches. When heated the radius increases to 6.2 inches. Estimate the change in the area.

Solution The area is

$$A = f(r) = \pi r^2.$$

When r changes by Δr the area changes by ΔA where

$$\Delta A \approx dA = f'(r_0)\Delta r$$
$$\Delta A \approx 2\pi r_0 \Delta r \approx 2\pi \cdot 6 \cdot (0.2) \approx 7.54 \text{ square inches.}$$

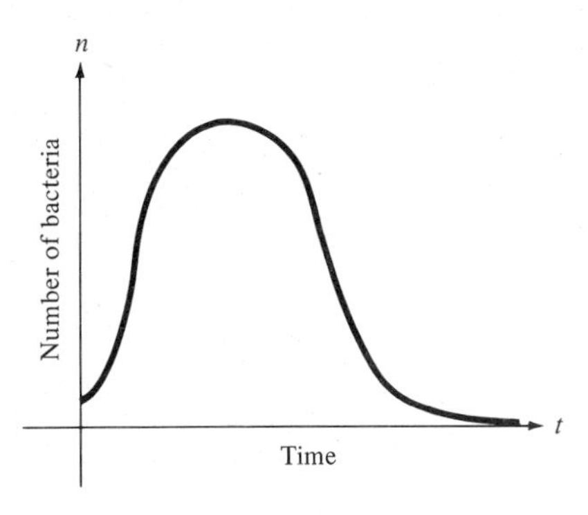

FIGURE 3.38 Population curve for bacteria when infection is treated with antibiotics.

Example 2 If a certain bacterial infection is treated with antibiotics the number of bacteria at time t follows a curve similar to the one in Figure 3.38 which can be approximated by a quartic polynomial over the period of infection.

A human bacterial infection has just been treated with a powerful antibiotic. The number of bacteria at time t $(0 \leq t \leq 200)$ is given by the formula

$$n = f(t) = \frac{t^4}{4} - 70t^3 + \frac{1625}{2}t^2 + 75{,}000t + 112{,}500{,}000,$$

where t is measured in hours.

(a) Use differentials to estimate the net change in the number of bacteria from $t = 10$ to $t = 10\frac{1}{2}$ hours.

(b) Estimate the net change from $t = 50$ to $t = 52$ hours.

Solution

(a) We use the formula

$$\Delta n \approx f'(t_0)\Delta t$$

with $t_0 = 10$ and $\Delta t = \frac{1}{2}$. Then

$$\begin{aligned}\Delta n &\approx (t_0^3 - 210t_0^2 + 1625t_0 + 75{,}000) \cdot \tfrac{1}{2} \\ &\approx (10^3 - 210 \cdot 10^2 + 1625 \cdot 10 + 75{,}000) \cdot \tfrac{1}{2} \approx 35{,}625.\end{aligned}$$

Since the formula is only approximate we would estimate that there will be a net increase of approximately 35,000 bacteria over the half-hour period.

(b) We use the formula

$$\Delta n \approx f'(t_0)\Delta t$$

with $t_0 = 50$ and $\Delta t = 2$. Then

$$\begin{aligned}\Delta n &\approx f'(t_0)\Delta t \approx (t_0^3 - 210t_0^2 + 1625t_0 + 75{,}000) \cdot 2 \\ &\approx (50^3 - 210 \cdot 50^2 + 1625 \cdot 50 + 75{,}000) \cdot 2 \\ &\approx (125{,}000) - 525{,}000 + 81{,}250 + 75{,}000) \cdot 2 \\ &\approx -487{,}500.\end{aligned}$$

There will be a net decrease of approximately one half million bacteria over the two-hour period.

The approximation formula $\Delta y \approx f'(x_0)\Delta x$ can be modified to give the approximate value of $f(x_0 + \Delta x)$. Since

$$\Delta y = f(x_0 + \Delta x) - f(x_0)$$

then

$$f(x_0 + \Delta x) - f(x_0) \approx f'(x_0)\Delta x$$

so that

Alternate form of the approximation formula

$$\underset{\text{new value of } f}{f(x_0 + \Delta x)} \approx \underset{\text{original value of } f}{f(x_0)} + \underset{\text{approximate change in } f}{f'(x_0)\Delta x}$$

Example 3 Let $f(x) = \sqrt{x}$. Use differentials (or increments) to approximate $f(64.04) = \sqrt{64.04}$.

Solution By the approximation formula

$$f(x_0 + \Delta x) \approx f(x_0) + f'(x_0)\Delta x$$

$$\sqrt{x_0 + \Delta x} \approx \sqrt{x_0} + \frac{1}{2\sqrt{x_0}} \cdot \Delta x.$$

We choose $x_0 = 64$ (so that $\sqrt{x_0}$ can be calculated easily) and $\Delta x = dx = 0.04$ (so that Δx is small and $f(x_0 + \Delta x)$ is the number $\sqrt{64.04}$ that we wish to compute). Then

$$\sqrt{64.04} \approx \sqrt{64} + \frac{1}{2\sqrt{64}} \cdot (0.04)$$

$$\sqrt{64.04} \approx 8 + \frac{1}{2 \cdot 8} \cdot (0.04) \approx 8 + \frac{0.04}{16}$$

$$\approx 8.0025.$$

Proportional Change Let $y = f(x)$ be a positive quantity. In most cases it is more important to know the proportional change in y than to know the actual change Δy. The proportional change, which is usually expressed as a percentage, is defined to be

Proportional change

$$\frac{\text{change in } y}{\text{original value of } y} = \frac{\Delta y}{y} \approx \frac{f'(x_0)\Delta x}{f(x_0)}.$$

Example 4 The profit function of the XYZ Block Company is

$$y = P(x) = -\tfrac{1}{3000}x^3 - \tfrac{1}{200}x^2 + 3x - 25,$$

where x is the number of production runs per week and y is measured in hundreds of dollars.

Calculate the proportional change in the net profit when output increases from $x = 40$ to $x = 45$ production runs.

Solution The proportional change is

$$\frac{\Delta y}{y} \approx \frac{P'(x_0)\Delta x}{P(x_0)}$$

where $x_0 = 40$ and $\Delta x = 5$. We calculate

$$P(x_0) = P(40) = -\frac{40^3}{3000} - \frac{40^2}{200} + 3 \cdot 40 - 25$$
$$\approx 65.7 \text{ hundred dollars,}$$
$$P'(x_0) = -\frac{40^2}{1000} - \frac{40}{100} + 3 = 1 \text{ hundred dollars.}$$

Therefore

$$\frac{\Delta y}{y} \approx \frac{P'(x_0)\Delta x}{P(x_0)} \approx \frac{1 \cdot 5}{65.7} \approx 0.076.$$

The company will have an increase in profit of approximately 7.6 percent if it increases production from 40 to 45 runs per week.

EXERCISES 3.11

1. Use the approximation formula

$$f(x_0 + \Delta x) \approx f(x_0) + f'(x_0)\Delta x$$

to approximate the following roots:

• (a) $\sqrt{100.1}$
(b) $\sqrt[3]{27.3}$
• (c) $\sqrt[3]{65}$
(d) $\sqrt{15.6}$

2. Use the approximation formula

$$\Delta y \approx f'(x_0)\Delta x$$

to estimate the change in y corresponding to the change in x.

(a) $y = 3x^2 - 2x + 7$, $x_0 = 2.3$, $\Delta x = .1$
• (b) $y = -5x^3 + 2x^2 - 7x + 1$, $x_0 = 2$, $\Delta x = -.3$
(c) $y = \sqrt{x^2 + 7}$, x changes from 3 to 3.1.

•3. Estimate the proportional change in y for each of the functions in Exercise 2.

4. The cost function for the H-2-OH Company is

$$C(x) = 300x^2 + 7000x + 800$$

dollars where x is the number of production runs per week.

(a) Calculate the approximate change in the cost if production increases from 20 to 25 runs per week.
• (b) Calculate the approximate proportional change in the cost if production increases from 20 to 25 runs per week.

4 THE INTEGRAL

4.1 ANTIDERIVATIVES

Introduction The calculus has two main divisions, one concerned with the *derivative*, the other with the *integral* (*antiderivative*). The first part of this book is devoted to the derivative. We now turn our attention to the integral.

In some cases it is easier to determine the derivative of a function than the function itself. The antiderivative of the derived function can then be used to determine the original function. For example, an automobile can be equipped with a device that makes a continuous recording of the velocity. Since the velocity is the derivative of the distance function then the distance is an antiderivative of the velocity. We should be able to use the information about the velocity to calculate the distance traveled over any part of a trip.

As a second example, consider the problem of determining the demand for a commodity as a function of price. Company analysts can vary the price by small amounts and determine the rate at which the demand changes with respect to price. In an ideal situation this knowledge about the rate of change (derivative of the demand function) can be used to obtain information about the actual demand function (antiderivative of the rate of change).

As we shall see, an antiderivative can be interpreted as an area under a curve. For centuries it was not realized that the determination of area is closely related to the study of tangent lines. Much of the credit for the development of the calculus goes to Newton and Liebniz for discovering just such a relationship.

The antiderivative

Antiderivatives Much of our work in the remainder of this book is concerned with finding *antiderivatives* of given functions. That is, given $f(x)$, we wish to find all functions $g(x)$ such that $g'(x) = f(x)$.

For example, since

$$\frac{d}{dx}(3x^2 - 2x + 5) = 6x - 2,$$

then $3x^2 - 2x + 5$ is an antiderivative of $f(x) = 6x - 2$. Observe that the antiderivative is not unique. We also have

$$3x^2 - 2x \text{ is an antiderivative of } 6x - 2,$$
$$3x^2 - 2x - \sqrt{17} \text{ is an antiderivative of } 6x - 2,$$

and so on. In general

$$3x^2 - 2x + C \text{ is an antiderivative of } 6x - 2$$

for any choice of the constant C.

We can show that all antiderivatives of $6x - 2$ are of this form. Let $g(x)$ be an arbitrary antiderivative of $6x - 2$. Then

$$g'(x) = \frac{d}{dx}(g(x)) = 6x - 2 = \frac{d}{dx}(3x^2 - 2x).$$

It follows from Theorem 3.8 that $g(x)$ and $3x^2 - 2x$ differ by at most a constant. Thus, there is a constant C such that

$$g(x) = 3x^2 - 2x + C.$$

The antidifferential

In actual practice the work involved with finding antiderivatives is simplified if we calculate *antidifferentials* rather than antiderivatives. If we use the differential notation we see that

$$3x^2 - 2x + C \text{ is an antidifferential of } (3x^2 - 2x)dx.$$

It is customary to denote a particular antiderivative of $f(x)$ (that is, a particular antidifferential of $f(x)\,dx$) by the symbol

$$\int f(x)\,dx.$$

Then by Theorem 3.8 every antiderivative of $f(x)$ is of form

$$\int f(x)\,dx + C,$$

where C is a constant. The expression

$$\int f(x)\,dx + C$$

The indefinite integral
Constant of integration

is called the (general) *antiderivative* or the *indefinite integral* of $f(x)$. The constant C *is called the constant of integration*. The process of calculating $\int f(x)dx + C$ is called *integration*.

One of the most useful of the formulas for integration is

The power rule

I1	$\int x^n dx = \dfrac{x^{n+1}}{n+1} + C \qquad (n \neq -1)$

This formula follows from the fact that

$$\frac{d}{dx}\left(\frac{x^{n+1}}{n+1}+C\right)=\frac{n+1}{n+1}x^{n+1-1}+0=x^n.$$

Other variables can be used in I1 rather than x. For example,

$$\int u^2du=\frac{u^{2+1}}{2+1}+C=\frac{u^3}{3}+C,$$

$$\int y^{12}dy=\frac{y^{12+1}}{12+1}+C=\frac{y^{13}}{13}+C,$$

and so on.

Example 1

(a) $$\int x^3dx=\frac{x^4}{4}+C.$$

(b) $$\int u^7du=\frac{u^8}{8}+C.$$

(c) $$\int dv=\int v^0dv=\frac{v^1}{1}+C=v+C.$$

(d) $$\int \sqrt{y}\,dy=\int y^{1/2}dy=\frac{y^{3/2}}{\frac{3}{2}}+C=\frac{2y\sqrt{y}}{3}+C.$$

Two convenient rules for computing indefinite integrals are listed below. These rules are analogous to the corresponding rules for derivatives. As we shall see, every derivative formula leads to a corresponding integral formula.

Theorem 4.1 Let f and g have antiderivatives; let k be a constant. Then

Basic integration formulas

I2 $$\int [f(x)\pm g(x)]\,dx=\int f(x)\,dx\pm\int g(x)\,dx+C,$$

I3 $$\int k\cdot f(x)\,dx=k\int f(x)\,dx+C.$$

PROOF OF **I2:** Let F be an antiderivative of f, G an antiderivative of g. Then

$$F'(x)=f(x),\qquad G'(x)=g(x).$$

Therefore,

$$\frac{d}{dx}(F(x)\pm G(x))=F'(x)\pm G'(x)=f(x)\pm g(x),$$

so that $F(x)\pm G(x)$ is an antiderivative of $f(x)\pm g(x)$. The general antiderivative of $f(x)\pm g(x)$ is obtained by adding a constant to this antiderivative:

$$\int (f(x)\pm g(x))\,dx=F(x)\pm G(x)+C$$

$$=\int f(x)\,dx\pm\int g(x)\,dx+C.$$

The proof of I3 is left for the reader (Exercise 4).

Example 2 Calculate $\int(3x^2 - 2x + 5)dx$.

Solution

$$\int (3x^2 - 2x + 5)dx = \int 3x^2dx - \int 2xdx + \int 5dx \qquad \text{(by I2)}$$

$$= 3\int x^2dx - 2\int xdx + 5\int dx \qquad \text{(by I3)}$$

$$= \frac{3x^3}{3} - \frac{2x^2}{2} + 5x + C \qquad \text{(by I1)}$$

$$= x^3 - x^2 + 5x + C.$$

Change of variable

The Substitution Principle The following examples illustrate how changes of variables can be handled in integration.

Example 3 Calculate $\int(2x - 5)^7 \cdot 2dx$.

Solution We make the substitution $u = 2x - 5$. Then $du = 2\,dx$. The integral becomes

$$\int (2x - 5)^7 2dx = \int u^7du = \frac{u^8}{8} + C \qquad \boxed{\begin{aligned} u &= 2x - 5 \\ du &= 2dx \end{aligned}}$$

$$= \frac{1}{8}(2x - 5)^8 + C.$$

Example 4 Calculate $\int(3x^2 + 8)^{1/2}\,x\,dx$.

Solution It would be natural to make the substitution $u = 3x^2 + 8$, $du = 6x\,dx$. Observe that we have $u^{1/2}$ as part of the function to be integrated, but, unfortunately, we do not have $du = 6x\,dx$. We have $x\,dx$ instead. We can transform the integral into the proper form by multiplying it by $\frac{1}{6} \cdot 6$ and taking the $\frac{1}{6}$ outside the integral sign:

$$\int (3x^2 + 8)^{1/2}x\,dx = \int (3x^2 + 8)^{1/2} \cdot \frac{1}{6} \cdot 6x\,dx$$

$$= \frac{1}{6}\int (3x^2 + 8)^{1/2}6x\,dx \qquad \text{(by I3)}$$

$$= \frac{1}{6}\int u^{1/2}du = \frac{1}{6} \cdot \frac{u^{3/2}}{\frac{3}{2}} + C \qquad \boxed{\begin{aligned} u &= 3x^2 + 8 \\ du &= 6x\,dx \end{aligned}}$$

$$= \frac{1}{6} \cdot \frac{2}{3}u^{3/2} + C = \frac{1}{9}(3x^2 + 8)^{3/2} + C.$$

The Substitution Principle, which we used above, can be stated in the following general form. The proof follows from the Chain Rule.

The Substitution Principle

Let f and h be functions which can be integrated. Let the substitution $u = g(x)$ transform the expression $f(x)\,dx$ into the expression $h(u)\,du$. Then the integrals of these expressions differ by at most a constant:

$$\int f(x)\,dx = \int h(u)\,du + C.$$

In applying the Substitution Principle we must transform the integral so that we have an exact fit with one of our integration formulas. This illustrates one of the major difficulties with integration as compared with differentiation. Rather than having a few general formulas which can be applied in succession we have a large number of very specialized formulas to which the integrals must be matched exactly.

We close with one further example.

Example 5 Calculate $\int (x + 2\sqrt{x})\left(1 + \frac{1}{\sqrt{x}}\right) dx$.

Solution Let $u = x + 2\sqrt{x}$. Then $du = (1 + 1/\sqrt{x})\, dx$. The integral becomes

$$\int (x + 2\sqrt{x})\left(1 + \frac{1}{\sqrt{x}}\right) dx = \int u\,du = \frac{u^2}{2} + C = \frac{(x + 2\sqrt{x})^2}{2} + C.$$

$$\boxed{\begin{aligned} u &= x + 2\sqrt{x} \\ du &= \left(1 + \frac{1}{\sqrt{x}}\right) dx \end{aligned}}$$

EXERCISES 4.1

1. Calculate the integrals:

- • (a) $\int 3x\, dx$
- (b) $\int (x^2 - x)\, dx$
- (c) $\int (5x^7 - 3x^6 + 2)\, dx$
- • (d) $\int 2\sqrt{x}\, dx$
- (e) $\int x^{-2}\, dx$
- • (f) $\int \left(x + \frac{1}{x^2}\right) dx$
- • (g) $\int (7x^{1/2} - 3x^{1/3})\, dx$
- (h) $\int (x + 1)^2\, dx$
- (i) $\int (x^3 + 3x^2 + 3x + 1)\, dx$
- • (j) $\int (15x^4 + 8x^3 - 3x^2 + 2)\, dx$
- • (k) $\int (\sqrt{x} + \sqrt[3]{x})\, dx$
- (l) $\int \left(\frac{3}{x^3} - \frac{2}{x^2} + \frac{1}{2\sqrt{x}}\right) dx$

2. Calculate the integrals. Use the Substitution Principle whenever possible.

- • (a) $\int (x - 1)^3\, dx$
- (b) $\int (x + 3)^4\, dx$
- • (c) $\int (2 - x)^2\, dx$
- (d) $\int (2x - 1)^{-2}\, dx$
- (e) $\int (1 + x)^4\, dx$
- • (f) $\int (\sqrt{x} + 1)\, dx$
- (g) $\int \frac{dx}{\sqrt{x + 1}}$
- (h) $\int \left(\frac{x - 1}{2}\right)^3 dx$
- • (i) $\int (x^2 - x - 1)^{-3}(2x - 1)\, dx$
- (j) $\int (-x^2 + 2x - 1)^2(x - 1)\, dx$
- (k) $\int (2x^2 + 4x + 1)^3(x + 1)\, dx$
- • (l) $\int x(x + 1)^2\, dx$
- (m) $\int (x^3 - 2x + 2)^4(3x^2 - 2)\, dx$
- • (n) $\int (3x^4 - 3x^2 - 6)^{-2}(12x^3 - 6x)\, dx$

3. Explain why $\int (x^3 + 1)^5\, dx \neq \frac{1}{3x^2} \int (x^3 + 1)^5 (3x^2)\, dx$.

4. Prove I3.

4.2 APPLICATIONS

In some cases it may be easier to determine the derivative of a function than the actual function itself. For example, it may be easier to determine the marginal profit function than the profit function. The unknown function can then be found as one of the antiderivatives of the known function.

Example 1 Statisticians of the La Phuma Cosmetics Company have determined from market surveys that the marginal profit on the perfume *La Phuma Number 8* is given by

$$MP(x) = -\frac{3}{1000}x^2 + \frac{x}{50} + 8,$$

where x is the number of production runs per year and $MP(x)$ is measured in thousands of dollars. Furthermore, they know that when $x = 10$ the profit on the perfume is \$50,000. Determine the profit function.

Solution Let $P(x)$ be the profit (also measured in thousands of dollars) for x production runs per year. Then the marginal profit is the derivative of the profit function so that

$$P'(x) = -\frac{3}{1000}x^2 + \frac{x}{50} + 8.$$

Then $P(x)$ is an antiderivative of the marginal profit so

$$\begin{aligned} P(x) &= \int \left(-\frac{3}{1000}x^2 + \frac{x}{50} + 8\right) dx \\ &= -\frac{3}{1000} \cdot \frac{x^3}{3} + \frac{1}{50} \cdot \frac{x^2}{2} + 8x + C \\ &= -\frac{x^3}{1000} + \frac{x^2}{100} + 8x + C, \end{aligned}$$

where C is a constant.

The above calculations give us all antiderivatives of $P'(x)$, only one of which is the actual profit function. To find it we must find the particular value of C that makes $P(x)$ equal 50 when $x = 10$. We find this value of C by substituting $x = 10$, $P(x) = 50$ in the above equation:

$$\begin{aligned} P(x) &= -\frac{x^3}{1000} + \frac{x^2}{100} + 8x + C \\ P(10) &= -\frac{1000}{1000} + \frac{100}{100} + 80 + C \\ 50 &= -1 + 1 + 80 + C \\ C &= -30. \end{aligned}$$

The profit function is

$$P(x) = -\frac{x^3}{1000} + \frac{x^2}{100} + 8x - 30.$$

An equation, such as

$$P'(x) = -\frac{3}{1000}x^2 + \frac{x}{50} + 8,$$

or

$$y' + 2xy = 3x^2,$$

that relates a function and its derivative is called a *differential equation*.

If a differential equation has the special form

$$y' = f(x)$$

where f is a function of x, then the solution can be obtained by integration:

$$y = \int f(x)\,dx + C.$$

In many cases we also are given an *initial condition*, the value of the unknown function at some particular value of x. (In Example 1 the initial condition is $P = 50$ when $x = 10$.) If we substitute the values given by the initial condition into the integrated function we can find the special value of C that solves the problem. The antiderivative that has this value of C is the particular one we need.

Distance, velocity, and acceleration

When an object moves along a straight path its velocity (v) is the derivative of the distance (s) and its acceleration (a) is the derivative of the velocity:

$$v(t) = s'(t) = \frac{ds}{dt}, \qquad a(t) = v'(t) = \frac{dv}{dt}.$$

It follows that the distance is an antiderivative of the velocity, which, in turn, is an antiderivative of the acceleration:

$$s(t) = \int v(t)dt + C_1 \quad \text{and} \quad v(t) = \int a(t)dt + C_2,$$

where the constants C_1 and C_2 can be determined from initial conditions.

Example 2 If an object is thrown straight up into the air the force of gravity subjects it to a constant acceleration[1] downward of 32 ft/sec². If the air resistance is negligible, then this force is the only one affecting the object.

A ball is thrown straight up from the ground with an initial velocity of 64 ft/sec. Derive the formula for the distance above the earth at time t. For what time range does the formula hold?

Solution We measure the distance above the ground as positive. Since the acceleration due to gravity is in the opposite direction, then it is a negative force:

$$a(t) = -32 \text{ ft/sec}^2.$$

[1] Actually, the acceleration is not quite constant. It depends on the distance from the object to the center of the earth. If the object is not thrown too high, the acceleration varies very little. Thus, the formula derived for the velocity is fairly accurate for a baseball thrown upward, but is very inaccurate for a rocket fired upward.

The velocity $v(t)$ is an antiderivative of the acceleration:

$$v(t) = \int a(t)dt = \int (-32)dt = -32t + C_1.$$

At time $t = 0$ the velocity is 64 ft/sec. Thus,

$$v(0) = -32 \cdot 0 + C_1 = 64$$

so that

$$C_1 = 64.$$

The velocity function is

$$v(t) = -32t + 64.$$

The distance $s(t)$ is an antiderivative of the velocity:

$$s(t) = \int v(t)dt = \int (-32t + 64)dt = -16t^2 + 64t + C_2.$$

At time $t = 0$ the distance is zero. Thus,

$$s(0) = -16 \cdot 0^2 + 64 \cdot 0 + C_2 = 0$$

so that

$$C_2 = 0.$$

The distance function is

$$s(t) = -16t^2 + 64t.$$

The formulas for distance and velocity hold from time $t = 0$ until the ball strikes the ground. To find the time of impact we set $s(t)$ equal to zero and solve for t by the quadratic formula

$$-16t^2 + 64t = 0$$
$$t = 0 \quad \text{or} \quad t = 4.$$

The formulas hold for $0 \leqslant t \leqslant 4$.

The formulas in Example 2 can be generalized for an object thrown straight up from an initial height of s_0 feet with an initial velocity of v_0 ft/sec. As in the example we have $a(t) = -32$ ft/sec^2. If we integrate this expression we obtain

$$v(t) = \int a(t)dt = -32t + C_1$$

where C_1 can be found from the initial condition $v(0) = v_0$ to be $C_1 = v_0$. Thus,

$$v(t) = -32t + v_0.$$

If we integrate this expression and use the initial condition $s(0) = s_0$ we obtain

$$s(t) = -16t^2 + v_0t + s_0.$$

Example 3 A man is driving along a straight road at 30 mph when the motor in his car stops running, causing a constant deceleration. At the end of 10 seconds he is moving at 15 mph. How far does the car move before it stops?

Solution Because of the small distances involved we convert to feet per second. The initial velocity is 44 ft/sec. Let a be the acceleration of the car after the motor stops. (Since the car is decelerating, a is negative.) If we measure time and distance from the point at which the motor stopped, we obtain the following formulas:

$$v(t) = at + v_0 = at + 44,$$
$$s(t) = \tfrac{1}{2}at^2 + 44t + s_0 = \tfrac{1}{2}at^2 + 44t.$$

When $t = 10$,

$$v(10) = 10a + 44 = 22 \qquad (15 \text{ mph} = 22 \text{ ft/sec})$$

and so

$$a = -2.2 = -\tfrac{11}{5} \text{ ft/sec}^2.$$

When we substitute this value of a, the formulas become

$$v(t) = -\tfrac{11}{5}\,t + 44,$$
$$s(t) = -\tfrac{11}{10}t^2 + 44t.$$

The car stops moving when the velocity becomes zero—that is, when

$$-\tfrac{11}{5}\,t + 44 = 0, \qquad t = 20 \text{ seconds}.$$

At the end of 20 seconds, when the car comes to a complete stop, it has traveled

$$s(20) = -\tfrac{11}{10}(20^2) + 44(20) = 440 \text{ feet}.$$

EXERCISES 4.2

1. Find the function $y = f(x)$ given the derivative y' and the initial condition.

   • (a) $y' = 3x^2 - 2x + 7$, $y = 3$ when $x = 0$
   (b) $y' = -2x^2 + x + 4$, $y = 8$ when $x = 1$
   • (c) $y' = \sqrt{x}$, $y = -3$ when $x = 4$
   (d) $y' = 2/\sqrt{x} - 3x + 2$, $y = 0$ when $x = 9$

2. The marginal cost for the production of x units is given below. Find the cost function $C(x)$ if $C(0) = 40$.

   • (a) $C'(x) = x/50 + 2$
   (b) $C'(x) = x^2/300 - x/40 + 7$
   • (c) $C'(x) = .001x^2 + .030x + 9$
   (d) $C'(x) = .0004x^3 + .0003x^2 + .0002x + 5$

3. If $a(t)$ is the acceleration function, v_0 and s_0 the velocity and distance at time $t = 0$, calculate $v(t)$ and $s(t)$.

   (a) $a(t) = t + 1$, $v_0 = 1$, $s_0 = 7$
   • (b) $a(t) = t^2 + 8$, $v_0 = -3$, $s_0 = 10$
   (c) $a(t) = \dfrac{1}{(t+1)^3} - t$, $v_0 = 2$, $s_0 = -4$
   • (d) $a(t) = (t + 2)^{-3}$, $v_0 = 1$, $s_0 = 2$

4. A particle is thrown upward from the earth with a velocity of 96 ft/sec. How high above the earth will the particle be when its velocity is zero?

•5. A ball is thrown upward from the ground and rises 64 feet before starting to fall. Find the total time until the ball returns to earth. At what times will it be 48 feet above the earth?

6. A vehicle operating at 20 mph can be stopped in 20 feet with a constant deceleration. What distance will be required to stop the vehicle at 40 mph with the same deceleration?

4.3 THE AREA PROBLEM

In the next few sections we take a temporary break from antiderivatives and consider problems related to the area under a curve. As we shall see later, the area problem is closely related to the calculation of antiderivatives.

How can we measure the area of a region in the xy-plane—for example, the area of the region pictured in Figure 4.1? Actually, two problems are involved: (1) What do we *mean* by the "area" of such a region? (2) How do we *calculate* its numerical value? Although most people have a good intuitive grasp of the concept of area, few are able to define it adequately.

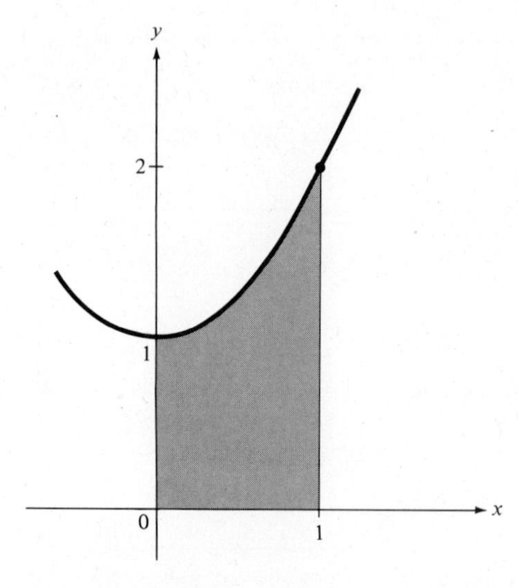

FIGURE 4.1 The area under the curve $y = x^2 + 1$, $0 \leq x \leq 1$.

The problem would be much easier to solve if the region were rectangular. From antiquity the area of a rectangle with base *b* and height *h* has been defined to be *bh*. Unfortunately, it is not easy to extend this definition to include a region with a curved boundary. The standard approach is to define the concept of "area" by using a limiting process involving rectangles.[2] That is, we try to fill up the region with rectangles and measure their areas. Since we cannot completely fill it up, no matter how many rectangles we use, we only obtain an approximation to the "area" of the region. We obtain better approximations by repeating the process and using more rectangles, each with smaller base than those used previously. If s_n is the sum of the areas of the rectangles when *n* rectangles are

---

[2] Except when discussing areas of rectangles, triangles, and circles, we shall use the word *area* in quotation marks until it is properly defined in Section 4.8.

used and if $\lim_{n\to\infty} s_n$ exists, it seems natural to *define* the "area" of the region to be the number obtained by this limiting process.

Figure 4.2 seems to verify these remarks. Obviously, s_8 is a better approximation to the "area" of the region than s_4. (Here we use our intuitive concept of "area.") s_{24} is a better approximation that s_8. It would appear that $\lim_{n\to\infty} s_n$ should exist and that this limit should be equal to the "area" of the region. It is also obvious that we must require all bases of the rectangles used in computing s_n to approach zero as n approaches infinity. If, for example, one of the rectangles is held fixed, the area of the rectangles will never approach the "area" of the region, no matter how "thin" the other rectangles are. (It is not necessary to require that all of the rectangles used in computing s_n have bases of the same length, although this is commonly done in order to simplify the computation.)

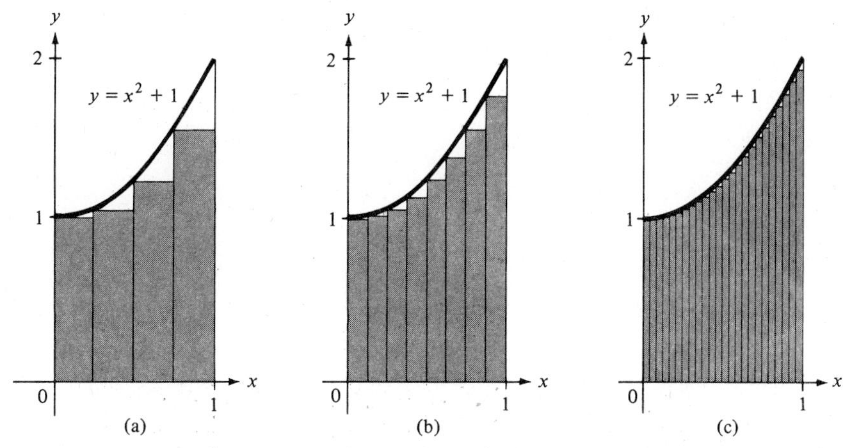

FIGURE 4.2 Approximations to the area under the curve $y = x^2 + 1$, $0 \leqslant x \leqslant 1$. (a) Four rectangles (area of rectangles $= s_4$); (b) eight rectangles (area of rectangles $= s_8$); (c) twenty-four rectangles (area of rectangles $= s_{24}$).

Example We now use the method outlined above to calculate the "area" of the region pictured in Figure 4.1—the "area" under the graph of $f(x) = x^2 + 1$, $0 \leqslant x \leqslant 1$.

The first step is to obtain a numerical value for s_n, the sum of the areas of the rectangles when n rectangles are used. In order to simplify the computation we partition the interval $[0, 1]$ into n equal subintervals and use these as the bases of the rectangle. Then (see Figure 4.3)

$x_0 = 0/n$ is the left-hand endpoint of the first subinterval,
$x_1 = 1/n$ is the left-hand endpoint of the second subinterval,
$x_2 = 2/n$ is the left-hand endpoint of the third subinterval,

$\vdots$

$x_{k-1} = (k-1)/n$ is the left-hand endpoint of the kth subinterval,

$\vdots$

$x_{n-1} = (n-1)/n$ is the left-hand endpoint on the nth subinterval.

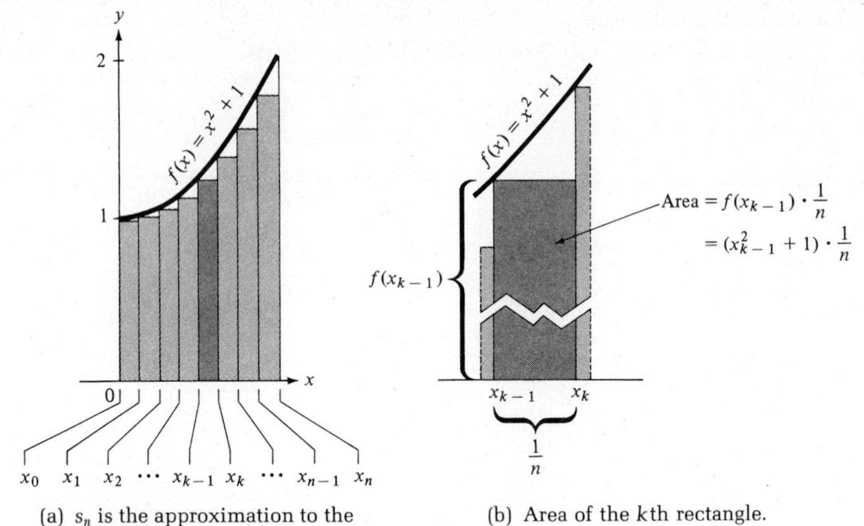

(a) s_n is the approximation to the area obtained by using n rectangles.

(b) Area of the kth rectangle.

FIGURE 4.3

We shall use the left-hand endpoint to determine the height of each rectangle. The left-hand endpoint of the kth subinterval is $x_{k-1} = (k-1)/n$. The height of the kth rectangle is

$$f(x_{k-1}) = \left(\frac{k-1}{n}\right)^2 + 1 = \frac{(k-1)^2}{n^2} + 1.$$

Since the length of the base is $1/n$, the area of the kth rectangle is

$$\left(\frac{(k-1)^2}{n^2} + 1\right)\frac{1}{n} = \frac{(k-1)^2}{n^3} + \frac{1}{n}.$$

Therefore s_n, the sum of the areas of the n rectangles; is equal to

$$\underbrace{\left[\frac{(1-1)^2}{n^3} + \frac{1}{n}\right]}_{\text{area of first rectangle}} + \underbrace{\left[\frac{(2-1)^2}{n^3} + \frac{1}{n}\right]}_{\text{area of second rectangle}} + \underbrace{\left[\frac{(3-1)^2}{n^3} + \frac{1}{n}\right]}_{\text{area of third rectangle}} + \cdots + \underbrace{\left[\frac{(n-1)^2}{n^3} + \frac{1}{n}\right]}_{\text{area of } n\text{th rectangle}}$$

$$s_n = \left[\frac{0^2}{n^3} + \frac{1}{n}\right] + \left[\frac{1^2}{n^3} + \frac{1}{n}\right] + \left[\frac{2^2}{n^3} + \frac{1}{n}\right] + \cdots + \left[\frac{(n-1)^2}{n^3} + \frac{1}{n}\right]$$

$$= \frac{1}{n^3}\,[0^2 + 1^2 + 2^2 + \cdots + (n-1)^2] + \underbrace{\left[\frac{1}{n} + \frac{1}{n} + \cdots + \frac{1}{n}\right]}_{n \text{ terms}}$$

$$= \frac{1}{n^3}\,[1^2 + 2^2 + \cdots + (n-1)^2] + 1.$$

This last sum is one of several that are considered in elementary algebra. There it is shown that if m is a positive integer, then

Formulas for sums

$$(1)\quad 1+2+3+\cdots+m=\frac{m(m+1)}{2}$$

$$(2)\quad 1^2+2^2+3^2+\cdots+m^2=\frac{m(m+1)(2m+1)}{6}$$

$$(3)\quad 1^3+2^3+3^3+\cdots+m^3=\frac{m^2(m+1)^2}{4}$$

We shall use these formulas without proof.

If we choose $m=n-1$ in the second formula, we obtain

$$1^2+2^2+\cdots+(n-1)^2=\frac{(n-1)n(2n-1)}{6}=\frac{2n^3-3n^2+n}{6}.$$

Therefore,

$$\begin{aligned} s_n &= \frac{1}{n^3}\left[\frac{2n^3-3n^2+n}{6}\right]+1 \\ &= \frac{2n^3}{6n^3}-\frac{3n^2}{6n^3}+\frac{n}{6n^3}+1 \\ &= \frac{1}{3}-\frac{1}{2n}+\frac{1}{6n^2}+1. \end{aligned}$$

We now let $n\to\infty$, obtaining

$$\begin{aligned} \text{area of region} = \lim_{n\to\infty} s_n &= \lim_{n\to\infty}\left[\frac{1}{3}-\frac{1}{2n}+\frac{1}{6n^2}+1\right] \\ &= \frac{1}{3}-0+0+1=\frac{4}{3}. \end{aligned}$$

The above example has been analyzed in considerable detail in order to acquaint the reader with certain basic concepts that will be developed in Section 4.5. The thoughtful reader will notice, however, that this method raises a new problem: *Suppose we construct the "approximating" rectangles in several different ways. What assurance do we have that we would obtain the same value for the "area" with each of these different constructions?*

For example, we could construct the rectangles so that they cover the region, rather than being contained within it, as in Figure 4.4a. We could use "horizontal" rectangles rather than vertical ones (that is, we could partition the y-axis rather than the x-axis), as in Figure 4.4b. We could use squares or rectangles "stacked" in some fashion, as in Figure 4.4c. We could even use geometrical figures different from rectangles—for example, circles, as in Figure 4.4d. It is not obvious that these approaches would lead to the same value for the "area" as that obtained above.

It is possible to reconcile these approaches and show that we would obtain the same "area" in each case. The proof, however, is very difficult and is far beyond the scope of this book.

As we will see, the limiting process used for the area problem can be used for many other problems as well. We will develop the basic theory for the more general concept, that of the *definite integral*. It can then be interpreted in terms of *area, volume, profit, work,* or some other quantity, according to the problem at hand.

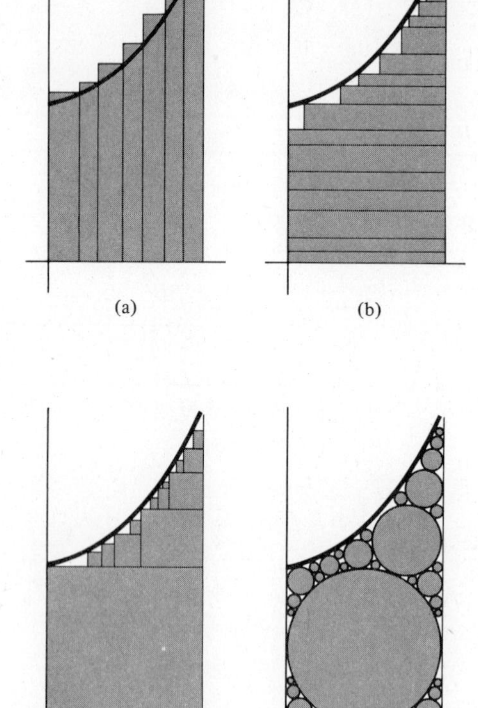

FIGURE 4.4 Alternative approaches to the area problem.

EXERCISES 4.3

1. Modify the example in the text to calculate the "areas" of the regions bounded by the following lines and curves. Sketch each region showing the rectangles for the approximating sum S_4.

   • (a) $y = x^2 + 1$, x-axis, lines $x = 1$, $x = 2$.
   (b) $y = 3x - x^2$, x-axis, lines $x = 0$, $x = 1$.
   • (c) $y = x^2 - 4$, x-axis, lines $x = 0$, $x = 2$.

2. Calculate the "area" of the region in Figure 4.1 by using rectangles that cover the region. (Use the same partition into n equal parts as in the example. Use the right-hand endpoint of each subinterval to determine the height of the rectangle. If S_n is the associated approximation to the area, show that $S_n = \frac{4}{3} + 1/2n + 1/6n^2$.)

4.4 SUMMATION NOTATION

In the preceding section we had to consider the sum $1^2 + 2^2 + \cdots + (n-1)^2$. Since we will encounter many such sums, we now adopt a more compact notation.

Σ notation

It is customary for mathematicians to use the Greek letter Σ (*sigma*) to indicate a sum. We write

$$\sum_{k=1}^{m} k^2 \qquad \text{(read "summation of } k^2 \text{ from 1 to } m\text{")}$$

as a shorthand notation for the sum $1^2 + 2^2 + \cdots + m^2$. Similarly,

$$\sum_{k=1}^{m} k^3 = 1^3 + 2^3 + \cdots + m^3.$$

In general, if f is a function defined on the integers $a, a+1, a+2, \ldots, b-1, b$, then

$$\sum_{k=a}^{b} f(k) = f(a) + f(a+1) + \cdots + f(b-1) + f(b).$$

The letter k, called a *dummy index*, does not actually occur in the expanded form of the sum. Hence, a different dummy index could be employed, say i, j, n, or x. Thus

$$\sum_{x=1}^{m} f(x) = \sum_{j=1}^{m} f(j) = \sum_{i=2}^{m+1} f(i-1) = \sum_{n=16}^{m+15} f(n-15).$$

Example 1 The following sums can be evaluated by use of the formulas for sums stated in Section 4.3.

$$\sum_{k=1}^{5} k^2 = 1^2 + 2^2 + 3^2 + 4^2 + 5^2 = \frac{5(5+1)(2\cdot 5+1)}{6} = \frac{5\cdot 6\cdot 11}{6} = 55,$$

$$\sum_{x=1}^{7} x = 1+2+3+4+5+6+7 = \frac{7(7+1)}{2} = 28,$$

$$\begin{aligned}\sum_{m=4}^{10} m &= 4+5+6+7+8+9+10 \\ &= (1+2+3)+(4+5+6+7+8+9+10)-(1+2+3) \\ &= \frac{10(10+1)}{2} - \frac{3(4)}{2} = 55 - 6 = 49.\end{aligned}$$

Using summation notation, the formulas in Section 4.3 can be expressed in the following compact form:

Formulas for sums

$$(1) \qquad \sum_{k=1}^{m} k = \frac{m(m+1)}{2}$$

$$(2) \qquad \sum_{k=1}^{m} k^2 = \frac{m(m+1)(2m+1)}{6}$$

$$(3) \qquad \sum_{k=1}^{m} k^3 = \frac{m^2(m+1)^2}{4}$$

We also need the following two special properties of sums:

$$(4) \qquad \sum_{k=1}^{m} cf(k) = c \sum_{k=1}^{m} f(k).$$

$$(5) \qquad \sum_{k=1}^{m} [f(k) \pm g(k)] = \left[\sum_{k=1}^{m} f(k)\right] \pm \left[\sum_{k=1}^{m} g(k)\right].$$

Property (4) is a generalization of the Distributive Law encountered by the reader in elementary algebra. This becomes obvious if we expand the sums involved:

$$\sum_{k=1}^{m} cf(k) = cf(1) + cf(2) + \cdot \cdot \cdot + cf(m)$$
$$= c[f(1) + f(2) + \cdot \cdot \cdot + f(m)] = c \sum_{k=1}^{m} f(k).$$

We can obtain property (5) by regrouping the terms and writing out new sums:

$$\sum_{k=1}^{m} [f(k) \pm g(k)]$$
$$= [f(1) \pm g(1)] + [f(2) \pm g(2)] + \cdot \cdot \cdot + [f(m) \pm g(m)]$$
$$= [f(1) + f(2) + \cdot \cdot \cdot + f(m)] \pm [g(1) + g(2) + \cdot \cdot \cdot + g(m)]$$
$$= \left[\sum_{k=1}^{m} f(k)\right] \pm \left[\sum_{k=1}^{m} g(k)\right].$$

Although property (5) is stated for two functions, f and g, the same result holds for three, four, or more functions.

On occasion we may encounter a sum involving a constant function. Observe that

$$(6) \qquad \sum_{k=1}^{m} c = \underbrace{c + c + \cdot \cdot \cdot + c}_{m \text{ terms}} = cm$$

Example 2 Simplify $\sum_{k=1}^{m} (6k^2 + 2k + 5)$.

Solution

$$\sum_{k=1}^{m} (6k^2 + 2k + 5)$$
$$= \sum_{k=1}^{m} 6k^2 + \sum_{k=1}^{m} 2k + \sum_{k=1}^{m} 5 \qquad \text{[by (5)]}$$
$$= 6 \sum_{k=1}^{m} k^2 + 2 \sum_{k=1}^{m} k + 5m \qquad \text{[by (4) and (6)]}$$
$$= 6 \cdot \frac{m(m+1)(2m+1)}{6} + 2 \cdot \frac{m(m+1)}{2} + 5m \qquad \text{[by (2) and (1)]}$$
$$= (2m^3 + 3m^2 + m) + (m^2 + m) + 5m$$
$$= 2m^3 + 4m^2 + 7m.$$

EXERCISES 4.4

1. Write out the following sums:

• (a) $\sum_{i=1}^{6} \frac{i}{3}$

• (b) $\sum_{j=0}^{4} (j+1)$

(c) $\sum_{k=2}^{5} 2k+1$

(d) $\sum_{j=1}^{6} (j+1) - \sum_{k=2}^{7} (3k-2)$

(e) $2\sum_{k=0}^{3} (2k^2+k-1)$

• (f) $\sum_{i=4}^{7} [i\, f(i)]^i$

(g) $\sum_{k=1}^{4} \frac{(k^2+1)(k+1)}{(2k-1)}$

2. Show that

$$\sum_{k=-n}^{n} k^2 = 2\sum_{m=0}^{n} m^2 = 2\sum_{j=1}^{n} j^2.$$

3. Let $f(x)$ be defined for the integers $x = 4$, 5, and 6. Write out the sums:

$$\sum_{j=4}^{4} f(j),\ \sum_{j=4}^{5} f(j),\ \sum_{j=4}^{6} f(j).$$

4. Verify formulas (1) to (5) for the special case $m = 5$.

5. Discuss the following "alternate definition" of the symbol $\Sigma_{k=1}^{n} f(k)$:

$$\begin{cases} (1) & \sum_{k=1}^{n} f(k) = f(1), \quad \text{if } n = 1. \\ (2) & \sum_{k=1}^{n} f(k) = f(n) + \sum_{k=1}^{n-1} f(k), \quad \text{if } n > 1. \end{cases}$$

(a) Use this definition to calculate $\Sigma_{k=1}^{n} f(k)$ for $n = 1$, 2, 3, 4, and 5.
(b) Is this definition equivalent to the one in the text?

6. Let a and b be two integers with $a < b$. If $f(x) = c$ for all x, show that

$$\sum_{x=a}^{b} f(x) = \sum_{x=a}^{b} c = (b-a+1)c.$$

7. Using the fact that $k^3 - (k-1)^3 = 3k^2 - 3k + 1$, establish that

$$\sum_{k=1}^{m} k^2 = \frac{m(m+1)(2m+1)}{6}.$$

(*Hint:* Sum both sides from $k = 1$ to $k = m$ and show that the left side is m^3. Use the result in the text for $\Sigma_{k=1}^{m} k$.)

4.5 THE DEFINITE INTEGRAL

In Section 4.3 we considered the problem of calculating the "area" of the region bounded by the graph of $f(x) = x^2 + 1$, the coordinate axes, and the vertical line $x = 1$. In this section we generalize the process used there.

Let f be continuous on the closed interval $[a, b]$. We partition the interval $[a, b]$ into n subintervals. To do this we choose numbers $x_0, x_1, x_2, \ldots, x_{n-1}, x_n$ such that

Partition of the interval $[a, b]$

$$a = x_0 < x_1 < x_2 < \cdots < x_{n-1} < x_n = b.$$

These numbers partition $[a, b]$ into the n subintervals $[x_0, x_1], [x_1, x_2], \ldots, [x_{n-1}, x_n]$. We next choose numbers $\xi_1, \xi_2, \ldots, \xi_n$ on these subintervals. That is,

$$\begin{aligned} x_0 &\leqslant \xi_1 \leqslant x_1, \\ x_1 &\leqslant \xi_2 \leqslant x_2, \\ &\vdots \\ x_{n-1} &\leqslant \xi_n \leqslant x_n. \end{aligned}$$

We now form the sum

The approximating sum S_n

$$S_n = \sum_{k=1}^{n} f(\xi_k)(x_k - x_{k-1}).$$

This sum is represented geometrically by the sum of the areas of the rectangles in Figure 4.5.

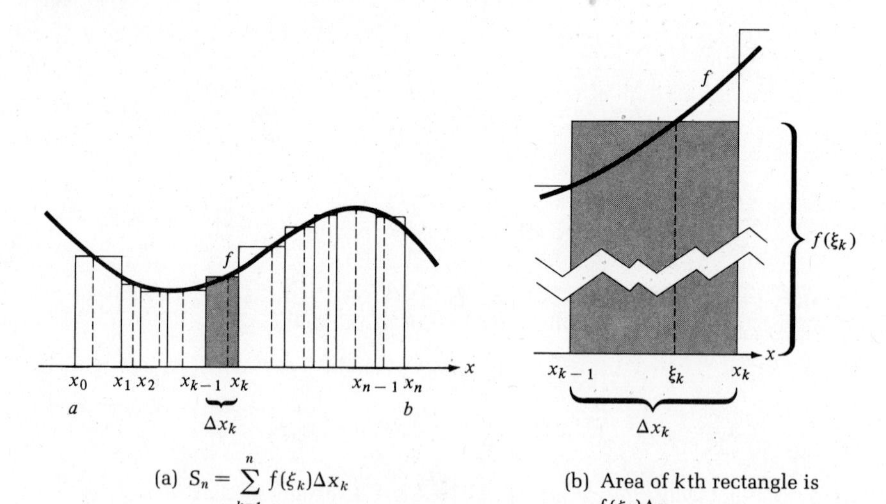

(a) $S_n = \sum_{k=1}^{n} f(\xi_k)\Delta x_k$

(b) Area of kth rectangle is $f(\xi_k)\Delta x_k$.

FIGURE 4.5

It is customary to use the following notation. Let

$$\begin{aligned} \Delta x_1 &= x_1 - x_0 && \text{(length of the first subinterval)} \\ \Delta x_2 &= x_2 - x_1 && \text{(length of the second subinterval),} \\ &\vdots \\ \Delta x_k &= x_k - x_{k-1} && \text{(length of the } k\text{th subinterval),} \\ &\vdots \\ \Delta x_n &= x_n - x_{n-1} && \text{(length of the } n\text{th subinterval).} \end{aligned}$$

Then

$$S_n = \sum_{k=1}^{n} f(\xi_k)\,\Delta x_k.$$

The number $\sum_{k=1}^{n} f(\xi_k)\,\Delta x_k$ depends on the partition $x_0, x_1, x_2, \ldots, x_n$ and on the numbers $\xi_1, \xi_2, \ldots, \xi_n$ as well as on the function f. In most cases, if any of these numbers is changed, the value of S_n also changes.

What happens to $\sum_{k=1}^{n} f(\xi_k)\,\Delta x_k$ as n gets large without bound and the length of each subinterval approaches zero? Several values are pictured in Figure 4.6. It can be shown that the sum tends to a limit as n gets large and each $\Delta x_k \to 0$. Furthermore, *this limit is independent of the partitions and the numbers* $\xi_1, \xi_2, \ldots, \xi_n$. This limit is called the *definite integral* of f over $[a, b]$. It is denoted by the symbol

The definite integral

$$\int_a^b f(x)\,dx.$$

(Read "integral from a to b of $f(x)\,dx$.") The formal definition follows.

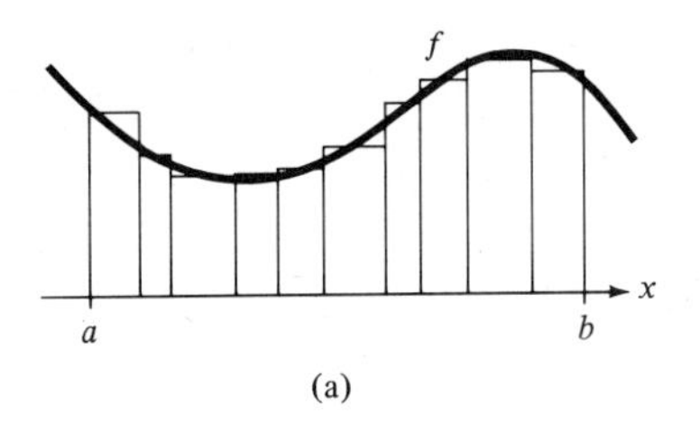

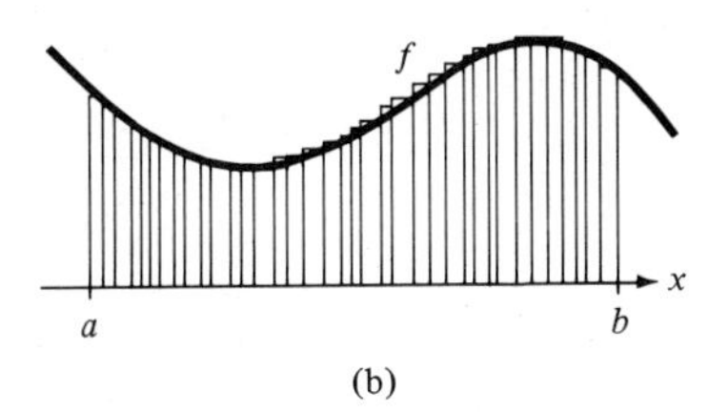

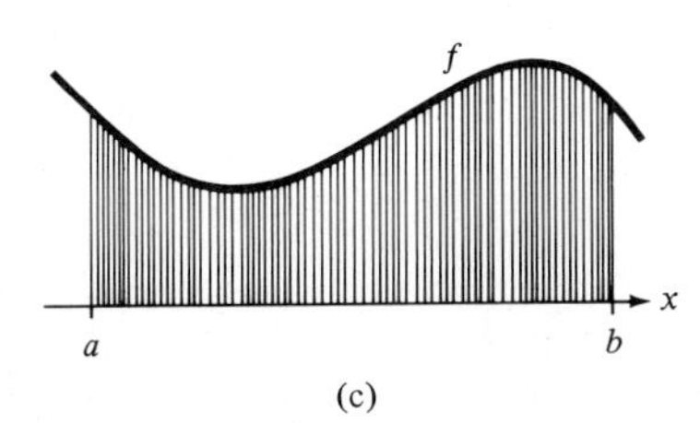

FIGURE 4.6 $S_n = \sum_{k=1}^{n} f(\xi_k)\Delta x_k$ tends to a limit as n gets large without bound.

Definition Let f be continuous on $[a, b]$. For each positive integer n let $a = x_0 < x_1 < \cdots < x_n = b$ be a partition of $[a, b]$ into n parts. For each k let $\Delta x_k = x_k - x_{k=1}$ and let ξ_k be a point on the kth subinterval. Then

$$\int_a^b f(x)\,dx = \lim_{\substack{n\to\infty \\ \Delta x_k \to 0}} \sum_{k=1}^{n} f(\xi_k)\,\Delta x_k.$$

The notation $\lim_{\substack{n\to\infty \\ \Delta x_k \to 0}}$ indicates that n gets large without bound and each $\Delta x_k \to 0$. The numbers a and b are called the *limits of integration*.

Remark The entire symbol $\int_a^b f(x)\,dx$ is used for a definite integral. The dx is a vestige of the Δx_k in the approximating sum and should not be confused with the differential. The reader will, of course, have noticed the similarity of the symbols for the antiderivative (indefinite integral) and the definite integral. In Section 4.7

we will show that this similarity is justified—that definite integrals can be calculated by the use of antiderivatives.

The proof that $\lim_{\substack{n\to\infty \\ \Delta x_k \to 0}} \sum_{k=1}^{n} f(\xi_k)\,\Delta x_k$ exists and is independent of the partitions used is comparatively difficult and can be found in more advanced texts.

Existence of the definite integral

As we mentioned above, $\int_a^b f(x)\,dx$ always exists if f is continuous on $[a, b]$. In certain cases the integral will exist even when f is not continuous on $[a, b]$. This is discussed in more detail in Section 4.10. The function f is said to be *integrable* on $[a, b]$ if $\int_a^b f(x)\,dx$ exists.

If we know that $\int_a^b f(x)\,dx$ exists, we have great freedom in choosing the partitions of $[a, b]$ that will be used to compute the integral. In most cases we choose all of the subintervals of equal length and let ξ_k be one of the endpoints of the kth subinterval. This can be done because any choice of subintervals and of the ξ_k leads to the same limit as any other choice, provided the length of each subinterval goes to zero as n gets large without bound.

$\int_a^b x\,dx = \frac{b^2 - a^2}{2}$

Example 1 Let $f(x) = x$.

(a) Partition $[a, b]$ into n equal subintervals.

(b) Calculate $\sum_{k=1}^{n} f(\xi_k)\,\Delta x_k$, using the right-hand endpoint of the kth subinterval for ξ_k.

(c) Show that $\int_a^b x\,dx = (b^2 - a^2)/2$.

Solution

(a) If $[a, b]$ is to be partitioned into n equal subintervals, then each must have length

$$\Delta x = \Delta x_k = \frac{b-a}{n}.$$

Thus,

$$\begin{aligned}
x_0 &= a \\
x_1 &= a + \Delta x \\
x_2 &= a + 2\,\Delta x \\
x_3 &= a + 3\,\Delta x \\
&\vdots \\
x_k &= a + k\,\Delta x \\
&\vdots \\
x_n &= a + n\,\Delta x = a + n \cdot \frac{b-a}{n} = b.
\end{aligned}$$

(See Figure 4.7.)

(b) If ξ_k is the right-hand endpoint of the kth subinterval, then

$$\xi_k = x_k = a + k\Delta x \qquad (k = 1, 2, \ldots, n).$$

Therefore, the nth approximating sum is

$$\begin{aligned}
S_n &= \sum_{k=1}^{n} f(\xi_k)\,\Delta x_k = \sum_{k=1}^{n} x_k\,\Delta x \\
&= \sum_{k=1}^{n} (a + k\,\Delta x)\,\Delta x = \sum_{k=1}^{n} (a\,\Delta x + k\,\Delta x^2)
\end{aligned}$$

$$= a\ \Delta x \sum_{k=1}^{n} 1 + \Delta x^2 \sum_{k=1}^{n} k$$

$$= a \cdot \Delta x \cdot n + \Delta x^2 \frac{n(n+1)}{2}$$

$$= a\left(\frac{b-a}{n}\right) n + \left(\frac{b-a}{n}\right)^2 \frac{n^2+n}{2}$$

$$= a(b-a) + \frac{(b-a)^2}{2} \frac{n^2+n}{n^2}$$

$$= ab - a^2 + \frac{b^2 - 2ab + a^2}{2}\left(1 + \frac{1}{n}\right).$$

(c)

$$\int_a^b x\ dx = \int_a^b f(x)\ dx = \lim_{n\to\infty} S_n$$

$$= \lim_{n\to\infty}\left[ab - a^2 + \frac{b^2 - 2ab + a^2}{2}\left(1 + \frac{1}{n}\right)\right]$$

$$= ab - a^2 + \frac{b^2 - 2ab + a^2}{2}(1+0)$$

$$= \frac{2ab - 2a^2 + b^2 - 2ab + a^2}{2} = \frac{b^2 - a^2}{2}.$$

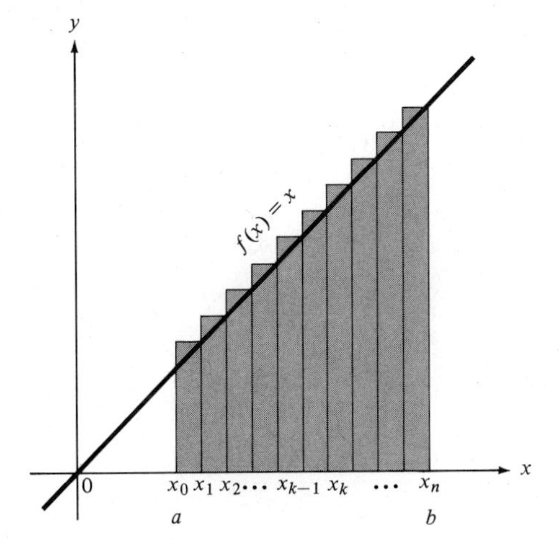

FIGURE 4.7 Example 1. $\int_a^b x\ dx = \frac{b^2 - a^2}{2}$

Example 2 Show that $\int_a^b x^2\,dx = (b^3 - a^3)/3$.

$$\int_a^b x^2 dx = \frac{b^3 - a^3}{3}$$

Solution We partition $[a, b]$ in the same way as in Example 1—into n subintervals of length

$$\Delta x = \Delta x_k = \frac{b-a}{n}$$

by choosing

$$x_0 = a,$$
$$x_1 = a + \Delta x,$$
$$x_2 = a + 2\ \Delta x,$$
$$\vdots$$
$$x_n = a + n\ \Delta x = b.$$

We also choose $\xi_1 = x_1, \xi_2 = x_2, \ldots, \xi_n = x_n$. The approximating sum is

$$
\begin{aligned}
S_n &= \sum_{k=1}^{n} f(\xi_k)\,\Delta x_k \\
&= \sum_{k=1}^{n} \xi_k^2\,\Delta x_k = \sum_{k=1}^{n} (a + k\,\Delta x)^2\,\Delta x \\
&= \sum_{k=1}^{n} [a^2\,\Delta x + 2ak\,\Delta x^2 + k^2\,\Delta x^3] \\
&= a^2\,\Delta x \sum_{k=1}^{n} 1 + 2a\,\Delta x^2 \sum_{k=1}^{n} k + \Delta x^3 \sum_{k=1}^{n} k^2 \\
&= a^2\,\Delta x \cdot n + 2a\,\Delta x^2\,\frac{n(n+1)}{2} + \Delta x^3\,\frac{n(n+1)(2n+1)}{6} \\
&= a^2\left(\frac{b-a}{n}\right) n + 2a\left(\frac{b-a}{n}\right)^2 \frac{n^2+n}{2} + \left(\frac{b-a}{n}\right)^3 \frac{2n^3+3n^2+n}{6} \\
&= a^2(b-a) + \frac{2a(b-a)^2}{2} \cdot \frac{n^2+n}{n^2} + \frac{(b-a)^3}{6} \cdot \frac{2n^3+3n^2+n}{n^3} \\
&= a^2(b-a) + a(b-a)^2\left(1+\frac{1}{n}\right) + \frac{(b-a)^3}{6}\left(2+\frac{3}{n}+\frac{1}{n^2}\right).
\end{aligned}
$$

Therefore,

$$
\begin{aligned}
\int_a^b x^2\,dx &= \lim_{\substack{n\to\infty \\ \Delta x\to 0}} \sum_{k=1}^{n} f(\xi_k)\,\Delta x_k = \lim_{n\to\infty} S_n \\
&= \lim_{n\to\infty}\left[a^2(b-a) + a(b-a)^2\left(1+\frac{1}{n}\right) + \frac{(b-a)^3}{6}\left(2+\frac{3}{n}+\frac{1}{n^2}\right)\right] \\
&= a^2(b-a) + a(b-a)^2(1+0) + \frac{(b-a)^3}{6}(2+0+0) \\
&= a^2b - a^3 + ab^2 - 2a^2b + a^3 + \frac{b^3 - 3ab^2 + 3a^2b - a^3}{3} \\
&= a^2b - a^3 + ab^2 - 2a^2b + a^3 + \frac{b^3}{3} - ab^2 + a^2b - \frac{a^3}{3} \\
&= \frac{b^3 - a^3}{3}.
\end{aligned}
$$

EXERCISES 4.5

1. Evaluate each of the following definite integrals. Make a careful sketch, showing the graph of the function being integrated and the rectangles for a typical approximating sum.

• (a) $\int_{-1}^{5} (x-1)dx$

(b) $\int_{-2}^{2} (-3x^2+4)dx$

• (c) $\int_{-1}^{3} (1-4x)dx$

(d) $\int_{0}^{2} (x^2-x)dx$

• (e) $\int_{2}^{5} (x+1)^2\,dx$

(f) $\int_{-1}^{1} (2x^2-1)dx$

2. Draw a careful sketch of $f(x) = x^3, -1 \leq x \leq 1$. Partition the interval $[-1, 1]$ into four subintervals. Form the approximating sum S_4 and represent it by rectangles in the sketch. Evaluate $\int_{-1}^{1} x^3 dx$.

   (a) Is S_4 the sum of the areas of the rectangles?

   • (b) Is $\int_{-1}^{1} x^3 dx$ equal to the area of the figure bounded by the graph of $y = x^3$, the lines $x = -1$, $x = 1$, and the x-axis?

3. Show that $\int_a^b x^0 dx = b - a$. Compare this result with those of Examples 1 and 2. What general rule would you conjecture?

$\int_a^b dx = b - a$

4. Define the function f by the following method: Let $f(x) = 0$ if x is an irrational number, $f(x) = 1$ if x is a rational number. Show that $\int_0^1 f(x)dx$ does not exist. (*Hint*: Show that for any partition the ξ_k can be chosen so that $S_n = 0$ or $S_n = 1$. Consequently, S_n has no limit as $n \to \infty$. You may use the fact that there are rational numbers and irrational numbers between any two real numbers.)

Example of nonintegrable function

4.6 DEFINITE INTEGRALS WITH VARIABLE UPPER LIMITS

We will find it convenient in the rest of this book to have the definite integral defined even when the upper limit of integration is less than or equal to the lower limit. To do this we define

$$\int_a^a f(x)\,dx = 0 \qquad \text{provided } f(a) \text{ exists}$$

$\int_a^b f(x)\,dx$ defined when $b \leq a$

and

$$\int_b^a f(x)\,dx = -\int_a^b f(x)\,dx \qquad \text{provided } a < b \text{ and } f \text{ is integrable on } [a, b].$$

For example,

$$\int_2^0 x\,dx = -\int_0^2 x\,dx = -\frac{2^2 - 0^2}{2} = -2.$$

We established in Section 4.5 that if $a < b$, then

$$\int_a^b dx = b - a, \qquad \int_a^b x\,dx = \frac{b^2 - a^2}{2}, \quad \text{and} \quad \int_a^b x^2\,dx = \frac{b^3 - a^3}{3}.$$

It is a simple matter to show that these rules also hold if $b \leq a$. We illustrate one proof in the following example.

Example 1 Show that the rule

$$\int_a^b x^2\,dx = \frac{b^3 - a^3}{3}$$

holds if $b \leq a$.

Solution If $b < a$, then

$$\int_b^a x^2\,dx = \frac{(a^3 - b^3)}{3}.$$

Therefore,

$$\int_a^b x^2\,dx = -\int_b^a x^2\,dx = -\frac{a^3 - b^3}{3} = \frac{b^3 - a^3}{3}.$$

If $a = b$, then

$$\int_a^b x^2\,dx = \int_a^a x^2\,dx = 0$$

and

$$\frac{b^3 - a^3}{3} = \frac{a^3 - a^3}{3} = 0.$$

Therefore,

$$\int_a^b x^2\,dx = 0 = \frac{b^3 - a^3}{3}.$$

The symbols x and dx in $\int_a^b f(x)\,dx$ do not appear in the final calculation. In a sense they act like placeholders. Thus, they can be replaced by other symbols, such as t and dt, w and dw, α and $d\alpha$, and so on. That is,

$$\int_a^b f(x)\,dx = \int_a^b f(t)\,dt = \int_a^b f(w)\,dw = \int_a^b f(\alpha)\,d\alpha = \cdots.$$

Functions defined by definite integrals with variable upper limits

Let f be integrable on an interval that contains a. We can define a new function F on this interval as follows. For each x on the interval let

$$F(x) = \int_a^x f(t)\,dt.$$

(See Figure 4.8.)

Since $\int_a^x f(t)\,dt$ is defined when $x < a$, $x = a$, and $x > a$, the above rule defines the function F for every x on the interval. If f is continuous everywhere, then it is integrable on every interval. In this case the rule

$$F(x) = \int_a^x f(t)\,dt$$

defines F for every x.

We use the notation $F(x) = \int_a^x f(t)\,dt$ rather than $\int_a^x f(x)\,dx$ to avoid confusion between x as the upper limit of integration and x as the placeholder symbol.

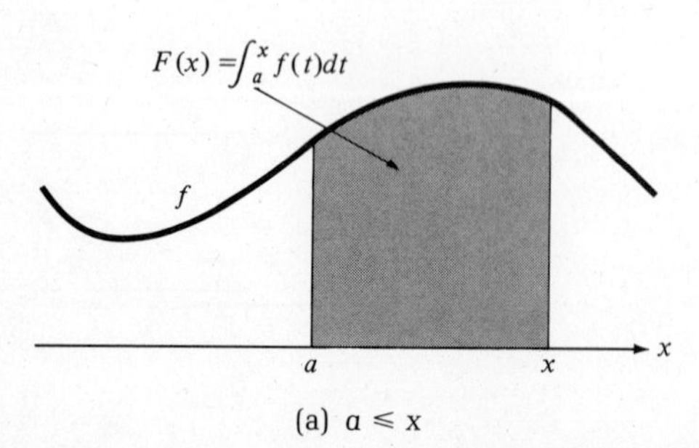

(a) $a \leq x$

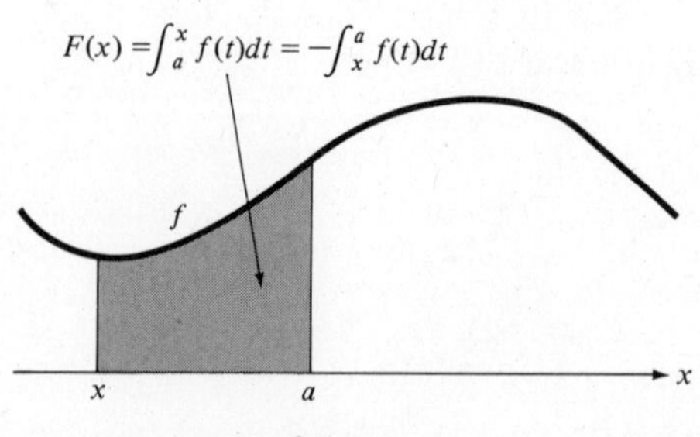

(b) $x < a$

FIGURE 4.8 The function F defined by an integral with a variable upper limit: $F(x) = \int_a^x f(t)\,dt$.

Example 2 Let $f(x) = x$. Define

$$F(x) = \int_1^x f(t)\,dt = \int_1^x t\,dt.$$

(a) Express F as a function of x by an algebraic rule. (b) Show that $F'(x) = f(x)$.

Solution

(a) Recall that

$$\int_a^b x\,dx = \int_a^b t\,dt = \frac{b^2 - a^2}{2}.$$

This rule holds for every choice of a and b. Therefore,

$$F(x) = \int_1^x t\,dt = \frac{x^2 - 1^2}{2} = \frac{x^2 - 1}{2}.$$

(b) $F'(x) = \dfrac{d}{dx}\left(\dfrac{x^2 - 1}{2}\right) = \dfrac{1}{2}(2x - 0) = x = f(x).$

Example 3 Let $f(x) = x^2$. Let

$$F(x) = \int_{-2}^x f(t)\,dt = \int_{-2}^x t^2\,dt.$$

(a) Express F as a function of x by an algebraic rule. (b) Show that $F'(x) = f(x)$.

Solution

(a) $F(x) = \displaystyle\int_{-2}^x t^2\,dt = \frac{x^3 - (-2)^3}{3} = \frac{x^3 + 8}{3}.$

(b) $F'(x) = \dfrac{d}{dx}\left(\dfrac{x^3 + 8}{3}\right) = \dfrac{1}{3}(3x^2 + 0) = x^2 = f(x).$

EXERCISES 4.6

1. Evaluate the following definite integrals. You may use the following results that hold when $a < b$:

$$\int_a^b dx = b - a, \quad \int_a^b x\,dx = \frac{b^2 - a^2}{2}, \quad \int_a^b x^2\,dx = \frac{b^3 - a^3}{3}.$$

(a) $\int_2^1 x\,dx$ (d) $\int_0^0 x^2\,dx$

• (b) $\int_7^7 x^2\,dx$ (e) $\int_3^6 x\,dx$

• (c) $\int_5^0 dx$ • (f) $\int_4^{-2} x^2\,dx$

2. Calculate $F(x) = \int_{-1}^x f(t)\,dt$ for each of the following functions $f(t)$. Show that $F'(x) = f(x)$ in each case.

(a) t
• (b) t^2
(c) 1

4.7 THE FUNDAMENTAL THEOREM OF CALCULUS

Thus far in this chapter we have had to use the definition for the calculation of definite integrals. This method is laborious, time-consuming, and inefficient—as evidenced by the very few definite integrals that we have calculated. In this section we discuss the *Fundamental Theorem of Calculus*. This theorem gives a close relationship between the definite integral and the derivative which can be used to evaluate many integrals with comparative ease.

A clue to the use of the Fundamental Theorem can be found in Examples 2 and 3 of the preceding section. We saw there that if $f(x) = x$ or $f(x) = x^2$ and if F is defined for each x by

$$F(x) = \int_a^x f(t)\, dt,$$

then

$$F'(x) = f(x).$$

Newton and Leibniz discovered that this relationship holds in general. This is the gist of the Fundamental Theorem of Calculus.[3]

The fundamental Theorem

Theorem 4.2 Let f be continuous on an interval containing a. Define the function F at each point x of the interval by

$$F(x) = \int_a^x f(t)\, dt.$$

Then F is differentiable and

$$F'(x) = f(x)$$

for each x on the interval.

There are several important aspects of the fundamental Theorem. First, the theorem tells us that any continuous function f has an *antiderivative*. That is, there exists a function F (defined by $F(x) = \int_a^x f(t)\, dt$) that has f as its derivative. This is the case even though f itself may not be differentiable. Thus, *any function that is continuous on an interval has an antiderivative on that interval.*

The main use that we will make of the Fundamental Theorem is in the evaluation of definite integrals. The following corollary to the theorem shows that we can evaluate $\int_a^b f(x)\, dx$ if we can find any antiderivative of f on the interval $[a, b]$. We do not actually need to determine the function $F(x) = \int_a^x f(x)\, dx$.

Definite integrals and antiderivatives

COROLLARY Let f be continuous on $[a, b]$. Let G be an antiderivative of f on $[a, b]$. Then

$$\int_a^b f(x)\, dx = G(b) - G(a).$$

PROOF Define F on the interval $[a, b]$ by

$$F(x) = \int_a^x f(t)\, dt.$$

---

[3] The proof of Theorem 4.2 is omitted. The interested reader can find the proof in A. Schwartz, *Calculus with Analytic Geometry*, 3rd ed. New York: Holt, Rinehart and Winston, 1974, pp. 137–139.

Then F and G are antiderivatives of f on $[a, b]$. By Theorem 3.8 there exists a constant C such that

$$F(x) = G(x) + C \qquad \text{for} \quad a \leq x \leq b.$$

To calculate C we evaluate $F(a)$:

$$\begin{aligned} F(a) &= \int_a^a f(t)\, dt = 0 \\ &= G(a) + C, \end{aligned}$$

so $C = -G(a)$ and

$$F(x) = \int_a^x f(t)\, dt = G(x) + C = G(x) - G(a).$$

For the special case $x = b$ we obtain

$$\int_a^b f(x)\, dx = \int_a^b f(t)\, dt = F(b) = G(b) - G(a).$$

The above corollary states one of the most useful results in mathematics, for it enables us to calculate complicated definite integrals by use of antiderivatives. As stated in the corollary we can use any antiderivative of the function that we wish to integrate. For simplicity we usually choose the antiderivative with the arbitrary constant C equal to zero.

Example 1 Calculate $\int_1^2 (2x - x^{-2/3})\, dx$.

Solution An antiderivative of $2x - x^{-2/3}$ is

$$G(x) = \frac{2x^2}{2} - \frac{x^{1/3}}{\frac{1}{3}} = x^2 - 3x^{1/3}.$$

Therefore,

$$\begin{aligned} \int_1^2 (2x - x^{-2/3})\, dx &= G(2) - G(1) \\ &= (2^2 - 3 \cdot 2^{1/3}) - (1^2 - 3 \cdot 1^{1/3}) \\ &= 4 - 3\sqrt[3]{2} - 1 + 3 \\ &= 6 - 3\sqrt[3]{2}. \end{aligned}$$

Notation Expressions such as $G(b) - G(a)$ occur with such frequency that we use a special shorthand symbol for them. We write

$$[G(x)]_a^b = G(b) - G(a).$$

$[G(x)]_a^b = G(b) - G(a)$

For example,

$$[3x^2]_{-1}^4 = 3 \cdot 4^2 - 3(-1)^2 = 3 \cdot 16 - 3 \cdot 1 = 45,$$

and

$$[\sqrt{x}]_2^3 = \sqrt{3} - \sqrt{2}.$$

By the corollary, if G is an antiderivative of f on $[a, b]$, then

$$\int_a^b f(x)\, dx = [G(x)]_a^b.$$

Example 2

(a) $\int_0^2 x\,dx = \left[\frac{x^2}{2}\right]_0^2 = \frac{2^2}{2} - \frac{0^2}{2} = 2.$

(b) $\int_{-1}^{1} (3x^2 - x^{1/3} + 5)\,dx$

$$= \left[\frac{3x^3}{3} - \frac{x^{4/3}}{4/3} + 5x\right]_{-1}^{1} = \left[x^3 - \frac{3x^{4/3}}{4} + 5x\right]_{-1}^{1}$$

$$= \left[1^3 - \frac{3 \cdot 1^{4/3}}{4} + 5 \cdot 1\right] - \left[(-1)^3 - \frac{3(-1)^{4/3}}{4} + 5(-1)\right]$$

$$= \left[1 - \frac{3}{4} + 5\right] - \left[-1 - \frac{3}{4} - 5\right]$$

$$= 1 - \frac{3}{4} + 5 + 1 + \frac{3}{4} + 5 = 12.$$

(c) $\int_0^4 \sqrt{x}\,dx = \int_0^4 x^{1/2}\,dx = \left[\frac{x^{3/2}}{3/2}\right]_0^4$

$$= \left[\frac{2}{3}x^{3/2}\right]_0^4 = \frac{2}{3} \cdot 4^{3/2} - \frac{2}{3} \cdot 0^{3/2}$$

$$= \frac{2}{3} \cdot 4\sqrt{4} - 0 = \frac{2}{3} \cdot 4 \cdot 2 = \frac{16}{3}.$$

EXERCISES 4.7

1. Use antiderivatives to calculate the following definite integrals:

(a) $\int_{-1}^{1} (3x^2 - x + 2)\,dx$

• (b) $\int_0^1 (x^2 + x - 1)\,dx$

(c) $\int_1^2 \frac{x^2 + 1}{x^2}\,dx$

• (d) $\int_1^4 \frac{x - 1}{\sqrt{x}}\,dx$

(e) $\int_1^3 |x|\,dx$ (be careful!)

• (f) $\int_{-4}^{-2} |x|\,dx$ (be careful!)

(g) $\int_0^2 \frac{3x^3 + 3x^2 + x + 1}{x + 1}\,dx$

• (h) $\int_{-1}^{0} (x - 1)^3\,dx$

(i) $\int_{-2}^{1} \frac{2x^3 - x^2 - 2x + 1}{2x - 1}\,dx$

• (j) $\int_1^3 \frac{x^2 - 1}{x - 1}\,dx$

2. It is possible to describe functions that are continuous everywhere but are so "bad" that they are not differentiable anywhere. Must such a function be the derivative of a function? Explain!

4.8 APPLICATIONS OF THE DEFINITE INTEGRAL

The theory of the definite integral is rich in a wide variety of applications. The development for a typical application has the following steps:

1. Find a sum of form $\Sigma_{k=1}^{n} f(x_k)\,\Delta x_k$ that approximates the quantity that we wish to define.

2. Define the desired quantity to be the limit of the sum in (1), which is equal to a definite integral:

$$\lim_{n\to\infty} \sum_{k=1}^{n} f(x_k)\,\Delta x_k = \int_a^b f(x)\,dx.$$

3. Use antiderivatives to calculate the values of the integral for particular functions $f(x)$ and particular numbers a and b.

The following applications should be studied carefully. They illustrate a few of the many and varied uses of the definite integral.

Area Let f be continuous on the closed interval $[a, b]$. We wish to define the *area* of the figure bounded by the graph of f, the x-axis, and the lines $x = a$ and $x = b$.

We first consider the case where f is *nonnegative* on the interval. Let $a = x_0, \ldots, x_{n-1}, x_n = b$ be a partition of $[a, b]$. Let $\xi_1, \xi_2, \ldots, \xi_n$ be points on the subintervals $[x_0, x_1], [x_1, x_2], \ldots, [x_{n-1}, x_n]$, respectively. Most people would agree that the sum

$$A_n = \sum_{k=1}^{n} f(\xi_k)\,\Delta x_k$$

is an approximation for the area of the figure. This is because the terms $f(\xi_1)\,\Delta x_1, f(\xi_2)\,\Delta x_2, \ldots, f(\xi_n)\,\Delta x_n$ are just the areas of the rectangles pictured in Figure 4.9 and it would appear that the sum of these areas should approximate the area of the figure. In fact, it would appear reasonable to expect the area to be equal to $\lim_{n\to\infty} A_n$. Since $\lim_{n\to\infty} A_n$ is equal to $\int_a^b f(x)\,dx$, it is customary for mathematicians to define the area of the figure to be this definite integral.

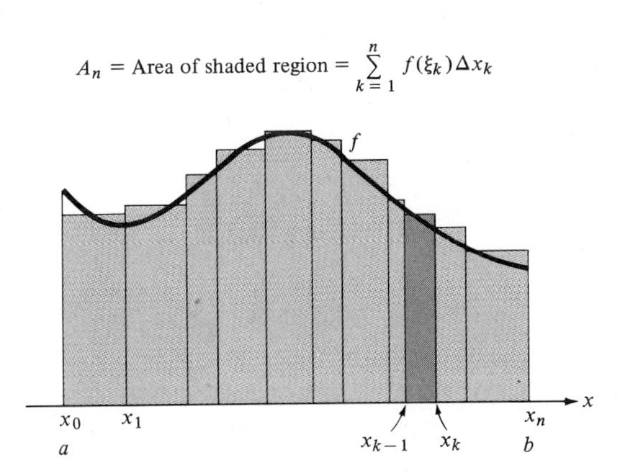

(a) $A_n = \sum_{k=1}^{n} f(\xi_k)\Delta x_k$

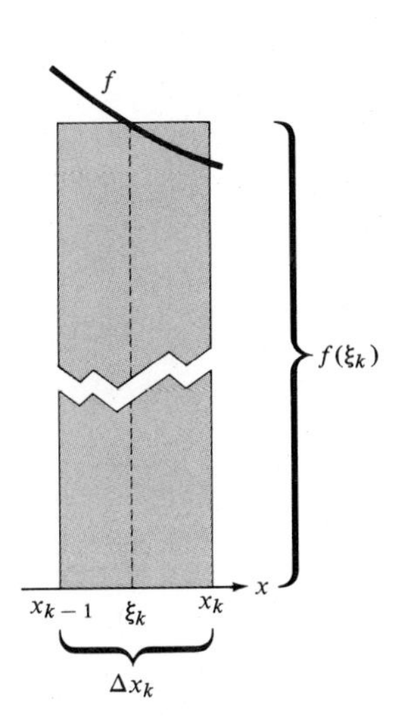

(b) Area of kth rectangle $= f(\xi_k)\Delta x_k$

FIGURE 4.9 $A = \int_a^b f(x)\,dx = \lim_{n\to\infty} \sum_{k=1}^{n} f(\xi_k)\Delta x_k$

Definition Let f be continuous and nonnegative on the closed interval $[a, b]$. The *area* of the portion of the plane bounded by the graph of $y = f(x)$, the x-axis, and the lines $x = a$, $x = b$ is defined to be

Area under a curve

$$A = \int_a^b f(x)\,dx.$$

Area between Two Curves How can we calculate the area of the region pictured in Figure 4.10? Here the graph of f is the upper boundary and the graph of g is the lower boundary.

We partition the interval $[a, b]$ into n parts by the numbers $a = x_0 < x_1 < x_2 < \cdots < x_{n-1} < x_n = b$. For each k we choose a point ξ_k in the interval $[x_{k-1}, x_k]$, and construct the rectangle with the number Δx_k as base and the number $f(\xi_k) - g(\xi_k)$ as height. The sum

$$A_n = \sum_{k=1}^{n} [f(\xi_k) - g(\xi_k)]\,\Delta x_k$$

is the sum of the areas of the rectangles pictured in Figure 4.10b.

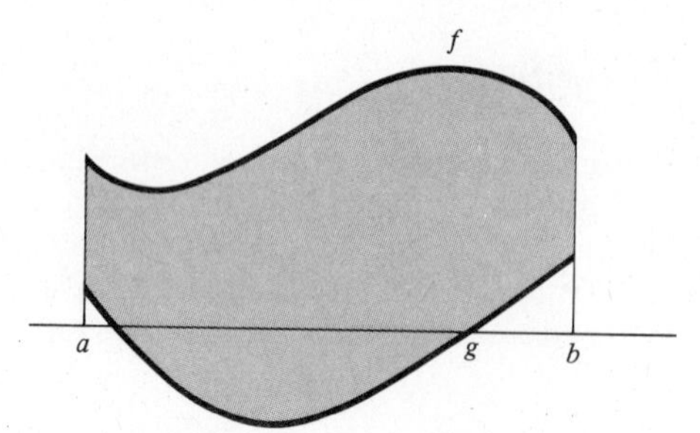

(a) Area between two curves:
$A = \int_a^b [f(x) - g(x)]\,dx$

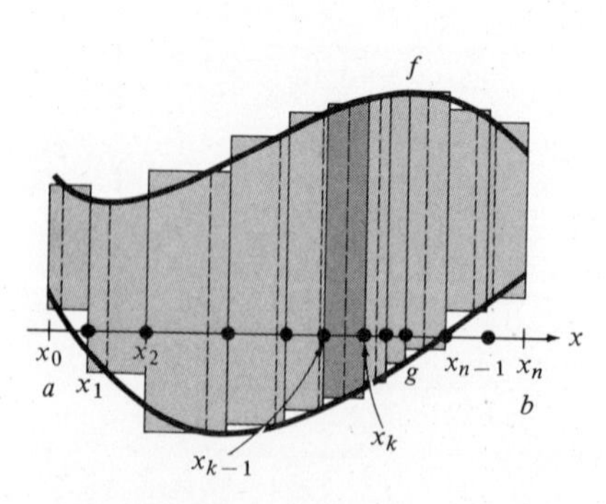

(b) An approximating sum for the area between two curves:

$A_n = \sum_{k=1}^{n} [f(\xi_k) - g(\xi_k)]\Delta x_k$

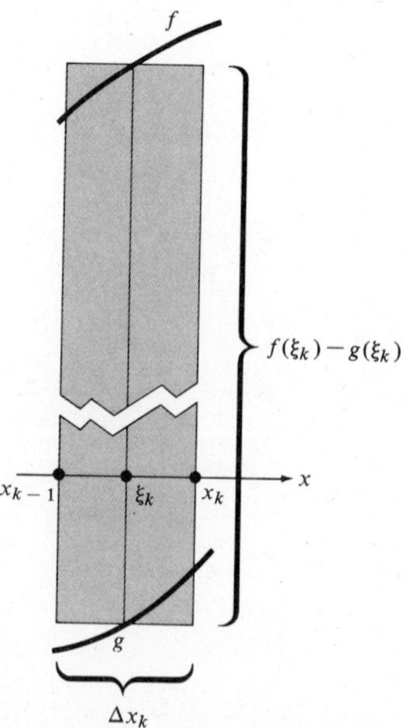

(c) Area of kth rectangle:
$A_k = [f(\xi_k) - g(\xi_k)]\Delta x_k$

FIGURE 4.10

If each Δx_k is small and n is large, it would appear that A_n should be a good approximation to the area of the region. Thus we define

$$\text{Area} = \lim_{\substack{n\to\infty \\ \Delta x_k \to 0}} A_n = \lim_{\substack{n\to\infty \\ \Delta x_k \to 0}} \sum_{k=1}^{n} [f(\xi_k) - g(\xi_k)]\, \Delta x_k.$$

But this limit is just the integral $\int_a^b [f(x) - g(x)]\, dx$. Thus

Area between two curves

$$\text{Area} = \int_a^b [f(x) - g(x)]\, dx.$$

Example 1 Calculate the area of the portion of the xy-plane bounded by the graphs of $f(x) = \sqrt{x}$ and $g(x) = x^2$.

Solution The region is pictured in Figure 4.11. Since $f(x) \geqslant g(x)$ for $0 \leqslant x \leqslant 1$, the area is

$$\begin{aligned} A &= \int_0^1 (f(x) - g(x))\, dx \\ &= \int_0^1 (\sqrt{x} - x^2)\, dx \\ &= \int_0^1 (x^{1/2} - x^2)\, dx = \left[\frac{x^{3/2}}{\frac{3}{2}} - \frac{x^3}{3}\right]_0^1 \\ &= \left[\frac{2}{3} x\sqrt{x} - \frac{x^3}{3}\right]_0^1 = \left[\frac{2}{3} \cdot 1\sqrt{1} - \frac{1^3}{3}\right] - [0 - 0] \\ &= \frac{2}{3} - \frac{1}{3} = \frac{1}{3}. \end{aligned}$$

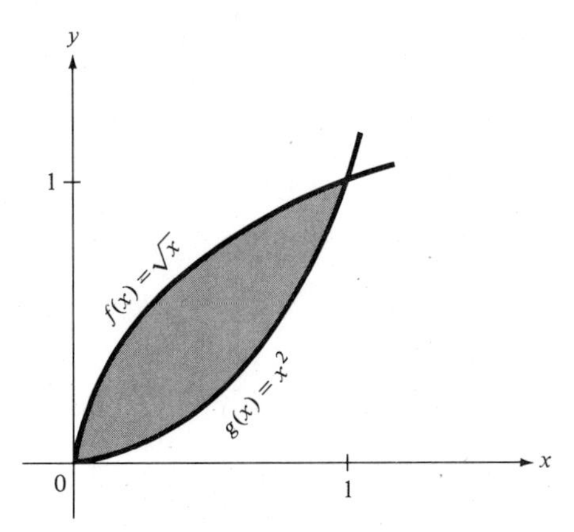

FIGURE 4.11 Example 1. Area of the region bounded by the graphs of $f(x) = \sqrt{x}$ and $g(x) = x^2$.

Example 2 Express the area of the region bounded by the graphs of $f(x) = x^3$ and $g(x) = x^3 + x^2 - x - 2$ as a definite integral.

Solution First we sketch the graphs (Figure 4.12). The points of intersection are $(-1, -1)$ and $(2, 8)$. Thus. The area of the region is

$$A = \int_{-1}^{2} (f(x) - g(x))\,dx = \int_{-1}^{2} (x^3 - (x^3 + x^2 - x - 2))\,dx$$
$$= \int_{-1}^{2} (-x^2 + x + 2)\,dx = \left[-\frac{x^3}{3} + \frac{x^2}{2} + 2x\right]_{-1}^{2}$$
$$= \left[-\frac{8}{3} + \frac{4}{2} + 4\right] - \left[-\frac{(-1)^3}{3} + \frac{(-1)^2}{2} + 2(-1)\right]$$
$$= 4\tfrac{1}{2}.$$

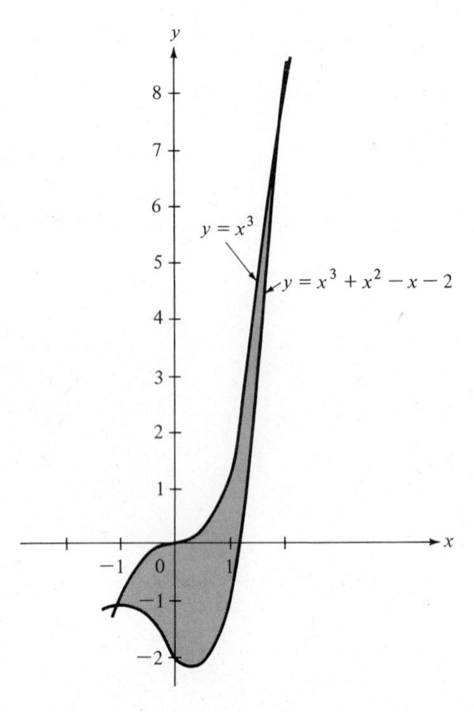

FIGURE 4.12 Example 2. Area of the region bounded by the graphs of $y = x^3$ and $y = x^3 + x^2 - x - 2$.

If $f(x) \leqslant g(x)$ over part of the interval $[a, b]$ and $g(x) \leqslant f(x)$ over another part, then we calculate the areas over the corresponding subintervals and add them to get the total area.

Example 3 Calculate the area of the closed figure on the interval $[-1, 1]$ bounded by the graphs of $y = x^3$ and $y = x$.

Solution The graphs are shown in Figure 4.13. Observe that

$$x^3 \geqslant x \quad \text{for } -1 \leqslant x \leqslant 0$$

and

$$x \geqslant x^3 \quad \text{for } \ 0 \leqslant x \leqslant 1.$$

Thus

$$\text{area} = \int_{-1}^{0} (x^3 - x)\,dx + \int_{0}^{1} (x - x^3)\,dx$$
$$= \left[\frac{x^4}{4} - \frac{x^2}{2}\right]_{-1}^{0} + \left[\frac{x^2}{2} - \frac{x^4}{4}\right]_{0}^{1}$$

$$= [0] - \left[\frac{(-1)^4}{4} - \frac{(-1)^2}{2}\right] + \left[\frac{1^2}{2} - \frac{1^4}{4}\right] - [0]$$

$$= -\left[\frac{1}{4} - \frac{1}{2}\right] + \left[\frac{1}{2} - \frac{1}{4}\right] = \frac{1}{2}.$$

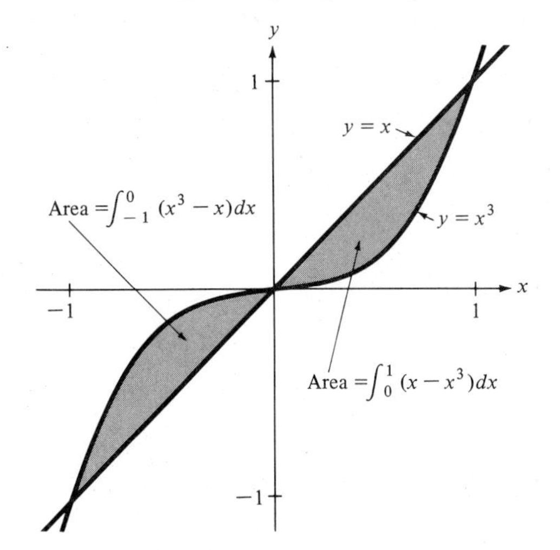

FIGURE 4.13 Example 3.
Area $= \int_{-1}^{0} (x^3 - x)\,dx + \int_{0}^{1} (x - x^3)\,dx$

Revenue Let $MR(x)$ be the marginal revenue of a company for a production level of x units. If $R(x)$ is the revenue function, then

$$MR(x) = R'(x),$$

Revenue and marginal revenue

so that $R(x)$ is one of the antiderivatives of the marginal revenue function. If a and b are two production levels, with $a < b$, then by the corollary to the Fundamental Theorem of Calculus

$$\int_a^b MR(x)\,dx = \int_a^b R'(x)\,dx = R(b) - R(a).$$

The difference $R(b) - R(a)$ is equal to the change in revenue from the production level of a units to the production level of b units. It follows that this change in revenue is equal to the integral of the marginal revenue function from $x = a$ to $x = b$. If $MR(x)$ is nonnegative for $a \leq x \leq b$, then the integral can be interpreted as the area under the marginal revenue curve from $x = a$ to $x = b$. (See Figure 4.14.)

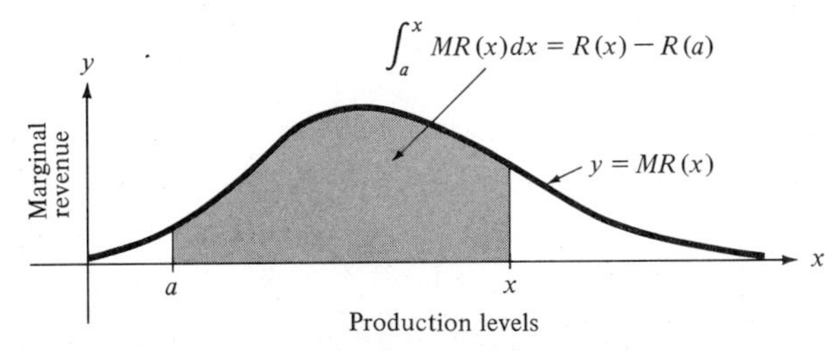

FIGURE 4.14 The area under the marginal revenue curve is equal to the difference in revenue at the two production levels:

$$\int_a^x MR(x)\,dx = R(x) - R(a)$$

If we know the marginal revenue function and the value of $R(a)$, then we can use the above relationship to find that

$$R(b) = R(a) + \int_a^b MR(x)\,dx.$$

More generally, the value of the revenue function $R(x)$ can be expressed as

$$R(x) = R(a) + \int_a^x MR(x)\,dx = R(a) + \int_a^x MR(t)\,dt.$$

Similar results hold for marginal cost and marginal profit functions.

Example 4 The marginal revenue function of the Substratum Casting Company is

$$MR(x) = -\frac{3x^2}{4000} + 8x - 2.$$

It is known that $R(20) = 500$. Express $R(x)$ as a function of x.

Solution

$$\begin{aligned} R(x) &= R(20) + \int_{20}^x MR(t)\,dt \\ &= R(20) + \int_{20}^x \left(-\frac{3t^2}{4000} + 8t - 2\right) dt \\ &= 500 + \left[\frac{-3t^2}{3 \cdot 4000} + \frac{8t^2}{2} - 2t\right]_{20}^x \\ &= 500 + \left[\frac{-t^3}{4000} + 4t^2 - 2t\right]_{20}^x \\ &= \frac{-x^3}{4000} + 4x^2 - 2x - 1058. \end{aligned}$$

Distance, Velocity, and Acceleration If $s(t)$ and $v(t)$ are the distance and velocity functions for straight line motion, then $s(t)$ is an antiderivative of the velocity function $v(t)$. It follows from the corollary to the Fundamental Theorem that

Velocity and distance

$$\int_a^b v(t)\,dt = s(b) - s(a),$$

so that the net distance traveled over the period $a \leqslant t \leqslant b$ is equal to the integral of the velocity function from a to b. If, furthermore, $v(t) \geqslant 0$ over this time interval then the integral is equal to the area under the graph of the velocity function. (See Figure 4.15.)

A similar result is concerned with the velocity function $v(t)$ and the acceleration function $a(t)$:

Acceleration and velocity

$$\int_a^b a(t)\,dt = v(b) - v(a).$$

Example 5 A plane flies in a straight line for two hours ($t = 0$ to $t = 2$) with velocity at time t (in hours) given by

$$v(t) = 500 + 100t.$$

How far does the plane travel over the two-hour period?

Solution The total distance traveled is equal to

$$\int_0^2 (500 + 100t)\, dt = \left[500t + \frac{100t^2}{2}\right]_0^2$$
$$= 500 \cdot 2 + 50 \cdot 2^2 - 0 + 0 = 1200 \text{ miles.}$$

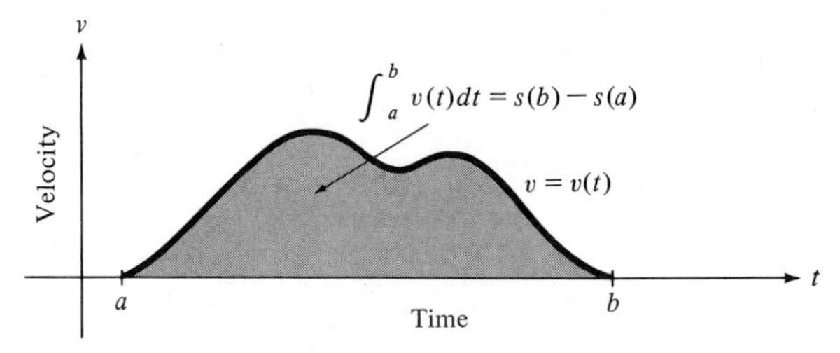

FIGURE 4.15 The area under the velocity curve is equal to the total distance traveled from time $t = a$ to $t = b$:

$$\int_a^b v(t)\, dt = s(b) - s(a)$$

Average Value The average value of n numbers $a_1, a_2, \ldots, a_n$, is defined to be

$$\frac{a_1 + a_2 + \cdots + a_n}{n}.$$

We will show how to extend this definition to calculate the average value of a continous function f over a closed interval $[a, b]$.

We first partition the interval into n equal subintervals, each of length $\Delta x = (b - a)/n$. This is accomplished by choosing the equally spaced points $a = x_0 < x_1 < x_2 < \cdots < x_n = b$.

Observe that the average value of f at the points $x_1, x_2, \ldots, x_n$ is

$$\frac{f(x_1) + f(x_2) + \cdots f(x_n)}{n}.$$

Recall that

$$\Delta x = \frac{b - a}{n}$$

so that

$$n = \frac{b - a}{\Delta x}.$$

If we substitute this value into the above expression we obtain

$$\frac{f(x_1) + f(x_2) + \cdots + f(x_n)}{n} = \frac{f(x_1) + f(x_2) + \cdots + f(x_n)}{\dfrac{b - a}{\Delta x}}$$
$$= \frac{f(x_1) + f(x_2) + \cdots + f(x_n)}{b - a} \cdot \Delta x$$
$$= \frac{\sum_{k=1}^{n} f(x_k)\, \Delta x}{b - a}.$$

The above sum is equal to the average value of f at the equally spaced points $x_1, x_2, \ldots, x_n$. The average value of f on $[a, b]$ is defined to be the limit of this sum as $n \to \infty$.

DEFINITION Let f be continuous on the closed interval $[a, b]$. The *average value of f on $[a, b]$ is equal to*

Average value

$$\text{average value} = \lim_{n\to\infty} \frac{1}{b-a} \cdot \sum_{k=1}^{n} f(x_k)/\Delta x = \frac{1}{b-a} \int_a^b f(x)\,dx.$$

This definition coincides with our intuitive notions of average value. For example, if $v(t)$ is a velocity function, then $\int_a^b v(t)\,dt$ is equal to the total distance traveled over the time interval $a \leq t \leq b$, and $b - a$ is equal to the length of the period of time. Then

$$\text{average value of } v(t) = \frac{1}{b-a} \int_a^b v(t)\,dt = \frac{\text{total distance}}{\text{elapsed time}},$$

which is how we originally defined the average velocity in Chapter 2.

Similarly, if $MP(x)$ is the marginal profit function at a production level of x units, then $\int_a^b MP(x)\,dx = P(b) - P(a)$ is equal to the change in profit from a production level of a units to a level of b units, and $b - a$ is equal to the change in production levels. Then

$$\begin{aligned} \text{average value of } MP(x) &= \frac{1}{b-a} \int_a^b MP(x)\,dx = \frac{P(b) - P(a)}{b-a} \\ &= \frac{\text{change in profit}}{\text{change in production}} \\ &= \text{the average change in profit corresponding to a change of production levels from } x = a \text{ to } x = b. \end{aligned}$$

EXERCISES 4.8

1. Calculate the area of the region with the given boundary. Sketch the region.

   • (a) $y = -x^2 + 5x - 4$, x-axis
   (b) $y = x^2 + 2$, $x = 2$, $x = -2$, x-axis
   • (c) The region in quadrant IV bounded by $y = x^2 - 1$, x-axis, y-axis
   (d) $y = x^3 + x - 1$, x-axis, y-axis, $x = 2$
   (e) $y = x^2 + \sqrt{x}$, x-axis, $x = 0$, $x = 4$
   • (f) $y = \sqrt{x} - x^2$, x-axis
   (g) $y = x^2$, $y = 2x$
   • (h) $y = x^2$, $y = x + 2$
   (i) Region in quadrant I bounded by $y = x^4$ and $y = x^3$
   • (j) $y = x/2$, $y = \sqrt{x}$
   (k) $y = |x|$, x-axis, $x = -3$, $x = 0$
   (l) $y = x^2$, $y = 8 - x^2$

2. Calculate the change in the profit function for a change in the production level from a to b units. ($MP(x)$ is the marginal profit function.)

   (a) $MP(x) = -\dfrac{x^2}{100} + 6x - 10$, $a = 50$, $b = 60$

• (b) $MP(x) = -\dfrac{4x^3}{10{,}000} - \dfrac{3x^2}{1000} + 50x - 20,\ a = 10,\ b = 20$

(c) $MP(x) = -\dfrac{3x^2}{1000} + 40x - 60,\ a = 10,\ b = 15$

• 3. Calculate the average marginal profit over the interval $[a, b]$ for each of the functions in Exercise 2.

4. Calculate the total distance traveled from $t = a$ to $t = b$ for each of the velocity functions. Then calculate the average velocity over the time interval $a \leq t \leq b$.

   (a) $v(t) = 3t^2 + 2t + 1,\ a = 0,\ b = 2$

   • (b) $v(t) = t - \sqrt{t},\ \ a = 4,\ b = 9$

   (c) $v(t) = 1 + \dfrac{1}{(t+1)^2},\ a = 0,\ b = 1$

   • (d) $v(t) = t + \sqrt{2t + 1},\ \ a = 4,\ b = 12$

5. Calculate the average value of f on the interval.

   (a) $f(x) = 2x,\ [1, 4]$
   • (b) $f(x) = x^2 + x - 1,\ [0, 1]$
   (c) $f(x) = x^2 - 4x,\ [0, 3]$
   • (d) $f(x) = x^3 + 1,\ [0, 2]$

6. Let f be a nonnegative function that is continuous on $[a, b]$. Let R be a rectangle with the interval $[a, b]$ as its base and with height equal to the average value of f on $[a, b]$. Show that

$$\text{area of } R = \int_a^b f(x)\,dx.$$

Illustrate with a sketch showing the rectangle R and the graph of f.

4.9 INFINITE LIMITS

In this section we take a temporary break from the theory of integration and consider some additional topics concerning limits.

What happens to $f(x)$ when x gets large without bound? We consider some examples (Figure 4.16):

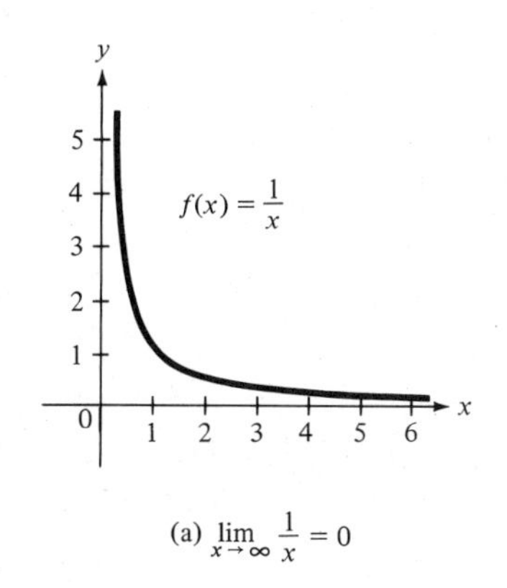

(a) $\lim_{x \to \infty} \frac{1}{x} = 0$

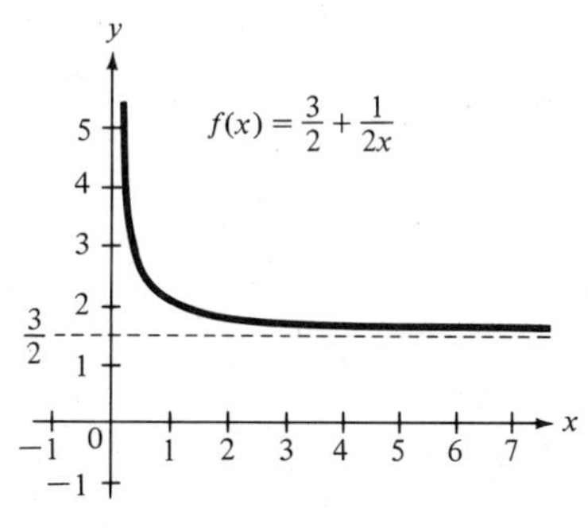

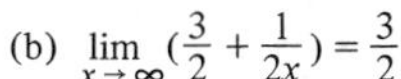

(b) $\lim_{x \to \infty} \left(\frac{3}{2} + \frac{1}{2x}\right) = \frac{3}{2}$

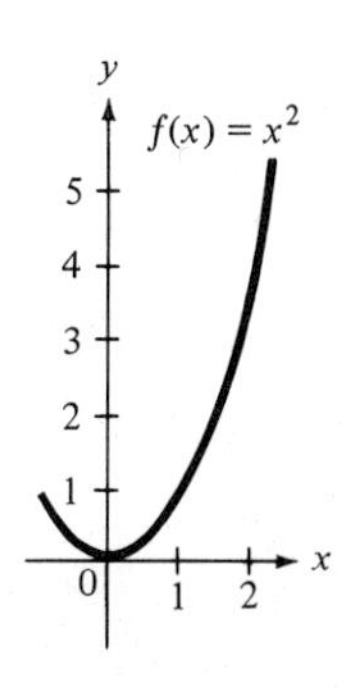

(c) $\lim_{x \to \infty} x^2 = \infty$

FIGURE 4.16

(a) If $f(x) = 1/x$, then $f(x)$ approaches zero.

(b) If $f(x) = \frac{3x+1}{2x} = \frac{3}{2} + \frac{1}{2x}$, then $f(x)$ approaches $\frac{3}{2} + 0 = \frac{3}{2}$.

(c) If $f(x) = x^2$, then $f(x)$ also gets large without bound.

In the first case above we write

$$\lim_{x\to\infty} \frac{1}{x} = 0$$

(read "limit of $1/x$ as x approaches infinity is zero"). In the second case we write

$$\lim_{x\to\infty} \frac{3x+1}{2x} = \lim_{x\to\infty} \left(\frac{3}{2} + \frac{1}{2x}\right) = \frac{3}{2} + 0 = \frac{3}{2},$$

in the third,

$$\lim_{x\to\infty} x^2 = \infty$$

(read "limit of x^2 as x approaches infinity is infinity").

These statements do not mean that we are considering a new number called infinity. The expression "$\lim_{x\to\infty} f(x) = L$" should be interpreted as a shorthand way of saying "$f(x)$ gets close to L as x gets large without bound." Similarly, "$\lim_{x\to\infty} f(x) = \infty$" means "$f(x)$ gets large without bound as x gets large without bound."

$\lim_{x\to\infty} f(x) = \infty$

If $f(x)$ gets large without bound through negative numbers as x gets large without bound through positive numbers, we write

$$\lim_{x\to\infty} f(x) = -\infty.$$

For example,

$$\lim_{x\to\infty} (1 - x^2) = -\infty.$$

There are various other combinations of limits of this type using $+\infty$, $-\infty$ and numbers. For example,

$$\lim_{x\to-\infty} (3x+1) = -\infty, \qquad \lim_{x\to-\infty} \frac{3x+1}{x} = 3, \qquad \lim_{x\to\infty} \frac{2}{x^2+1} = 0.$$

The meanings of these expressions should be clear if they are expressed in words. For example, the first of these limits states that "$3x + 1$ gets large without bound through negative values as x gets large without bound through negative values."

We will give the formal definition for only one of the limits of this type. The interested reader can find the others in most of the standard calculus texts.[4]

DEFINITION Let f be defined for all x greater than some fixed number n. We say that

$\lim_{x\to\infty} f(x) = L$

$$\lim_{x\to\infty} f(x) = L$$

[4] See, for example, A. Schwartz, *Calculus with Analytic Geometry*, 3rd ed. New York: Holt, Rinehart and Winston, 1974, pp. 490–500.

provided: For each $\epsilon > 0$ there is a number N (which depends on ϵ) such that

$$\text{if } x > N, \qquad \text{then } \quad |f(x) - L| < \epsilon.$$

In other words, we can keep $f(x)$ within any preassigned distance of L by choosing sufficiently large values of x. Geometrically, this means that the horizontal line $y = L$ is an asymptote for the graph of f as x gets large without bound. (See Figure 4.17.) Similarly, if $\lim_{x\to-\infty} f(x) = L$, then the line $y = L$ is an asymptote as x gets large without bound through negative values.

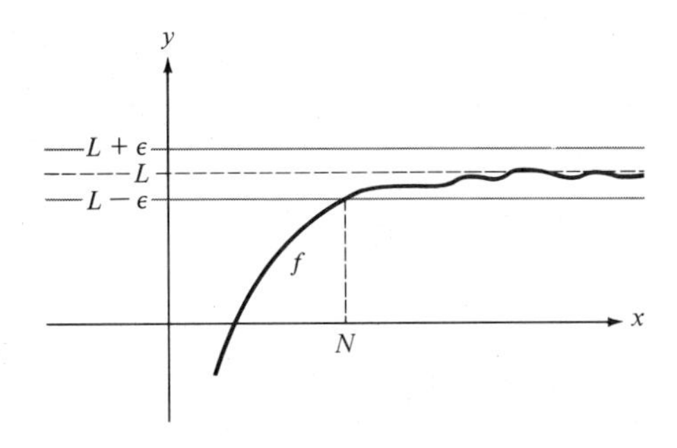

FIGURE 4.17 $\lim_{x\to\infty} f(x) = L$. If $x > N$, then $f(x)$ is within the ϵ-distance of L.

The Limit Theorem (Theorem 2.1) can be extended to cover certain of these limits.

Theorem 4.3 (*The Extended Limit Theorem*) If

$$\lim_{x\to\infty} f(x) = L \quad \text{and} \quad \lim_{x\to\infty} g(x) = M,$$

where L and M are numbers (neither is equal to $+\infty$ or $-\infty$), then

L1 $\lim_{x\to\infty} [f(x) \pm g(x)] = L \pm M = \lim_{x\to\infty} f(x) \pm \lim_{x\to\infty} g(x),$

L2 $\lim_{x\to\infty} [f(x) \cdot g(x)] = LM = \lim_{x\to\infty} f(x) \cdot \lim_{x\to\infty} g(x),$

L3 $\lim_{x\to\infty} \dfrac{f(x)}{g(x)} = \dfrac{L}{M} = \dfrac{\lim_{x\to\infty} f(x)}{\lim_{x\to\infty} g(x)}$ provided $M \neq 0$,

L4 $\lim_{x\to\infty} \sqrt[n]{f(x)} = \sqrt[n]{L} = \sqrt[n]{\lim_{x\to\infty} f(x)}$ provided $L > 0$ if n is even.

Similar properties hold if $\lim_{x\to-\infty} f(x) = L$ and $\lim_{x\to-\infty} g(x) = M$.

The situation is more complicated if one or both of the limits is infinite. Then it may be best to examine the behavior of the various functions involved to determine the values of the limits. In general the sum, difference, and product cause little trouble. It is the quotient that is most difficult.

There is one important case when we can determine the values of the limit of a quotient without difficulty. If both the numerator and denominator are polynomials, we can calculate the limit of the quotient by the technique illustrated in the following example.

Example 1 Calculate $\lim_{x\to\infty} \frac{2x^2 - x + 1}{3x^2 + x - 7}$.

Solution We cannot apply the Limit Theorem, because both numerator and denominator become infinite. To change the function to a more usable form we multiply both numerator and denominator by $1/x^2$. Then

$$\frac{2x^2 - x + 1}{3x^2 + x - 7} = \frac{(2x^2 - x + 1)\frac{1}{x^2}}{(3x^2 + x - 7)\frac{1}{x^2}} = \frac{2 - \frac{1}{x} + \frac{1}{x^2}}{3 + \frac{1}{x} - \frac{7}{x^2}}.$$

In this form both the numerator and denominator have finite limits. Therefore,

$$\lim_{x\to\infty} \frac{2x^2 - x + 1}{3x^2 + x - 7} = \lim_{x\to\infty} \frac{2 - \frac{1}{x} + \frac{1}{x^2}}{3 + \frac{1}{x} - \frac{7}{x^2}} = \frac{\lim_{x\to\infty}\left(2 - \frac{1}{x} + \frac{1}{x^2}\right)}{\lim_{x\to\infty}\left(3 + \frac{1}{x} - \frac{7}{x^2}\right)}$$

$$= \frac{2 - \lim_{x\to\infty}\frac{1}{x} + \lim_{x\to\infty}\frac{1}{x^2}}{3 + \lim_{x\to\infty}\frac{1}{x} - \lim_{x\to\infty}\frac{7}{x^2}} = \frac{2 - 0 + 0}{3 + 0 - 0} = \frac{2}{3}.$$

Example 2 Calculate $\lim_{x\to\infty} \frac{x + 1}{2x^2 - 3}$.

Solution Multiply numerator and denominator by $1/x^2$. Then

$$\frac{x + 1}{2x^2 - 3} = \frac{\frac{1}{x} + \frac{1}{x^2}}{2 - \frac{3}{x^2}}.$$

As $x \to \infty$, the numerator approaches 0 while the denominator approaches 2. Thus,

$$\lim_{x\to\infty} \frac{x + 1}{2x^2 - 3} = \lim_{x\to\infty} \frac{\frac{1}{x} + \frac{1}{x^2}}{2 - \frac{3}{x^2}} = \frac{0}{2} = 0.$$

Example 3 (*The Inhibited Growth Law*) We will show in Chapter 9 that under certain assumptions the size $P = P(t)$ of a population at time t is given by

$$P(t) = \frac{m}{1 + \frac{c}{e^{at}}}$$

where m, c, and a are positive constants which depend on the population and $e \approx 2.718$ (a positive constant that will be discussed in Chapter 5). The growth law is derived from the assumption that inhibiting factors tend to limit the size of the population when it gets too large.

Because a is positive and $e > 1$ it follows that $e^a > 1$ so that $e^{at} \to \infty$ as $t \to \infty$. Then $c/e^{at} \to 0$ as $t \to \infty$ and

$$\lim_{t\to\infty} P(t) = \lim_{t\to\infty} \frac{m}{1 + \dfrac{c}{e^{at}}} = \frac{m}{1+0} = m.$$

Thus, the population tends to stabilize at the value m after a long period of time.

Example 4 (*Accumulation of Capital*) Assume that a certain segment of a society has a tendency to accumulate capital. If $F(t)$ is the amount that has been accumulated at time t, then $\lim_{t\to\infty} F(t)$ describes the maximum amount that can be accumulated over a long period of time.

In theoretical studies of the distribution of capital it is customary to let $f(t)\, dt$ be the amount of new capital that this segment obtains between times t and $t + dt$. (The actual function f depends on the particular society being studied.) Then the amount accumulated at time T is

$$F(T) = F(0) + \int_0^T f(t)\, dt.$$

Thus, the maximum amount that can be accumulated over a very long period of time is

$$\lim_{T\to\infty} F(T) = F(0) + \lim_{T\to\infty} \int_0^T f(t)\, dt.$$

This type of limit will be studied in the next section.

The Limit $\lim_{x\to a} f(x) = \infty$. There is another type of infinite limit: If $f(x)$ gets large without bound as $x \to a$ we say

$$\lim_{x\to a} f(x) = \infty.$$

$\lim_{x\to a} f(x) = \infty$

As an example consider the function $f(x) = 1/x^2$ (Figure 4.18). As $x \to 0$ from either side, $f(x)$ gets large without bound so

$$\lim_{x\to 0} f(x) = \infty.$$

Geometrically, this means that if we draw a horizontal line through $(0, M)$ there exists a positive number δ such that, excepting the point $(a, f(a))$, the entire portion of the graph between $a - \delta$ and $a + \delta$ is above the horizontal line.

One-sided limits

A function will sometimes approach one limit as x approaches a through values greater than a and another limit as x approaches a through values less than a. For example, if $f(x) = 1/(x - 1)$, then (see Figure 4.19)

$$f(x) \to +\infty \qquad \text{as } x \to 1 \text{ through values greater than 1,}$$

and

$$f(x) \to -\infty \qquad \text{as } x \to 1 \text{ through values less than 1.}$$

We write

$$\lim_{x\to 1^+} f(x) = +\infty \quad \text{and} \quad \lim_{x\to 1^-} f(x) = -\infty$$

(read "the limit as x approaches 1 from the right of $f(x)$ is $+\infty$" and "the limit as x approaches 1 from the left of $f(x)$ is $-\infty$").

Observe that the graph of f has the line $x = a$ as a vertical asymptote if the limit as x approaches a from either side becomes infinite. (See Figures 4.18, 4.19, and 4.20.)

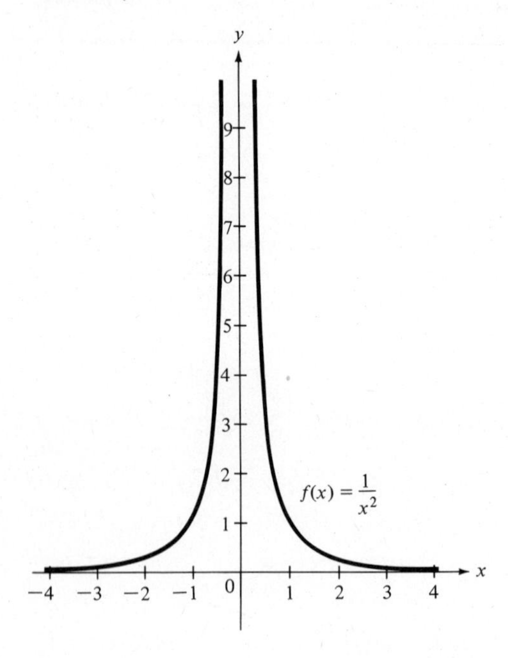

FIGURE 4.18 $\lim_{x \to 0} \frac{1}{x^2} = \infty$

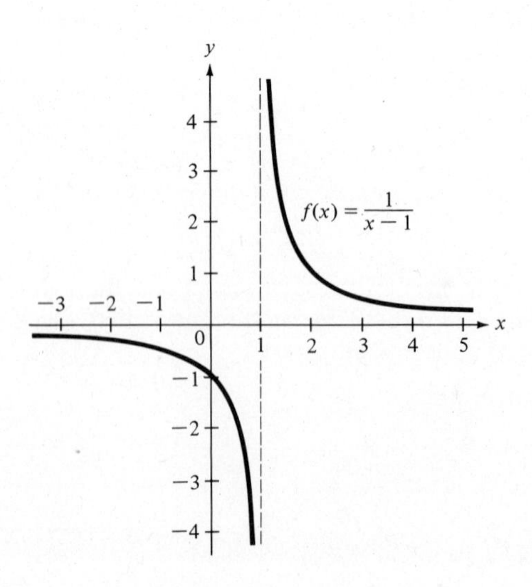

FIGURE 4.19

$\lim_{x \to 1^+} \frac{1}{x-1} = \infty,$

$\lim_{x \to 1^-} \frac{1}{x-1} = -\infty.$

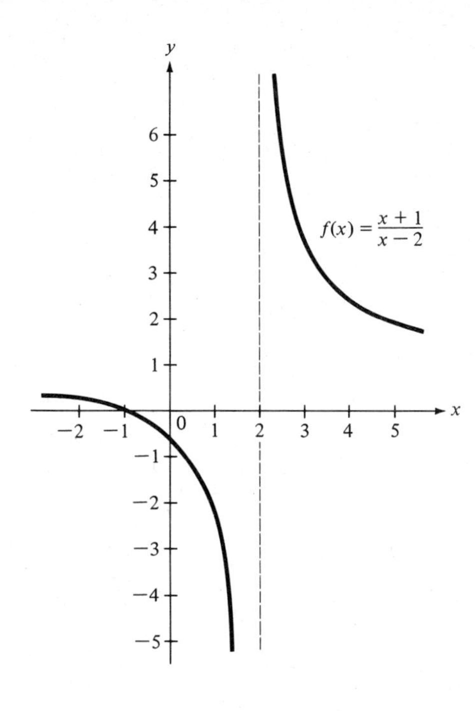

FIGURE 4.20 Example 5.

$$\lim_{x\to 2^+} \frac{x+1}{x-1} = \infty,$$

$$\lim_{x\to 2^-} \frac{x+1}{x-1} = -\infty.$$

Example 5 Calculate

$$\lim_{x\to 2^+} \frac{x+1}{x-2} \quad \text{and} \quad \lim_{x\to 2^-} \frac{x+1}{x-2}.$$

Solution

$$\lim_{x\to 2^+} (x+1) = \lim_{x\to 2^-} (x+1) = \lim_{x\to 2} (x+1) = 3.$$

As $x \to 2^+$, $(x - 2)$ approaches zero through positive values. Therefore, $(x + 1)/(x - 2)$ gets numerically large without bound through positive values as $x \to 2^+$, and so

$$\lim_{x\to 2^+} \frac{x+1}{x-2} = \infty.$$

On the other hand, $x - 2$ approaches zero through negative values as $x \to 2^-$. Therefore, $(x + 1)/(x - 2)$ gets numerically large without bound through negative values as $x \to 2^-$, and so

$$\lim_{x\to 2^-} \frac{x+1}{x-2} = -\infty.$$

(See Figure 4.20.)

EXERCISES 4.9

1. Evaluate the following limits:

• (a) $\lim_{x\to\infty} 2\sqrt{x}$

• (b) $\lim_{x\to\infty} \frac{4x^2 - 3x + 5}{6x^2 + 2x - 1}$

• (c) $\lim_{x\to\infty} \frac{4x^4 + 6x^3 - 8x^2 + 2x - 1}{x^3 + 7x^2 - 10x + 6}$

(d) $\lim_{x\to\infty} \frac{4x^3 - 6x^2 + 2x + 7}{2x^4 - 4x^2 + 3}$

• (e) $\lim_{x\to\infty} x^{-1/3}$

(f) $\lim_{x\to\infty} \frac{6x^2 - 3x + 1}{x^2 - 1}$

• (g) $\lim_{x\to\infty} \frac{3}{|x|}$

(h) $\lim_{x\to\infty} \frac{6 - 2x + 3x^2}{7 - x}$

2. Evaluate the following limits. Illustrate with a sketch.

(a) $\lim_{x\to -1} \frac{x^2 - 1}{x + 1}$

• (b) $\lim_{x\to 1^+} \frac{x + 1}{x^2 - 1}$

(c) $\lim_{x\to 3^+} \frac{x^2 + 9}{x^2 - 9}$, $\lim_{x\to 3^-} \frac{x^2 + 9}{x^2 - 9}$

• (d) $\lim_{x\to 1^-} \frac{(x + 1)(x - 2)}{x^2 - 1}$

(e) $\lim_{x\to 1^+} \frac{(x + 1)(x - 2)}{x^2 - 1}$

• (f) $\lim_{x\to 2} \frac{x^2 + 3x - 10}{x^2 - 4}$

4.10 IMPROPER INTEGRALS

An integral sometimes can be defined over an interval that is not finite in length. Such an integral is called an *improper integral.*

Suppose that f is continuous for all $x \geqslant a$. We can calculate $\int_a^b f(x)\,dx$ as a function of b for any $b \geqslant a$. We define

$\int_a^\infty f(x)\,dx$

$$\int_a^\infty f(x)\,dx = \lim_{b\to\infty} \int_a^b f(x)\,dx \qquad \text{provided a finite limit exists.}$$

(See Figure 4.21.) If a finite limit does not exist we say that the integral does not exist.

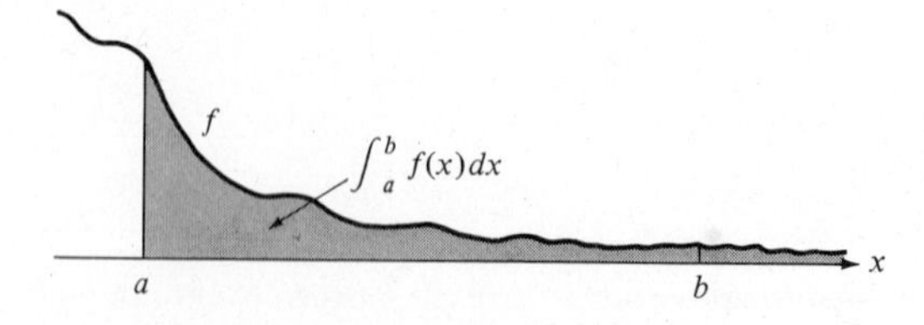

FIGURE 4.21 $\int_a^\infty f(x)\,dx = \lim_{b\to\infty} \int_a^b f(x)\,dx$, provided the limit exists.

Example 1 Calculate $\int_1^\infty (1/x^2)\,dx$.

Solution Let $b \geqslant 1$. Then

$$\int_1^b \frac{1}{x^2}\,dx = \int_1^b x^{-2}\,dx = \left[-\frac{1}{x}\right]_1^b = 1 - \frac{1}{b}.$$

$$\lim_{b\to\infty} \int_1^b \frac{1}{x^2}\,dx = \lim_{b\to\infty} \left(1 - \frac{1}{b}\right) = 1.$$

Therefore,

$$\int_1^\infty \frac{1}{x^2}\,dx = \lim_{b\to\infty}\int_1^b \frac{1}{x^2}\,dx = 1.$$

If $f(x) > 0$ for all $x > a$, then $\int_a^b f(x)\,dx$ represents the area under the graph of f for $a < x < b$. If $\lim_{b\to\infty} \int_a^b f(x)\,dx = L$, then the total area under the curve is finite.

Example 2 In Example 4 of Section 4.9 we considered the accumulation of capital by a certain segment of society. If $f(t)\,dt$ is the amount that is accumulated between times t and $t + dt$ then the total amount owned after a very large period of time is approximately equal to

$$F(0) + \lim_{T\to\infty}\int_0^T f(t)\,dt = F(0) + \int_0^\infty f(t)\,dt,$$

where $F(0)$ is the amount that had been accumulated at time $t = 0$.

If $f(t) > 0$ for all t, then the above expression gives an upper limit for the amount at any time. Since the society will only last a finite length of time, say until time $t = t_0$, then the maximum amount that can be accumulated is

$$F(0) + \int_0^{t_0} f(t)\,dt.$$

Unfortunately, it is not possible to predict the value of t_0 in advance. Thus, we use the expression

$$F(0) + \int_0^\infty f(t)\,dt$$

as a convenient approximation if t_0 is large.

Other improper integrals can be defined similarly. For example, if f is continuous for all $x \leqslant b$, we define

$$\int_{-\infty}^b f(x)\,dx = \lim_{a\to-\infty}\int_a^b f(x)\,dx$$

$\int_{-\infty}^b$ f(x) dx

provided a finite limit exists.

If f is continuous for all x, we define

$$\int_{-\infty}^\infty f(x)\,dx = \int_{-\infty}^0 f(x)\,dx + \int_0^\infty f(x)\,dx$$

$\int_{-\infty}^\infty$ f(x) dx

provided both of the integrals on the right-hand side of the equation exist. (See Figure 4.22.) There is nothing special about zero in this definition; we could have used any real number a rather than zero.

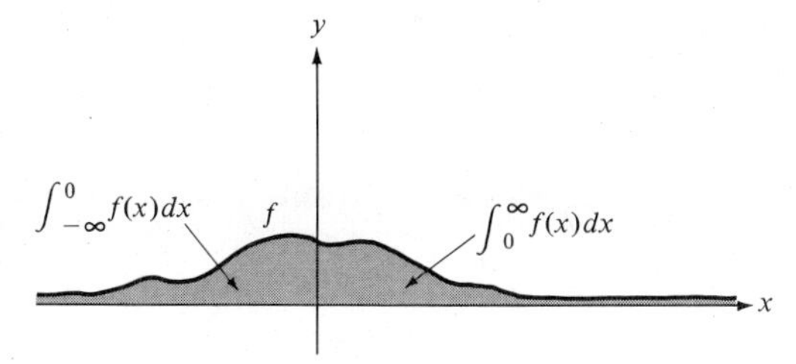

FIGURE 4.22 $\int_{-\infty}^\infty f(x)\,dx = \int_{-\infty}^0 f(x) \int_0^\infty f(x)\,dx$, provided both of the integrals on the right-hand side of the equation exist.

Example 3 Show that $\int_{-\infty}^{\infty} x\,dx$ does not exist.

Solution $\int_{-\infty}^{\infty} x\,dx = \int_{-\infty}^{0} x\,dx + \int_{0}^{\infty} x\,dx$ provided both integrals on the right exist. The integral $\int_{-\infty}^{0} x\,dx = \lim_{a\to-\infty} \int_{a}^{0} x\,dx$ provided the limit is finite. We compute

$$\lim_{a\to-\infty} \int_a^0 x\,dx = \lim_{a\to-\infty} \left[\frac{x^2}{2}\right]_a^0 = \lim_{a\to-\infty} \left[\frac{0^2 - a^2}{2}\right] = \lim_{a\to-\infty} \left[-\frac{a^2}{2}\right] = -\infty.$$

Since the limit is infinite the integral $\int_{-\infty}^{0} x\,dx$ does not exist. Therefore, $\int_{-\infty}^{\infty} x\,dx$ does not exist.

There is one other type of improper integral. It may be possible to define $\int_a^b f(x)\,dx$ when f is not continuous on $[a, b]$. We first consider the simplest case—the case where f is continuous on $[a, b]$ except at the left-hand endpoint a. If m is any number such that $a < m \leq b$, then $\int_m^b f(x)\,dx$ exists (owing to the fact that f is continuous on $[m, b]$). We consider this integral to be function of m and take the limit as m approaches a through values greater than a. (See Figure 4.23.) If a finite limit exists, we define $\int_a^b f(x)\,dx$ to be this limit:

$\int_a^b f(x)\,dx$, f discontinuous at a

$$\int_a^b f(x)\,dx = \lim_{m\to a^+} \int_m^b f(x)\,dx \qquad \text{provided a finite limit exists.}$$

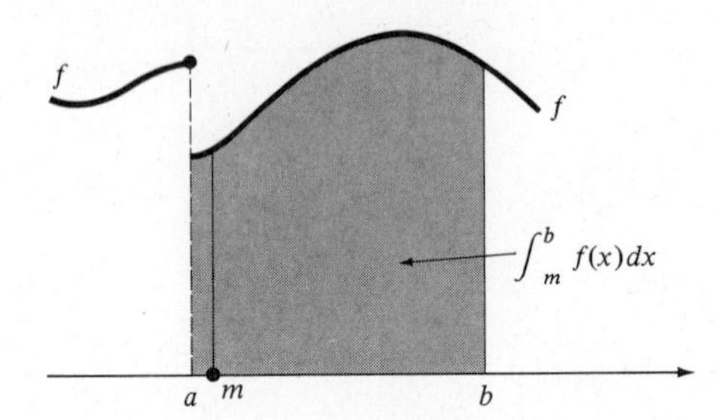

FIGURE 4.23 If f is discontinuous at a then $\int_a^b f(x)\,dx = \lim_{m\to a^+} \int_m^b f(x)\,dx$, provided the limit exists.

Example 4 Calculate $\int_0^1 \frac{1}{\sqrt{x}}\,dx$.

Solution If $0 < m \leq 1$, then $1/\sqrt{x}$ is continuous on $[m, 1]$. Then

$$\int_m^1 \frac{1}{\sqrt{x}}\,dx = \int_m^1 x^{-1/2}\,dx = \left[\frac{x^{1/2}}{\frac{1}{2}}\right]_m^1 = [2\sqrt{x}]_m^1$$
$$= 2\sqrt{1} - 2\sqrt{m} = 2 - 2\sqrt{m},$$

$$\int_a^b \frac{1}{\sqrt{x}}\,dx = \lim_{m\to0^+} \int_m^1 \frac{1}{\sqrt{x}}\,dx = \lim_{m\to0^+} [2 - 2\sqrt{m}\,] = 2 - 2\cdot 0 = 2.$$

If f is continuous on $[a, b]$ except at b, we define

$\int_a^b f(x)\,dx$, f discontinuous at b

$$\int_a^b f(x)\,dx = \lim_{n\to b^-} \int_a^n f(x)\,dx \qquad \text{provided a finite limit exists.}$$

(See Figure 4.24.)

If f is continuous on $[a, b]$ except at the two endpoints, we define

$$\int_a^b f(x)\,dx = \lim_{\substack{m\to a^+\\ n\to b^-}} \int_m^n f(x)\,dx$$

$$= \lim_{m\to a^+}\left(\lim_{n\to b^-}\int_m^n f(x)\,dx\right) \qquad \text{provided a finite limit exists.}$$

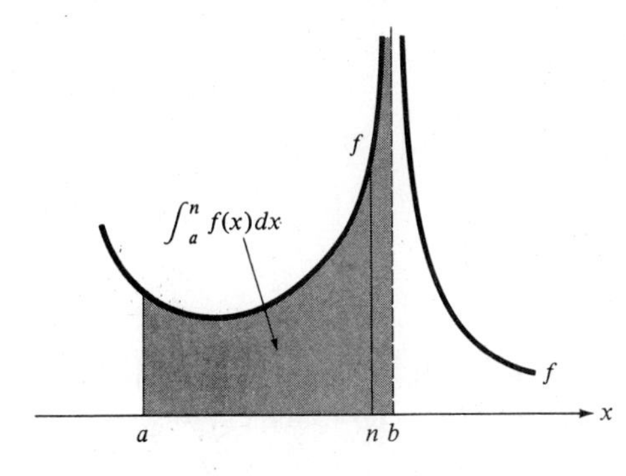

FIGURE 4.24 If f is discontinuous at b then $\int_a^b f(x)\,dx = \lim_{n\to b^-}\int_a^n f(x)\,dx$, provided the limit exists.

If f is continuous on $[a, b]$ except at a finite number of points, say at the points $c_1, c_2, \ldots, c_n$, where $a \leqslant c_1 < c_2 < \cdots < c_n \leqslant b$, we define

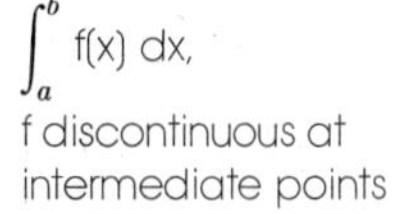

$\int_a^b$ f(x) dx,
f discontinuous at intermediate points

$$\int_a^b f(x)\,dx = \int_a^{c_1} f(x)\,dx + \int_{c_1}^{c_2} f(x)\,dx + \cdots + \int_{c_n}^b f(x)\,dx$$

provided all the integrals on the right-hand side of the equation exist. (See Figure 4.25.)

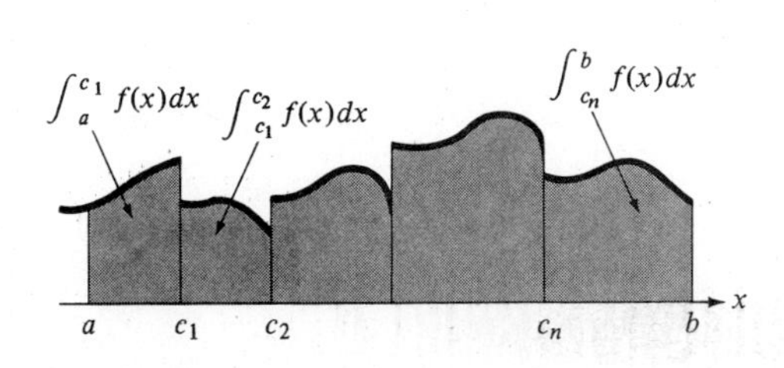

FIGURE 4.25 $\int_a^b f(x)\,dx = \int_a^{c_1} f(x)\,dx + \int_{c_1}^{c_2} f(x)\,dx + \cdots + \int_{c_n}^b f(x)\,dx$, provided all of the integrals on the right-hand side of the equation exist.

Example 5 Calculate $\int_0^3 [x]\,dx$.

Solution [x] is the greatest integer not greater than x (Section 2.5). This function is continuous everywhere except at the integers. Therefore,

$$\int_0^3 [x]\,dx = \int_0^1 [x]\,dx + \int_1^2 [x]\,dx + \int_2^3 [x]\,dx$$

$$= \lim_{\substack{m\to 0^+\\ m\to 1^-}} \int_m^n [x]\,dx + \lim_{\substack{m\to 1^+\\ n\to 2^-}} \int_m^n [x]\,dx + \lim_{\substack{m\to 2^+\\ n\to 3^-}} \int_m^n [x]\,dx$$

$$= \lim_{\substack{m\to 0^+\\ m\to 1^-}} \int_m^n 0\cdot dx + \lim_{\substack{m\to 1^+\\ n\to 2^-}} \int_m^n 1\cdot dx + \lim_{\substack{m\to 2^+\\ n\to 3^-}} \int_m^n 2\cdot dx$$

$$= \lim_{\substack{m\to 0^+\\ n\to 1^-}} [0\cdot(n-m)] + \lim_{\substack{m\to 1^+\\ n\to 2^-}} [1\cdot(n-m)] + \lim_{\substack{m\to 2^+\\ n\to 3^-}} [2\cdot(n-m)],$$

$$= 0\cdot(1-0) + 1\cdot(2-1) + 2\cdot(3-2) = 3.$$

(See Figure 4.26.)

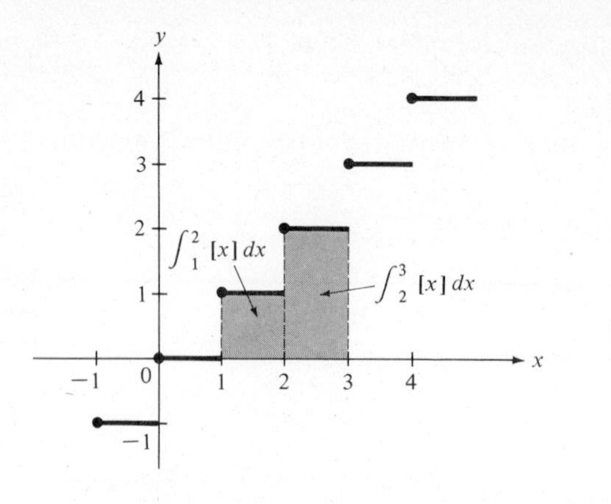

FIGURE 4.26 Example 5. $\int_0^3 [x]\,dx = \int_0^1 [x]\,dx + \int_1^2 [x]\,dx + \int_2^3 [x]\,dx = 0 + 1 + 2 = 3.$

The two basic types of improper integrals can be combined in many ways. For example, if f is continuous for all $x > a$ but is discontinuous at a, we define

$$\int_a^\infty f(x)\,dx = \lim_{n\to\infty}\left(\lim_{m\to a^+}\int_m^n f(x)\,dx\right),$$

provided both limits exist and are finite. If f is continuous for all x except c_1, $c_2, \ldots, c_n$, where $c_1 < c_2 \cdots < c_n$, we define

$$\int_{-\infty}^\infty f(x)\,dx = \int_{-\infty}^{c_1} f(x)\,dx + \int_{c_1}^{c_2} f(x)\,dx + \cdots + \int_{c_n}^\infty f(x)\,dx,$$

provided all of the integrals on the right-hand side of the equation exist. (See Figure 4.27.)

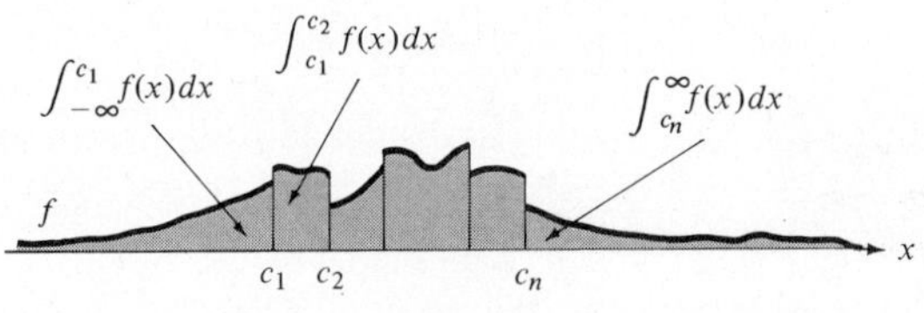

FIGURE 4.27 $\int_{-\infty}^\infty f(x)\,dx = \int_{-\infty}^{c_1} f(x)\,dx + \int_{c_1}^{c_2} f(x)\,dx + \cdots + \int_{c_n}^\infty f(x)\,dx$, provided all of the integrals on the right-hand side of the equation exist.

Example 6 Let $f(x) = \begin{cases} 1/\sqrt{x} & \text{if } 0 < x \le 1 \\ 2/x^2 & \text{if } x > 1. \end{cases}$

Calculate $\int_0^\infty f(x)\,dx$.

Solution

$$\int_0^\infty f(x)\,dx = \int_0^1 f(x)\,dx + \int_1^\infty f(x)\,dx \quad \text{(see Figure 4.28)}$$

$$= \int_0^1 \frac{1}{\sqrt{x}}\,dx + 2\int_1^\infty \frac{1}{x^2}\,dx.$$

These integrals can be evaluated (see Examples 1 and 4) as

$$\int_0^1 \frac{1}{\sqrt{x}}\,dx = 2, \qquad \int_1^\infty \frac{1}{x^2}\,dx = 1$$

so that

$$\int_0^\infty f(x)\,dx = 2 + 2 \cdot 1 = 4.$$

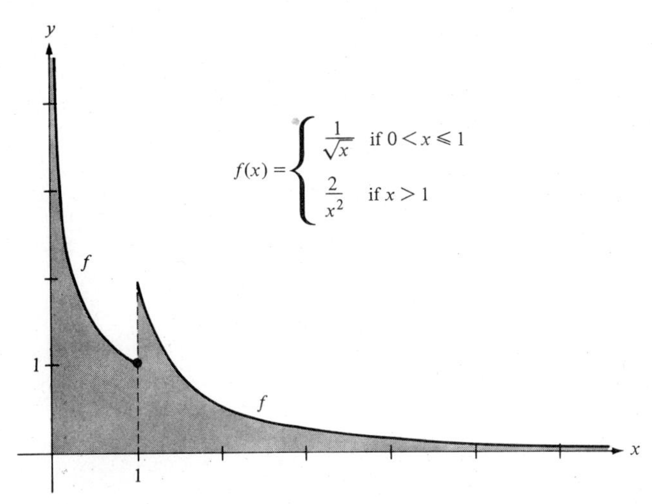

FIGURE 4.28 Example 6. $\int_0^\infty f(x)\,dx = \int_0^1 \frac{dx}{\sqrt{x}} + \int_1^\infty \frac{2}{x^2}\,dx.$

EXERCISES 4.10

1. Examine the following integrals to see if they exist. If one fails to exist, explain why. If it exists, calculate its value.

   (a) $\int_0^1 x^{-1/2}\,dx$

   • (b) $\int_4^\infty x^{-1/2}\,dx$

   • (c) $\int_{-1}^1 x^{-1/3}\,dx$

   (d) $\int_1^\infty (x^2+2)^{-2}\,x\,dx$

   (e) $\int_0^3 \frac{x\,dx}{(x^2-1)^{1/3}}$

   • (f) $\int_2^5 \frac{x\,dx}{(9-x^2)^2}$

   (g) $\int_0^\infty \frac{x\,dx}{(x^2+1)^3}$

   • (h) $\int_{-\infty}^\infty 2x(x^2+1)^{-3/2}\,dx$

2. Show that there is no value of $w > 1$ such that

$$\int_1^w x^{-2}\,dx = 3.$$

3. Students frequently wish to define $\int_{-\infty}^\infty f(x)\,dx$ by the rule

$$\text{“}\int_{-\infty}^\infty f(x)\,dx = \lim_{n\to\infty} \int_{-n}^n f(x)\,dx.\text{”}$$

This exercise will demonstrate that the "students definition" is not equivalent to the definition given in the book.

(a) Show that if $\int_{-\infty}^{\infty} f(x)\,dx$ exists, then

$$\lim_{n\to\infty} \int_{-n}^{n} f(x)\,dx = \int_{-\infty}^{\infty} f(x)\,dx.$$

(b) By an example show that $\lim_{n\to\infty} \int_{-n}^{n} f(x)\,dx$ may exist even though $\int_{-\infty}^{\infty} f(x)\,dx$ fails to exist. (*Hint:* Consider the function defined by $f(x) = 4x^3$.)

4. Calculate the areas of the regions bounded by the following curves and lines. Make a sketch of each region.

- (a) $y = [x]x$ $(1 \leq x \leq 3)$, the x-axis (greatest integer function)

(b) $y = 1/x^2$ $(2 \leq x)$, the x-axis

- (c) $y = 1/\sqrt{x}$ $(0 \leq x \leq 4)$, the x-axis

(d) $y = \dfrac{[x]}{x^2}$ $(1 \leq x \leq 3)$, the x-axis (greatest integer function)

- (e) $y = x - [x]$ $(0 \leq x \leq 2)$, the x-axis (greatest integer function)

(f) Region bounded above by $y = 1$, bounded below by $y = 1/x^2$ $(x > 0)$.
(g) Region bounded above by $y = 1/x^2$, bounded below by $y = x^2$ $(x > 0)$.

4.11 PROBABILITY DENSITY FUNCTIONS

Roughly speaking, the probability that an event will have a particular outcome is equal to the likelihood of that outcome occuring. For example, if we toss a fair coin, then the probability of *heads* is $\frac{1}{2}$, indicating that approximately one half of the tosses will be *heads* and one half will be *tails*. This type of result can be established by an elementary argument concerning events with a finite number of possible outcomes, all of which are equally likely. The probability of a particular type of outcome is

Probability of an event

$$p = \frac{\text{number of favorable outcomes}}{\text{number of possible outcomes}}.$$

(The word "favorable" refers to outcomes that yield the result in which we are interested.)

In our coin-tossing example there are two possible outcomes, equally likely to occur. One is *heads*, the other is *tails*. Thus, the probability of *heads* is

$$p = \frac{\text{number of favorable outcomes}}{\text{number of possible outcomes}} = \frac{1}{2}.$$

Many probability problems involve an infinite number of possible outcomes. For example, if we throw a dart at a dartboard there is an infinity of possible locations at which the dart can strike the board. (Here we assume that the tip of the dart has no thickness.) Obviously, the probability that any given point will be struck is zero. We might ask, however, for the probability that the dart will strike the board within a certain distance of the center of the bullseye. Such problems are best handled by probability density functions.

Probability density function

A *probability density function* is an integrable function f defined on the entire x-axis such that

(1) $f(x) \geqslant 0$ for all x,

(2) $\int_{-\infty}^{\infty} f(x)\,dx = 1.$

Probability of an event

If an event has outcomes which can be described by numerical values, then the *probability that the outcome will be in the range* $a \leqslant x \leqslant b$ *is*

$$p = \int_a^b f(x)\,dx.$$

(See Figure 4.29.)

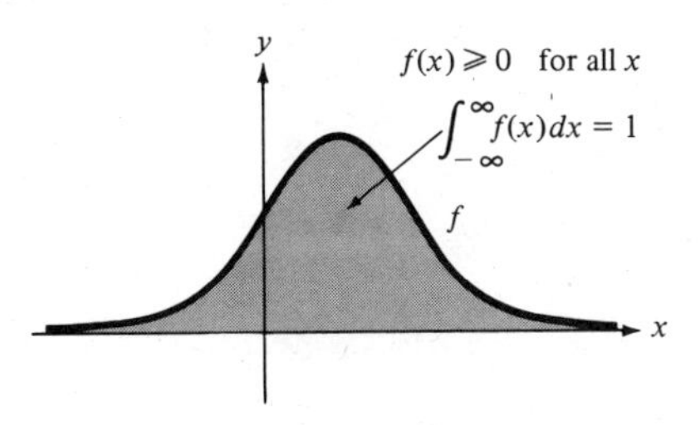

(a) A typical probability density function.

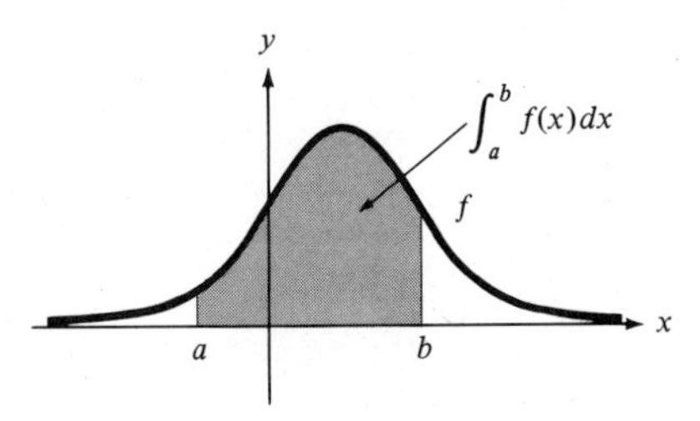

(b) The probability that x is between a and b is $p = \int_a^b f(x)\,dx$.

FIGURE 4.29 Probability density functions have the properties that:
(1) $f(x) \geqslant 0$ for all x
(2) $\int_{-\infty}^{\infty} f(x)\,dx = 1$

The particular probability density function is determined by the problem under consideration. One type of problem will involve a probability density function of a particular type, another type of problem will involve a completely different type of density function. Much of the work in a theoretical course in statistics involves determining probability density functions for particular types of problems.

In our work in this section we give the probability density functions for certain problems without proof.

Example 1 An event consists of randomly choosing a point on the closed interval [1, 4]. The probability density function is

$$f(x) = \begin{cases} \frac{1}{3} & \text{if } 1 \leqslant x \leqslant 4 \\ 0 & \text{if } x < 1 \text{ or } x > 4. \end{cases}$$

(See Figure 4.30.)

(a) Verify that f is a probability density function.

(b) Calculate the probability that the selected point is between 3.0 and 3.2 (inclusive).

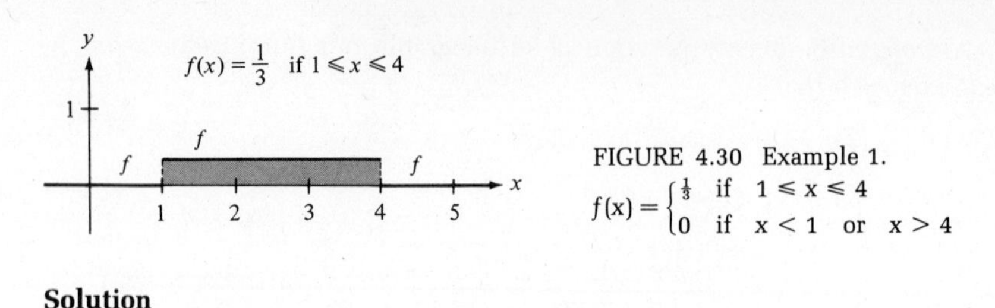

FIGURE 4.30 Example 1.
$$f(x) = \begin{cases} \frac{1}{3} & \text{if } 1 \leqslant x \leqslant 4 \\ 0 & \text{if } x < 1 \text{ or } x > 4 \end{cases}$$

Solution

(a) We must show that $f(x) \geqslant 0$ for all x and that

$$\int_{-\infty}^{\infty} f(x)\,dx = 1.$$

The first fact follows directly from the definition of f. To calculate the integral we note that since $f(x) = 0$ if x is not on the interval [1, 4], then

$$\int_{-\infty}^{\infty} f(x)\,dx = \int_{1}^{4} f(x)\,dx = \int_{1}^{4} \frac{1}{3}\,dx = \frac{1}{3}\,[x]_1^4 = \frac{1}{3}\,[4-1]$$
$$= \frac{1}{3} \cdot 3 = 1.$$

Since both of the necessary conditions are met, then f is a probability density function.

(b) The probability that the point will be between 3.0 and 3.2 is

$$p = \int_{3.0}^{3.2} f(x)\,dx = \int_{3.0}^{3.2} \frac{1}{3}\,dx = \frac{1}{3}\,[x]_{3.0}^{3.2} = \frac{1}{3}\,[3.2 - 3.0]$$
$$= \tfrac{1}{3}[.2] = \tfrac{1}{3} \cdot \tfrac{1}{5} = \tfrac{1}{15}.$$

Thus, on the average, we expect a point to be chosen on the interval [3.0, 3.2] once out of every 15 times.

The following example is typical of the problems solved by this method.

Example 2 The Broken China Company manufactures 1000 china plates each day. The probability density function for the proportion of plates that pass inspection is

$$f(x) = \begin{cases} 13 \cdot 14x^{12}\,(1-x) & \text{if } 0 \leqslant x \leqslant 1 \\ 0 & \text{if } x < 0 \text{ or } x > 1. \end{cases}$$

(a) Verify that f is a probability density function.

(b) Find the probability that between 500 and 800 plates will pass inspection on a given day.

Solution

(a) If x is between 0 and 1, then x^{12} and $1 - x$ both are nonnegative, so that

$$f(x) = 13 \cdot 14x^{12}\,(1-x) \geqslant 0.$$

By definition $f(x) \geqslant 0$ (since $f(x) = 0$) for any other value of x. Thus, $f(x) \geqslant 0$ for all x.

To calculate $\int_{-\infty}^{\infty} f(x)\,dx$ we observe that since $f(x) = 0$ when $x < 0$ or $x > 1$, then

$$\int_{-\infty}^{\infty} f(x)\,dx = \int_0^1 f(x)\,dx = \int_0^1 13 \cdot 14x^{12}\,(1-x)\,dx$$

$$= 13 \cdot 14 \int_0^1 (x^{12} - x^{13})\,dx = 13 \cdot 14 \left[\frac{x^{13}}{13} - \frac{x^{14}}{14}\right]_0^1$$

$$= 13 \cdot 14 \left[\frac{1}{13} - \frac{1}{14}\right] - 13 \cdot 14 \cdot 0$$

$$= 13 \cdot 14 \left[\frac{14-13}{13 \cdot 14}\right] = 13 \cdot 14 \cdot \frac{1}{13 \cdot 14} = 1.$$

Thus f is a probability density function.

(b) If between 500 and 800 plates pass inspection on a given day then x, the *proportion* of plates that pass, is between 0.5 and 0.8. The probability that x is within this range is

$$p = \int_{0.5}^{0.8} f(x)\,dx = \int_{0.5}^{0.8} 13 \cdot 14x^{12}\,(1-x)\,dx$$

$$= 13 \cdot 14 \int_{0.5}^{0.8} (x^{12} - x^{13})\,dx = 13 \cdot 14 \left[\frac{x^{13}}{13} - \frac{x^{14}}{14}\right]_{0.5}^{0.8}$$

$$= 13 \cdot 14 \left[\frac{14x^{13} - 13x^{14}}{13 \cdot 14}\right]_{0.5}^{0.8} = [14x^{13} - 13x^{14}]_{0.5}^{0.8}$$

$$= [14(0.8)^{13} - 13(0.8)^{14}] - [14(0.5)^{13} - 13(0.5)^{14}]$$

$$\approx 0.615.$$

EXERCISES 4.11

1. Verify that f is a probability density function; then calculate the probability that x is between 1 and 2.

(a) $f(x) = \begin{cases} \frac{1}{2} & \text{if } 0 \leqslant x \leqslant 2 \\ 0 & \text{if } x < 0 \text{ or } x > 2 \end{cases}$

(c) $f(x) = \begin{cases} 0 & \text{if } x < 0 \\ \dfrac{3\sqrt{x}}{16} & \text{if } 0 \leqslant x \leqslant 4 \\ 0 & \text{if } x > 4 \end{cases}$

• (b) $f(x) = \begin{cases} 0 & \text{if } x < 1 \\ 1/x^2 & \text{if } x \geqslant 1 \end{cases}$

• (d) $f(x) = \begin{cases} 0 & \text{if } x < 0 \\ \frac{1}{2}x & \text{if } 0 \leqslant x \leqslant 2 \\ 0 & \text{if } x > 2 \end{cases}$

2. (See Example 2.)
   (a) Find the probability that at least 60 percent of the plates manufactured in a single day pass inspection.
   • (b) Find the probability that no more than 60 percent of the plates manufactured in a single day pass inspection.

•3. Professor Crankcase attempts to arrive at his college between 9:20 and 9:30 each morning so that he can walk directly into his 9:30 class. Because of varying traffic conditions the actual time of arrival is between 9:10 and 9:40

and is described by the following probability density function (where t is the number of minutes after 9:00):

$$f(t) = \begin{cases} 0 & \text{if } t < 10 \text{ or } t > 40 \\ \dfrac{1}{15^2}(t-10) & \text{if } 10 \leq t \leq 25 \\ \dfrac{-1}{15^2}(t-40) & \text{if } 25 \leq t \leq 40 \end{cases}$$

Find the probability that the professor arrives between 9:20 and 9:30 on a given morning.

LOGARITHMIC AND EXPONENTIAL FUNCTIONS

5.1 EXPONENTS AND LOGARITHMS

In Section A.5 of the Appendix we define b^n for each positive integer n by

$$b^n = b \cdot b \cdot b \cdot \cdot \cdot b \qquad (n \text{ factors}).$$

In addition, if $b \neq 0$, we define

$$b^0 = 1 \qquad (\textit{zero exponent})$$

and

$$b^{-n} = \frac{1}{b^n} \qquad (\textit{negative exponent}).$$

With these definitions in mind the following laws of exponents can be established:

Laws of exponents

E1 $$b^m \cdot b^n = b^{m+n},$$

E2 $$\frac{b^m}{b^n} = b^{m-n},$$

E3 $$(b^m)^n = b^{mn},$$

E4 $$(ab)^n = a^n b^n.$$

Fractional exponents $b^{m/n}$ are defined as roots (for $n > 0$):

$$b^{1/n} = \sqrt[n]{b} \qquad \text{(the principal } n\text{th root of } b\text{)}$$

and

$$b^{m/n} = (\sqrt[n]{b})^m.$$

Fractional exponents

These fractional exponents are defined provided b is greater than or equal to zero when n is even.

In advanced courses in mathematics it is shown that if $b > 0$, then b^x can be defined for every real number x. Thus, such expressions as

$$3^{\sqrt{2}}, \ 5^{\pi}, \quad \text{and} \quad (\sqrt{13})^{\sqrt{17}}$$

are meaningful. These numbers can be defined as limits. For example,

$$3^{\sqrt{2}} = \lim_{x \to \sqrt{2}} 3^x.$$

It can be proved that the rules of exponents E1 to E4 hold when the exponents are any real numbers. Thus, it makes sense to define a function f by

$$f(x) = b^x,$$

where b is a fixed positive number. It can be shown that this function f is continuous and differentiable for all x. The graph of $f(x) = b^x$ (for $b > 1$) is shown in Figure 5.1.

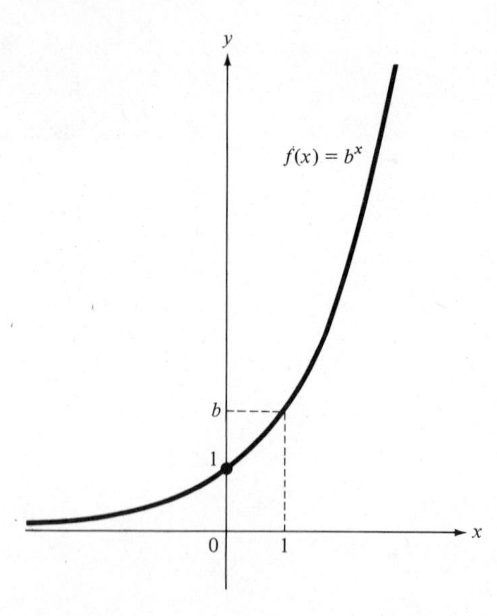

FIGURE 5.1 The graph of $f(x) = b^x$ $(b > 0)$.

Logarithms Let $b > 1$. The exponential function $f(x) = b^x$ has the following properties:

Properties of the exponential function

1. f is continuous and differentiable.
2. f is an increasing function (indicating a positive derivative).
3. $\lim_{x\to\infty} b^x = \infty$ and $\lim_{x\to-\infty} b^x = 0$.

It follows from these properties that if we draw a horizontal line through any point $(0, M)$ on the positive part of the y-axis, this line will intersect the

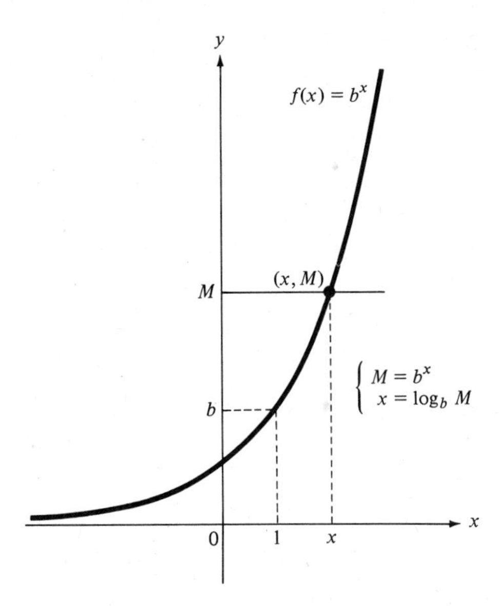

FIGURE 5.2 $x = \log_b M$ if and only if $M = b^x$.

graph of $f(x) = b^x$ in exactly one point (x, M). (This is pictured in Figure 5.2.) In other words, if $M > 0$, there is a unique number x such that $M = b^x$. This number x is called the *logarithm of M to the base b*. It is denoted by $x = \log_b M$. That is,

Logarithm

$$x = \log_b M \qquad \text{if and only if} \qquad M = b^x.$$

For example, since $1000 = 10^3$, then $\log_{10} 1000 = 3$.

Since logarithms are nothing more than exponents written without the base, then the rules of exponents can be applied to them. For example, the rule

$$b^x \cdot b^y = b^{x+y}$$

can be used to show that

$$\log_b (MN) = \log_b M + \log_b N.$$

Other rules follow similarly. The three most important rules for logarithms are stated in the following theorem.

Properties of logarithms

Theorem 5.1 Let M and N be positive numbers, $b > 1$. Then

(a) $\log_b MN = \log_b M + \log_b N$.

(b) $\log_b \left(\dfrac{M}{N}\right) = \log_b M - \log_b N$.

(c) $\log_b M^k = k \log_b M$ for any real number k.

PROOF OF (a) Let $\log_b M = m$ and $\log_b N = n$. Then $M = b^m$, $N = b^n$, and

$$MN = b^m \cdot b^n = b^{m+n}.$$

But $\log_b (MN)$ is the only real number x such that $MN = b^x$. Therefore, since $MN = b^{m+n}$, it follows that

$$\log_b MN = m + n = \log_b M + \log_b N.$$

Similar arguments can be used to establish (b) and (c).

Example 1 Table IV at the end of this book is a table of logarithms to the base e ($e \approx 2.71828$). Use Table IV and Theorem 5.1 to calculate the following logarithms:

(a) $\log_e (8.31)(7.62)$

(b) $\log_e \dfrac{8.31}{7.62}$

(c) $\log_e (5.50)^3$

Solution

(a)
$$\log_e ((8.31)(7.62) = \log_e 8.31 + \log_e 7.62 = 2.1175 + 2.0308 = 4.1483.$$

(b)
$$\log_e \left(\frac{8.31}{7.62}\right) = \log_e 8.31 - \log_e 7.62 = 2.1175 - 2.0308 = 0.0867.$$

(c)
$$\log_e (5.50)^3 = 3 \log_e (5.5) = 3 \cdot (1.7047) = 5.1141.$$

Example 2 Use Table IV and Example 1 to calculate the following numbers:

(a) (8.31)(7.62)

(b) 8.31/7.62

(c) $(5.50)^3$

Solution

(a)
$$\log_e(8.31)(7.62) = 4.1483 \qquad \text{(from Example 1)},$$

which is approximately equal to $\log_e 63.3$. Thus

$$\log_e(8.31)(7.62) \approx \log_e (63.3),$$

so that

$$(8.31)(7.62) \approx 63.3.$$

(b)
$$\log_e \left(\frac{8.31}{7.62}\right) = 0.0867 \qquad \text{(from Example 1)}$$
$$\approx \log_e (1.09) \qquad \text{(by Table IV)}.$$

Therefore,

$$\frac{8.31}{7.62} \approx 1.09.$$

(c)
$$\log_e (5.50)^3 = 5.1141 \qquad \text{(from Example 1)}.$$

This number is larger than any contained in Table IV. Observe, however, that $\log_e 100 = 4.6052$. Thus,

$$\log_e (5.50)^3 = 5.1141 = 4.6052 + 0.5089 = \log_e 100 + 0.5089.$$

Now 0.5089 is approximately equal to $\log_e 1.66$. Thus,

$$\begin{aligned}\log_e (5.50)^3 &= \log_e 100 + 0.5089\\ &\approx \log_e 100 + \log_e 1.66\\ &\approx \log_e (100)(1.66) \quad \text{(by Theorem 5.1a)}\\ &\approx \log_e 166.\end{aligned}$$

Since $\log_e (5.50)^3 \approx \log_e 166$, then

$$(5.50)^3 \approx 166.$$

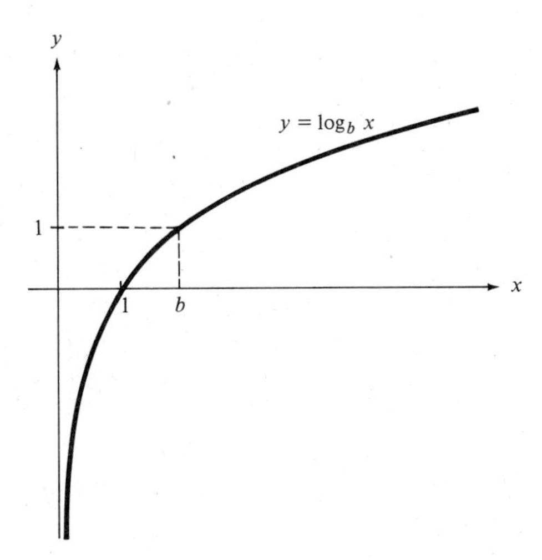

FIGURE 5.3 Graph of the logarithmic function $y = \log_b x$ $(b > 0)$.

The graph of $L(x) = \log_b x$ $(b > 1)$ is shown in Figure 5.3. The domain of this function is the set of all positive real numbers. The range is the set of all real numbers. Observe that

$$\log_b 1 = 0 \quad \text{and} \quad \log_b b = 1.$$

The function L is continuous and differentiable. It is an increasing function (indicating a positive derivative).

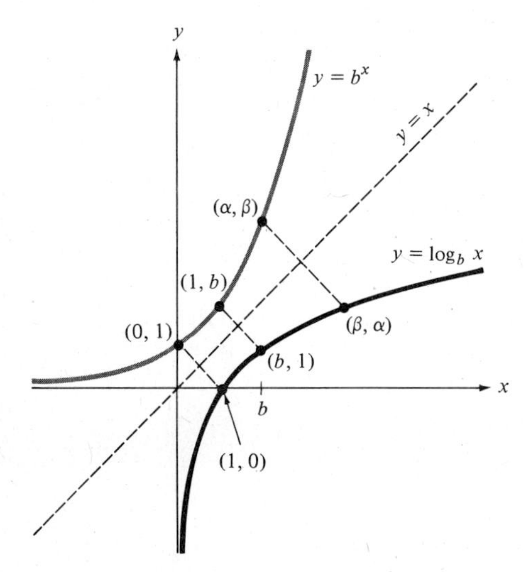

FIGURE 5.4 The graph of $y = \log_b x$ is the reflection of the graph of $y = b^x$ in the line $y = x$.

Exponential and logarithmic functions

There is a remarkable similarity between the graphs of $y = b^x$ and $y = \log_b x$. This can be seen if the two graphs are plotted on the same axis system (Figure 5.4). Observe that one is the reflection of the other though the 45° line $y = x$. This is equivalent to the fact that the point (α, β) is on the graph of $y = b^x$ if and only if (β, α) is on the graph of $y = \log_b x$. But this follows from the fact that (α, β) is on the graph of $y = b^x$ if and only if $\beta = b^\alpha$, which is true if and only if $\log_b \beta = \alpha$, which is true if and only if (β, α) is on the graph of $y = \log_b x$.

EXERCISES 5.1

1. Use the rules of exponents to simplify the following expressions:

   (a) $5^{1/3} \cdot 5^{1/2}$
   • (b) $(7^{1/3})^{1/2}$
   (c) $(3^{-1/2} \cdot 3^{2/3})^6$
   (d) $(5^{-1/2} + 5^{1/2})^2$
   • (e) $3^a \cdot 3^b/3^c$
   (f) $(5^{1/6} \cdot 10^{1/2})^0$

2. Use the rules of logarithms to calculate the following logarithms. A table of logarithms to the base e ($e \approx 2.71828$) can be found in Table IV.

   (a) $\log_3 (3^2 \cdot 3^{2/3})$
   • (b) $\log_5 (5^{2/3}/5^{1/2})$
   • (c) $\log_{10} (\sqrt[3]{10^{12}})$
   (d) $\log_e (5^2 \cdot 7^{1/2})$
   • (e) $\log_e (3^{12} \cdot 5^{14}/8^{25})$
   (f) $\log_e (7^2 \cdot 5^{-3})^6$

3. Use logarithms from Table IV to calculate the following numbers. (In (e) write $171 = (100)(1.71)$.)

   • (a) $(5.32)^2$
   • (b) $\sqrt[7]{95.5}$
   (c) $(1.42)(9.67)$
   • (d) $85.6/23.1$
   • (e) $\sqrt{171}$
   (f) $(63.5)(92.4)$

4. Use the relationship $\log_b y = x$ if and only if $y = b^x$ to write the following equations in a different form:

   (a) $\log_e 35.2 = 3.5610$
   • (b) $\log_e (7.08) = 1.9573$
   (c) $10^2 = 100$
   • (d) $10^{-3} = 1/1000$

5. Use Table IV to make an accurate graph of $y = \log_e x$, $1/10 \leqslant x \leqslant 10$. Sketch tangent lines at several points on the graph by placing a ruler so that it just touches the graph. Verify that the slope of the tangent line at the point (x, y) is equal to $1/x$, the reciprocal of the x-coordinate of the point. (It will be shown later that $d(\log_e x)/dx = 1/x$.) (Use the rules of logarithms to calculate needed values of $\log_e x$ for $1/10 \leqslant x \leqslant 1$. For example, $\log_e 1/5 = \log_e 1 - \log_e 5 = -\log_e 5$.)

6. Use Table V at the end of the book to make an accurate sketch of $y = e^x$, $-2 \leqslant x \leqslant 2$. Sketch tangent lines at several points on the graph by placing a ruler so that it just touches the graph. Verify that the slope of the tangent line at the point (x, y) is equal to the y-coordinate of the point. (It will be shown that $d(e^x)/dx = e^x$.)

7. Complete the proof of Theorem 5.1 by proving that

   (a) $\log_b \dfrac{M}{N} = \log_b M - \log_b N$

   (b) $\log_b M^k = k \log_b M$

5.2 DERIVATIVES OF LOGARITHMIC FUNCTIONS. THE NUMBER e

Let $L(x) = \log_b x$. We wish to calculate $L'(a)$, the derivative of L at a. It can be shown that the function L is both differentiable and continuous.

We assume these facts without further proof and proceed to calculate the value of the derivative.

$$L'(a) = \lim_{\Delta x \to 0} \frac{L(a + \Delta x) - L(a)}{\Delta x} = \lim_{\Delta x \to 0} \frac{\log_b (a + \Delta x) - \log_b a}{\Delta x}$$

$$= \lim_{\Delta x \to 0} \frac{1}{\Delta x} \log_b \left(\frac{a + \Delta x}{a}\right) \qquad \text{(by Theorem 5.1)}$$

$$= \frac{1}{a} \cdot \lim_{\Delta x \to 0} \frac{a}{\Delta x} \cdot \log_b \left(\frac{a + \Delta x}{a}\right) \quad \left(\text{multiplying by } \frac{a}{a}\right)$$

$$= \frac{1}{a} \cdot \lim_{\Delta x \to 0} \log_b \left(1 + \frac{\Delta x}{a}\right)^{a/\Delta x} \qquad \text{(by Theorem 5.1).}$$

It is not at all obvious that this last limit exists. It is proved in more advanced courses, however, that

$$\lim_{\Delta x \to 0} \left(1 + \frac{\Delta x}{a}\right)^{a/\Delta x}$$

exists and is independent of the number a. This limit is equal to an irrational number between 2 and 3, called e. To be more exact

The number e

$$\lim_{\Delta x \to 0} \left(1 + \frac{\Delta x}{a}\right)^{a/\Delta x} = e \approx 2.71828.$$

Furthermore, since the logarithm function is continuous at e, then

$$\lim_{\Delta x \to 0} \left[\log_b \left(1 + \frac{\Delta x}{a}\right)^{a/\Delta x}\right] = \log_b \left[\lim_{\Delta x \to 0} \left(1 + \frac{\Delta x}{a}\right)^{a/\Delta x}\right]$$

$$= \log_b e.$$

Therefore

$$L'(a) = \frac{1}{a} \cdot \lim_{\Delta x \to 0} \left[\log_b \left(1 + \frac{\Delta x}{a}\right)^{a/\Delta x}\right]$$

$$= \frac{1}{a} \cdot \log_b e.$$

If we replace a by x, we have the following theorem:

Theorem 5.2 $\dfrac{d}{dx} (\log_b x) = \dfrac{1}{x} \cdot \log_b e.$

For example, $\log_{10} e = 0.43429$ (approximately). Therefore

$$\frac{d}{dx} (\log_{10} x) = \frac{0.43429}{x} \quad \text{for each } x > 0.$$

If we examine the derivative formula,

$$\frac{d}{dx}(\log_b x) = \frac{1}{x}\log_b e,$$

we see that it can be simplified if we choose e as the logarithm base. Since

$$\log_e e = 1,$$

we obtain the formula

$$\frac{d}{dx}(\log_e x) = \frac{1}{x}\log_e e = \frac{1}{x}.$$

Because of this formula and others which also are simplified, it is customary to use base e logarithms in all of higher mathematics.

Natural logarithm

Logarithms to the base e are called *natural logarithms*. They are denoted by the special symbol

$$\ln x.$$

That is,

$$\ln x = y \qquad \text{if and only if} \qquad x = e^y.$$

(See Figure 5.5.)

Using this notation, the formula for the derivative of the natural logarithm function is

$$\frac{d}{dx}(\ln x) = \frac{1}{x}.$$

This can be extended by the Chain Rule to

Derivative of ln x

D9	$\frac{d}{dx}(\ln u) = \frac{1}{u} \qquad (u(x) > 0),$

where u is a differentiable function of x.

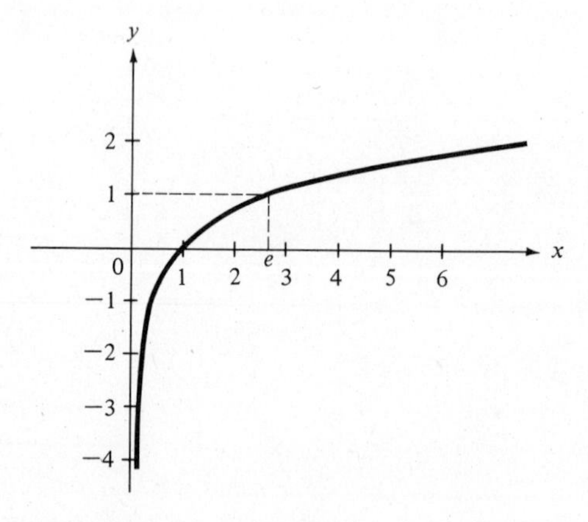

FIGURE 5.5 Graph of $y = \ln x$.

Example 1

(a) $\frac{d}{dx}\ln(3x^2+2) = \frac{1}{3x^2+2}\cdot\frac{d}{dx}(3x^2+2) = \frac{6x}{3x^2+2}.$

(b) $$\frac{d}{dx}(\ln(2x+1))^3 = 3(\ln(2x+1))^2 \cdot \frac{d}{dx}(\ln(2x+1))$$
$$= 3(\ln(2x+1))^2 \cdot \frac{1}{2x+1} \cdot \frac{d}{dx}(2x+1)$$
$$= \frac{6}{2x+1}(\ln(2x+1))^2.$$

(c) $$\frac{d}{dx}\ln(2x+1)^3 = \frac{1}{(2x+1)^3} \cdot \frac{d}{dx}(2x+1)^3$$
$$= \frac{1}{(2x+1)^3} \cdot 3(2x+1)^2 \frac{d}{dx}(2x+1) = \frac{6}{2x+1}.$$

Example 2 Calculate the equation of the line tangent to the graph of $y = \ln x$ at the point (1, 0).

Solution $\frac{d}{dx}(\ln x) = \frac{1}{x}$. When $x = 1$, the value of the derivative is 1. Therefore, the equation of the tangent line is

$$y - 0 = 1 \cdot (x - 1), \qquad y = x - 1.$$

(See Figure 5.6.)

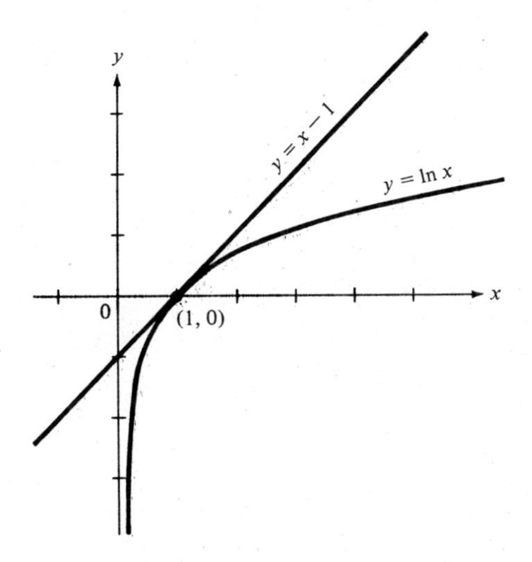

FIGURE 5.6 Example 2. The tangent line to the graph of $y = \ln x$ at the point (1, 0) is

$$y = x - 1.$$

The rules of logarithms can be used to simplify the calculation of many derivatives. These rules can be restated for positive functions $u(x)$ and $v(x)$ as

$$\ln(uv) = \ln u + \ln v,$$
$$\ln\left(\frac{u}{v}\right) = \ln u - \ln v,$$
$$\ln(u^k) = k \ln u.$$

Example 3 Calculate

$$\frac{d}{dx}\ln\frac{(3x^2+1)^{13}(x^4+7)^8}{x^2+2}.$$

Solution We first use the rules of logarithms to simplify the function:

$$\frac{d}{dx}\ln\frac{(3x^2+1)^{13}\cdot(x^4+7)^8}{x^2+2}$$
$$=\frac{d}{dx}[13\ln(3x^2+1)+8\ln(x^4+7)-\ln(x^2+2)]$$
$$=13\cdot\frac{d}{dx}[\ln(3x^2+1)]+8\cdot\frac{d}{dx}[\ln(x^4+7)]-\frac{d}{dx}[\ln(x^2+2)]$$
$$=13\cdot\frac{1}{3x^2+1}\cdot 6x+8\cdot\frac{1}{x^4+7}\cdot 4x^3-\frac{1}{x^2+2}\cdot 2x$$
$$=\frac{78x}{3x^2+1}+\frac{32x^3}{x^4+7}-\frac{2x}{x^2+2}.$$

EXERCISES 5.2

1. Calculate the derivatives:

• (a) $\ln(3x)$
(b) $\ln(x^2-x-1)$
(c) $\ln x^3$
• (d) $x\ln x$
(e) $(\ln x)^3$
• (f) $\dfrac{\ln x}{x^2+1}$
(g) $(x-1)^2\ln[(x-1)^2]$
• (h) $[\ln(x-1)]^{10}$
(i) $\ln(\ln x)$
(j) $\ln(x-\sqrt{x^2+1})$
• (k) $\ln(x-\sqrt{x^2+1})^3$
• (l) $\dfrac{x\ln x}{(x-1)^6}$

2. Locate all local maxima and minima and discuss the concavity. Sketch the graph

• (a) $y=x\ln x$
(b) $y=\ln x$

3. Calculate the derivatives of the following functions:

(a) $\log_3 x^2$, $x>0$
• (b) $\log_8\sqrt{x^2+1}$
• (c) $[\log_3(x^2-x-1)]^3$
(d) $\dfrac{\log_{10}x}{x+1}$, $x>0$

4. Calculate dy/dx. (*Hint:* First use the laws of logarithms to simplify the expressions.)

• (a) $y=\ln\left[\dfrac{(x^3+1)^5\sqrt[3]{x^2+5}}{(x^2+3)^{8/9}}\right]$

(b) $y=\ln\left[\dfrac{(x^3+x+1)^{1/4}(x^4-x+1)^6}{(x^2-1)^3(x^6+7x^3+x+3)}\right]$

5. Find the equation of the tangent line to the graph of $y=\ln x$ $(x>0)$ at the point $(e, 1)$ and at the point (x_0, y_0).

6. Find the equation of a tangent line to the curve $y=\ln x^2$ that is

• (a) Parallel to $2x-6y+7=0$.
(b) Parallel to $x-2y-8=0$.

5.3 ANTIDERIVATIVES

Since $d(\ln x)/dx = 1/x$, then $\ln x$ is an antiderivative of $1/x$. Therefore

$$\int \frac{1}{x}\,dx = \ln x + C \qquad (x > 0).$$

This last formula fills a gap in our integration formulas. Recall that

$$\int x^n\,dx = \frac{x^{n+1}}{n+1} + C \qquad \text{if } n \neq -1.$$

The lack of a comparable formula for the case $n = -1$ was not an oversight. As we see above, the integral $\int x^{-1}\,dx = \int (1/x)\,dx$ is not a polynomial function.

By using the integration formula above in conjunction with the Substitution Principle, we can calculate any integral that can be reduced to the form $\int du/u$, where $u(x) > 0$.

Example 1 Calculate $\displaystyle\int \frac{2x\,dx}{x^2+1}$.

Solution Let $u = x^2 + 1$. Then $du = 2x\,dx$. Since $u > 0$ for all x, then

$$\int \frac{2x\,dx}{x^2+1} = \int \frac{du}{u} = \ln u + C$$

$$= \ln (x^2 + 1) + C.$$

$u = x^2 + 1$
$du = 2x\,dx$

We are limited in our use of the integration formula

$$\int \frac{du}{u} = \ln u + C, \qquad u > 0$$

because it is applicable only where $u(x) > 0$. A modification will extend the formula's usefulness. It can be shown (see Exercise 4) that

$$\frac{d}{dx}(\ln |x|) = \frac{1}{x} \qquad (x \neq 0).$$

(See Figure 5.7.) Then $\ln |x|$ is an antiderivative of $1/x$ and

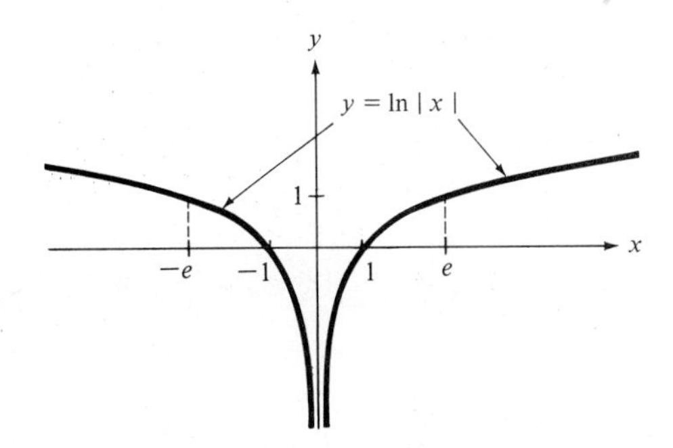

FIGURE 5.7 The graph of $y = \ln|x|$. $\dfrac{d}{dx}(\ln|x|) = \dfrac{1}{x}$.

I4
$$\begin{cases} \displaystyle\int \frac{1}{x}\,dx = \ln |x| + C & (x \neq 0). \\ \displaystyle\int \frac{1}{u}\,du = \ln |u| + C & (u(x) \neq 0). \end{cases}$$

In applying 14 we must be certain that $u(x) \neq 0$ at any point we are considering. Thus, for example, we can use I4 to calculate

$$\int_a^b \frac{1}{x}\,dx$$

if a and b are both positive or both negative, but not in other cases.

Example 2 Calculate $\displaystyle\int \frac{x\,dx}{x^2 - 1}$.

Solution Let $u = x^2 - 1$, $du = 2x\,dx$. Then

$$\begin{aligned} \int \frac{x\,dx}{x^2 - 1} &= \frac{1}{2}\int \frac{2x\,dx}{x^2 - 1} \\ &= \frac{1}{2}\int \frac{du}{u} = \frac{1}{2}\ln |u| + C \\ &= \frac{1}{2}\ln |x^2 - 1| + C. \end{aligned}$$

$u = x^2 - 1$
$du = 2x\,dx$

This formula can be applied over any interval on which $u(x) = x^2 - 1$ is not zero. Since $x^2 - 1$ is zero only when $x = \pm 1$, then the formula can be used over any interval not containing $x = -1$ or $x = 1$.

Example 3 Calculate $\displaystyle\int \frac{3x^2 + 4x + 1}{x - 1}\,dx$.

Solution We divide $x - 1$ into $3x^2 + 4x + 1$ by long division, obtaining

$$\frac{3x^2 + 4x + 1}{x - 1} = 3x + 7 + \frac{8}{x - 1}.$$

Then

$$\begin{aligned} \int \frac{3x^2 + 4x + 1}{x - 1}\,dx &= \int \left(3x + 7 + \frac{8}{x - 1}\right) dx \\ &= 3\int x\,dx + 7\int dx + 8\int \frac{dx}{x - 1} \\ &= \frac{3x^2}{2} + 7x + 8\ln |x - 1| + C. \end{aligned}$$

Integrals of rational functions

Any rational function of form $(a_n x^n + \cdots + a_1 x + a_0)/(ax + b)$ can be integrated by the technique illustrated in Example 3.

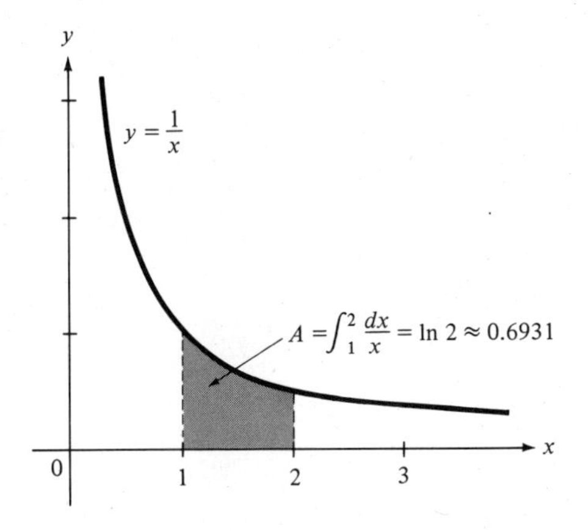

FIGURE 5.8 Example 4.

Example 4 Calculate the area of the region bounded by the graph of $xy = 1$, the x-axis, and the vertical lines $x = 1$, $x = 2$. (See Figure 5.8.)

Solution The area is

$$A = \int_1^2 \frac{dx}{x} = [\ln |x|]_1^2 = \ln |2| - \ln |1| = \ln 2 \approx 0.6931.$$

One additional integral is needed for completeness. On occasion we may need to integrate the logarithmic function $y = \ln x$. It can be shown that

I5	$\int \ln x \, dx = x \ln x - x + C \qquad (x > 0).$

The proof of this result is left for the reader. (See Exercise 3.)

EXERCISES 5.3

1. Calculate the following integrals. Over what intervals do the integrals exist?

(a) $\int \frac{2}{x} \, dx$

• (b) $\int \left(7x^2 - \frac{5}{3x}\right) dx$

(c) $\int \frac{dx}{x + 2}$

• (d) $\int \frac{dx}{x - 3}$

(e) $\int \frac{dx}{2x + 3}$

• (f) $\int \frac{(\ln x)^2}{x} \, dx$

(g) $\int \frac{1}{\ln x} \cdot \frac{dx}{x}$

• (h) $\int [(\ln x)^2 + 3(\ln x) + 7] \cdot \frac{dx}{x}$

• (i) $\int \frac{4x^2 + 3x + 1}{x^2} \, dx$

(j) $\int \frac{7x^4 + 5x^3 - 3x^2 - 8x + 1}{x - 1} \, dx$

2. Calculate the following definite integrals *if they exist*. If an integral is improper, explain why. (For (e), see Exercise 3.)

(a) $\int_0^1 \frac{dx}{x + 2}$

• (b) $\int_0^2 \frac{dx}{x - 2}$

(c) $\int_0^1 \frac{dx}{2x-3}$

(e) $\int_1^e \ln x\, dx$

• (d) $\int_0^1 \frac{dx}{3x-2}$

• (f) $\int_0^2 \frac{x}{x^2+1}\, dx$

3. Prove that $d(x \ln x)/dx = 1 + \ln x$. By integrating both sides of this equation show that

I5 $$\int \ln x\, dx = x \cdot \ln x - x + C \qquad (x > 0).$$

4. Prove that $d(\ln |x|)/dx = 1/x$ if $x \neq 0$. (*Hint:* If $x < 0$, let $u = -x = |x|$ and use the Chain Rule.) Use this result to show that

$$\int \frac{dx}{x} = \ln |x| + C \qquad (x \neq 0).$$

Show that this formula reduces to

$$\int \frac{dx}{x} = \ln x + C$$

when $x > 0$.

•5. Calculate the area of the region bounded by the graph of $xy = 2$, the x-axis, and the lines $x = -1$ and $x = -6$. (*Hint:* First sketch the curve.)

6. Show that the improper integral

$$\int_0^1 \frac{dx}{x}$$

is not equal to a finite number. This justifies the fact that $u(x)$ must not be zero when we apply I4. (*Hint:* First use Theorem 5.1c to show that $\lim_{x\to\infty} \ln x = \infty$.)

5.4 THE EXPONENTIAL FUNCTION $y = e^x$

It was stated in Section 5.1 that if b is a positive number, then the function $f(x) = b^x$ is differentiable. We now show how to calculate the derivative of the exponential function when $b = e$.

Theorem 5.3 $\frac{de^x}{dx} = e^x$.

PROOF Let $y = e^x$. Then

$$\ln y = x.$$

We calculate the derivative of each side of this last equation using D9:

$$\frac{d}{dx}(\ln y) = \frac{dx}{dx},$$

$$\frac{1}{y} \cdot \frac{dy}{dx} = 1,$$

$$\frac{dy}{dx} = y.$$

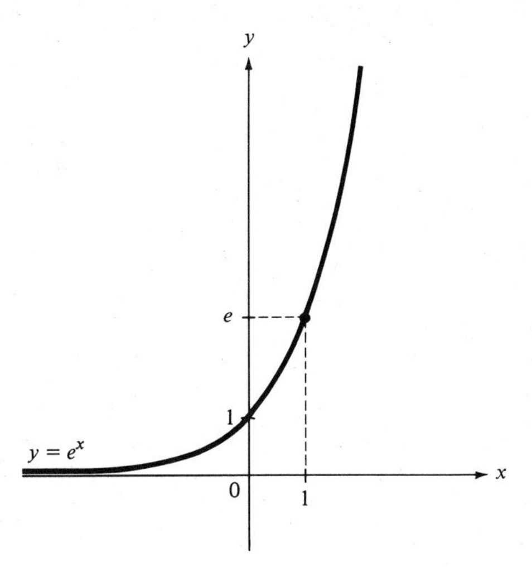

FIGURE 5.9 The exponential function $y = e^x$. At each point on the graph the slope of the tangent line is equal to the y-coordinate of the point:

$$\frac{d}{dx}(e^x) = e^x.$$

Since $y = e^x$, then

$$\frac{d}{dx}(e^x) = e^x.$$

(Figure 5.9.)

If we apply the Chain Rule to this formula, we obtain the more general formula

D10	$$\frac{d}{dx}(e^u) = e^u \cdot \frac{du}{dx}.$$

Derivative of e^u

Example 1

(a) $\frac{d}{dx}(e^{2x}) = e^{2x} \cdot \frac{d}{dx}(2x) = 2e^{2x}$

(b) $\frac{d}{dx}(3e^{x^2}) = 3e^{x^2} \cdot \frac{d}{dx}(x^2) = 3e^{x^2} \cdot 2x = 6xe^{x^2}$

(c) $\frac{d}{dx}(e^{1/x}) = e^{1/x} \cdot \frac{d}{dx}\left(\frac{1}{x}\right) = e^{1/x}\left(-\frac{1}{x^2}\right) = -\frac{e^{1/x}}{x^2}$

Since e^x is its own derivative, then it is also its own antiderivative. Therefore, the corresponding integration formula is

I6	$$\int e^x\,dx = e^x + C.$$

Integral of e^x

Example 2 Calculate $\int_{-1}^{1} e^{2x}\,dx$.

Solution We first calculate the indefinite integral. Let $u = 2x$, $du = 2\,dx$. Then

$$\int e^{2x}\,dx = \frac{1}{2}\int e^{2x} \cdot 2\,dx = \frac{1}{2}\int e^u\,du$$

$u = 2x$
$du = 2\,dx$

$$= \frac{1}{2}e^u + C = \frac{1}{2}e^{2x} + C.$$

Therefore

$$\int_{-1}^{1} e^{2x}\,dx = \frac{1}{2}\,[e^{2x}]_{-1}^{1} = \frac{1}{2}\,[e^{2\cdot 1} - e^{2\cdot(-1)}]$$
$$= \frac{1}{2}\,(e^{2} - e^{-2}).$$

Example 3 Calculate the area bounded by the graph of $y = e^x$, the x-axis, and the vertical lines $x = 0$ and $x = 1$ (Figure 5.10).

Solution The area is equal to

$$A = \int_{0}^{1} e^{x}\,dx = [e^{x}]_{0}^{1} = e^{1} - e^{0} = e - 1$$
$$\approx 2.718 - 1 \approx 1.718.$$

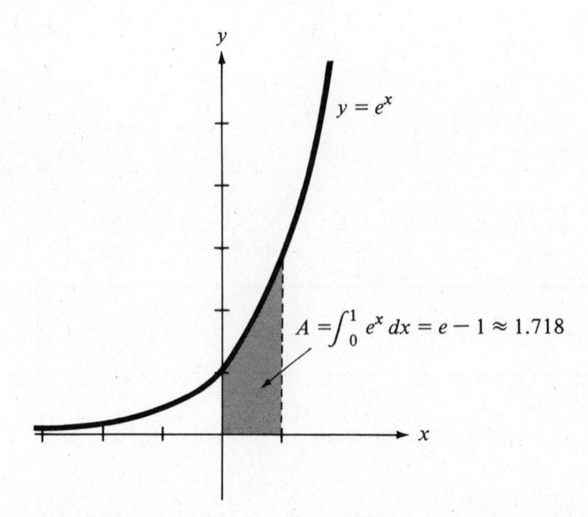

FIGURE 5.10 Example 3.

Example 4 Calculate the area of the region in quadrant I bounded by the graph of $y = e^{-x}$ and the x-axis (Figure 5.11).

Solution The area of the region is defined to be

$$A = \int_{0}^{\infty} e^{-x}\,dx$$

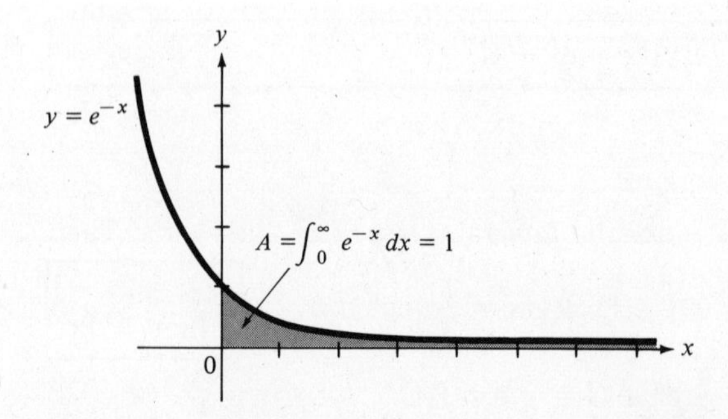

FIGURE 5.11 Example 4.

provided the improper integral has a finite value. The indefinite integral

$$\int e^{-x}\,dx$$

can be calculated by the substitution $u = -x$, $du = -dx$:

$$\int e^{-x}\,dx = -\int e^{-x}(-dx) = -\int e^{u}\,du = -e^{u} + C \qquad \boxed{\begin{aligned} u &= -x \\ du &= -dx \end{aligned}}$$

$$= -e^{-x} + C.$$

Therefore

$$\begin{aligned} A &= \int_0^\infty e^{-x}\,dx = \lim_{b\to\infty}\left[\int_0^b e^{-x}\,dx\right] \\ &= \lim_{b\to\infty}\left[-e^{-x}\right]_0^b = \lim_{b\to\infty}\left[(-e^{-b}) - (-e^{-0})\right] \\ &= \lim_{b\to\infty}\left[-\frac{1}{e^b} + 1\right] = 0 + 1 = 1. \end{aligned}$$

Thus the region, although unbounded in extent, has an area of one unit.

Other Bases Derivatives and integrals of exponential functions of form

Change of base

$$y = b^x \qquad (b > 0)$$

can be calculated by a simple device. Recall that

$$b = e^u \qquad \text{if and only if} \qquad u = \ln b.$$

Therefore

$$b = e^{\ln b} \quad \text{and} \quad y = b^x = (e^{\ln b})^x = e^{\ln b \cdot x}.$$

Since this function is now written in the form $y = e^u$, we can apply D10 or I6.

Application. The Normal Probability Density Function A set of numbers (*data set*) is said to be *distributed normally* if the corresponding probability density function is

The normal probability density function

$$f(x) = \frac{1}{\sqrt{2\pi}\sigma}\,e^{-(x-\mu)^2/2\sigma^2}$$

where μ and σ are constants, $\sigma > 0$.

The graph of the normal density function f is symmetric about the vertical line $x = \mu$ (Figure 5.12). The number μ is the *mean* (average) and σ is the *standard deviation* of the original set. In every case approximately 68 percent of the data items are within one standard deviation of the mean (between $\mu - \sigma$ and $\mu + \sigma$), approximately 95 percent of them are within two standard deviations of the mean (between $\mu - 2\sigma$ and $\mu + 2\sigma$), and so on.

A very great number of data sets are distributed normally. For example, such biological traits as weight and height within a species are distributed normally, as are psychological measurements such as I.Q. scores, S.A.T. scores, and so on. One of the original studies of the normal curve involved errors of measurement. It was found that the recorded measurements are normally distributed around the "true" value of the quantity being measured.

If data are distributed normally with mean μ and standard deviation σ,

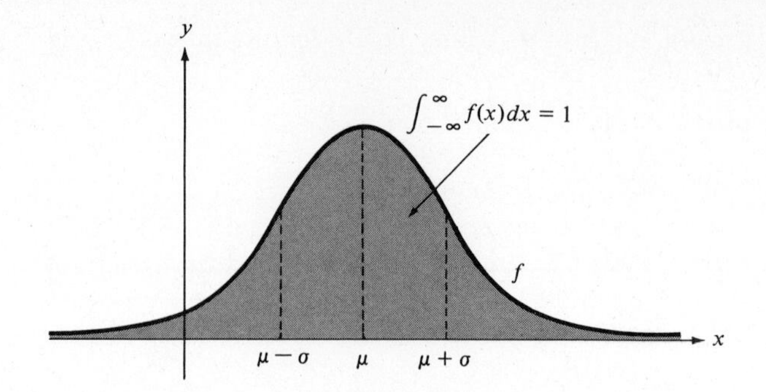

FIGURE 5.12 The normal probability density function:

$$f(x) = \frac{1}{\sqrt{2\pi}\sigma} e^{-(x-\mu)^2/2\sigma^2}$$

(μ = mean, σ = standard deviation). Approximately 68 percent of the data items are between $\mu - \sigma$ and $\mu + \sigma$.

then the probability that a given data item x is between a and b is equal to the area under the graph of the density function between a and b:

Probability that x is between a and b

$$\text{probability} = \frac{1}{\sqrt{2\pi}\sigma} \int_a^b e^{-(x-\mu)^2/2\sigma^2}\, dx.$$

(See Figure 5.13.)

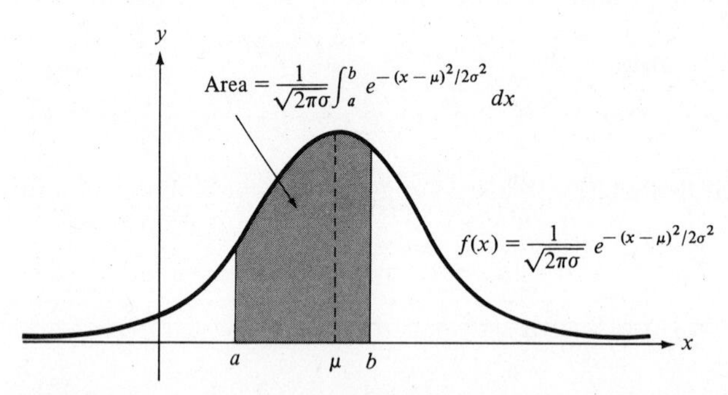

FIGURE 5.13 The normal probability density function. The probability that x is between a and b is

$$p = \frac{1}{\sqrt{2\pi}\sigma} \int_a^b e^{-(x-\mu)^2/2\sigma^2}\, dx.$$

The above integral is not one of the elementary integrals considered in this book. It cannot be calculated by antiderivatives. For this reason elaborate tables have been calculated for the area under various parts of the normal density function. These tables express the area from μ (the mean) in multiples of σ (the standard deviation). A small part of one of these tables is reproduced in Table 5.1.

Table 5.1 lists areas under the normal density curve from μ to $\mu + x\sigma$ where $x \geq 0$. To find the area from $\mu - x\sigma$ to μ we use the symmetry about $x = \mu$. The area from $\mu - x\sigma$ to μ is equal to the area from μ to $\mu + x\sigma$.

TABLE 5.1 Area under Normal Probability Density Curve from μ to $\mu + x\sigma$

x	Area	x	Area
.00	.0000	2.00	.4773
.25	.0987	2.25	.4878
.50	.1915	2.50	.4938
.75	.2734	2.75	.4970
1.00	.3413	3.00	.4987
1.25	.3944	3.25	.4994
1.50	.4332	3.50	.4998
1.75	.4599	3.75	.4999
2.00	.4773	4.00	.5000

Example 5 The area from μ to $\mu + .5\sigma$ is .1915. This number corresponds to the entry $x = .5$ in Table 5.1. It is equal to the probability that a given data item has its value between μ and $\mu + .5\sigma$. The probability that a data item is between $\mu - .5\sigma$ and μ also is .1915. The probability that it is between $\mu - .5\sigma$ and $\mu + .5\sigma$ is $.1915 + .1915 = .3830$.

Example 6 I.Q. scores are distributed normally with mean $\mu = 100$ and standard deviation $\sigma = 15$. (The raw scores are adjusted so that the final scores have this particular distribution.) What is the probability that a given person has an I. Q. score greater than 130?

Solution The probability that a given data item (I. Q. score) is greater than $\mu + 2\sigma$ (130) is equal to the area under the normal density curve from $\mu + 2\sigma$ to ∞.

The total area under the curve from μ to ∞ is $\frac{1}{2}$. (This is one half of the total area under the curve, which must be 1 for any probability density function.) From Table 5.1 the area from μ to $\mu + 2\sigma$ is .4773. Thus

$$\begin{aligned} \text{probability} &= \text{area from } \mu + 2\sigma \text{ to } \infty \\ &= (\text{area from } \mu \text{ to } \infty) - (\text{area from } \mu \text{ to } \mu + 2\sigma) \\ &= 0.5000 - 0.4773 = 0.0227. \end{aligned}$$

Thus slightly more than 2 percent of the population have I.Q. scores greater than 130.

EXERCISES 5.4

1. Calculate the derivatives of the following functions:

• (a) $2e^x + 1$
(b) $(e^x + 3)^2$
(c) e^{3x}
• (d) $\dfrac{e^{2x} - 1}{e^{2x} + 1}$
(e) xe^x
• (f) e^{x^2}
• (g) $\ln (e^x + 1)$
(h) $\ln (\ln (e^{2x} - 3))$

2. Let $b > 0$, $b \neq 1$. Show that

$$\frac{d(b^x)}{dx} = \ln b \cdot b^x \quad \text{and} \quad \frac{d(b^u)}{dx} = \ln b \cdot b^u \cdot \frac{du}{dx}$$

(*Hint:* Write $b = e^{\ln b}$.)

3. Using the results obtained in Exercise 2, calculate the derivative of each of the following functions:

(a) 3^x
• (b) $2^{x+1}3^{x-1}$
• (c) $3^{x+1/2x^{3^x}}$
(d) $5^{x^2}e^{2^x}$

4. Calculate the following integrals:

• (a) $\int e^{3x}\,dx$
(b) $\int xe^{x^2}\,dx$
• (c) $\int (x-1)e^{x^2-2x+1}\,dx$
• (d) $\int_0^1 \frac{dx}{e^{2x}}$
(e) $\int \frac{e^x}{e^x - 1}\,dx$
(f) $\int \frac{e^{\sqrt{x}}}{\sqrt{x}}\,dx$
(g) $\int \frac{(e^x+1)^2}{e^x}\,dx$
• (h) $\int \frac{e^{x^{-2}}}{x^3}\,dx$
• (i) $\int \frac{e^{2x}}{1 - 2e^{2x} + e^{4x}}\,dx$
(j) $\int (e^x + e^{-x})^2\,dx$

5. Calculate the area of the figure bounded by the curve $y = e^x$, the coordinate axes, and the line $x = -2$.

•6. Calculate the area of the figure bounded by the curve $y = e^x$, the positive y-axis, and the negative x-axis.

7. The function $y = e^{-x^2}$ is closely related to the normal density function.

(a) Determine the following properties of its graph: (1) relative extrema, (2) asymptotes, (3) points of inflection, (4) symmetry.
(b) Show that the rectangle of maximum area that has its base on the x-axis and two vertices on the graph of $y = e^{-x^2}$ has these vertices at the points of inflection.

8. S.A.T. scores are distributed normally with mean $\mu = 500$ and standard deviation $\sigma = 100$. What is the probability that a student has his S.A.T. score in each of the following ranges?

(a) $400 \leq x \leq 500$
• (b) $500 \leq x \leq 700$
(c) $450 \leq x \leq 550$
• (d) $750 \leq x \leq 800$

9. Weights of college women are distributed normally with mean $\mu = 127$ pounds and standard deviation $\sigma = 17$ pounds. What is the probability that a randomly chosen college woman will have her weight in each of the following ranges?

(a) Between 110 and 144 pounds
• (b) Greater than 144 pounds
• (c) Less than 93 pounds
(d) Greater than 110 pounds

5.5 DIFFERENTIAL EQUATIONS

In the study of population growth it is assumed that the factors which influence growth do not change over a few generations. If this assumption holds, the result is a short-term situation in which *the rate of change of the population with respect to time is proportional to the size of the population*. For example, if the population increases by 15 percent one year it also will increase by 15 percent the next, and the next, and so on. This proportional increase will continue until there is a change in the factors which influence growth.

If $P = P(t)$ is the population at time t, then the assumption that the rate of change is proportional to the size of the population can be expressed as the equation

$$\frac{dP}{dt} = kP,$$

where k is the constant of proportionality.[1]

An equation, such as the one above, that relates a function and its derivative is called a *differential equation*. Examples of differential equations are

Differential equation

$$y' = 3x^2 + 2x,$$

$$\frac{dy}{dx} = -3x^2y^2,$$

$$\frac{dP}{dt} = kP,$$

and so on. By a *solution* of a differential equation we mean any function that when substituted for the unknown function makes the differential equation a true statement.

Example 1 The solution of the differential equation

$$y' = 3x^2 + 2x$$

can be obtained by integration:

$$\frac{dy}{dx} = 3x^2 + 2x,$$

$$dy = (3x^2 + 2x)\,dx,$$

$$\int dy = \int (3x^2 + 2x)\,dx,$$

$$y = \frac{3x^3}{3} + \frac{2x^2}{2} + C = x^3 + x^2 + C.$$

The solution is any function of form

$$y = x^3 + x^2 + C,$$

where C is a constant.

[1] In strict terms this assumption is not realistic. Over a very short period of time (say one tenth of a second) the population probably will not change at all. Thus, the "instantaneous" rate of change should be zero most of the time and nonzero at a few isolated moments. In order to make sense of the assumption we must consider it as describing the rate of change at a mythical "average" time.

Observe that the process of solving a differential equation must involve integration at some step. Thus a constant of integration will appear in the final answer. The solution which involves an arbitrary constant is called the *general solution*. Solutions which are obtained by choosing particular numbers for the constant are called *particular solutions*.

General and particular solutions

It follows from Example 1 that the general solution of

$$y' = 3x^2 + 2x$$

is

$$y = x^3 + x^2 + C.$$

Examples of particular solutions of this differential equation are

$$\begin{aligned} y &= x^3 + x^2, \\ y &= x^3 + x^2 - 12, \\ y &= x^3 + x^2 + 13e, \\ y &= x^3 + x^2 + \sqrt{17}, \end{aligned}$$

and so on.

Many of the applications of the calculus involve differential equations. The usual procedure for solving a problem by the use of differential equations is as follows:

Steps in solving a problem by differential equations

1. Show that a function to be calculated, such as population size, satisfies a certain differential equation.
2. Find the general solution of the differential equation.
3. Among the infinite number of solutions given by the general solution find the particular solution that solves the original problem. This involves the use of supplementary information, such as the population size at a fixed time, to find the particular value of the arbitrary constant that solves the problem.

Example 2 Find the particular solution of the differential equation

$$y' = 3x^2 + 2x$$

that satisfies the condition $y = 2$ when $x = -2$.

Solution We found in Example 1 that the general solution of the differential equation is

$$y = x^3 + x^2 + C.$$

If $y = 2$ when $x = -2$, then

$$2 = (-2)^3 + (-2)^2 + C = -8 + 4 + C,$$

so that

$$C = 6.$$

The particular solution is

$$y = x^3 + x^2 + 6.$$

Separation of Variables There is a general method that can be used to solve a large number of differential equations, such as

Separation of variables

$$\frac{dy}{dx} = -3x^2y^2.$$

We consider the derivative dy/dx to be a quotient of two differentials dy and dx and separate the variables, writing all of the terms with y on one side of the equals sign and all of the terms involving x on the other. We then integrate both sides of the differential equation in order to get an equation involving x and y.

Example 3

(a) Solve the differential equation

$$\frac{dy}{dx} = -3x^2y^2.$$

(b) Find the particular solution that passes through the point (2, 1).

Solution

(a)
$$\frac{dy}{dx} = -3x^2y^2.$$

$$\frac{dy}{y^2} = -3x^2dx \qquad \text{(variables are separated).}$$

$$\int y^{-2}\,dy = \int (-3x^2)\,dx.$$

$$\frac{y^{-1}}{-1} = -\frac{3x^3}{3} + C = -x^3 + C.$$

$$-\frac{1}{y} = -x^3 + C.$$

$$\frac{1}{y} = x^3 - C.$$

$$y = \frac{1}{x^3 - C}.$$

(b) For the particular solution we want $y = 1$ when $x = 2$. We substitute into the equation for y and solve for C:

$$1 = \frac{1}{2^3 - C} = \frac{1}{8 - C}$$

$$8 - C = 1$$

$$C = 7.$$

The particular solution is

$$y = \frac{1}{x^3 - 7}.$$

The Differential Equation $\frac{dP}{dt} = kP$ We now return to the differential equation that describes the unrestricted growth of populations.

The differential equation $\frac{dP}{dt} = kP$

Theorem 5.4 The general solution of the differential equation

$$\frac{dP}{dt} = kP,$$

where k is a given constant, is

$$P = Ce^{kt}.$$

PROOF We separate the variables and integrate:

$$\frac{dP}{dt} = kP,$$

$$\frac{dP}{P} = kdt, \qquad \text{(variables are separated)}$$

$$\int \frac{dP}{P} = \int kdt,$$

$$\ln |P| = kt + C_1.$$

Recall that $\ln y = x$ if and only if $y = e^x$. It follows from this relationship that

$$y = e^x = e^{\ln y}$$

for every positive y. Thus

$$\ln |P| = kt + C_1,$$

$$e^{\ln |P|} = e^{kt+C_1},$$

$$|P| = e^{kt}e^{C_1},$$

$$P = \pm e^{C_1}e^{kt},$$

so that

$$P = Ce^{kt},$$

where $C = \pm e^{C_1}$.

Example 4 At time $t = 0$, five million bacteria were living in George's lung. Three hours later the number had increased to nine million. Assuming that the conditions for growth do not change over the next few hours find the number of bacteria at the end of (a) 12 hours; (b) 36 hours.

Solution The population size at time t is

$$P = P(t) = Ce^{kt},$$

where C and k are constants. To evaluate the constants we use the facts that

$$P(0) = 5$$

and

$$P(3) = 9,$$

where $P(t)$ is measured in millions of bacteria. Since $P(0) = 5$, then

$$5 = Ce^{k \cdot 0} = C \cdot e^0 = C \cdot 1$$

so that

$$C = 5.$$

Thus, the population function is

$$P(t) = 5e^{kt}.$$

Since $P(3) = 9$, then

$$9 = 5e^{k \cdot 3} = 5 \cdot e^{3k}$$

$$\tfrac{9}{5} = e^{3k}$$

$$3k = \ln \tfrac{9}{5} = \ln (1.8) = 0.5878 \qquad \text{(by Table IV)}$$

$$k = \tfrac{1}{3}(0.5878) = 0.1959.$$

The population function is

$$P = 5e^{0.1959t}.$$

(a) When $t = 12$

$$P = 5e^{(0.1959)\cdot 12} = 5e^{2.3508}$$
$$\approx 5 \cdot (10.486) \approx 52.43 \qquad \text{(by Table V)}.$$

Thus the population after 12 hours is in excess of 52 million.

(b) When $t = 36$

$$P = 5e^{(0.1959)36} = 5e^{7.0524}$$
$$\approx 5 \cdot (1153) \approx 5765 \qquad \text{(by Table V)}.$$

The population after 36 hours is in excess of five billion.

Remark It is obvious that the situation described in Example 4 cannot continue indefinitely. Eventually the growth rate must slow down—either George will die or the defense system of his body will produce enough antibodies to kill the bacteria. A similar situation must prevail in any population growth situation described by the differential equation

$$\frac{dP}{dt} = kP.$$

This law of growth can only continue while the factors influencing growth do not change. Eventually, if nothing else happens, the sheer size of the population will make the acquisition of food more difficult and thereby slow down the rate of increase. Thus, the function

$$P(t) = Ce^{kt}$$

describes a law of unrestricted growth which can only occur over a few generations. Formulas for inhibited growth can be derived from other assumptions about the growth rate. These formulas for inhibited growth are much more realistic over a long period of time than is formula $P = Ce^{kt}$.

EXERCISES 5.5

1. Use separation of variables to solve the following differential equations:

(a) $\dfrac{dy}{dx} = \dfrac{x+1}{y-1}$

(b) $\dfrac{dy}{dx} = \dfrac{y-1}{x+1}$

• (c) $\dfrac{dy}{dx} = (x+1)(y-1)$

(d) $y'(x+1) = 2xy$

(e) $y' = \dfrac{2xy}{y^2+1}$

• (f) $y' = x - 4xy$

(g) $y' = 3y$

• (h) $y' + 7y = 0$

2. Solve each of the following differential equations. Then find a particular solution satisfying the additional restriction.

(a) $dP/dt = -P$, $P = -e^2$ when $t = 1$

• (b) $5(dP/dt) - P = 0, P(0) = -1$

(c) $dP/dt = P \ln 4$, $P(2) = \frac{1}{2}$

• (d) $dy/dx = (x + 1)y$, $y = 3$ when $x = 0$

(e) $dx/dt = 5x^2$, $x = 7$ when $t = 3$

•3. A culture of bacteria contained seven million bacteria at 3:00 P.M. Four hours later the number had increased to 14 million. Assume that the conditions for growth had not changed over the four-hour interval. How many bacteria were in the culture at 5:00 P.M.?

4. The population of the United States was approximately 151 million in 1950 and approximately 179 million in 1960. Assume that the law of unrestricted growth held for the United States population from 1950 to 1970. Calculate (from the formula) the population to be expected in 1970. Compare this figure with the true population in 1970 as found in an almanac. What, if anything, does this comparison show?

•5. A roast with bacterial contamination of 2 percent was placed in a home freezer on June 1. By July 1 the contamination had increased to 3 percent. Assuming that 4 percent is the maximum allowable contamination for human consumption, what is the deadline for safe use of the roast? (Assume that the rate of change of the population of bacteria is proportional to the size of the population.)

5.6 APPLICATIONS. I

Radioactive Decay The rate at which a radioactive material changes to lead is proportional to the amount of the material present. In symbols, if $A = A(t)$ is the amount present at time t, then

$$\frac{dA}{dt} = kA,$$

where k is the constant of proportionality. Since this differential equation is, except for notation, the same as the one in Section 5.5, then

$$A(t) = Ce^{kt},$$

where C is a constant.

Example 1 Radioactive carbon (carbon-14) has a half-life of 5568 years. (At the end of that period of time one half of the original amount remains.) Derive a formula for the amount remaining after t years.

Solution Let $A(t)$ be the amount present at time t. We measure time from the beginning of the period in which we are interested. Then

$$A(t) = Ce^{kt},$$

where C is a constant. At the beginning of the period of time the amount present was $A(0)$. Thus

$$A(0) = Ce^{0 \cdot t} = C$$

and the formula is

$$A(t) = A(0)e^{kt}.$$

At the end of 5568 years the amount present is $A(0)/2$. Thus

$$\frac{A(0)}{2} = A(5568) = A(0)e^{5568k}$$

and

$$\tfrac{1}{2} = e^{5568k}.$$

If we take the logarithm of both sides we obtain

$$\ln \tfrac{1}{2} = 5568k, \qquad k = \tfrac{1}{5568} \ln \tfrac{1}{2}.$$

Thus

$$A(t) = A(0)e^{\ln(1/2)\cdot(t/5568)}.$$

Traces of carbon-14 exist in many living organisms. By analyzing the carbon-14 in the remains we can determine the approximate date at which the organism lived.

Example 2 An analysis of an ancient campsite used by North American Indians reveals that one tenth of the carbon-14 present in the original ashes has decomposed. Use this information to date the campsite.

Solution We determined above that the amount of carbon-14 remaining after t years is $A(t) = A(0)e^{\ln(1/2)\cdot(t/5568)}$. In this case nine tenths of the original carbon-14 remains, and so

$$\frac{9A(0)}{10} = A(0)e^{\ln(1/2)\cdot(t/5568)}, \qquad \frac{9}{10} = e^{\ln(1/2)\cdot(t/5568)}.$$

Taking logarithms of both sides,

$$\ln \frac{9}{10} = \ln \frac{1}{2} \cdot \frac{t}{5568}$$

$$t = 5568 \cdot \frac{\ln \dfrac{9}{10}}{\ln \dfrac{1}{2}} = 5568 \left(\frac{\ln 9 - \ln 10}{\ln 1 - \ln 2} \right)$$

$$\approx 5568 \left(\frac{2.1972 - 2.3026}{0 - 0.6931} \right) \approx 847.$$

To allow for errors of measurement we date the campsite at approximately 850 years.

Investments It is frequently assumed that the rate at which an investment grows in value is proportional to the value of the investment. Consider the following example:

Example 3 Mr. Smith purchased \$100,000 worth of bank bonds that pay 5 percent interest. He has owned the bonds for seven months and wishes to sell them now. (They cannot be redeemed for another five months.) What is the current market value?

Solution Mr. Smith assumes that the bonds have been increasing in value continuously and that the rate of change is proportional to the value. If $V(t)$ is the value at the time t (measured in years), then

$$\frac{dV}{dt} = kV$$

and so

$$V(t) = Ce^{kt} \qquad (0 \leq t \leq 1),$$

where C and k are constants. We evaluate C by

$$V(0) = Ce^{k \cdot 0} = C = 100{,}000.$$

When $t = 1$, we have

$$\begin{aligned} V(1) &= 100{,}000e^k = 105{,}000,\\ e^k &= 1.05,\\ k &= \ln(1.05). \end{aligned}$$

Therefore

$$\begin{aligned} V(t) &= 100{,}000e^{\ln(1.05)\cdot t} = 100{,}000(e^{\ln(1.05)})^t\\ &= 100{,}000(1.05)^t. \end{aligned}$$

At the present time seven months have elapsed and the value is

$$V(\tfrac{7}{12}) = 100{,}000(1.05)^{7/12}.$$

To evaluate this number we use logarithms:

$$\begin{aligned} \ln V(\tfrac{7}{12}) &= \ln[100{,}000(1.05)^{7/12}] = \ln 100{,}000 + \tfrac{7}{12}\ln(1.05)\\ &\approx \ln 100{,}000 + \tfrac{7}{12}(0.0488) \approx \ln 100{,}000 + 0.0285\\ &\approx \ln 100{,}000 + \ln(1.03) = \ln 103{,}000.\\ V(\tfrac{7}{12}) &\approx \$103{,}000. \end{aligned}$$

Example 4 Ten years ago Mr. Smith purchased a tract of land for \$10,000 and immediately planted it in timber at a cost of \$5000. In another 15 years the timber can be marketed for an estimated \$40,000. At that time the land is expected to have a value of \$20,000 (due primarily to inflation). Mr. Smith has received an offer of \$30,000 for the land and timber. Is this a reasonable offer?

Solution Mr. Smith considers that there are two ways in which he can determine the present value. (I) He can consider the land and timber together, assuming an initial investment of \$15,000. (II) He can consider the two as separate investments. To be on the safe side he computes the present value both ways.

Method I Assume an initial investment of \$15,000 that increases continuously for 25 years, obtaining a final value of \$60,000 (\$40,000 for timber and \$20,000 for land). We find, as in the previous analyses, that the formula for the value at time t is of form

$$V(t) = 15{,}000e^{kt} \qquad (0 \leq t \leq 25).$$

To find k we calculate

$$\begin{aligned} V(25) &= 15{,}000e^{25k} = 60{,}000,\\ e^{25k} &= 4, \end{aligned}$$

$$25k = \ln 4,$$
$$k = \tfrac{1}{25} \ln 4.$$

The formula is

$$V(t) = 15{,}000e^{(t/25)\ln 4}.$$

The present value is

$$V(10) = 15{,}000e^{(10/25)\ln 4} \approx \$26{,}100.$$

Method II Assume initial investments of \$10,000 for land and \$5000 for timber. Let $V_L(t)$ and $V_T(t)$ be the values of the land and timber at time t, respectively. It follows that

$$V_L(t) = 10{,}000e^{(t/25)\ln 2} \quad \text{and} \quad V_T(t) = 5000e^{(t/25)\ln 8}.$$

At present time ($t = 10$) the total value is

$$V = V_L(10) + V_T(10) = 10{,}000e^{(10/25)\ln 2} + 5000e^{(10/25)\ln 8}$$
$$\approx \$13{,}200 + \$11{,}500 = \$24{,}700.$$

By either method of computation the offer is greater than the present value and should be considered.

Other methods are available for solving Examples 3 and 4. The answers obtained from these methods differ by small amounts from the ones that we obtained.

Depreciation The assumption that the rate of change in value is proportional to the value is also reasonable when computing depreciation.

Example 5 A refrigerator was purchased five years ago for \$300. In another five years it will have a junk value of \$30. Determine its present value.

Solution If $V(t)$ is the value at time t, then

$$\frac{dV}{dt} = kV$$

and so

$$V(t) = Ce^{kt} \qquad (0 \leq t \leq 10).$$

Computing the value of V when $t = 0$ and $t = 10$, we find that $C = 300$, $k = \frac{1}{10} \ln (\frac{1}{10})$. Thus,

$$V(t) = 300e^{\ln(1/10)\cdot(t/10)} = 300(\tfrac{1}{10})^{t/10}.$$

Using logarithms we can compute that the present value is

$$V(5) \approx \$95.$$

Most household appliances are made to last ten years. We assume that a ten-year-old appliance has a salvage value of one tenth the original purchase price. Then the value at time t is

$$V(t) = V_0e^{\ln(1/10)\cdot(t/10)} = V_0(\tfrac{1}{10})^{t/10} \qquad (0 \leq t \leq 10),$$

where V_0 is the original purchase price. Observe that $(\frac{1}{10})^{3/10} \approx \frac{1}{2}$. (This can be verified by logarithms.) Thus the value of an appliance has a "half-life" of approximately three years. For example, if the original value is $300, the value at three years is approximately $150, at six years approximately $75, and at nine years approximately $37.

Remark A large number of methods are used to compute depreciation. These methods lead to estimates of the true value that vary greatly from each other. The method illustrated in Example 5 gives one of the best estimates for the true value, but is one of the most difficult to apply. For this reason simpler, but less accurate, methods are normally used.

EXERCISES 5.6

•1. The half-life of isotopic radium is approximately 2000 years. Beginning with 200 grams of this radium, find a formula that will give the amount remaining after t years. After how many years will one third of the original 200 grams remain?

2. A sample of a radioactive element has a mass of 1000 milligrams. At the end of one year 100 milligrams remain. What is the half-life? How much will remain after ten years?

•3. A radioactive element decays in such a way that there is a 40 percent loss after a two-year period. How much will remain at the end of five years? What is the half-life?

4. The town of Beeburgh had a population of 1800 in 1955. Ten years later the population had increased to 2500. Assuming the law of exponential growth is applicable, what will be the population in 1980?

•5. A culture of RSV virus weighed 1.0 milligram at 9:00 A.M. At 11:00 A.M. the culture weighed 0.8 milligram. How much is this culture expected to weigh at 3:00 P.M.?

6. The population of a city increases at a constant rate of 5 percent a year. In how many years will the population double? Triple?

•7. Three years ago George purchased a new Gasdrinker Super Sport automobile for $7500. In seven more years the value is expected to be $1500. Find its current value.

8. To commence operation Maggie's Restaurant purchased a grill for $600 and a coffee urn for $150. These appliances are expected to last fifteen years and have a salvage value of $30 each. After five years of operation, Maggie is forced to sell this equipment. What is the value of the two appliances at the time of the sale?

•9. The Mountain Investment Company purchased a parcel of land in downtown Ceeburgh for $100,000 and invested $1,000,000 in a modern ten-story office building. In ten years the land is expected to be worth $150,000 and the building $1,850,000. What is the value of the real estate after five years? Use two methods.

5.7 APPLICATIONS. II

We close this chapter with a number of diverse examples which show how logarithmic and exponential functions arise in practical applications.

Relative Rates of Change In many growth problems the relative rate of change is more important than the rate of change itself. If a business has an increase in revenue of $5000 per year this may be good or bad, depending on the size of the revenue. If the revenue is $25,000 this is a very good increase (20 percent), but if the revenue is $100,000,000, then the increase is poor (0.005 percent).

The *relative rate of change* of a function $y = y(t)$ is defined to be the ratio

$$\frac{\text{rate of change of } y}{\text{value of } y} = \frac{dy/dt}{y}.$$

In our example, if $y = R(t) = 25{,}000$ and $dy/dt = R'(t) = 5000$, then the relative rate of change is

$$\frac{dy/dt}{y} = \frac{5000}{25{,}000} = \frac{1}{5} = 0.20,$$

which corresponds to an increase in revenue of 20 percent.

The relative rate of change of y with respect to t can be calculated most easily by noting that it is the derivative of the *natural logarithm of y:*

$$\frac{d}{dt}(\ln y) = \frac{1}{y} \cdot \frac{dy}{dt} = \frac{dy/dt}{y} = \text{relative rate of change of } y.$$

A number of differential equations are set up by using a relative rate of change. The law of unrestricted growth

$$P = Ce^{kt},$$

for example, was derived from the assumption that

$$\frac{dP/dt}{P} = k,$$

the relative rate of change of the population is a constant.

Allometric Relationships (*Application to Biology*) During childhood different parts of the human body grow at different rates. The eyes, for example, grow much slower than the legs. Biologists who have studied these growth rates have found that the different rates are *allometric*. That is, the relative rates of growth are proportional.

As an illustration, let x be the measure of one part of a body (say the diameter of an eyeball) and y the measure of another part (say length of a leg). Then there exists a constant k such that

$$\textit{relative rate of change of } y = k \cdot (\textit{relative rate of change of } x)$$

$$\frac{dy/dt}{y} = k \cdot \frac{dx/dt}{x}.$$

This differential equation can be solved by integration after multiplying it by the differential dt:

$$\frac{dy}{y} = k\frac{dx}{x}$$

$$\int \frac{dy}{y} = k \int \frac{dx}{x}$$

$$\ln y = k \ln x + C_1$$

$$e^{\ln y} = e^{k \ln x + C_1} = e^{C_1} e^{\ln x \cdot k} = e^{C_1}(e^{\ln x})^k$$
$$y = Cx^k \qquad (\text{where } C = e^{C_1}).$$

The value of y is proportional to a power of the value of x.

Similar results hold for any organism. The specific constants C and k depend on the species, the race, and the sex and individual characteristics of the organism.

Present Value (*Application to Economics*) If we invest $x at an interest rate of r per year, compounded continuously (the rate of change of the value is proportional to the value), then after t years the value is

$$y = xe^{rt}.$$

If we know the value of y and the rate r, we can solve for x

$$x = \frac{y}{e^{rt}} = ye^{-rt}.$$

This value x is called the *present value of* y. It is the amount of money we must invest in order to get a return of $y in t years.

Example 1 How much money should we invest at 8 percent per year, compounded continuously, in order to have $20,000 in 15 years.

Solution The present value of $20,000 in 15 years is

$$\begin{aligned} x &= 20{,}000e^{-(0.08)(15)} = 20{,}000e^{-1.2} \\ &= (20{,}000)(0.30119) \qquad (\text{by Table V}) \\ &= \$6023.80. \end{aligned}$$

Example 2 George plans to sell his doughnut shop, which makes an average profit of $5000 per year. He will invest the money in real estate at the equivalent of 8 percent per year, compounded continuously. He decides that the fair price would be the present value of the future earnings over the next 20 years. Compute this quantity.

Solution Assume that the profit would be earned continuously over the years. Over the period of time from t to t + dt he would earn

$$5000\, dt$$

dollars. This money has a present value of

$$(5000\, dt)e^{-0.08t}.$$

We can determine the present value of the total earnings over the 20-year period by integrating this expression from t = 0 to t = 20:

$$\begin{aligned} \text{present value} &= \int_0^{20} 5000e^{-0.08t}\, dt \\ &= 5000 \int_0^{20} e^{-0.08t}\, dt = \frac{5000}{-.08}[e^{-0.08t}]_0^{20} \\ &= (-62{,}500)[e^{-1.6} - e^0] \\ &= (-62{,}500)[0.20190 - 1] = \$49{,}881.25. \end{aligned}$$

Example 3 (See Example 2.) Calculate the present value of all of the future earnings of the doughnut shop.

Solution If we assume that the doughnut shop will last forever, then the present value of all future earnings is given by

$$\begin{aligned}\text{present value} &= \int_0^\infty 5000e^{-0.08t}\,dt = 5000\int_0^\infty e^{-0.08t}\,dt \\ &= \frac{5000}{-0.08}\,[e^{-0.08t}]_0^\infty = \frac{5000}{-0.08}\cdot\lim_{b\to\infty}\,[e^{-0.08t}]_0^b \\ &= \frac{5000}{-0.08}\cdot\lim_{b\to\infty}\left[\frac{1}{e^{0.08b}} - e^0\right] \\ &= (-62{,}500)\cdot[0-1] = \$62{,}500.\end{aligned}$$

Autocatalytic Reactions (*Application to Chemistry*) In a chemical reaction one or more substances change to form a new substance. Certain of these reactions are autocatalytic—the new substance acts as a catalyst to cause additional reactions.

In a simple autocatalytic reaction substance A changes into substance B, each molecule of A changing into a molecule of B. Let a be the number of molecules of A and b the number of molecules of B at time $t = 0$ when the reaction begins. If $x = x(t)$ is the number of molecules that had changed at time t, then *the rate of change of x is proportional to the amount of A and the amount of B at time t.*

At a given instant there are $a - x$ molecules of A and $b + x$ molecules of B. Therefore

$$\frac{dx}{dt} = k(a-x)(b+x),$$

where k is a constant.

This differential equation can be rewritten as

$$\frac{dx}{(a-x)(b+x)} = k\,dt.$$

The left-hand side of the new equation is not similar to any that we have encountered. We can integrate it by observing first that

$$\frac{1}{a-x} + \frac{1}{b+x} = \frac{(b+x)+(a-x)}{(a-x)(b+x)} = \frac{a+b}{(a-x)(b+x)}$$

so that

$$\frac{1}{(a-x)(b+x)} = \frac{1}{a+b}\left(\frac{1}{a-x} + \frac{1}{b+x}\right).$$

If we substitute this expression for $\dfrac{1}{(a-x)(b+x)}$ we obtain

$$\frac{1}{(a-x)(b+x)}\,dx = k\,dt$$

$$\frac{1}{a+b}\left(\frac{1}{a-x} + \frac{1}{b+x}\right)dx = k\,dt$$

$$\frac{1}{a+b}\int\left(\frac{1}{a-x}+\frac{1}{b+x}\right)dx = k\int dt$$

$$\frac{1}{a+b}[-\ln(a-x)+\ln(b+x)] = kt + C.$$

This equation can be used to compute x in a particular example or can be changed to an equivalent equation involving the exponential function.

EXERCISES 5.7

1. Calculate the relative rates of change for the following profit functions when $x = 10$:

   • (a) $P(x) = -x^2 + 500x + 5$

   (b) $P(x) = \dfrac{5000}{1+e^{-x}}$

2. Use the identity

$$\frac{1}{(a-x)(b+x)} = \frac{1}{a+b}\left(\frac{1}{a-x}+\frac{1}{b+x}\right)$$

   to solve the following differential equations:

   (a) $dx = (2-x)(1+x)\,dt$

   • (b) $dx/dt = (5+x)(5-x)$

   • (c) $dx/dt = 7(x+1)(x-1)$

   (d) $dy/dx = 2y(y+1)$

•3. In an autocatalytic reaction each molecule of substance A is transformed into a molecule of substance B. At the beginning of the reaction there are 100 molecules of A and 10 molecules of B. Twenty seconds later there are 55 molecules of A and 55 of B. Use the differential equation

$$\frac{dx}{dt} = k(100-x)(10+x)$$

   to determine an equation relating x and t, where $x = x(t)$ is the number of molecules that have been transformed at time t. Calculate the value of x when $t = 40$.

4. Calculate the present value of the following amounts of money invested at continuous compounding of interest for the stated number of years.

   • (a) \$5000, 6 percent, 10 years

   (b) \$10,000, 4 percent, 30 years

   (c) \$100,000, 5 percent, 40 years

•5. An author will receive \$3000 per year in royalties for the next five years and nothing thereafter. Find the present value of his future royalties assuming continuous compounding of interest at 5 percent per year.

THE TRIGONOMETRIC FUNCTIONS

6.1 RADIAN MEASURE

Remedial Section The dawn of ancient civilization brought the need for accurate surveying of land. Legend correlates this need with the annual flooding of the Nile valley and the need to re-establish boundary lines after the flood waters had receded. Although the origins of this art are lost in antiquity, we know that it was developed to a high degree by the surveyors in the Babylonian and Egyptian empires.

Undoubtedly, surveying was at first purely a descriptive process based on measurement, but it soon developed along mathematical lines. The key to this development was the discovery that ratios of corresponding sides of similar triangles are equal. Tables of these ratios made it possible to compute distances and areas rather than measure them. The ancient Babylonians computed fairly accurate tables of these trigonometric functions, and at about the same time discovered many properties of triangles. For example, many centuries before Pythagoras they knew the truth of the Pythagorean Theorem and used it in the computation of the tables.

Degree measure

The first requirement of a system for measuring angles is a basic unit of measure. The standard unit is the *degree,* subdivided into *minutes* and *seconds.* To define a degree we first divide a circle into 360 equal arcs. Next we construct an angle with vertex at the center of the circle that intercepts one of these arcs at its two endpoints. (Such an angle is called a *central angle.*) This central angle is

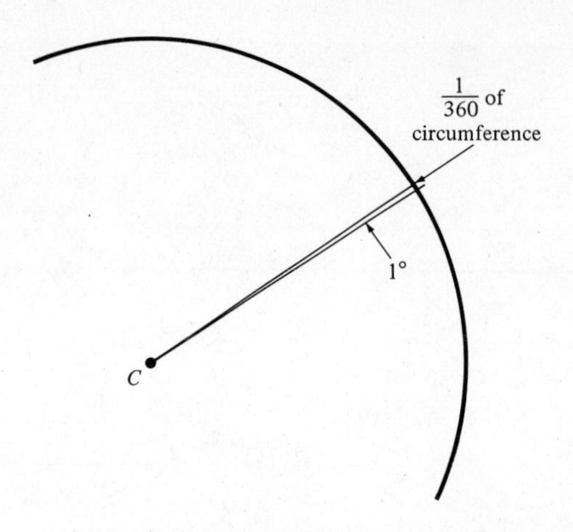

FIGURE 6.1 One degree.

defined to have a measure of one degree (1°). (See Figure 6.1.) One *minute* (1′) has one sixtieth the measure of a degree; one *second* (1″) has one-sixtieth the measure of a minute.

The degree-minute-second system is quite adequate for all of the practical purposes of measurement. There have been a few attempts to replace it, but the advantages of having the number 360 divisible by most of the small integers, coupled with the disadvantages of a transition period when two systems are being used, have assured the survival of this system.

The situation is quite different when we attempt to apply the principles of calculus to trigonometric functions. There is no mathematical reason for the choice of 360 as a basis for the system. This problem is similar to the one we faced with the logarithmic function. Base 10 is very convenient when logarithms are used for computation, but base e is natural for the calculus. When we apply calculus to the trigonometric functions, we use a different unit of measure: the *radian*.

Radian measure

As with the definition of a degree we begin with a circle. We mark off an arc equal in length to the radius. The central angle that intersects the two end-points of the arc is defined to have a measure of *one radian* (Figure 6.2.)

It is a simple matter to correlate radian measure with degree measure. Observe that the complete circle has arc length $2\pi r$. Thus, the corresponding central

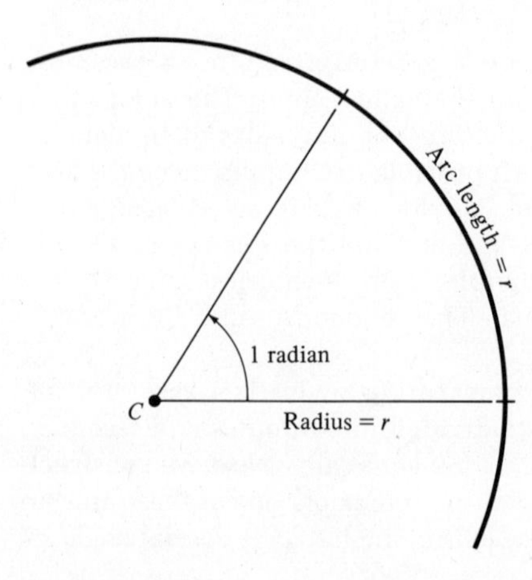

FIGURE 6.2 One radian.

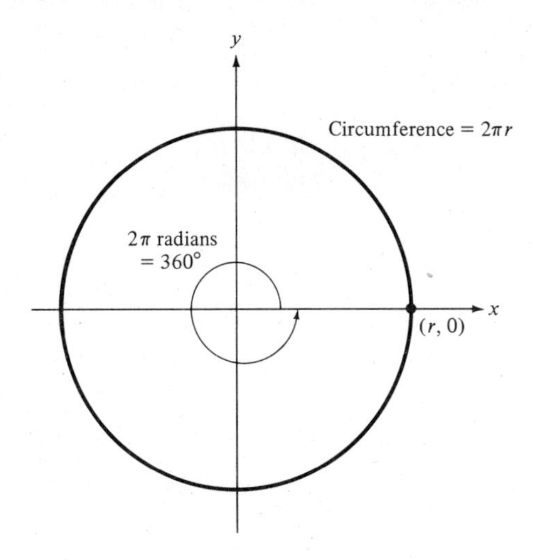

FIGURE 6.3 2π radians $= 360°$, π radians $= 180°$, 1 radian $= \dfrac{180°}{\pi} \approx 57°17'45''$.

angle has a measure of 2π radians. In the degree system this angle measures 360°. Thus

Conversion to radian measure

$$\begin{aligned} 2\pi \text{ radians} &= 360°, \\ \pi \text{ radians} &= 180°, \\ 1 \text{ radian} &= \frac{180°}{\pi} \approx 57°17'45'', \\ 1° &= \frac{\pi}{180} \text{ radians.} \end{aligned}$$

In radian measure the "standard" angles of 30°, 45°, 60°, 90° are measured as

$$30° = \frac{\pi}{6} \text{ radians,}$$

$$45° = \frac{\pi}{4} \text{ radians,}$$

$$60° = \frac{\pi}{3} \text{ radians,}$$

$$90° = \frac{\pi}{2} \text{ radians.}$$

Example 1

(a) $$27° = 27 \cdot \frac{\pi}{180} = \frac{3\pi}{20} \approx 0.4712 \text{ radian.}$$

(b) $$1 \text{ radian} = \frac{180°}{\pi} \approx 57.2958° \approx 57°17'44.806''.$$

(c) $$2.3 \text{ radians} = (2.3)\frac{180°}{\pi} \approx 131.78°.$$

Area of a Sector The use of radian measure enables us to compute the area of the sector with central angle α. Observe that if $\alpha = 2\pi$, the sector is the entire

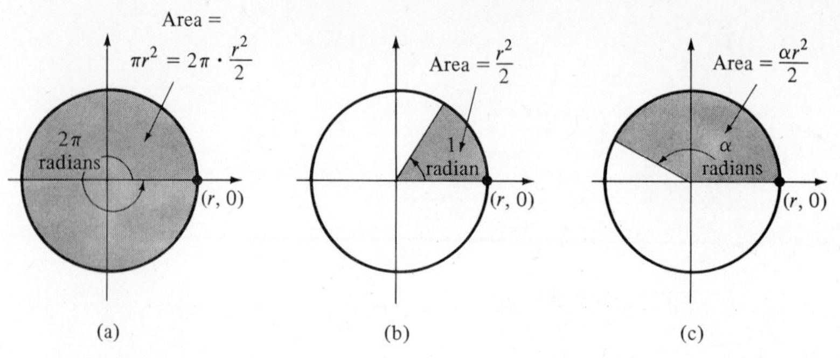

FIGURE 6.4 The sector with central angle α has area $\alpha r^2/2$.

circular disk and thus has area πr^2 (Figure 6.4a). If we divide this result by 2π, we have the area of the sector with central angle of one radian:

Area of sector

$$\text{area of sector with central angle 1 radian} = \frac{\pi r^2}{2\pi} = \frac{r^2}{2}$$

$$\text{area of sector with central angle } \alpha \text{ radians} = \frac{\alpha r^2}{2} \quad (0 \leqslant \alpha \leqslant 2\pi).$$

Example 2 A circle has radius $r = 3$ feet. A sector with central angle $\alpha = 60°$ ($\pi/3$ radians) has area

$$A = \frac{\pi}{3} \cdot \frac{3^2}{2} \approx 4.712 \text{ square feet.}$$

EXERCISES 6.1

1. Convert to radian measure (express in multiples of π):
   - • (a) 15°
   - (b) 200°
   - (c) 80°
   - (d) 270°
   - • (e) 310°
   - (f) 210°
   - • (g) 36°
   - (h) 135°

2. Convert to degree measure:
   - (a) $\pi/8$
   - • (b) $\pi/12$
   - (c) $7\pi/10$
   - • (d) $4\pi/3$
   - (e) $5\pi/12$
   - • (f) $2\pi/5$
   - (g) $\pi/6$
   - (h) 5π

3. Calculate the areas of the following sectors. (The central angle and radius are given.)
   - (a) 2 radians, 9 inches
   - • (b) 144°, 5 centimeters
   - (c) 120°, 12 feet
   - • (d) 45°, 8 inches
   - (e) $2\pi/5$ radians, 3 feet
   - • (f) 30°, 3 miles

4. Use the definition of radian measure to show that if a central angle of θ radians subtends an arc of length s in a circle of radius r, then $s = r\theta$.

5. Using the notation and result of Exercise 4, complete the missing entries in the following table:

r	θ	s
6 feet	30°	—
42 feet	—	7 feet
—	2.5 radians	5 inches

6.2 THE TRIGONOMETRIC FUNCTIONS

Remedial Section Traditionally the trigonometric functions have been defined for acute angles by using the sides of right triangles. For example, referring to Figure 6.5,

$$\sin \alpha = \frac{\text{length of side opposite angle } \alpha}{\text{length of hypotenuse}} = \frac{A}{C},$$

$$\cos \alpha = \frac{\text{length of side adjacent to angle } \alpha}{\text{length of hypotenuse}} = \frac{B}{C}.$$

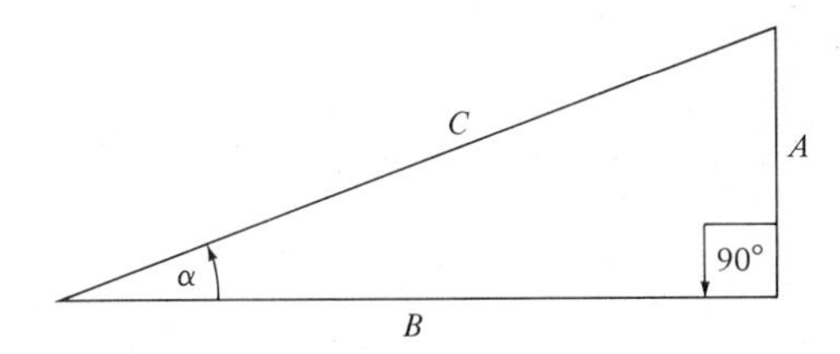

FIGURE 6.5 Classical definitions of the sine and cosine:

$$\sin \alpha = A/C, \qquad \cos \alpha = B/C.$$

The major drawback to this approach is that it defines the trigonometric functions only for angles between 0° and 90° (0 radians and $\pi/2$ radians). When we extend the results of the calculus to trigonometric functions, it becomes necessary to define these functions for *numbers* rather than angles and to have their domains of definition as large as possible.

There is no problem in changing the domains of definition of the trigonometric functions from angles to numbers. We define the *cosine of the number* x to be equal to the *cosine of the angle of* x *radians*. The *sine of the number* x is equal to the *sine of the angle of* x *radians*, and so on. For example, since $45° = \pi/4$ radians, then

$$\cos \frac{\pi}{4} = \cos \left(\frac{\pi}{4} \text{ radians} \right) = \cos 45° = \frac{1}{\sqrt{2}},$$

$$\sin \frac{\pi}{4} = \sin \left(\frac{\pi}{4} \text{ radians} \right) = \sin 45° = \frac{1}{\sqrt{2}}.$$

We now show how to enlarge the domains of these functions so that they are defined for numbers outside of the interval $0 \leqslant x \leqslant \pi/2$. First, we construct a circle of radius 1 and center at the origin. (The equation of this circle is $x^2 + y^2 = 1$.) Next, we construct the central angle of α radians with initial side on the positive x-axis. The terminal side is found by measuring the angle counterclockwise if $\alpha \geqslant 0$, clockwise is $\alpha < 0$. (Since the circle has radius 1 we can locate the terminal side of the angle by measuring an arc of length $|\alpha|$ from the point (1, 0) on the positive x-axis. We measure the arc in a counterclockwise direction if $\alpha \geqslant 0$ and a clockwise direction if $\alpha < 0$.) Observe that this process can be used to measure an angle of α radians for any real number α. (See Figure 6.6.)

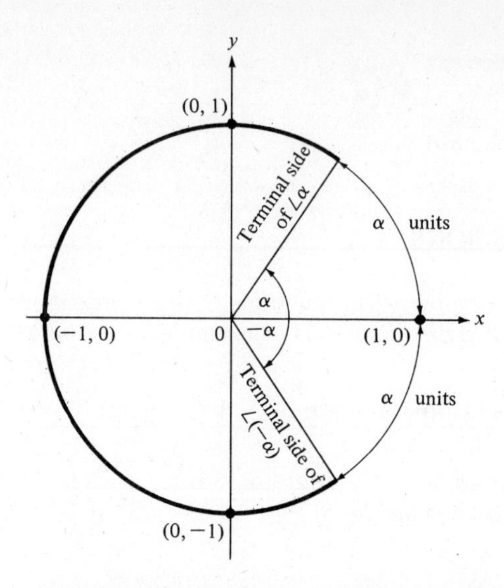

FIGURE 6.6 Standard positions of positive and negative angles. ($\alpha > 0$.)

We now let $P_\alpha(x_\alpha, y_\alpha)$ be the point of intersection of the terminal side of α and the circle $x^2 + y^2 = 1$ (Figure 6.7). The *cosine* and *sine* of α are defined in terms of the coordinates of P_α

sin α
cos α

$$\cos \alpha = x_\alpha \qquad \sin \alpha = y_\alpha.$$

Thus, the point P_α on the unit circle has coordinates $(\cos \alpha, \sin \alpha)$.

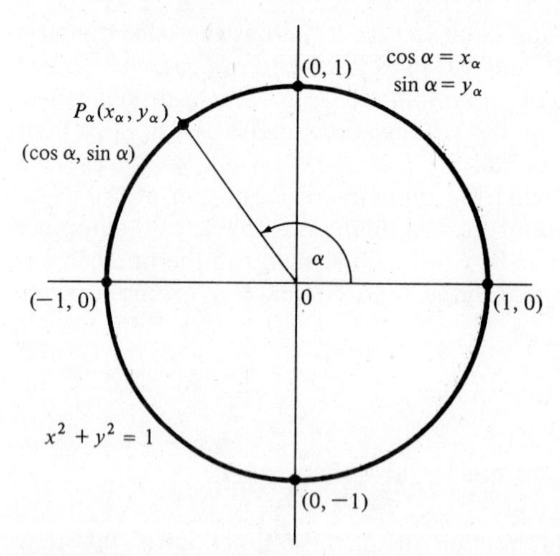

FIGURE 6.7 The cosine and sine of α: $\cos \alpha = x_\alpha$, $\sin \alpha = y_\alpha$.

Example 1

(a) If $\alpha = 0$, then P_α is the point (1, 0). Therefore

$$\cos 0 = 1 \quad \text{and} \quad \sin 0 = 0.$$

(b) If $\alpha = \pi/6$ ($\alpha = 30°$), then the x-coordinate of P_α is $\sqrt{3}/2$ and the y-coordinate is 1/2. Thus,

$$\cos \frac{\pi}{6} = \frac{\sqrt{3}}{2},$$

$$\sin \frac{\pi}{6} = \frac{1}{2}.$$

(See Figure 6.8.)

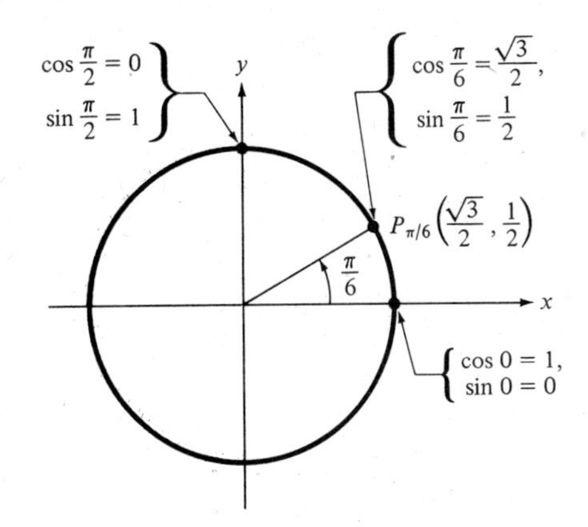

FIGURE 6.8 Example 1.

The other trigonometric functions are defined as follows:

tangent: $\tan \alpha = \dfrac{\sin \alpha}{\cos \alpha} = \dfrac{y_\alpha}{x_\alpha}$ (provided $\cos \alpha \neq 0$), tan α

cotangent: $\cot \alpha = \dfrac{\cos \alpha}{\sin \alpha} = \dfrac{x_\alpha}{y_\alpha}$ (provided $\sin \alpha \neq 0$), cot α

secant: $\sec \alpha = \dfrac{1}{\cos \alpha} = \dfrac{1}{x_\alpha}$ (provided $\cos \alpha \neq 0$), sec α

cosecant: $\csc \alpha = \dfrac{1}{\sin \alpha} = \dfrac{1}{y_\alpha}$ (provided $\sin \alpha \neq 0$). csc α

For example, since $P_{\pi/2}$ is the point (0, 1), then

$$\cos \frac{\pi}{2} = 0,$$

$$\sin \frac{\pi}{2} = 1,$$

$$\cot \frac{\pi}{2} = \frac{\cos \frac{\pi}{2}}{\sin \frac{\pi}{2}} = \frac{0}{1} = 0,$$

$$\csc \frac{\pi}{2} = \frac{1}{\sin \frac{\pi}{2}} = \frac{1}{1} = 1,$$

while sec $\pi/2$ and tan $\pi/2$ are not defined (Figure 6.8).

The trigonometric functions are now defined for all values of x. For example, since 2π radians is the angle of one complete rotation in a counterclockwise direction and

$$\frac{19\pi}{4} = 2\pi + 2\pi + \frac{3\pi}{4},$$

then $P_{19\pi/4}$ is the point obtained by an angle of two complete rotations followed by an angle of $3\pi/4$ radians (Figure 6.9). Thus,

$$P_{19\pi/4} = P_{3\pi/4} = \left(-\frac{1}{\sqrt{2}}, \frac{1}{\sqrt{2}}\right).$$

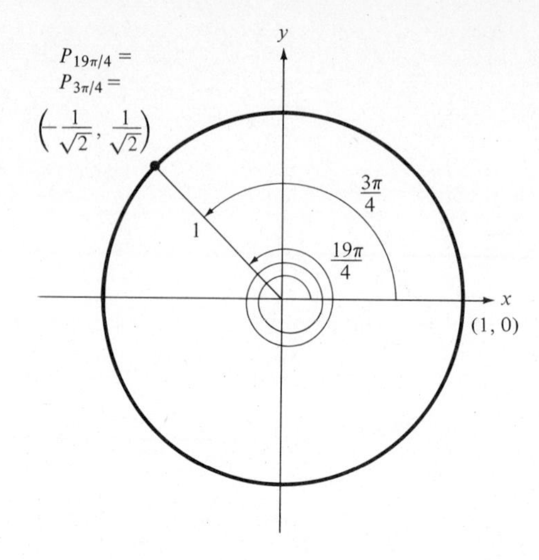

FIGURE 6.9
$\cos 19\pi/4 = \cos 3\pi/4 = -1/\sqrt{2}$,
$\sin 19\pi/4 = \sin 3\pi/4 = 1/\sqrt{2}$.

Consequently,

$$\sin\frac{19\pi}{4} = \sin\frac{3\pi}{4} = \frac{1}{\sqrt{2}}, \qquad \cot\frac{19\pi}{4} = \frac{\cos\frac{19\pi}{4}}{\sin\frac{19\pi}{4}} = -1,$$

$$\cos\frac{19\pi}{4} = \cos\frac{3\pi}{4} = -\frac{1}{\sqrt{2}}, \qquad \sec\frac{19\pi}{4} = \frac{1}{\cos\frac{19\pi}{4}} = \frac{1}{-\frac{1}{\sqrt{2}}} = -\sqrt{2},$$

$$\tan\frac{19\pi}{4} = \frac{\sin\frac{19\pi}{4}}{\cos\frac{19\pi}{4}} = -1, \qquad \csc\frac{19\pi}{4} = \frac{1}{\sin\frac{19\pi}{4}} = \frac{1}{\frac{1}{\sqrt{2}}} = \sqrt{2}.$$

Computation of the Trigonometric Functions It is necessary to know the values of the trigonometric functions only for first-quadrant angles. We can compute the values for angles in the other quadrants from them by using reference angles. First construct the angle α in standard position, using the unit circle, then measure the *smallest angle* formed by the terminal side of α and the x-axis. This angle, β, is called the *reference angle corresponding to* α. (See Figure 6.10.)

Reference angle

Observe that the coordinates of the points P_α and P_β are equal in absolute value and that those of P_β are nonnegative. Thus,

$$|\cos\alpha| = |x_\alpha| = |x_\beta| = x_\beta = \cos\beta$$

and

$$|\sin\alpha| = |y_\alpha| = |y_\beta| = y_\beta = \sin\beta.$$

Therefore,

$$\cos\alpha = \pm\cos\beta,$$
$$\sin\alpha = \pm\sin\beta,$$
$$\tan\alpha = \frac{\sin\alpha}{\cos\alpha} = \pm\frac{\sin\beta}{\cos\beta} = \pm\tan\beta,$$

the sign being determined by the quadrant in which the terminal side of α is located.

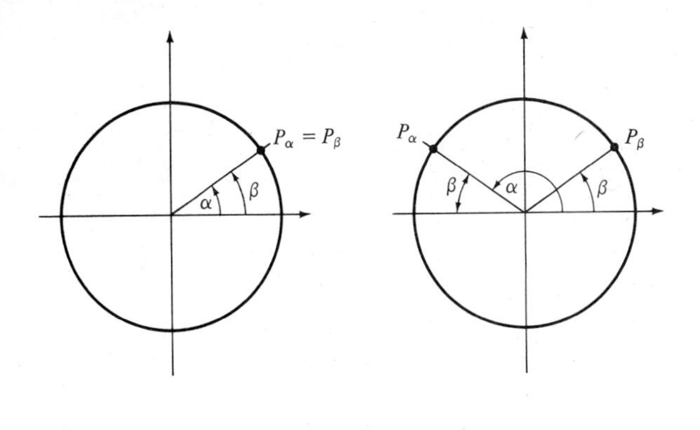

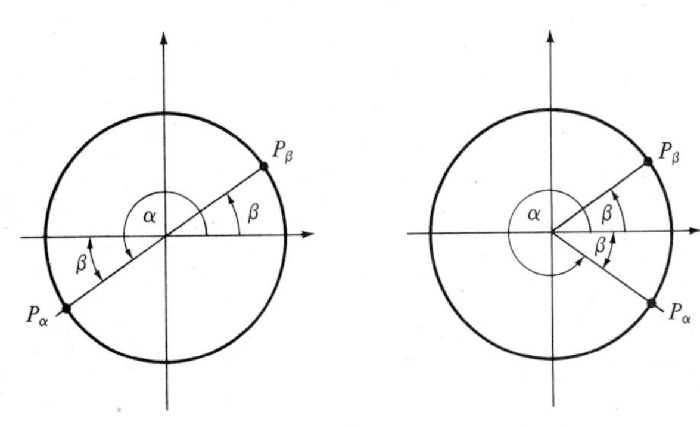

FIGURE 6.10 Reference angles: $\sin \alpha = \pm \sin \beta$, $\cos \alpha = \pm \cos \beta$.

Example 2 Use the fact that $P_{\pi/4}$ is the point $(1/\sqrt{2},\ 1/\sqrt{2})$ to compute (a) $\tan (3\pi/4)$, (b) $\sin (5\pi/4)$, (c) $\cos (7\pi/4)$.

Solution In each case the reference angle is $\pi/4$. (See Figure 6.11.)

(a) $3\pi/4$ is a quadrant II angle. Since the sine is positive and the cosine is negative in this quadrant, then

$$\sin \frac{3\pi}{4} = \sin \frac{\pi}{4} = \frac{1}{\sqrt{2}},$$

$$\cos \frac{3\pi}{4} = -\cos \frac{\pi}{4} = -\frac{1}{\sqrt{2}},$$

$$\tan \frac{3\pi}{4} = \frac{\sin \frac{3\pi}{4}}{\cos \frac{3\pi}{4}} = \frac{\frac{1}{\sqrt{2}}}{-\frac{1}{\sqrt{2}}} = -1.$$

(b) $5\pi/4$ is a quadrant III angle. In quadrant III the sine and cosine are both negative. Therefore

$$\sin \frac{5\pi}{4} = -\sin \frac{\pi}{4} = -\frac{1}{\sqrt{2}}.$$

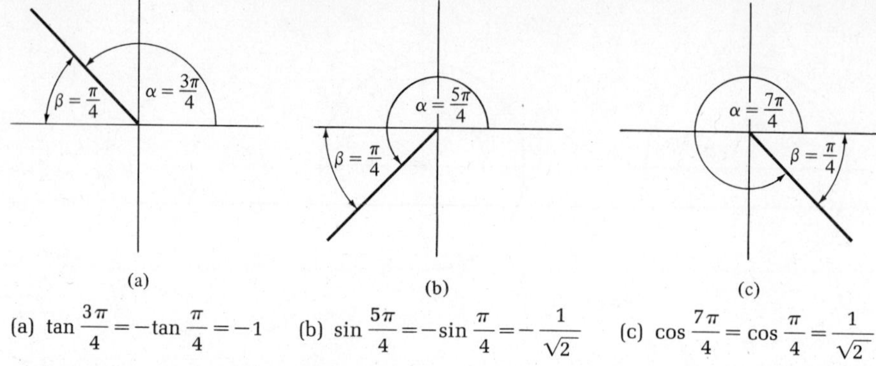

(a) $\tan \frac{3\pi}{4} = -\tan \frac{\pi}{4} = -1$ (b) $\sin \frac{5\pi}{4} = -\sin \frac{\pi}{4} = -\frac{1}{\sqrt{2}}$ (c) $\cos \frac{7\pi}{4} = \cos \frac{\pi}{4} = \frac{1}{\sqrt{2}}$

FIGURE 6.11 Example 2. β is the reference angle for α.

(c) $7\pi/4$ is a quadrant IV angle. In quadrant IV the cosine is positive and the sine is negative. Therefore

$$\cos \frac{7\pi}{4} = \cos \frac{\pi}{4} = \frac{1}{\sqrt{2}}.$$

The angles 0, $\pi/6$, $\pi/4$, $\pi/3$, $\pi/2$ are called *standard angles* because their trigonometric functions can be computed easily. The functions of all multiples of these angles can be computed by use of reference angles, as in Example 1. Table 6.1 lists the functions of the standard angles.

TABLE 6.1 Trigonometric Functions of Standard Angles

α	P_α	$\cos \alpha$	$\sin \alpha$	$\tan \alpha$
0	(1, 0)	1	0	0
$\frac{\pi}{6}$	$\left(\frac{\sqrt{3}}{2}, \frac{1}{2}\right)$	$\frac{\sqrt{3}}{2}$	$\frac{1}{2}$	$\frac{1}{\sqrt{3}}$
$\frac{\pi}{4}$	$\left(\frac{1}{\sqrt{2}}, \frac{1}{\sqrt{2}}\right)$	$\frac{1}{\sqrt{2}}$	$\frac{1}{\sqrt{2}}$	1
$\frac{\pi}{3}$	$\left(\frac{1}{2}, \frac{\sqrt{3}}{2}\right)$	$\frac{1}{2}$	$\frac{\sqrt{3}}{2}$	$\sqrt{3}$
$\frac{\pi}{2}$	(0, 1)	0	1	–

Table II at the end of the book lists approximations to the first-quadrant angles in increments of 0.01 radian.

In conjunction with the use of tables we must know the signs of the trigonometric functions in each quadrant. Figure 6.12 diagrams this information. The signs of the other trigonometric functions can be obtained by taking reciprocals of those given.

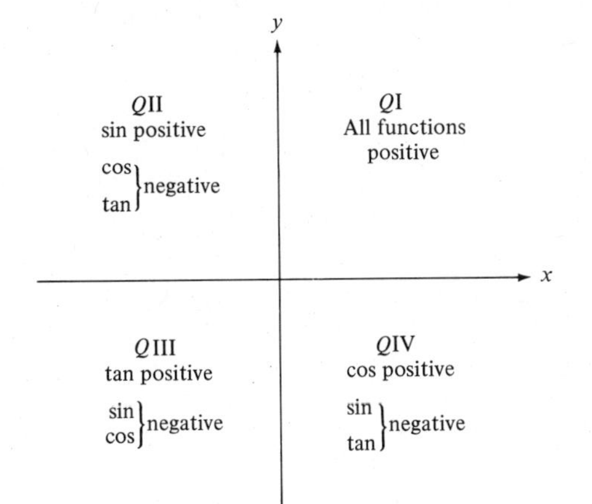

FIGURE 6.12 Signs of the trigonometric functions in the four quadrants.

Example 3 Use Table II to approximate (a) cos 2.91, (b) tan 4.21.

Solution

(a) 2.91 radians is a quadrant II angle. (See Figure 6.13.) Using the approximation $\pi \approx 3.14$, we see that the reference angle is

$$\beta = \pi - 2.91 \approx 3.14 - 2.91 = 0.23.$$

Thus,

$$\cos 2.91 \approx -\cos 0.23 \approx -0.97367.$$

(b) 4.21 radians is a quadrant III angle. The reference angle is

$$\beta = 4.21 - \pi \approx 4.21 - 3.14 = 1.07.$$

Therefore,

$$\tan 4.12 \approx \tan 1.07 \approx 1.8270.$$

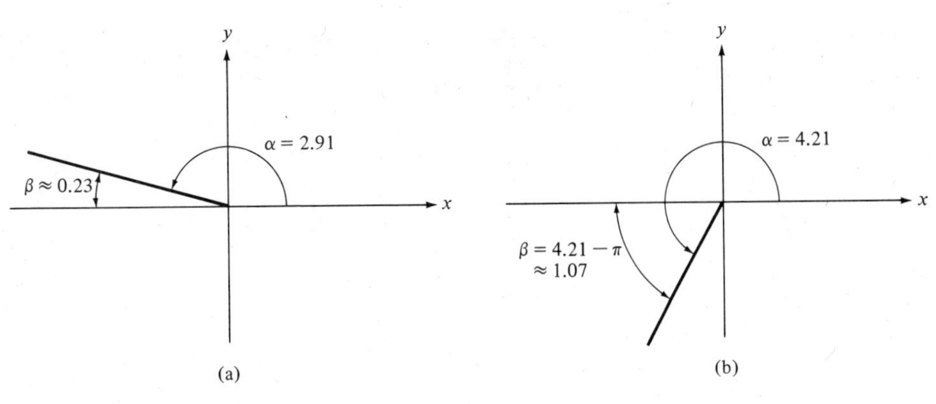

FIGURE 6.13 Example 3.

EXERCISES 6.2

1. Construct each of the following angles in standard position. Construct the reference angle β and give the measure of β. (If α is not expressed as a multiple of π, use the approximation $\pi \approx 3.14$ when calculating β.)

(a) $\alpha = 3\pi/4$
• (b) $\alpha = 5\pi/4$
• (c) $\alpha = \pi/2$
(d) $\alpha = -3\pi/4$
(e) $\alpha = 25\pi/6$
• (f) $\alpha = 3.61$
(g) $\alpha = 6.43$
• (h) $\alpha = -2.37$
(i) $\alpha = 9.2$
• (j) $\alpha = -15.2$

2. (a) to (j). State the coordinates of P_α and P_β for the numbers in Exercise 1.

3. Use Table 6.1 or Table II to compute or approximate each of the following numbers. Make a sketch showing both α and the reference angle β. For (h) to (n) use the approximation $\pi \approx 3.14$.

(a) $\cos(3\pi/2)$
• (b) $\tan(-3\pi/4)$
(c) $\sec(3\pi/4)$
(d) $\cot(-5\pi/3)$
• (e) $\sin(3\pi/4)$
(f) $\cos(-\pi)$
• (g) $\tan(-2\pi/3)$
• (h) $\sin 5.41$
(i) $\cos 1.35$
• (j) $\cot 3.12$
(k) $\tan(-4.30)$
• (l) $\sin(-2.98)$
(m) $\sin(2.98)$
(n) $\cos(-9.82)$

6.3 PERIODIC FUNCTIONS

Certain phenomena occur in cycles with astonishing regularity. The average monthly temperature of New York City, for example, is approximately the same each January, each February, each March, and so on, throughout the year. (See Figure 6.14.) If $f(t)$ is the average temperature of month t, then

$$f(t + 12) \approx f(t)$$

for each value of t.

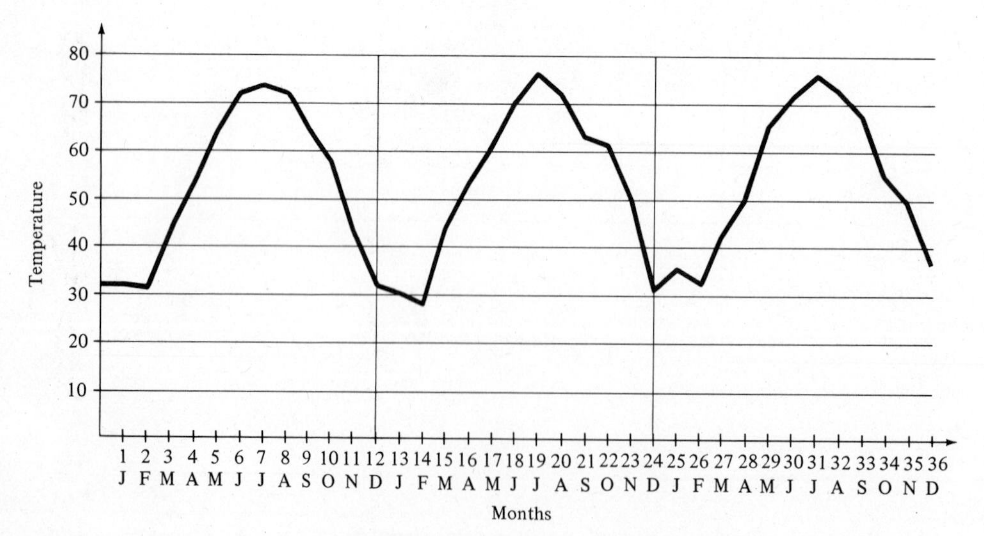

FIGURE 6.14 The average monthly temperature of New York City over a three-year period. The temperature graph is almost periodic:

$$f(t + 12) \approx f(t) \quad \text{for every } t$$

Period

A function f is said to be *periodic* if there exists a positive number p such that

$$f(x+p) = f(x)$$

for every value of x. The smallest such number p is called the *period* of the function.

Observe that if f is periodic with period p, then

$$f(x+2p) = f((x+p)+p) = f(x+p) = f(x),$$
$$f(x+3p) = f((x+2p)+p) = f(x+2p) = f(x),$$

and so on. Thus, the values of the function f are determined completely by the values on the interval [0, p].

Fundamental cycle

The portion of the graph over the interval [0, p] is called the *fundamental cycle* of the function. The portions over the intervals [p, 2p], [2p, 3p], [3p, 4p], and so on, are duplicates of the fundamental cycle. (See Figure 6.15.)

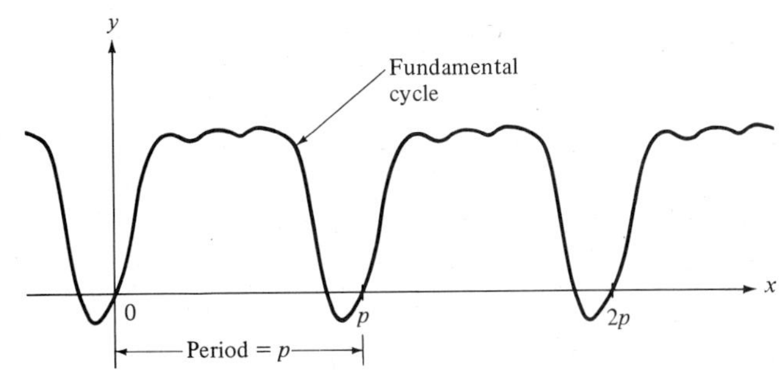

FIGURE 6.15 A periodic function of period p. The fundamental cycle is the portion of the graph over the interval [0, p].

It can be shown that continuous periodic functions can be approximated to any desired degree of accuracy by adding together the proper combinations of the sine and cosine functions. We will return to this discussion after we consider the graphs of the basic trigonometric functions.

The Sine and Cosine Functions Let P_α be the point on the terminal side of angle α that is on the unit circle. (See Figure 6.16.) P_α has coordinates $x = \cos\alpha$, $y = \sin\alpha$. Observe that the point $P_{\alpha+2\pi}$ is equal to the point P_α since $\alpha + 2\pi$ is the angle obtained by adding one complete revolution to the angle α. Therefore

$$\cos(\alpha+2\pi) = \text{x-coordinate of } P_{\alpha+2\pi}$$
$$= \text{x-coordinate of } P_\alpha = \cos\alpha$$

and, similarly,

$$\sin(\alpha+2\pi) = \sin\alpha.$$

It can be shown that 2π is the smallest number p such that

$$\cos(\alpha+p) = \cos\alpha$$

and

$$\sin(\alpha+p) = \sin\alpha$$

for every α. Therefore, the sine and cosine functions are periodic with period $p = 2\pi$.

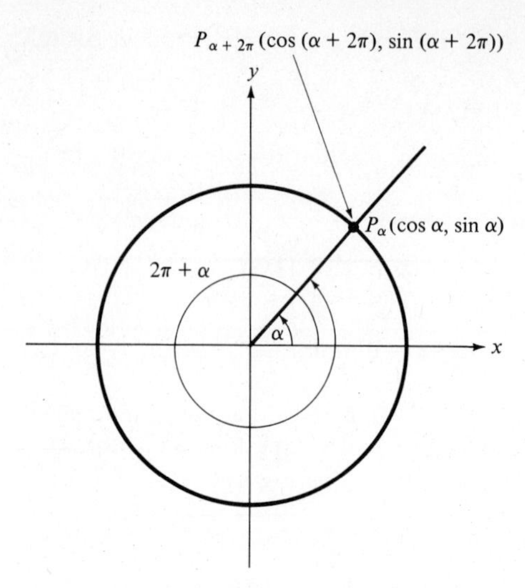

FIGURE 6.16 Because $P_\alpha = P_{\alpha+2\pi}$ the trigonometric functions are periodic:
$\cos\alpha =$ x-coordinate of $P_\alpha = \cos(\alpha + 2\pi)$
$\sin\alpha =$ y-coordinate of $P_\alpha = \sin(\alpha + 2\pi)$

The graphs of $y = \sin x$ and $y = \cos x$ are shown in Figure 6.17. The fundamental cycle for each graph is defined on the interval $[0, 2\pi]$. The similarity of the two graphs is not accidental. We will show later that the graph of $y = \cos x$ can be obtained by shifting the graph of $y = \sin x$ a distance of $\pi/2$ to the left.

Amplitude

The sine and cosine functions have values which range from -1 to $+1$. One half of this range is called the *amplitude* of the functions. Thus

$$\text{amplitude of the sine function} = \tfrac{1}{2}[1 - (-1)] = 1.$$

and

$$\text{amplitude of the cosine function} = \tfrac{1}{2}[1 - (-1)] = 1.$$

(See Figure 6.17.)

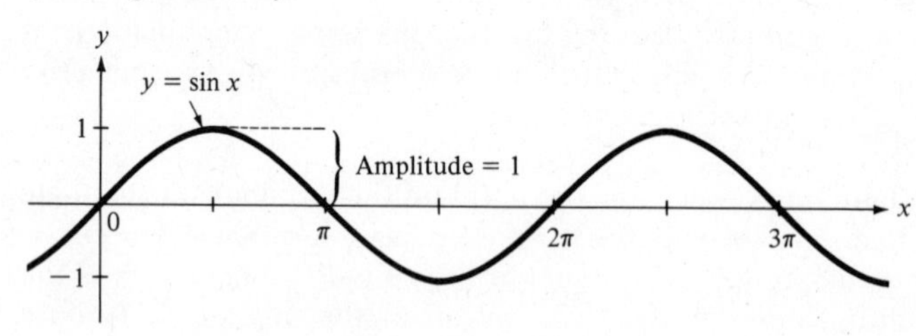

(a) Graph of $y = \sin x$

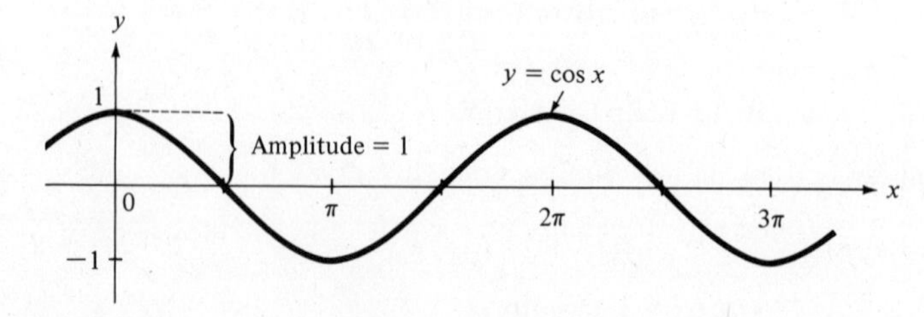

(b) Graph of $y = \cos x$

FIGURE 6.17 The sine and cosine functions are periodic with period $p = 2\pi$. Each graph has amplitude equal to 1.

Graph of $y = \sin bx$

The graphs of $y = a \sin bx$ and $y = a \cos bx$ The graph of $y = \sin bx$ is obtained by a horizontal compression of the graph of $y = \sin x$. This compression is by a factor of b. (If $b > 1$ the graph is compressed, if $0 < b < 1$ it is stretched.) Thus, the graph of $y = \sin 2x$ has a period of $2\pi/2 = \pi$, the graph of $y = \sin 3x$ has a period of $2\pi/3$, and so on. The fundamental cycle of the function $y = \sin bx$ is defined over the interval $[0, 2\pi/b]$. On this interval the graph has the basic shape of the sine curve. (See Figure 6.18a.)

Graph of $y = a \sin bx$

The graph of $y = a \sin bx$ can be obtained from the graph of $y = \sin bx$ by a vertical "stretch." (The y-coordinate of each point on the graph of $y = \sin bx$ is multiplied by the number a.) Thus, the amplitude of $y = a \sin bx$ is a and the period is $2\pi/b$. The graphs of $y = \sin x$, $y = \sin 3x$, and $y = 2 \sin 3x$ are shown for comparison in Figure 6.18b.

In a similar way the graph of $y = a \cos bx$ can be obtained from the graph of $y = \cos x$ by a horizontal compression and a vertical "stretch."

The functions $y = \sin x$ and $y = \cos x$ are differentiable (and therefore continuous) everywhere. The other trigonometric functions can be obtained from these two by taking quotients and reciprocals. Consequently the other trigonomet-

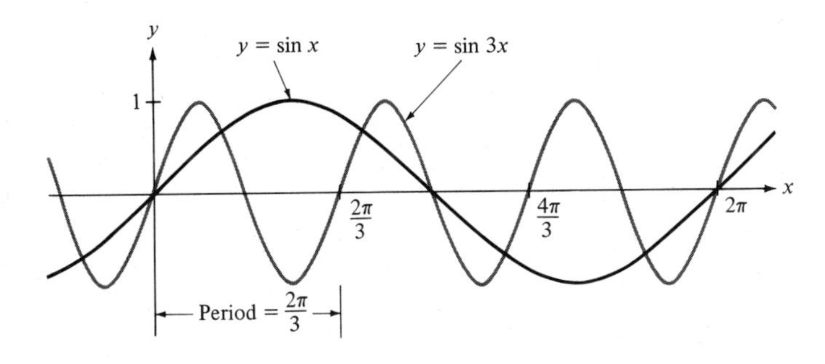

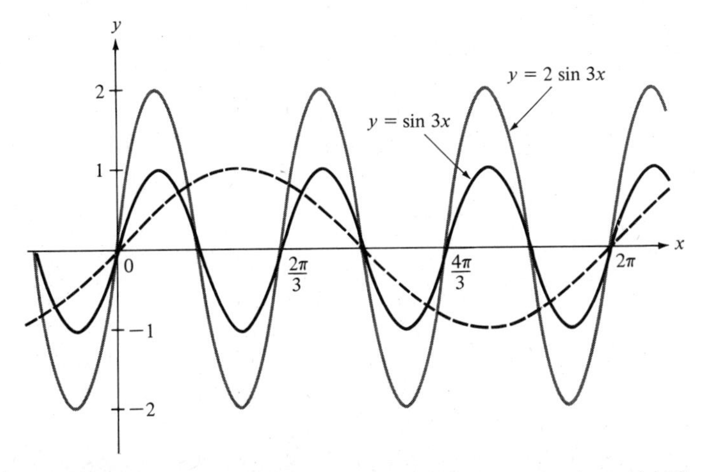

FIGURE 6.18 The graphs of $y = \sin x$, $y = \sin 3x$, $y = 2 \sin 3x$. (a) The graph of $y = \sin 3x$ is obtained by compressing the graph of $y = \sin x$ horizontally. The period is $2\pi/3$, the amplitude is 1. (b) The graph of $y = 2 \sin 3x$ is obtained by "stretching" the graph of $y = \sin 3x$ vertically. The period is $2\pi/3$, the amplitude is 2.

ric functions are differentiable (and continuous) everywhere except where the denominators are zero. Their graphs have vertical asymptotes at these points of discontinuity. The graphs of the tangent, cotangent, secant, and cosecant functions are shown in Figures 6.19 and 6.20.

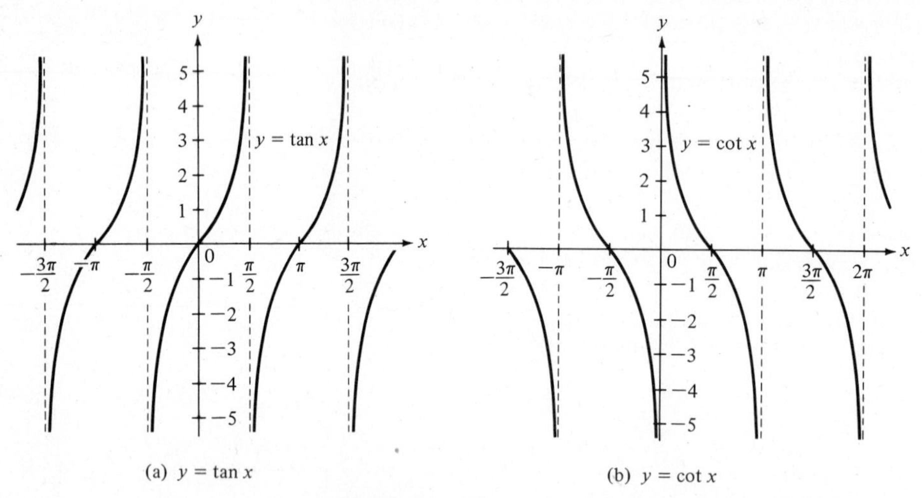

FIGURE 6.19 The graphs of the tangent and cotangent functions.

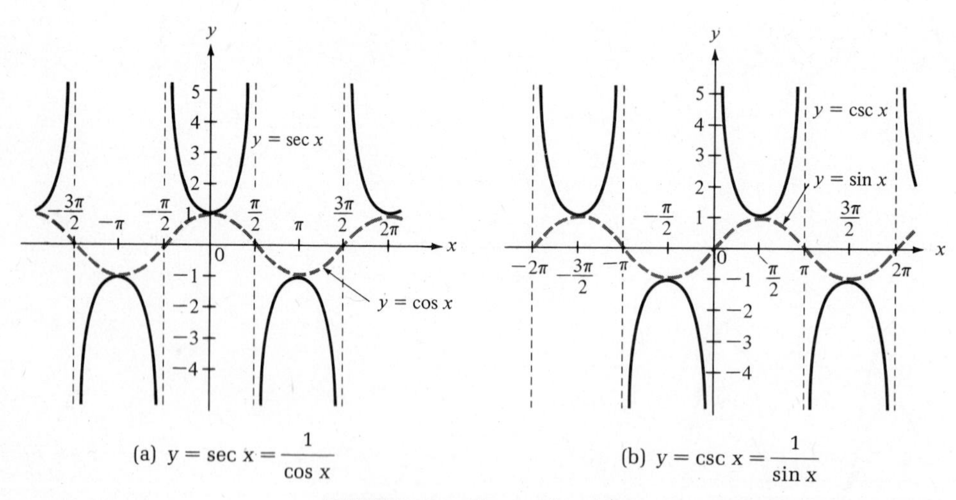

FIGURE 6.20 The graphs of the secant and cosecant functions.

Periodic Functions Various combinations of sin x, cos x, sin 2x, cos 2x, sin 3x, cos 3x, and so on, can be formed. For example, we can consider the functions

$$y = 2 \sin x + \tfrac{1}{2} \sin 3x + \tfrac{1}{4} \cos 4x$$

and

$$y = \tfrac{1}{2} - \sin x - \tfrac{1}{2} \sin 2x - \tfrac{1}{3} \sin 3x - \tfrac{1}{4} \sin 4x - \tfrac{1}{5} \sin 5x - \tfrac{1}{6} \sin 6x.$$

These two graphs are shown in Figure 6.21. Both graphs are periodic with period 2π, but have almost no resemblance to the original sine and cosine functions.

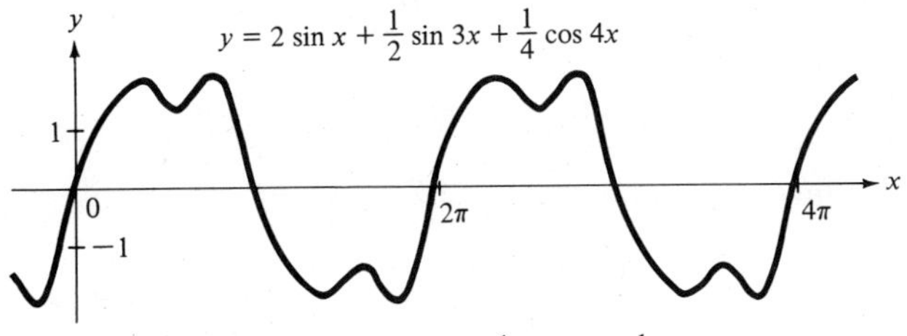

(a) Graph of $y = 2 \sin x + \frac{1}{2} \sin 3x + \frac{1}{4} \cos 4x$

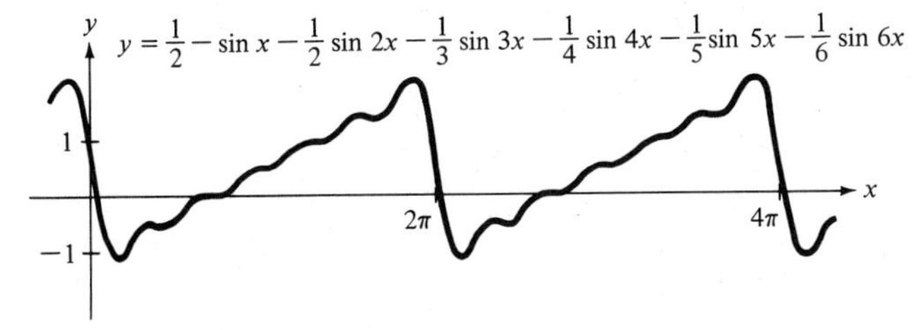

(b) Graph of $y = \frac{1}{2} - \sin x - \frac{1}{2} \sin 2x - \frac{1}{3} \sin 3x - \frac{1}{4} \sin 4x - \frac{1}{5} \sin 5x - \frac{1}{6} \sin 6x$

FIGURE 6.21

As we mentioned earlier, every continuous periodic function can be approximated to any degree of accuracy by adding various sine and cosine functions. Figure 6.22 shows an approximation to the temperature graph of Figure 6.14. The approximating function can be expressed in terms of sines and cosines.

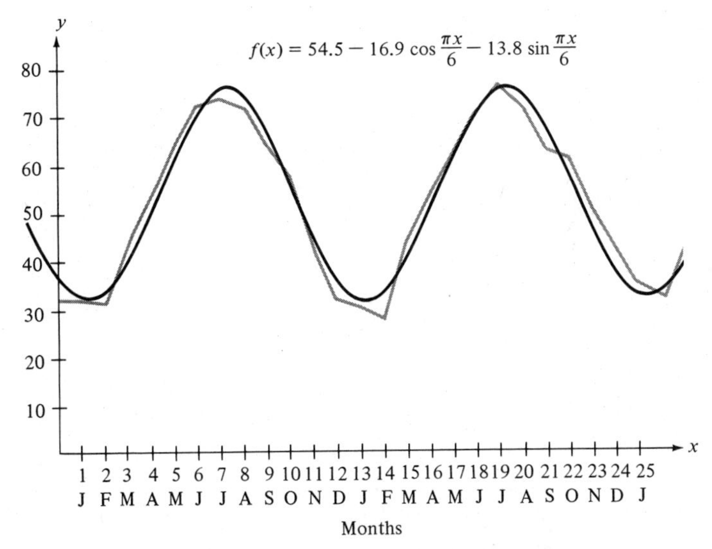

FIGURE 6.22 The graph of a periodic function f that approximates the average monthly temperature of New York City. (See Figure 6.14.)

Actually, if we allow an infinite number of terms, then every continuous periodic function f with period 2π can be written in the form

$$(1) \quad f(x) = a_0 + a_1 \cos x + b_1 \sin x + a_2 \cos 2x + b_2 \sin 2x + a_3 \cos 3x + b_3 \sin 3x + \cdots.$$

The approximating functions that we mentioned above are obtained by keeping the first few terms of the infinite "sum" (which dominate the values of f) and dropping the final terms (which have an insignificant influence on the values of f). The branch of mathematics that deals with such problems is called *Fourier analysis*. Functions of form (1) are called *Fourier series*.

EXERCISES 6.3

1. Sketch the graphs of the following functions. State the period and amplitude of each function.

   (a) $y = \sin x$
   • (b) $y = \sin 2x$
   • (c) $y = 2 \sin 2x$
   (d) $y = \cos x$
   (e) $y = \cos 3x$
   • (f) $y = 2 \cos 3x$

2. Show that $f(x + 2\pi) = f(x)$ for each of the following functions. Sketch the portion of the graph on $[0, 2\pi]$. Use the periodicity of the function to sketch it over the interval $[-2\pi, 4\pi]$.

   (a) $f(x) = \sin x + \cos x$
   (b) $f(x) = \sin x + \sin 2x$
   (c) $f(x) = \sin x + \frac{1}{4} \sin 2x$
   (d) $f(x) = 2 \cos x + \frac{1}{5} \sin 2x$

3. For each of the following quantities let $f(t)$ be the average value for month t. Which functions f are approximately periodic with "period" $t = 12$?

   (a) Dow-Jones average
   (b) Number of marriages in the United States per 1000 persons
   (c) Number of births in the United States per 1000 persons
   (d) Rainfall on Kansas
   (e) Amount of ice cream sold in the United States

6.4 IDENTITIES

A large number of identities relate the trigonometric functions. Some of the most important are listed in Table VIII (to which the reference numbers refer).

Pythagorean identities

Pythagorean Identities The most basic identity is the *Pythagorean identity:*

T5 $$\cos^2 \alpha + \sin^2 \alpha = 1.$$

This identity is a consequence of the fact that the point P_α $(\cos \alpha, \sin \alpha)$ is on the unit circle $x^2 + y^2 = 1$. Other forms of the Pythagorean identity are obtained by dividing both sides of T5 by $\cos^2 \alpha$ and $\sin^2 \alpha$, respectively:

T6 $$1 + \tan^2 \alpha = \sec^2 \alpha,$$

T7 $$\cot^2 \alpha + 1 = \csc^2 \alpha.$$

The Sum and Difference Identities The sum and difference identities express the functions of the sum and difference of two angles in terms of the original angles. These are

Sum and difference identities

T8 $$\sin (\alpha \pm \beta) = \sin \alpha \cos \beta \pm \cos \alpha \sin \beta,$$

T9 $$\cos (\alpha \pm \beta) = \cos \alpha \cos \beta \mp \sin \alpha \sin \beta.$$

The symbol "$\mp$" in T9 is read "minus-or-plus." Identity T9 represents the two identities

$$\cos(\alpha+\beta) = \cos\alpha\cos\beta - \sin\alpha\sin\beta,$$
$$\cos(\alpha-\beta) = \cos\alpha\cos\beta + \sin\alpha\sin\beta.$$

A large number of special identities can be derived from these two basic identities. For example, we can obtain the formula for $\tan(\alpha\pm\beta)$ by writing

$$\tan(\alpha\pm\beta) = \frac{\sin(\alpha\pm\beta)}{\cos(\alpha\pm\beta)}$$

and using the two sum and difference identities. If we choose special values for α and β we get other identities (see Example 2). These identities can be used to compute the functions of angles obtained as sums or differences of standard angles.

Example 1 Use the fact that $15° = 45° - 30°$ to calculate $\sin 15°$ and $\cos 15°$.

Solution We use the table of functions of standard angles in Section 6.2:

$$\sin 15° = \sin(45° - 30°) = \sin 45°\cos 30° - \cos 45°\sin 30° \quad \text{(by T8)}$$
$$= \frac{1}{\sqrt{2}}\cdot\frac{\sqrt{3}}{2} - \frac{1}{\sqrt{2}}\cdot\frac{1}{2} = \frac{\sqrt{3}-1}{2\sqrt{2}} \approx 0.2588,$$
$$\cos 15° = \cos(45° - 30°) = \cos 45°\cos 30° + \sin 45°\sin 30° \quad \text{(by T9)}$$
$$= \frac{1}{\sqrt{2}}\cdot\frac{\sqrt{3}}{2} + \frac{1}{\sqrt{2}}\cdot\frac{1}{2} = \frac{\sqrt{3}+1}{2\sqrt{2}} \approx 0.9659.$$

Example 2 Show that $\sin x = \cos(x - \pi/2)$. Interpret this result graphically.

Solution We use identity T9:

$$\cos\left(x - \frac{\pi}{2}\right) = \cos x\cos\frac{\pi}{2} + \sin x\sin\frac{\pi}{2}$$
$$= \cos x\cdot 0 + \sin x\cdot 1 = \sin x.$$

This result shows that the graph of $y = \sin x$ can be obtained by shifting the graph of $y = \cos x$ a distance of $\pi/2$ units to the right (see Section 1.5). The two graphs are pictured in Figure 6.23.

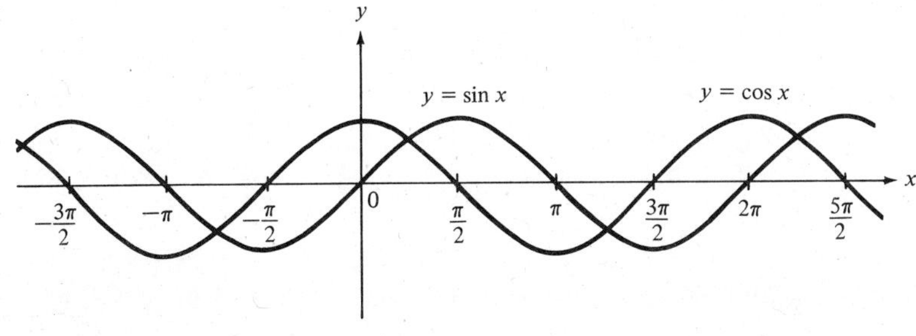

FIGURE 6.23 $\sin x = \cos\left(x - \frac{\pi}{2}\right)$. The graph of $y = \cos x$ is a translate of the graph of $y = \sin x$.

The Double-Angle Identities If we take $\alpha = \beta$ in T8 and T9, we obtain

$$\sin 2\alpha = \sin(\alpha + \alpha) = \sin\alpha\cos\alpha + \cos\alpha\sin\alpha = 2\sin\alpha\cos\alpha,$$
$$\cos 2\alpha = \cos(\alpha + \alpha) = \cos\alpha\cos\alpha - \sin\alpha\sin\alpha = \cos^2\alpha - \sin^2\alpha.$$

This last identity can be changed to other useful forms by use of the Pythagorean identity (T5):

$$\begin{aligned}\cos 2\alpha &= \cos^2\alpha - \sin^2\alpha = \cos^2\alpha - (1 - \cos^2\alpha)\\ &= 2\cos^2\alpha - 1,\\ \cos 2\alpha &= \cos^2\alpha - \sin^2\alpha = (1 - \sin^2\alpha) - \sin^2\alpha = 1 - 2\sin^2\alpha.\end{aligned}$$

We have thus proved the following four identities:

Double-angle identities

T12 $\sin 2\alpha = 2\sin\alpha\cos\alpha.$

T13 $\cos 2\alpha = \cos^2\alpha - \sin^2\alpha.$

T14 $\cos 2\alpha = 2\cos^2\alpha - 1.$

T15 $\cos 2\alpha = 1 - 2\sin^2\alpha.$

These identities can be used in conjunction with T8 and T9 to derive identities for $\sin n\alpha$ and $\cos n\alpha$, where n is any given positive integer.

Example 3 Express $\sin 3\alpha$ in terms of $\sin\alpha$.

Solution

$$\begin{aligned}\sin 3\alpha &= \sin(2\alpha + \alpha) = \sin 2\alpha\cos\alpha + \cos 2\alpha\sin\alpha\\ &= (2\sin\alpha\cos\alpha)\cos\alpha + (\cos^2\alpha - \sin^2\alpha)\sin\alpha\\ &= 2\sin\alpha\cos^2\alpha + \sin\alpha\cos^2\alpha - \sin^3\alpha\\ &= 3\sin\alpha\cos^2\alpha - \sin^3\alpha = 3\sin\alpha(1 - \sin^2\alpha) - \sin^3\alpha\\ &= 3\sin\alpha - 4\sin^3\alpha.\end{aligned}$$

EXERCISES 6.4

1. Simplify by using the indicated identity from Table VIII. Then sketch the graph. (*Hint* for (e): $\cos(\pi/4) = \sin(\pi/4) = 1/\sqrt{2}$.)

   • (a) $y = \sin x\cos x$ (T12)
   (b) $y = \cos^2 x - \sin^2 x$ (T13)
   • (c) $y = \cos\left(\dfrac{3\pi}{4} - x\right)$ (T9)
   (d) $y = \sin^2 x + \cos^2 x$ (T5)
   • (e) $y = \dfrac{\sin x}{\sqrt{2}} + \dfrac{\cos x}{\sqrt{2}}$ (T8)
   (f) $y = \sqrt{\dfrac{\cos x + 1}{2}}$, $0 \leq x \leq \pi$ (T17)

2. Use the indicated identities to calculate the trigonometric functions of the following angles. (*Hint:* In (a) write $15° = 30°/2$; in (c) write $75° = 30° + 45°$.)

   • (a) $\sin 15°$ (T16)
   (b) $\cos 22.5°$ (T17)
   (c) $\sin 75°$ (T8)
   • (d) $\cos 165°$ (T9)
   (e) $\tan 67.5°$ (T9 and T17)
   • (f) $\cot 105°$ (T8 and T9)

•3. Calculate the trigonometric functions of α given that P_α is in quadrant III and $\sin\alpha = -4/5$.

4. If P_α is in quadrant I with $\sin\alpha = 3/5$ and P_β is in quadrant III with $\tan \beta = 5/12$, calculate:

- • (a) $\sin(\alpha + \beta)$
- (b) $\cos(\alpha - \beta)$
- (c) $\sin 2\beta$
- • (d) $\cos 2\alpha$
- • (e) $\sin \beta/2$
- (f) $\cos \alpha/2$
- • (g) $\cot 2\alpha$
- (h) $\tan(\alpha + \beta)$

5. Use identities T1 to T17 to show that the following are true:

(a) $(1 - 2\sin^2\theta)^2 = 1 - \sin^2 2\theta$
(b) $\cos^4\theta - \sin^4\theta = \cos 2\theta$
(c) $\cos(\pi/3 - \theta) - \sin(\pi/6 - \theta) = \sqrt{3}\sin\theta$
(d) $\cos 3x = 4\cos^3 x - 3\cos x$
(e) $\frac{1}{2}[1 + \cos 2\theta] = \cos^2\theta$
(f) $\sin(\theta + 3\pi/2) = -\cos\theta = \sin(3\pi/2 - \theta)$

6.5 DERIVATIVES OF THE TRIGONOMETRIC FUNCTIONS

When we attempt to differentiate the functions sin x and cos x, we are faced with the necessity of evaluating two basic limits:

$$\lim_{\Delta x \to 0} \frac{\sin \Delta x}{\Delta x} \quad \text{and} \quad \lim_{\Delta x \to 0} \frac{\cos \Delta x - 1}{\Delta x}.$$

To see how these limits are involved, we begin the calculation of the derivative of sin x. Let $f(x) = \sin x$. Then

$$\frac{d}{dx}(\sin x) = \lim_{\Delta x \to 0} \frac{f(x + \Delta x) - f(x)}{\Delta x}.$$

We modify the form of the difference quotient $[f(x + \Delta x) - f(x)]/\Delta x$ by use of identity T8:

$$\begin{aligned}\frac{f(x + \Delta x) - f(x)}{\Delta x} &= \frac{\sin(x + \Delta x) - \sin x}{\Delta x} \\ &= \frac{\sin x \cos \Delta x + \cos x \sin \Delta x - \sin x}{\Delta x} \\ &= \sin x \cdot \frac{\cos \Delta x - 1}{\Delta x} + \cos x \cdot \frac{\sin \Delta x}{\Delta x}.\end{aligned}$$

Since sin x and cos x are independent of Δx, it follows that

$$\begin{aligned}\frac{d}{dx}(\sin x) &= \lim_{\Delta x \to 0} \frac{f(x + \Delta x) - f(x)}{\Delta x} \\ &= \sin x \cdot \lim_{\Delta x \to 0} \frac{\cos \Delta x - 1}{\Delta x} + \cos x \cdot \lim_{\Delta x \to 0} \frac{\sin \Delta x}{\Delta x}.\end{aligned}$$

Before proceeding further, we evaluate these two limits.

Evaluation of the Limits To simplify the notation we replace Δx by k. We first consider the case where k approaches zero through positive values. Since we are interested in small values of k (the values near zero), we restrict ourselves to the

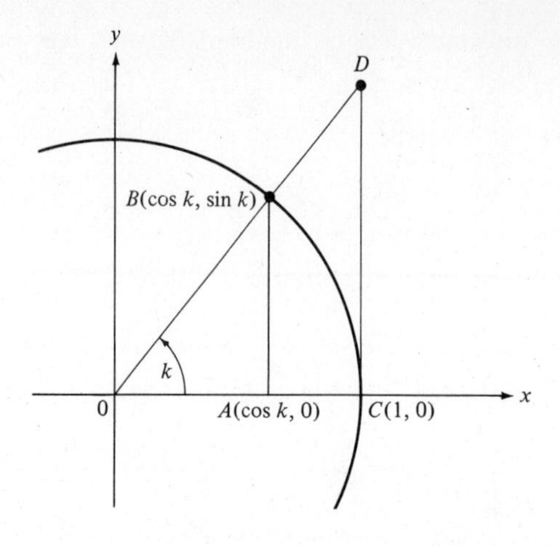

FIGURE 6.24

range $0 < k < \pi/2$. We construct the unit circle and the angle k in standard position. We also construct the two right triangles shown in Figure 6.24. It is obvious that

$$\text{area of } \triangle OAB < \text{area of sector } OCB < \text{area of } \triangle OCD.$$

Observe that $|O\,C| = |O\,B| = 1$, $|O\,A| = \cos k$ and $|A\,B| = \sin k$. (These last two relationships hold because B is the point $(\cos k, \sin k)$.) Thus, the area of triangle OAB is $(\cos k \cdot \sin k)/2$.

To calculate the area of triangle OCD, note that it is similar to triangle OAB, so that ratios of corresponding sides are equal. In particular,

$$\frac{|C\,D|}{|O\,C|} = \frac{|A\,B|}{|O\,A|} = \frac{\sin k}{\cos k}.$$

Since $|O\,C| = 1$, then $|C\,D| = \sin k/\cos k$. It follows that the area of OCD is

$$\frac{1}{2}\,|O\,C| \cdot |C\,D| = \frac{1}{2} \cdot 1 \cdot \frac{\sin k}{\cos k} = \frac{\sin k}{2\cos k}.$$

Recall from Section 6.1 that the area of sector OCB is $(k \cdot 1^2)/2 = k/2$. Therefore,

$$\text{area } \triangle OAB < \text{area sector } OCB < \text{area } \triangle OCD,$$

$$\frac{\cos k \sin k}{2} < \frac{k}{2} < \frac{\sin k}{2\cos k}.$$

We multiply these inequalities by $2/\sin k$, obtaining

$$\cos k < \frac{k}{\sin k} < \frac{1}{\cos k}.$$

and take reciprocals, obtaining

$$\frac{1}{\cos k} > \frac{\sin k}{k} > \cos k.$$

As k approaches 0 through positive values, both $\cos k$ and $1/\cos k$ approach 1. Thus, $(\sin k)/k$ is "squeezed in" between two functions, both of which approach

1. It follows that $(\sin k)/k$ also approaches 1 as k approaches 0 through positive values. Therefore,

$$\lim_{k\to 0^+} \frac{\sin k}{k} = 1.$$

We now consider the case where $k \to 0^-$. If k is negative, we can write $k = -K$, where $K > 0$. Then

$$\begin{aligned}\lim_{k\to 0^-} \frac{\sin k}{k} &= \lim_{-K\to 0^-} \frac{\sin(-K)}{-K} = \lim_{K\to 0^+} \frac{\sin(-K)}{-K} \\ &= \lim_{K\to 0^+} \frac{-\sin K}{-K} \qquad \text{(By T10)} \\ &= \lim_{K\to 0^+} \frac{\sin K}{K} = 1.\end{aligned}$$

Therefore

$$\lim_{k\to 0^+} \frac{\sin k}{k} = \lim_{k\to 0^-} \frac{\sin k}{k} = 1$$

and so

$$\lim_{k\to 0} \frac{\sin k}{k} = 1.$$

This limit can now be used to evaluate $\lim_{k\to 0} [(\cos k - 1)/k]$, avoiding another tedious argument based on geometry. (We also need the more elementary facts that $\lim_{k\to 0} \sin k = 0$ and $\lim_{k\to 0} (\cos k + 1) = 2$.) Observe that

$$\begin{aligned}\frac{\cos k - 1}{k} &= \frac{\cos k - 1}{k} \cdot \frac{\cos k + 1}{\cos k + 1} = \frac{\cos^2 k - 1}{k(\cos k + 1)} \\ &= \frac{-\sin^2 k}{k(\cos k + 1)} \qquad \text{(by T5)} \\ &= (-1)\,\frac{\sin k}{k} \cdot \frac{\sin k}{\cos k + 1}.\end{aligned}$$

We now take the limit as k approaches zero:

$$\begin{aligned}\lim_{k\to 0} \frac{\cos k - 1}{k} &= (-1) \lim_{k\to 0} \frac{\sin k}{k} \cdot \frac{\lim_{k\to 0} \sin k}{\lim_{k\to 0} (\cos k + 1)} \\ &= (-1) \cdot 1 \cdot \frac{0}{2} = 0.\end{aligned}$$

We have proved the following theorem.

Basic Limits

Theorem 6.1 $\lim_{k\to 0} \dfrac{\sin k}{k} = 1$ and $\lim_{k\to 0} \dfrac{\cos k - 1}{k} = 0.$

Theorem 6.1 can be used to evaluate other types of limits involving the trigonometric functions.

Example 1 Show that $\lim_{x\to 0} \frac{\sin 3x}{x} = 3$.

Solution Let $k = 3x$. Then

$$\lim_{x\to 0} \frac{\sin 3x}{x} = \lim_{x\to 0} \frac{\sin 3x}{x} \cdot \frac{3}{3}$$

$$= 3 \lim_{x\to 0} \frac{\sin 3x}{3x} = 3 \lim_{k\to 0} \frac{\sin k}{k} = 3 \cdot 1 = 3.$$

Example 2 Show that $\lim_{x\to 0} \frac{x \csc x}{\sec x} = 1$.

Solution

$$\lim_{x\to 0} \frac{x \csc x}{\sec x} = \lim_{x\to 0} \frac{x \cdot \frac{1}{\sin x}}{\frac{1}{\cos x}}$$

$$= \lim_{x\to 0} \frac{x}{\sin x} \cdot \cos x = \lim_{x\to 0} \frac{1}{\frac{\sin x}{x}} \cdot \cos x$$

$$= \frac{1}{\lim_{x\to 0} \frac{\sin x}{x}} \cdot \lim_{x\to 0} \cos x = \frac{1}{1} \cdot 1 = 1.$$

Derivatives of sin x and cos x We now return to the problem of finding the derivative of the sine function.

Theorem 6.2 $\frac{d}{dx}(\sin x) = \cos x$.

PROOF

$$\frac{d}{dx}(\sin x) = \lim_{\Delta x\to 0} \frac{\sin (x + \Delta x) - \sin x}{\Delta x}$$

$$= \lim_{\Delta x\to 0} \frac{\sin x \cos \Delta x + \cos x \sin \Delta x - \sin x}{\Delta x}$$

$$= \sin x \cdot \lim_{\Delta x\to 0} \frac{\cos \Delta x - 1}{\Delta x} + \cos x \cdot \lim_{\Delta x\to 0} \frac{\sin \Delta x}{\Delta x}$$

$$= \sin x \cdot 0 + \cos x \cdot 1 \quad \text{(by Theorem 6.1)}$$

$$= \cos x.$$

Applying the Chain Rule, we obtain the more general formula

D11 $$\frac{d}{dx}(\sin u) = \cos u \cdot \frac{du}{dx}.$$

Example 3 Calculate $\frac{d}{dx}(\sin \sqrt{x})$.

Solution Using D11 with $u = \sqrt{x} = x^{1/2}$, we obtain

$$\frac{d}{dx}(\sin \sqrt{x}) = \cos \sqrt{x} \cdot \frac{d}{dx}(x^{1/2}) = (\cos \sqrt{x}) \cdot \tfrac{1}{2}x^{-1/2}$$
$$= \frac{\cos \sqrt{x}}{2\sqrt{x}}.$$

Example 4 Calculate $\frac{d}{dx}(e^{\sin x^2})$.

Solution Recall that $\frac{d}{dx}(e^u) = e^u \cdot \frac{du}{dx}$. Therefore,

$$\frac{d}{dx}(e^{\sin x^2}) = e^{\sin x^2} \cdot \frac{d}{dx}(\sin x^2)$$
$$= e^{\sin x^2}(\cos x^2)\frac{d}{dx}(x^2)$$
$$= 2x \cos x^2 e^{\sin x^2}.$$

The derivative of the cosine function can be calculated by an argument similar to the one for the sine function. We give a simpler proof, however, that is based on the two reduction identities (T11):

$$\cos x = \sin\left(\frac{\pi}{2} - x\right),$$
$$\sin x = \cos\left(\frac{\pi}{2} - x\right).$$

Theorem 6.3 $\frac{d}{dx}(\cos x) = -\sin x.$

PROOF

$$\frac{d}{dx}(\cos x) = \frac{d}{dx}\left(\sin\left(\frac{\pi}{2} - x\right)\right) \qquad \text{(by T11)}$$
$$= \cos\left(\frac{\pi}{2} - x\right) \cdot \frac{d}{dx}\left(\frac{\pi}{2} - x\right) \qquad \text{(by D11)}$$
$$= \sin x \cdot (-1) \qquad \text{(by T11)}$$
$$= -\sin x.$$

Applying the Chain Rule, we obtain

D12 $$\frac{d}{dx}(\cos u) = -\sin u \cdot \frac{du}{dx}.$$

Derivatives of the other trigonometric functions

Example 5 Prove that $\frac{d}{dx}(\tan x) = \sec^2 x$.

Solution We write $\tan x = \sin x/\cos x$ and use the formula for the derivative of the quotient (D7):

$$\tan x = \frac{d}{dx}\left(\frac{\sin x}{\cos x}\right) = \frac{\cos x \cdot \frac{d}{dx}(\sin x) - \sin x \cdot \frac{d}{dx}(\cos x)}{\cos^2 x}$$

$$= \frac{\cos x \cdot \cos x - \sin x\,(-\sin x)}{\cos^2 x}$$

$$= \frac{\cos^2 x + \sin^2 x}{\cos^2 x}$$

$$= \frac{1}{\cos^2 x} \qquad \text{(by T5)}$$

$$= \sec^2 x.$$

The derivatives of the other basic trigonometric functions can be obtained by using the derivatives of sin x and cos x as in Example 5. The formulas are listed below.

Basic derivative formulas

D11	$\frac{d}{dx}(\sin u) = \cos u \cdot \frac{du}{dx}.$
D12	$\frac{d}{dx}(\cos u) = -\sin u \cdot \frac{du}{dx}.$
D13	$\frac{d}{dx}(\tan u) = \sec^2 u \cdot \frac{du}{dx}.$
D14	$\frac{d}{dx}(\cot u) = -\csc^2 u \cdot \frac{du}{dx}.$
D15	$\frac{d}{dx}(\sec u) = \sec u \tan u \cdot \frac{du}{dx}.$
D16	$\frac{d}{dx}(\csc u) = -\csc u \cot u \cdot \frac{du}{dx}.$

Example 6 Show that the slope of a tangent line to the graph of $y = \sin x$ can be no greater than 1 and no less than -1.

Solution At each number x the slope of the tangent line is $d(\sin x)/dx = \cos x$. Since $-1 \leq \cos x \leq 1$ for all x, the proposition is proved.

Example 7 Show that the tangent function is an increasing function for $-\pi/2 < x < \pi/2$.

Solution $d(\tan x)/dx = \sec^2 x > 0$ for all x between $-\pi/2$ and $\pi/2$. Since the derivative is positive at each x between $-\pi/2$ and $\pi/2$, the function is increasing on the interval.

EXERCISES 6.5

1. Calculate the value (if any) of each of the following limits. If a limit fails to exist, see whether (possibly infinite) one-sided limits exist.

(a) $\lim_{x\to 0} \frac{\tan x}{x}$

• (b) $\lim_{x\to 0} \frac{\cos 2x}{x}$

• (c) $\lim_{x\to 0} \frac{\sin 3x}{2x}$

(d) $\lim_{x\to \pi/2} \frac{\sin(2x-\pi)}{x-\frac{\pi}{2}}$

• (e) $\lim_{x\to 0} \frac{\sin^2 x}{x}$

• (f) $\lim_{x\to 0} x \cot x$

(g) $\lim_{x\to 0} \frac{\sin x}{x^2}$

(h) $\lim_{x\to 0} \frac{1-\cos x}{x^2}$

2. Calculate the derivative of each of the following functions:

(a) $\cos 5x$

(b) $2 \sin\left(\frac{x}{3}\right)$

(c) $\tan 2x$

• (d) $\sin(x^2+5)$

• (e) $\cos(x^2-x-2)$

• (f) $\sin^2(3x)$

(g) $\sin(\cos x)$

• (h) $\cot(e^x)$

(i) $\ln(\sec x + \tan x)$

• (j) $e^{\sin(x^2)}$

3. Determine all local maxima and minima for the following functions in the range $0 \leqslant x \leqslant 2\pi$. Sketch the graphs.

• (a) $\cos^2 x$

(b) $\sin^2 x$

• (c) $\cos x - \sin x$

(d) $\sin x + \cos x$

4. Prove the following derivative formulas:

(a) $\frac{d}{dx}(\sec x) = \sec x \tan x$

(b) $\frac{d}{dx}(\csc x) = -\csc x \cot x$

(c) $\frac{d}{dx}(\cot x) = -\csc^2 x$

5. Use the derivative formulas of this section to establish the following properties of the trigonometric functions:

(a) The cosine function decreases from 1 to 0 as x increases from 0 to $\pi/2$.
(b) The sine function is increasing for $-\pi/2 < x < \pi/2$ and decreasing for $\pi/2 < x < 3\pi/2$.
(c) The sine function has a local minimum at $x = 3\pi/2$.
(d) The cosine function has local maxima at $x = 2n\pi$ $(n = 0, \pm 1, \pm 2, \ldots)$.

(e) The secant function has local minima at $x = 2n\pi$ ($n = 0, \pm 1, \pm 2, \ldots$).
(f) The cotangent function is decreasing for $0 < x < \pi$.
(g) The minimum value of the slope of the tangent line to the graph of $y = \tan x$ is 1.
(h) The cosecant function has local maxima at $x = 3\pi/2 + 2n\pi$ ($n = 0, \pm 1, \pm 2, \ldots$), and at no other points.

6. Locate all points of inflection for the graphs of the following functions. Where are the graphs concave upward? Concave downward?

• (a) $\cos x$
(b) $\sin x$
(c) $\sec x$
• (d) $\tan x$

6.6 INTEGRATION FORMULAS

Each differentiation formula has its corresponding integration formula. The integrals corresponding to the derivatives listed in the preceding section are listed below:

Basic integration formulas

I7	$\int \cos u \, du = \sin u + C.$
I8	$\int \sin u \, du = -\cos u + C.$
I9	$\int \sec^2 u \, du = \tan u + C.$
I10	$\int \csc^2 u \, du = -\cot u + C.$
I11	$\int \sec u \tan u \, du = \sec u + C.$
I12	$\int \csc u \cot u \, du = -\csc u + C.$

Example 1

$$\begin{aligned}\int_0^{\pi/2} \sin 2x \, dx &= \frac{1}{2}\int_0^{\pi/2} \sin 2x \, 2dx = \frac{1}{2}\,[-\cos 2x]_0^{\pi/2} \\ &= \frac{1}{2}\,[(-\cos \pi) - (-\cos 0)] \\ &= \frac{1}{2}\,[-(-1) + 1] = 1.\end{aligned}$$

Since $d(\sin x) = \cos x \, dx$, then integrals of form

$\int f(\sin x)\, dx$

$$\int f(\sin x) \cos x \, dx,$$

where f is integrable, can be integrated by the substitution

$$u = \sin x, \qquad du = \cos x \, dx.$$

Similarly, integrals of form $\int f(\cos x) \sin x\, dx$ can be integrated by the substitution $u = \cos x$, $du = -\sin x\, dx$.

Example 2 Integrate $\int \sin^3 x \cos x\, dx$.

Solution Let $u = \sin x$, $du = \cos x\, dx$. Then

$$\int \sin^3 x \cos x\, dx = \int u^3\, du = \frac{u^4}{4} + C = \frac{\sin^4 x}{4} + C.$$

$u = \sin x$
$du = \cos x\, dx$

$\int \sin^m x\, dx$ (m odd)

Any integral of form $\int \sin^m x\, dx$ or $\int \cos^m x\, dx$ where *m is an odd integer* can be evaluated by first using the Pythagorean identity $\sin^2 x + \cos^2 x = 1$. The method is illustrated in the following example:

Example 3 Integrate $\int \sin^5 x\, dx$.

Solution We first write the integral in the form

$$\int f(\cos x) \sin x\, dx$$

by making the substitution $\sin^2 x = 1 - \cos^2 x$:

$$\begin{aligned}\int \sin^5 x\, dx &= \int (\sin^2 x)^2 \sin x\, dx \\ &= \int (1 - \cos^2 x)^2 \sin x\, dx \\ &= \int (1 - 2\cos^2 x + \cos^4 x) \sin x\, dx \\ &= \int (1 - 2u^2 + u^4)(-du) \\ &= -u + \frac{2u^3}{3} - \frac{u^5}{5} + C \\ &= -\cos x + \frac{2\cos^3 x}{3} - \frac{\cos^5 x}{5} + C.\end{aligned}$$

$u = \cos x$
$du = -\sin x\, dx$

Of the six basic trigonometric functions, only the sine and cosine can be integrated directly. The others require various techniques of integration. As an example we integrate the tangent function.

Example 4 Show that $\int \tan x\, dx = \ln |\sec x| + C$.

Solution $\int \tan x\, dx = \int \frac{\sin x}{\cos x}\, dx$. We make the substitution $u = \cos x$, $du = -\sin x\, dx$. The integral becomes

$$\int \tan x\, dx = \int \frac{\sin x\, dx}{\cos x} = \int \frac{-du}{u} = -\int \frac{du}{u}$$

$$= -\ln|u| + C = -\ln|\cos x| + C$$

$$= \ln\left(\frac{1}{|\cos x|}\right) + C = \ln|\sec x| + C.$$

$u = \cos x$
$du = -\sin x\, dx$

$\int \sin^m x\, dx$ (m even)

Integrals of Even Powers of sin x and cos x The technique illustrated in Example 3 does not help us to integrate even powers of the sine and cosine functions. If we tried to use it to integrate $\sin^2 x$, for example, we would change $\sin^2 x$ to $1 - \cos^2 x$, then change $\cos^2 x$ to $1 - \sin^2 x$, giving us the original integral. However, we can integrate even powers of sin x and cos x by using the Double-Angle Identities (T14 and T15) to reduce the power. The technique is illustrated below.

Example 5 Integrate $\int \sin^2 x\, dx$.

Solution We write the Double-Angle Identity (T15) in the form

$$\sin^2 x = \frac{1 - \cos 2x}{2}$$

and substitute in the integral

$$\int \sin^2 x\, dx = \int \frac{1 - \cos 2x}{2}\, dx = \frac{1}{2}\int dx - \frac{1}{2}\int \cos 2x\, dx$$

$$= \frac{x}{2} - \frac{1}{2} \cdot \frac{1}{2}\int \cos 2x \cdot 2\, dx$$

$$= \frac{x}{2} - \frac{1}{4}\int \cos u\, du$$

$u = 2x$
$du = 2\, dx$

$$= \frac{x}{2} - \frac{1}{4}\sin u + C$$

$$= \frac{x}{2} - \frac{1}{4}\sin 2x + C.$$

We can integrate higher powers by the technique of Example 5. For example, to integrate $\int \sin^4 x\, dx$ we first use T15 and then T14:

$$\sin^4 x = (\sin^2 x)^2 = \left(\frac{1 - \cos 2x}{2}\right)^2 = \frac{1}{4}(1 - 2\cos 2x + \cos^2 2x)$$

$$= \frac{1}{4}\left(1 - 2\cos 2x + \left[\frac{1 + \cos 4x}{2}\right]\right) \qquad \text{(by T14)}$$

$$= \frac{1}{4}\left(1 - 2\cos 2x + \frac{1}{2} + \frac{1}{2}\cos 4x\right) = \frac{3}{8} - \frac{1}{2}\cos 2x + \frac{1}{8}\cos 4x.$$

Then

$$\int \sin^4 x\, dx = \int \left(\frac{3}{8} - \frac{1}{2}\cos 2x + \frac{1}{8}\cos 4x\right) dx$$

$$= \frac{3x}{8} - \frac{1}{2}\int \cos 2x\, dx + \frac{1}{8}\int \cos 4x\, dx$$

$$= \frac{3x}{8} - \frac{1}{2} \cdot \frac{1}{2} \int \cos 2x \cdot 2\,dx + \frac{1}{8} \cdot \frac{1}{4} \int \cos 4x \cdot 4\,dx$$
$$= \frac{3x}{8} - \frac{1}{4} \sin 2x + \frac{1}{32} \sin 4x + C.$$

EXERCISES 6.6

1. Integrate:
   - (a) $\int \cos (2x - 1)\,dx$
   - • (b) $\int \sin (3x + 3)\,dx$
   - • (c) $\int x \sin 2x^2\,dx$
   - (d) $\int \tan 3x\,dx$
   - • (e) $\int e^{\cos x} \sin x\,dx$
   - • (f) $\int \sin^3 4x\,dx$
   - (g) $\int \cos^2 4x\,dx$
   - • (h) $\int \sin^2 x \cos^2 x\,dx$
   - (i) $\int \sec 3x \tan 3x\,dx$
   - • (j) $\int \csc^2 2x\,dx$
   - (k) $\int \cot (2x + 1)\,dx$
   - • (l) $\int \cos^2 2x \sin 2x\,dx$
   - • (m) $\int \frac{\sin x}{1 + \cos x}\,dx$
   - (n) $\int (1 + \tan x)^2\,dx$
   - • (o) $\int (\cos x) \sqrt{1 + 2 \sin x}\,dx$

2. Calculate the area bounded by the x-axis, the lines $x = 0$ and $x = \pi/4$, and the graphs of the following functions. Make a sketch.
   - (a) $y = \cos x$
   - • (b) $y = \sin 2x$
   - • (c) $y = \tan x$

•3. Calculate the area of one of the closed regions bounded by the graphs of $y = \sin x$ and $y = \cos x$. (See Figure 6.23.)

*6.7 INVERSE TRIGONOMETRIC FUNCTIONS

If x is any number in the range $-1 \leq x \leq 1$ there is a unique number y in the range $-\pi/2 \leq y \leq \pi/2$ such that

$$\sin y = x.$$

This number y is called the *sine-inverse* or *arcsine* of x. Since a unique value of y exists for every x in the interval, then y is defined as a function of x—the *sine-inverse function*. We write

$$y = \text{Sin}^{-1}\,x.$$

$\text{Sin}^{-1}\,x$

Thus, $y = \text{Sin}^{-1}\,x$ if and only if $\sin y = x$ where $-\pi/2 \leq y \leq \pi/2$. The graph of the sine-inverse function is shown in Figure 6.25a.

For example, $\pi/4$ is the only value of y in the range $-\pi/2 \leq y \leq \pi/2$ such that $\sin y = 1/\sqrt{2}$. It follows that

$$\text{Sin}^{-1}\left(\frac{1}{\sqrt{2}}\right) = \text{the number whose sine is } \frac{1}{\sqrt{2}} = \frac{\pi}{4}.$$

To calculate the values of the sine-inverse function we use the trigonometric tables.

Example 1 Calculate
(a) $\text{Sin}^{-1}\,(0.48818)$
(b) $\text{Sin}^{-1}\,(-0.74464)$

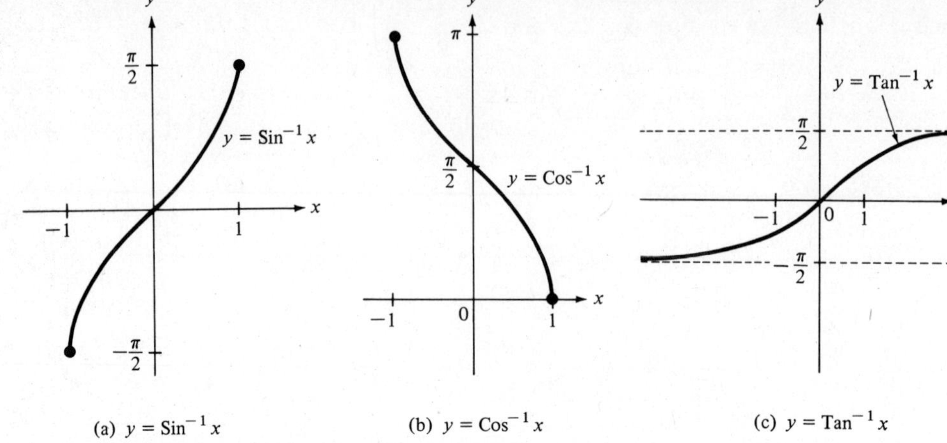

FIGURE 6.25 Graphs of three inverse trigonometric functions.

Solution

(a) Since

$$\sin (0.51) \approx 0.48818 \qquad \text{(by Table II)},$$

then

$$\text{Sin}^{-1} (0.48818) \approx 0.51.$$

(b) Let $\text{Sin}^{-1} (-0.74464) = y$. Then

$$\sin y = -0.74464 \quad \text{and} \quad -\frac{\pi}{2} \leqslant y \leqslant \frac{\pi}{2}.$$

Since $\sin y$ is negative it follows that y is a quadrant IV angle. Let β be the reference angle. (See Figure 6.26.) Then

$$\sin \beta = 0.74464$$

so that

$$\beta \approx 0.84 \qquad \text{(by Table II)}.$$

Observe in Figure 6.26 that $\beta = -y$ Therefore,

$$y \approx -0.84.$$
$$\text{Sin}^{-1} (-0.74464) \approx -0.84.$$

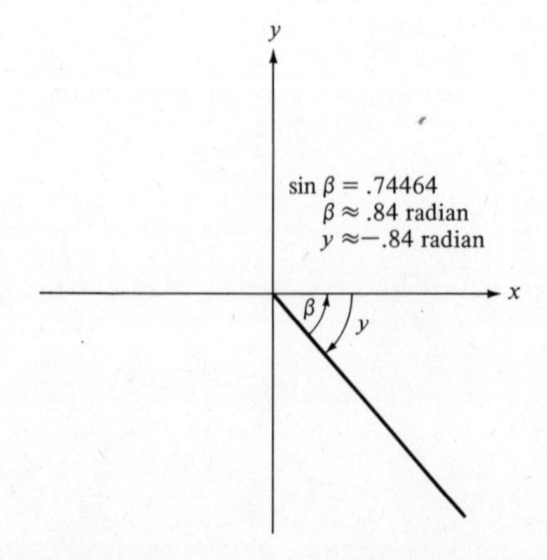

FIGURE 6.26 Example 1b

For each trigonometric function there is a corresponding inverse trigonometric function. For example,

$$\text{Cos}^{-1}\,x = y \qquad \text{if and only if } \cos y = x \text{ where } 0 \le y \le \pi$$

$\text{Cos}^{-1}\,x$

and

$$\text{Tan}^{-1}\,x = y \qquad \text{if and only if} \qquad \tan y = x \text{ where } -\frac{\pi}{2} \le y \le \frac{\pi}{2}.$$

$\text{Tan}^{-1}\,x$

(See Figure 6.25b and c.) We shall not consider the other inverse trigonometric functions in this book.

Application to Integration The reader who has examined Table VII at the end of the book will have observed that several of the integration formulas involve inverse trigonometric functions. In the remainder of this section we indicate how these formulas are derived.

Example 2 Show that

I13	$\displaystyle\int \frac{1}{x^2+1}\,dx = \text{Tan}^{-1}\,x + C.$

$\displaystyle\int \frac{1}{x^2+1}\,dx$

Solution Let $u = \text{Tan}^{-1}\,x$. Then

$$x = \tan u \quad \text{and} \quad dx = \sec^2 u\,du.$$

If we substitute these values into the integral we obtain

$$\int \frac{1}{x^2+1}\,dx = \int \frac{1}{\tan^2 u + 1}\sec^2 u\,du$$

$u = \text{Tan}^{-1}\,x$
$x = \tan u$
$dx = \sec^2 u\,du$

$$= \int \frac{1}{\sec^2 u}\cdot \sec^2 u\,du = \int du = u + C \qquad \text{(by T6)}$$
$$= \text{Tan}^{-1}\,x + C.$$

Example 3 Show that

I14	$\displaystyle\int \frac{1}{\sqrt{1-x^2}}\,dx = \text{Sin}^{-1}\,x + C.$

$\displaystyle\int \frac{1}{\sqrt{1-x^2}}\,dx$

Solution Let $u = \text{Sin}^{-1}\,x$. Then

$$x = \sin u, \qquad dx = \cos u\,du.$$

When we substitute these values into the integral we obtain

$$\int \frac{1}{\sqrt{1-x^2}}\,dx = \int \frac{1}{\sqrt{1-\sin^2 u}}\cos u\,du.$$

$u = \text{Sin}^{-1}\,x$
$x = \sin u$
$dx = \cos u\,du$

Now

$$1 - \sin^2 u = \cos^2 u \qquad \text{(by T5)},$$

so that

$$\sqrt{1-\sin^2 u} = \sqrt{\cos^2 u} = |\cos u|.$$

Since $u = \text{Sin}^{-1} x$, then $-\pi/2 \leqslant u \leqslant \pi/2$, which implies that $\cos u \geqslant 0$. Thus,

$$\sqrt{1 - \sin^2 u} = |\cos u| = \cos u.$$

Returning to the integral we have

$$\int \frac{1}{\sqrt{1 - x^2}}\, dx = \int \frac{1}{\sqrt{1 - \sin^2 u}} \cos u\, du = \int \frac{1}{\cos u} \cos u\, du$$
$$= \int du = u + C = \text{Sin}^{-1} x + C.$$

EXERCISES 6.7

1. Use Table II to calculate the following numbers. For (d) to (f) draw a figure showing the angle and the reference angle.

   (a) $\text{Sin}^{-1}(.29552)$
   • (b) $\text{Tan}^{-1}(.87707)$
   (c) $\text{Cos}^{-1}(.26750)$
   • (d) $\text{Sin}^{-1}(-.89121)$
   (e) $\text{Tan}^{-1}(-.09024)$
   • (f) $\text{Cos}^{-1}(-.94604)$

2. Solve the following equations for x in terms of y:

   (a) $y = \cos x$
   • (b) $y = \sin(2x)$
   (c) $2y = 1 - \cos 3x$
   • (d) $y = \text{Cos}^{-1}(3x)$
   (e) $y = \text{Tan}^{-1}(x) + 2$
   • (f) $2y + 1 = \text{Sin}^{-1}(x + 2)$

3. Use the formulas in Examples 2 and 3 to calculate the integrals.

   • (a) $\displaystyle\int_0^1 \frac{dx}{1 + x^2}$
   (b) $\displaystyle\int \frac{dx}{1 + (2x)^2}$ (let $v = 2x$)
   (c) $\displaystyle\int_0^{\frac{\sqrt{3}}{2}} \frac{dx}{\sqrt{1 - x^2}}$
   • (d) $\displaystyle\int \frac{2\, dx}{\sqrt{1 - 9x^2}}$ (let $v = 3x$)

7 TECHNIQUES OF INTEGRATION

In general integration is more difficult than differentiation. The basic reason is that the integration formulas are specialized whereas the derivative formulas are quite general. We frequently can use a derivative formula to break a difficult derivative problem into several parts in which simpler derivatives can be calculated. With integration, on the other hand, we must have an exact fit to a formula before it can be applied. Thus, for example, the "complicated" integral $\int e^x \sqrt{e^x + 1}\, dx$ can be integrated easily by the substitution $u = e^x$, while the "simple" integral $\int e^{\sqrt{x}}\, dx$ cannot be integrated by any elementary method at all.

Because of the difficulty of integration many techniques have been devised to aid in the integration of special classes of functions. In this chapter we consider three of the most important of these techniques—*substitution, partial fractions,* and *integration-by-parts*. We then turn our attention to numerical methods of approximate integration.

In light of Theorem 4.2 the reader is apt to think that any continuous function can be integrated by use of the proper method. In a certain sense this is true. The function has an integral, but, unfortunately, we may not be able to express it in terms of any previously known function. It is conceivable that such an integral could only be approximated by some special method. The situation is analogous to that in which we might try to calculate $\int_1^x (dt/t)$ without any prior knowledge of the logarithmic function. The integral would define a completely new function, which we would then have to study. As a matter of fact, many of the important functions studied in higher mathematics are defined as integrals of more elementary functions.

7.1 THE SUBSTITUTION METHOD

The method of *substitution*, the most basic of all of the techniques of integration, has been used many times thus far in this book. It consists of substituting a function of a new variable for the original one, thereby changing the integral to one that can be evaluated more easily. For example, to integrate $\int e^{2x} 2\,dx$ we let $2x = u$, changing the integral to $\int e^u\,du$. The value of this last integral is $e^u + C$, which we rewrite as $e^{2x} + C$.

In general, let $\int f(x)\,dx$ be the original integral, where f is continuous over the interval under consideration. The method rests on finding functions g and h (g differentiable and h continuous) such that $f(x)\,dx = h(g(x))g'(x)\,dx$. We then make the substitution $g(x) = u$, changing the original integral to $\int h(u)\,du$, which we integrate. The validity of this method is established in the following theorem.

The substitution principle

Theorem 7.1 Let g be a differentiable function of x. Suppose the substitution $u = g(x)$, $du = g'(x)\,dx$ changes $f(x)\,dx$ into $h(u)\,du$ where h is continuous. If H is an antiderivative of h, then

$$\int f(x)\,dx = H(g(x)) + C = H(u) + C = \int h(u)\,du + C.$$

PROOF We must show that $H(g(x))$ is an antiderivative of $f(x)$. Recall that $h(u)\,du = f(x)\,dx$, so that

$$h(u) \cdot \frac{du}{dx} = f(x).$$

It now follows from the Chain Rule that

$$\begin{aligned} \frac{d}{dx}(H(g(x))) &= \frac{d}{dx}(H(u)) = \frac{d}{du}(H(u)) \cdot \frac{du}{dx} \\ &= H'(u) \cdot \frac{du}{dx} = h(u) \cdot \frac{du}{dx} = f(x). \end{aligned}$$

Since $H(g(x))$ is an antiderivative of $f(x)$, then

$$\int f(x)\,dx = H(g(x)) + C = H(u) + C = \int h(u)\,du + C.$$

The Substitution Principle can be used with definite as well as indefinite integrals.

Theorem 7.2 Let f be continuous and g differentiable on the closed interval $[a, b]$. Suppose the substitution $u = g(x)$, $du = g'(x)\,dx$ changes $f(x)\,dx$ into $h(u)\,du$. Let $u = \alpha$ when $x = a$ and $u = \beta$ when $x = b$. If h is continuous on the closed interval with endpoints α, β, then

Substitution with definite integrals

$$\int_a^b f(x)\,dx = \int_\alpha^\beta h(u)\,du.$$

PROOF Observe that $g(a) = \alpha$ and $g(b) = \beta$. Since h is continuous on the closed interval with endpoints α and β, then h has an antiderivative H on that interval. Furthermore, by Theorem 7.1,

$$\int f(x)\,dx = \int h(u)\,du = H(u) + C = H(g(x)) + C.$$

Therefore,

$$\int_a^b f(x)\,dx = [H(g(x)) + C]_a^b = [H(g(b)) + C] - [H(g(a)) + C]$$
$$= H(g(b)) - H(g(a)) = H(\beta) - H(\alpha)$$
$$= \int_\alpha^\beta h(u)\,du.$$

This result is very useful, for it enables us to evaluate the integral obtained by the substitution without returning to the original function or the original limits of integration.

Example 1 Calculate $\int_e^{e^2} \frac{1}{x \ln x}\,dx$.

Solution

$$\int_e^{e^2} \frac{1}{x \ln x}\,dx = \int_e^{e^2} \frac{1}{\ln x} \cdot \frac{dx}{x}.$$

If we let $u = \ln x$, then $du = dx/x$ and

$$\frac{1}{\ln x}\frac{dx}{x} = \frac{1}{u}\,du$$

$u = \ln x$ $du = \frac{1}{x}\,dx$

When $x = e$, $u = \ln e = 1$, and when $x = e^2$, $u = \ln e^2 = 2$. Therefore,

$$\int_e^{e^2} \frac{1}{\ln x} \cdot \frac{dx}{x} = \int_1^2 \frac{1}{u}\,du$$
$$= [\ln u]_1^2 = \ln 2 - \ln 1 = \ln 2.$$

Example 2 Calculate $\int_0^{\pi/2} \sin^3 x \cos x\,dx$.

Solution Let $u = \sin x$, $du = \cos x\,dx$. When $x = 0$, $u = 0$, when $x = \pi/2$, $u = 1$. Therefore,

$$\int_0^{\pi/2} \sin^3 x \cos x\,dx = \int_0^1 u^3\,du = \left[\frac{u^4}{4}\right]_0^1 = \frac{1}{4}.$$

$u = \sin x$ $du = \cos x\,dx$

Example 3 Calculate $\int_0^1 \frac{e^x}{1 + e^x}\,dx$.

Solution Let $u = 1 + e^x$, $du = e^x\,dx$. When $x = 0$, $u = 2$, when $x = 1$, $u = 1 + e$. Therefore,

$$\int_0^1 \frac{e^x}{1 + e^x}\,dx = \int_2^{1+e} \frac{du}{u} = [\ln u]_2^{1+e}$$
$$= \ln(1 + e) - \ln 2 = \ln\left(\frac{1 + e}{2}\right).$$

$u = 1 + e^x$ $du = e^x\,dx$

EXERCISES 7.1

1. Integrate:

• (a) $\int e^{\sqrt{t}} \cdot \frac{dt}{\sqrt{t}}$

(b) $\int x^2 \cos x^3 \, dx$

(c) $\int_0^{\pi/2} \sin x \cdot e^{\cos x} \, dx$

• (d) $\int_0^3 \frac{x^2}{x^3+4} \, dx$

(e) $\int \frac{1}{\sqrt{x}-1} \cdot \frac{1}{\sqrt{x}} \, dx$

• (f) $\int (x - \ln x)\left(1 - \frac{1}{x}\right) dx$

(g) $\int_1^{e^2} \frac{\ln x}{x} \, dx$

(h) $\int \frac{e^{-x}}{1+e^{-x}} \, dx$

• (i) $\int_0^1 \frac{e^x}{\sqrt{1+e^x}} \, dx$

(j) $\int \csc^2 x \cdot e^{\cot x} \, dx$

2. Integrate:

• (a) $\int_{e^{\pi}}^{e^{2\pi}} \frac{\sin(\ln x)}{x} \, dx$

• (b) $\int_0^1 \frac{e^{2x}+2e^x}{e^x+1} \, dx$

(c) $\int_3^9 \frac{2x \, dx}{\sqrt{x^2-1}}$

(d) $\int_0^{27} \frac{x^{-2/3}}{2+x^{1/3}} \, dx$

• (e) $\int \frac{dx}{x(\ln x)^2}$

(f) $\int \frac{\cos x}{\sin x} \, dx$

7.2 INTEGRATION BY PARTS

One of the most important techniques of integration is based on the formula for the derivative of a product:

D6 $$\frac{d}{dx}(uv) = u \cdot \frac{dv}{dx} + v \cdot \frac{du}{dx}.$$

If we integrate both sides of D6 with respect to x, recalling that $\int \frac{d}{dx}(f(x)) \, dx = f(x) + C$, we obtain

$$\int \frac{d}{dx}(uv) \, dx = \int \left(u \frac{dv}{dx}\right) dx + \int \left(v \frac{du}{dx}\right) dx$$
$$uv + C = \int u \, dv + \int v \, du.$$

Solving this equation for $\int u \, dv$, we have the *integration-by-parts* formula:

Integration-by-parts formula

I15 $$\int u \, dv = uv - \int v \, du + C.$$

To apply the formula to the integration of a product, say $\int f(x)g(x) \, dx$, we choose one factor, say $f(x)$, as u and the remaining factor $g(x) \, dx$ as dv. It usually is best to try to choose the factors so that $v \, du$ can be integrated more easily than the original integral. In some of the simpler cases, at least, it works well to choose u and dv so that du is simpler than u and v is not much more complicated than dv.

Example 1 Calculate $\int xe^x\,dx$.

Solution Let $u = x$, $dv = e^x\,dx$. Then $du = dx$, $v = e^x$. (Actually we should have an arbitrary constant added to $v = e^x$, but it would be canceled out later. See Exercise 2.) Then

$u = x$ $du = dx$	$dv = e^x\,dx$ $v = e^x$

$$\int xe^x\,dx = \int u\,dv = uv - \int v\,du + C$$
$$= xe^x - \int e^x\,dx + C$$
$$= xe^x - e^x + C.$$

As one would suspect from a careful study of the example above, integration by parts is especially useful for integrating functions of form $xf(x)$, where $f(x)$ can be integrated twice. Thus, it is the most natural technique for such integrals as

$$\int xe^x\,dx, \qquad \int x\sin x\,dx, \qquad \int x\cos x\,dx.$$

Reduction formulas

More sophisticated techniques can be based on this method. For example, certain *reduction formulas* can be derived quite easily. These are formulas in which a given integral is expressed in terms of similar integrals of simpler form.

Example 2 Let n be a positive integer. Use integration by parts to derive the reduction formula

$$\boxed{\int x^n e^x\,dx = x^n e^x - n\int x^{n-1}e^x\,dx + C.}$$

Solution Let $u = x^n$, $dv = e^x\,dx$. Then $du = nx^{n-1}\,dx$, $v = e^x$.

$u = x^n$ $du = nx^{n-1}\,dx$
$dv = e^x\,dx$ $v = e^x$

$$\int x^n e^x\,dx = \int u\,dv = uv - \int v\,du + C$$
$$= x^n e^x - \int e^x(nx^{n-1}\,dx) + C$$
$$= x^n e^x - n\int x^{n-1}e^x\,dx + C.$$

To illustrate the use of the reduction formula we calculate $\int x^n e^x\,dx$ for $n = 1, 2, 3$.

$n = 1$: $$\int xe^x\,dx = xe^x - \int e^x\,dx = xe^x - e^x + C.$$

$n = 2$: $$\int x^2e^x\,dx = x^2e^x - 2\int xe^x\,dx$$
$$= x^2e^x - 2\{xe^x - e^x\} + C \quad \text{(by case } n = 1\text{)}$$
$$= x^2e^x - 2xe^x + 2e^x + C.$$

$$n = 3: \quad \int x^3e^x\,dx = x^3e^x - 3\int x^2e^x\,dx$$

$$= x^3e^x - 3\{x^2e^x - 2xe^x + 2e^x\} + C \quad \text{(by case } n = 2)$$

$$= x^3e^x - 3x^2e^x + 6xe^x - 6e^x + C.$$

In certain cases it may be necessary to apply the integration-by-parts formula twice.

Example 3 Calculate $\int e^x \sin x\,dx$.

Solution: We let $u = \sin x$, $dv = e^x\,dx$. Then $du = \cos x\,dx$, $v = e^x$.

$$\int e^x \sin x\,dx = \int u\,dv = uv - \int v\,du$$

$u = \sin x$ $du = \cos x\,dx$	$dv = e^x\,dx$ $v = e^x$

$$= e^x \sin x - \int e^x \cos x\,dx + C_1.$$

We apply the formula again with $U = \cos x$, $dV = e^x\,dx$, $dU = -\sin x\,dx$, $V = e^x$:

$$\int e^x \sin x\,dx = e^x \sin x - \int e^x \cos x\,dx + C_1$$

$$= e^x \sin x - \int U\,dV + C_1$$

$$= e^x \sin x - \left\{UV - \int V\,dU + C_2\right\} + C_1$$

$U = \cos x$ $dU = -\sin x\,dx$	$dV = e^x\,dx$ $V = e^x$

$$= e^x \sin x - e^x \cos x + \int e^x(-\sin x)\,dx + (C_1 - C_2),$$

$$\int e^x \sin x\,dx = e^x \sin x - e^x \cos x - \int e^x \sin x\,dx + (C_1 - C_2).$$

This equation is of form $X = Y - X$ (where $X = \int e^x \sin x\,dx$). If we solve the equation for X, we have

$$2\int e^x \sin x\,dx = e^x \sin x - e^x \cos x + (C_1 - C_2),$$

$$\int e^x \sin x\,dx = e^x\left(\frac{\sin x - \cos x}{2}\right) + C,$$

where $C = (C_1 - C_2)/2$.

EXERCISES 7.2

1. Use integration by parts:

(a) $\int x^2e^{2x}\,dx$

• (b) $\int x \ln x\,dx$

(c) $\int x\sqrt{x+1}\,dx$

• (d) $\int x^2 \ln x\,dx$

• (e) $\int \ln x\,dx$

(f) $\int (\ln x)^2\,dx$

(g) $\int x^3 \ln x\,dx$

• (h) $\int e^x \cos x\,dx$

2. This exercise refers to the solution of Example 1. Show that if we choose $v = e^x + K$, where K is a constant, we get the same answer as if we choose $v = e^x$. Show that this is true in general.

3. Use the reduction formula of Example 2 to calculate $\int x^4e^x\,dx$ and $\int x^5e^x\,dx$.

4. Use integration by parts to obtain the following reduction formulas:

$$\int x^n \cos x\,dx = x^n \sin x - n\int x^{n-1} \sin x\,dx$$

$$\int x^n \sin x\,dx = -x^n \cos x + n\int x^{n-1} \cos x\,dx$$

5. Use the reduction formulas of Exercise 4 to evaluate the following integrals:

• (a) $\int x^3 \sin x\,dx$

(b) $\int x^4 \cos x\,dx$

(c) $\int x^4 \sin x\,dx$

(d) $\int x^5 \cos x\,dx$

7.3 PARTIAL FRACTIONS

When we add fractions we add the numerators over a common denominator. For example,

$$\frac{1}{x} + \frac{2}{x+1} + \frac{3}{x+2} = \frac{(x+1)(x+2) + 2x(x+2) + 3x(x+1)}{x(x+1)(x+2)}$$

$$= \frac{(x^2+3x+2) + (2x^2+4x) + (3x^2+3x)}{x(x+1)(x+2)}$$

$$= \frac{6x^2+10x+2}{x(x+1)(x+2)}.$$

The method of partial fractions enables us to reverse the steps in the above process and decompose fractions, such as

$$\frac{6x^2+10x+2}{x(x+1)(x+2)},$$

into "partial" fractions, such as

$$\frac{1}{x} + \frac{2}{x+1} + \frac{3}{x+2},$$

which can more easily be integrated.

Remark In our work we will consider rational functions (quotients of polynomials) in which the numerator has lower degree than the denominator. If this condition is not met we first carry out the long division process, dividing denominator into numerator, until we reduce the problem to an equivalent one involving a fraction in which the numerator has lower degree than the denominator.

For example, if our original fraction is

$$\frac{2x^2 + x + 3}{x^2 - 1} \qquad \text{(numerator has same degree as denominator)}$$

we divide $x^2 - 1$ into $2x^2 + x + 3$, obtaining

$$\frac{2x^2 + x + 3}{x^2 - 1} = 2 + \frac{x + 5}{x^2 - 1} = 2 + \frac{x + 5}{(x - 1)(x + 1)}.$$

We now have only to decompose $\frac{x + 5}{(x - 1)(x + 1)}$ into partial fractions.

We will not consider the more complicated problems involving partial fractions. Instead we will restrict ourselves to the case where the denominator is a product of distinct linear terms, say $(x - a)(x - b)(x - c)$. . . . In that case there is a partial fraction corresponding to each of the linear terms. The decomposition is

$$\frac{f(x)}{(x - a)(x - b)(x - c) \cdots} = \frac{A}{x - a} + \frac{B}{x - b} + \frac{C}{x - c} + \cdots$$

where $A, B, C, \ldots$ are constants which must be determined. The following example illustrates the method.

Example 1

(a) Write $\frac{x + 5}{(x - 1)(x + 1)}$ as a sum of partial fractions.

(b) Integrate $\int \frac{2x^2 + x + 3}{x^2 - 1}\, dx$.

Solution

(a) $\frac{x + 5}{(x - 1)(x + 1)}$ can be written as a sum of two partial fractions

$$\frac{x + 5}{(x - 1)(x + 1)} = \frac{A}{x - 1} + \frac{B}{x + 1}.$$

To calculate A and B we add the two fractions on the right:

$$\frac{x + 5}{(x - 1)(x + 1)} = \frac{A}{x - 1} + \frac{B}{x + 1} = \frac{A(x + 1) + B(x - 1)}{(x - 1)(x + 1)}.$$

Since this equation is an identity the numerators must be equal functions:

$$x + 5 = A(x + 1) + B(x - 1).$$

To calculate A and B we substitute the values $x = -1$ and $x = 1$, obtaining

$$x = -1: \quad -1 + 5 = A \cdot 0 + B \cdot (-2),$$
$$4 = -2B,$$
$$B = -2.$$

$$x = 1: \quad 1 + 5 = A \cdot 2 + B \cdot 0,$$
$$6 = 2A,$$
$$A = 3.$$

The partial fraction decomposition is

$$\frac{x+5}{(x-1)(x+1)} = \frac{3}{x-1} - \frac{2}{x+1}.$$

(b) Recall that

$$\frac{2x^2 + x + 3}{x^2 - 1} = 2 + \frac{x+5}{(x-1)(x+1)}.$$

Therefore

$$\int \frac{2x^2 + x + 3}{x^2 - 1}\,dx = \int \left[2 + \frac{x+5}{(x-1)(x+1)}\right] dx$$
$$= \int \left[2 + \frac{3}{x-1} - \frac{2}{x+1}\right] dx$$
$$= 2x + 3\ln|x-1| - 2\ln|x+1| + C.$$

Example 2 Integrate $\int \frac{x^3 + 2x^2 + 5x + 7}{x^2 + 4x + 3}\,dx.$

Solution

1. Since the numerator has degree greater than that of the deonominator we divide it by the denominator:

$$\frac{x^3 + 2x^2 + 5x + 7}{x^2 + 4x + 3} = x - 2 + \frac{10x + 13}{x^2 + 4x + 3}$$
$$= x - 2 + \frac{10x + 13}{(x+1)(x+3)}.$$

2. We decompose $\frac{10x + 13}{(x+1)(x+3)}$ into partial fractions:

$$\frac{10x + 13}{(x+1)(x+3)} = \frac{A}{x+1} + \frac{B}{x+3}.$$

To find A and B we add the fractions on the right:

$$\frac{10x + 13}{(x+1)(x+3)} = \frac{A(x+3) + B(x+1)}{(x+1)(x+3)}.$$
$$10x + 13 = A(x+3) + B(x+1) \quad \text{(equating the numerators).}$$

We now substitute the values $x = -3$ and $x = -1$ into the equation

$$10x + 13 = A(x+3) + B(x+1).$$

$$x = -3: \quad -30 + 13 = A \cdot 0 + B \cdot (-2).$$
$$-17 = -2B.$$
$$B = \tfrac{17}{2}.$$
$$x = -1: \quad -10 + 13 = A \cdot 2 + B \cdot 0$$
$$3 = 2A$$
$$A = \tfrac{3}{2}.$$

The partial fraction decomposition is

$$\frac{10x + 13}{(x+1)(x+3)} = \frac{3/2}{x+1} + \frac{17/2}{x+3}$$

3. $\displaystyle\int \frac{x^3 + 2x^2 + 5x + 7}{x^2 + 4x + 3}\,dx$

$$= \int \left[x - 2 + \frac{10x + 13}{(x+1)(x+3)} \right] dx = \int \left[x - 2 + \frac{3/2}{x+1} + \frac{17/2}{x+3} \right] dx$$

$$= \frac{x^2}{2} - 2x + \frac{3}{2} \ln |x+1| + \frac{17}{2} \ln |x+3| + C.$$

Our work on partial fractions can be extended to cover denominators which are products of powers of linear and quadratic functions. The interested reader can find an explanation of the technique in almost any introductory calculus book written for engineering students.[1]

EXERCISES 7.3

1. Use the methods of this section to integrate the following functions:

(a) $\displaystyle\int \frac{3x + 3}{(x-1)(x+2)}\,dx$

• (b) $\displaystyle\int \frac{2x + 1}{(x-2)(x-3)}\,dx$

(c) $\displaystyle\int \frac{6}{(x+5)(x+1)}\,dx$

• (d) $\displaystyle\int \frac{3x - 1}{x(x+7)}\,dx$

(e) $\displaystyle\int \frac{3x^2 + 10x - 4}{x(x-1)(x+2)}\,dx$

• (f) $\displaystyle\int \frac{5x + 7}{x^2 + 3x + 2}\,dx$

(g) $\displaystyle\int \frac{8x^3 + 2x^2 - 5x + 7}{x^2 - 4x + 3}\,dx$

• (h) $\displaystyle\int \frac{dx}{(x-a)(x-b)}$

7.4 TABLES OF INTEGRALS

The calculation of integrals can be quite laborious and may depend as much on luck as on skill. For this reason extensive tables of integrals have been published. The standard mathematics reference books all contain tables of moderate length. For example, the Rinehart tables[2] list 494 integrals, and the CRC tables[3] give 463 integrals. While these lists are adequate for most elementary work, more extensive tables listing thousands of integrals are available for the specialist.

The usual table has integrals classified according to type: exponential forms, logarithmic forms, forms involving $\sqrt{a^2 - x^2}$, forms involving $\sqrt{ax + b}$, and so on. To avoid trivial duplication, only certain basic integrals are given. Related integrals must be reduced to one of these forms by appropriate substitu-

---

[1] See, for example, A. Schwartz, *Calculus with Analytic Geometry*, 3rd ed. New York: Holt, Rinehart and Winston, 1974, pp. 437–448.

[2] H. D. Larsen, *Rinehart Mathematical Tables, Formulas and Curves*. New York: Holt, Rinehart and Winston, 1953.

[3] C. D. Hodgman, *CRC Standard Mathematical Tables*, 12th ed. Cleveland, Ohio: Chemical Rubber Publishing Co., 1959.

tions or other methods. Thus, a table may list many forms involving $\sin x$ or $\sin ax$, but no forms involving $\sin 7x$.

Table VII at the end of the book contains a list of integrals selected from the Rinehart tables. Observe that the constant of integration is omitted. Natural logarithms are denoted by "ln" as in this book, rather than by "log" as is customary in most tables and advanced books.

The following examples illustrate the use of Table VII. (The reference numbers refer to Table VII.)

Example 1 Evaluate $\int \sin^3[2x - \pi/2]\, dx$.

Solution Let $u = 2x - \pi/2$. Then $du = 2\, dx$, and the integral becomes

$$\int \sin^3\left(2x - \frac{\pi}{2}\right) dx = \int \sin^3 u \cdot \frac{du}{2} = \frac{1}{2}\int \sin^3 u\, du$$

$$= \frac{1}{2}\left[\frac{\cos^3 u}{3} - \cos u\right] + C \quad (\text{by } \mathbf{43})$$

$$= \frac{\cos^3\left(2x - \frac{\pi}{2}\right)}{6} - \frac{\cos\left(2x - \frac{\pi}{2}\right)}{2} + C.$$

$$\boxed{\begin{aligned} u &= 2x - \frac{\pi}{2} \\ du &= 2\, dx \end{aligned}}$$

Example 2 Evaluate $\int x^3 \sqrt{2x + 3}\, dx$.

Solution The integral is of form $\int x^m \sqrt{ax + b}\, dx$. We use reduction formula **20** with $a = 2$, $b = 3$, $m = 3$.

$$\int x^3 \sqrt{2x + 3}\, dx$$

$$= \frac{2}{2(2 \cdot 3 + 3)}\left[x^3 \sqrt{(2x + 3)^3} - 3 \cdot 3 \int x^2 \sqrt{2x + 3}\, dx\right] \quad (\text{by } \mathbf{20})$$

$$= \frac{1}{9} x^3 \sqrt{(2x + 3)^3} - \int x^2 \sqrt{2x + 3}\, dx$$

$$= \frac{1}{9} x^3 \sqrt{(2x + 3)^3} - \left[\frac{2(15 \cdot 2^2 x^2 - 12 \cdot 2 \cdot 3x + 8 \cdot 3^2}{105 \cdot 2^3} \sqrt{(2x + 3)^3}\right] + C \quad (\text{by } \mathbf{19})$$

$$= \frac{1}{9} x^3 \sqrt{(2x + 3)^3} - \left[\frac{5x^2 - 6x + 6}{35}\right] \sqrt{(2x + 3)^3} + C$$

$$= \frac{35x^3 - 45x^2 + 54x - 54}{315} \sqrt{(2x + 3)^3} + C.$$

EXERCISES 7.4

1. Use Table VII to calculate each of the following integrals. State the reference number of each formula each time that it is used.

   • (a) $\int \frac{\sqrt{16 + x^2}}{x^2}\, dx$ (b) $\int (2x^2 + x + 1)^2\, dx$

• (c) $\int \frac{dx}{(2x^2 + x + 1)}$

(d) $\int x^2 (5x - 7)^{-5}\, dx$

(e) $\int \frac{7x}{(x + 8)^4}\, dx$

• (f) $\int \sqrt{16 + x^2}\, dx$

(g) $\int x^2 \ln (2x)\, dx$

(h) $\int \frac{\sqrt{25 - x^2}}{x^3}\, dx$

• (i) $\int x^2 \sqrt{9 - x^2}\, dx$

(j) $\int \sin^4 3x\, dx$

• (k) $\int \frac{dx}{x \ln 2x}$

(l) $\int x^3 (\ln x)^2\, dx$

(m) $\int \frac{e^{5x}}{x^3}\, dx$

• (n) $\int x \ln x\, dx$

(o) $\int x^3 e^{5x}\, dx$

• (p) $\int \frac{dx}{1 + 3e^{5x}}$

(q) $\int \frac{e^x}{3 + e^x}\, dx$

(r) $\int \frac{dx}{x \ln (2x)}$

7.5 THE TRAPEZOIDAL RULE

If f is a function that has no elementary antiderivative, then the methods discussed previously cannot be used to calculate $\int_a^b f(x)\, dx$. Rather we must use a numerical method to approximate it. The simplest such numerical method is to use one of the approximating sums. That is, we partition the interval $[a, b]$ into n equal parts and form the approximating sum

$$[f(x_1) + f(x_2) + \cdots + f(x_n)]\ \Delta x,$$

where $\Delta x = (b - a)/n$ and $x_k = a + k\ \Delta x$. The value of this sum is approximately equal to the value of the integral $\int_a^b f(x)\, dx$. If f is nonnegative on the interval $[a, b]$, the approximating sum above represents the sum of the areas of the rectangles shown in Figure 7.1.

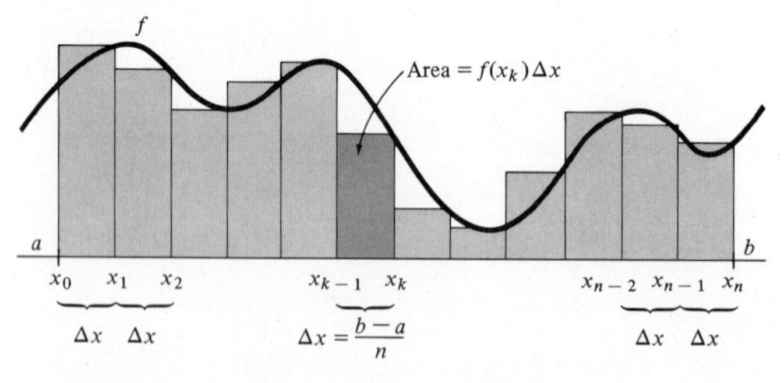

FIGURE 7.1 The area under the curve is $\int_a^b f(x)\, dx$.

The sum of the areas of the rectangles is $\sum_{k=1}^{n} f(x_k)\ \Delta x$ $\left(\text{where } \Delta x = \frac{b - a}{n}\right)$.

$$\int_a^b f(x)\, dx \approx \sum_{k=1}^{n} f(x_k)\ \Delta x \quad \text{if } n \text{ is large.}$$

A rather large error usually results if an approximating sum is used to estimate the integral. We can reduce much of this error by using trapezoids rather than rectangles in setting up the approximation. We form the kth trapezoid by connecting the points

$$(x_{k-1}, f(x_{k-1})) \quad \text{and} \quad (x_k, f(x_k))$$

with a line segment. (See Figure 7.2.)

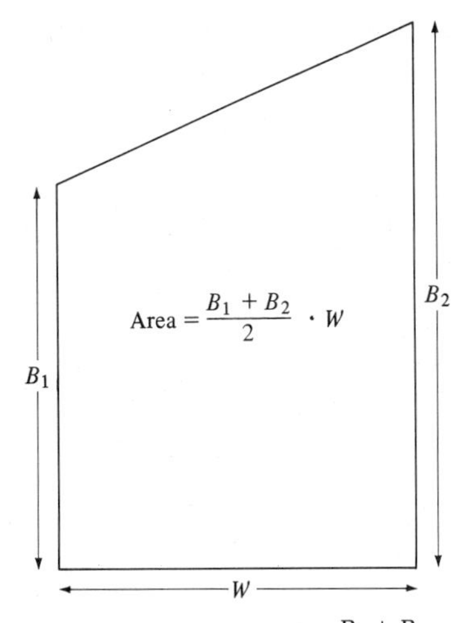

(a) Area of a trapezoid: $A = \dfrac{B_1 + B_2}{2} \cdot W$.

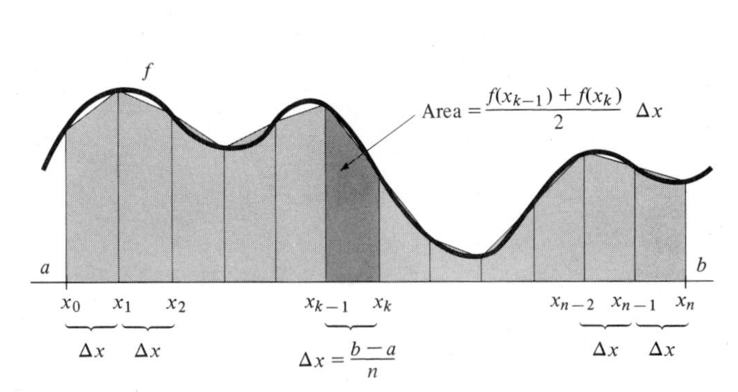

(b) $\displaystyle\int_a^b f(x)\,dx \approx \frac{f(x_0) + f(x_1)}{2}\Delta x + \frac{f(x_1) + f(x_2)}{2}\Delta x + \cdots + \frac{f(x_{n-1}) + f(x_n)}{2}\Delta x$

$\displaystyle\approx \left[\frac{f(x_0)}{2} + f(x_1) + f(x_2) + \cdots + f(x_{n-1}) + \frac{f(x_n)}{2}\right]\Delta x.$

FIGURE 7.2 The Trapezoidal Rule.

Recall that the area of a trapezoid with bases B_1 and B_2 and width W is

$$\frac{B_1 + B_2}{2} \cdot W.$$

(See Figure 7.2a.) Therefore the area of the kth trapezoid used in the approximation is

$$\frac{f(x_{k-1}) + f(x_k)}{2} \cdot \Delta x.$$

It follows that

$$\int_a^b f(x)\,dx \approx \left[\frac{f(x_0) + f(x_1)}{2} + \frac{f(x_1) + f(x_2)}{2} + \frac{f(x_2) + f(x_3)}{2} + \cdots + \frac{f(x_{n-1}) + f(x_n)}{2}\right]\Delta x.$$

(See Figure 7.2b.)

If we collect like terms and simplify, we obtain the *Trapezoidal Rule:*

$$\int_a^b f(x)\,dx \approx \left[\frac{f(x_0)}{2} + f(x_1) + f(x_2) + \cdots + f(x_{n-1}) + \frac{f(x_n)}{2}\right]\Delta x.$$

Trapezoidal Rule

Observe that the formula in the Trapezoidal Rule differs from an approximating sum only in that $f(x_n)$ is replaced by $[f(x_0) + f(x_n)]/2$, the average of the values of f at the two end points.

Example 1 Use the Trapezoidal Rule with $n = 4$ to approximate $\int_1^3 \ln x\,dx$.

Solution $\Delta x = (3-1)/4 = \frac{1}{2}$. Therefore,

$$\begin{aligned}\int_1^3 \ln x\,dx &\approx \left[\frac{\ln 1}{2} + \ln 1.5 + \ln 2.0 + \ln 2.5 + \frac{\ln 3}{2}\right]\cdot\frac{1}{2}\\ &\approx \left[\frac{0}{2} + 0.4055 + 0.6931 + 0.9163 + \frac{1.0986}{2}\right]\cdot\frac{1}{2}\\ &= \frac{2.5642}{2} = 1.2821.\end{aligned}$$

Example 2 Use the Trapezoidal Rule with five subintervals to approximate $\int_1^2 e^{-x^2}\,dx$. (The integral $\int e^{-x^2}\,dx$, which is important in statistics, cannot be calculated by elementary methods.)

Solution $\Delta x = (2-1)/5 = \frac{1}{5} = 0.2$. We use Table V:

$$\begin{aligned}\int_1^2 e^{-x^2}\,dx &\approx \left[\frac{e^{-(1.0)^2}}{2} + e^{-(1.2)^2} + e^{-(1.4)^2} + e^{-(1.6)^2} + e^{-(1.8)^2} + \frac{e^{-(2.0)^2}}{2}\right](0.2)\\ &\approx \left[\frac{e^{-1.00}}{2} + e^{-1.44} + e^{-1.96} + e^{-2.56} + e^{-3.24} + \frac{e^{-4.0}}{2}\right](0.2)\\ &\approx \left[\frac{0.36788}{2} + 0.23693 + 0.14086 + 0.07731\right.\\ &\qquad\qquad \left. + 0.03916 + \frac{0.01832}{2}\right](0.2)\\ &\approx (0.68736)(0.2) \approx 0.1375.\end{aligned}$$

Approximate integration can be used to make numerical tables. For example, since $1/x$ can be converted into a decimal equivalent with little trouble, we can approximate $\ln a = \int_1^a (1/x)\,dx$ by the trapezoidal rule.

Example 3 Use the Trapezoidal Rule to approximate $\ln 2 = \int_1^2 (1/x)\,dx$. Use a partition of [1, 2] into ten parts.

Solution $\Delta x = \frac{1}{10}$. Using Table I, we obtain

$$\begin{aligned}\ln 2 &= \int_1^2 \frac{1}{x}\,dx\\ &\approx \left[\frac{1}{2}\cdot\frac{1}{1.0} + \frac{1}{1.1} + \frac{1}{1.2} + \frac{1}{1.3} + \frac{1}{1.4} + \frac{1}{1.5} + \frac{1}{1.6}\right.\\ &\qquad \left. + \frac{1}{1.7} + \frac{1}{1.8} + \frac{1}{1.9} + \frac{1}{2}\cdot\frac{1}{2.0}\right]\cdot\frac{1}{10}\end{aligned}$$

$\approx [0.50000 + 0.90909 + 0.83333 + 0.76923 + 0.71429$
$+ 0.66667 + 0.62500 + 0.58824 + 0.55556 + 0.52632$
$+ 0.25000] \cdot (0.1) \approx 0.6938.$

EXERCISES 7.5

1. Use the Trapezoidal Rule to approximate the following integrals:

(a) $\int_0^5 \sqrt{x^3 + 2x}\, dx,\ n = 5$

• (b) $\int_0^2 \sqrt{2 + x^4}\, dx,\ n = 6$

(c) $\int_0^4 \sqrt{x}\, dx,\ n = 8$

• (d) $\int_1^2 \frac{dx}{1 + x},\ n = 5$

(e) $\int_0^1 e^{-x^2}\, dx,\ n = 10$

• (f) $\int_0^2 (1 + x^3)^{1/3}\, dx,\ n = 6$

(g) $\int_1^3 \sqrt{4 + x^2}\, dx,\ n = 4$

• (h) $\int_1^4 \ln 2x\, dx,\ n = 6$

7.6 SIMPSON'S RULE

The Trapezoidal Rule involves the use of line segments to approximate the function f. This means that the concavity of the graph of f is not taken into account. Another method of approximate integration partially utilizes the concavity. This method, known as *Simpson's Rule,* involves the approximation of the function f by parabolic arcs rather than line segments. Our first task is to calculate the area under an arc of a parabola $y = ax^2 + bx + C$.

In our setup we will use three noncollinear points, $P_0(x_0, y_0)$, $P_1(x_1, y_1)$, $P_2(x_2, y_2)$, with nonnegative y-coordinates that have their x-coordinates equally spaced. If $x_0 < x_1 < x_2$ and $\Delta x = x_1 - x_0 = x_2 - x_1$, we can represent the points as $P_0(x_1 - \Delta x, y_0)$, $P_1(x_1, y_1)$ and $P_2(x_1 + \Delta x, y_2)$.

Area under parabolic arc

Lemma Let $P_0(x_1 - \Delta x, y_0)$, $P_1(x_1, y_1)$ and $P_2(x_1 + \Delta x, y_2)$ be three noncollinear points with nonnegative y-coordinates. The area under the parabolic arc $y = ax^2 + bx + c$ that connects the three points is

$$A = \frac{\Delta x}{3}(y_0 + 4y_1 + y_2).$$

PROOF Let $y = ax^2 + bx + c$ be the equation of the parabola through the three points. (See Figure 7.3.) The area under the parabolic arc that connects the three points is

$$A = \int_{x_1 - \Delta x}^{x_1 + \Delta x} (ax^2 + bx + c)\, dx = \left[\frac{ax^3}{3} + \frac{bx^2}{2} + cx\right]_{x_1 - \Delta x}^{x_1 + \Delta x}$$
$$= \frac{a(x_1 + \Delta x)^3}{3} + \frac{b(x_1 + \Delta x)^2}{2} + c(x_1 + \Delta x)$$
$$- \frac{a(x_1 - \Delta x)^3}{3} - \frac{b(x_1 - \Delta x)^2}{2} - c(x_1 - \Delta x).$$

After simplification, this reduces to

$$A = \Delta x\left(2ax_1{}^2 + \frac{2a\,\Delta x^2}{3} + 2bx_1 + 2c\right).$$

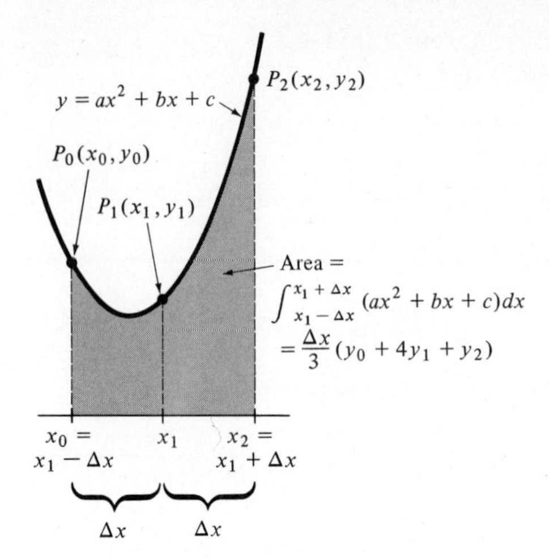

FIGURE 7.3 Area under the parabola $= \frac{\Delta x}{3} (y_0 + 4y_1 + y_2)$.

We now show that $(y_0 + 4y_1 + y_2) \cdot \frac{\Delta x}{3}$ has the same value. Since P_0, P_1, and P_2 are on the graph of the parabola, their coordinates satisfy the equation

$$y = ax^2 + bx + c.$$

Therefore,

$$\begin{aligned} y_0 &= a(x_1 - \Delta x)^2 + b(x_1 - \Delta x) + c = ax_1^2 - 2ax_1\,\Delta x + a\,\Delta x^2 + bx_1 - b\,\Delta x + c, \\ y_1 &= \qquad\qquad\qquad\qquad\qquad\quad\;\; = ax_1^2 \qquad\qquad\qquad\qquad\quad + bx_1 \qquad\quad + c, \\ y_2 &= a(x_1 + \Delta x)^2 + b(x_1 + \Delta x) + c = ax_1^2 + 2ax_1\,\Delta x + a\,\Delta x^2 + bx_1 + b\,\Delta x + c. \end{aligned}$$

If we multiply the middle equation by 4, add, divide by 3, and simplify, we have

$$\frac{1}{3} (y_0 + 4y_1 + y_2) = 2ax_1^2 + \frac{2a\,\Delta x^2}{3} + 2bx_1 + 2c.$$

Therefore,

$$A = \Delta x \left(2ax_1^2 + \frac{2a\,\Delta x^2}{3} + 2bx_1 + 2c \right) = \frac{\Delta x}{3} (y_0 + 4y_1 + y_2).$$

Example 1 Use the lemma to calculate the area bounded by the parabola $y = 3x^2 - 2x + 7$, the x-axis, and the vertical lines $x = 1$ and $x = 3$.

Solution Let $x_0 = 1$, $x_2 = 3$. Then $x_1 = 2$, the midpoint between x_0 and x_2, and $\Delta x = 1$. We calculate

$$\begin{aligned} y_0 &= 3 \cdot 1^2 - 2 \cdot 1 + 7 = 8, \\ y_1 &= 3 \cdot 2^2 - 2 \cdot 2 + 7 = 15, \\ y_2 &= 3 \cdot 3^2 - 2 \cdot 3 + 7 = 28. \end{aligned}$$

$$A = \frac{\Delta x}{3} (y_0 + 4y_1 + y_2) = \frac{1}{3} (8 + 4 \cdot 15 + 28) = \frac{96}{3} = 32.$$

Simpson's Rule We now are ready to establish Simpson's Rule for the approximate integration of $\int_a^b f(x)\,dx$.

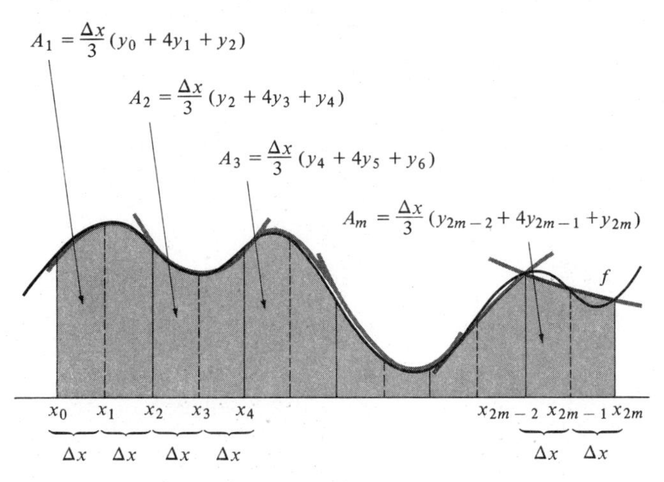

FIGURE 7.4 Simpson's Rule:

$$\int_a^b f(x)\,dx \approx \frac{\Delta x}{3}(y_0 + 4y_1 + 2y_2 + 4y_3 + 2y_4 + \cdots + 2y_{2m-2} + 4y_{2m-1} + y_{2m}).$$

First we partition the interval $[a, b]$ into an *even* number of equal subintervals of length Δx by the partition $a = x_0 < x_1 < x_2 < \cdots < x_{2m} = b$. (See Figure 7.4.) Observe that each pair of subintervals is determined by three points in the plane with equally spaced x-coordinates. (The first pair is determined by x_0, x_1, x_2, the second pair by x_2, x_3, x_4, and so on.) Now we approximate the curve with a parabolic arc over each pair of subintervals, the parabola being determined by the three points mentioned above. The areas under the parabolic arcs are

$$A_1 = \frac{\Delta x}{3}\{f(x_0) + 4f(x_1) + f(x_2)\} \qquad \text{(first pair of subintervals)},$$

$$A_2 = \frac{\Delta x}{3}\{f(x_2) + 4f(x_3) + f(x_4)\} \qquad \text{(second pair of subintervals)},$$

$$A_3 = \frac{\Delta x}{3}\{f(x_4) + 4f(x_5) + f(x_6)\} \qquad \text{(third pair of subintervals)},$$

$$A_m = \frac{\Delta x}{3}\{f(x_{2m-2}) + 4f(x_{2m-1}) + f(x_{2m})\} \qquad (m\text{th pair of subintervals}).$$

The total area under the parabolic arcs is $A_1 + A_2 + \cdots + A_m$. If we add these terms, factor out the common factor $\Delta x/3$, and collect like terms we have Simpson's Rule:

$$\int_a^b f(x)\,dx \approx \frac{\Delta x}{3}\{f(x_0) + 4f(x_1) + 2f(x_2) + 4f(x_3) + 2f(x_4) + \cdots + 2f(x_{2m-2}) + 4f(x_{2m-1}) + f(x_{2m})\}.$$

Simpson's Rule

Remark In applying Simpson's Rule we must have an *even* number of equal subintervals. To emphasize this fact we use $2m$ rather than n to represent the number of subintervals. The value of Δx is, of course,

$$\Delta x = \frac{b-a}{2m}.$$

Example 2 Use Simpson's Rule with ten subintervals to approximate

$$\ln 2 = \int_1^2 \frac{1}{x}\, dx.$$

Solution $\Delta x = \frac{1}{10}$, $x_k = 1 + k\,\Delta x = 1 + k/10$, and $f(x_k) = 1/x_k$. Using Table I, we obtain

$$\begin{aligned}
\ln 2 &= \int_1^2 \frac{1}{x}\, dx \\
&\approx \frac{\frac{1}{10}}{3}\left(\frac{1}{1.0} + \frac{4}{1.1} + \frac{2}{1.2} + \frac{4}{1.3} + \frac{2}{1.4} + \frac{4}{1.5}\right. \\
&\quad \left. + \frac{2}{1.6} + \frac{4}{1.7} + \frac{2}{1.8} + \frac{4}{1.9} + \frac{1}{2.0}\right) \\
&\approx \frac{1}{30}(1.00000 + 3.63636 + 1.66667 + 3.07692 + 1.42857 \\
&\quad + 2.66667 + 1.25000 + 2.35294 + 1.11111 + 2.10526 + 0.50000) \\
&\approx \frac{1}{30}(20.79450) \approx 0.69315.
\end{aligned}$$

This approximation is much better than the one obtained with the Trapezoidal Rule in Example 3 of Section 7.6.

Example 3 Use Simpson's Rule with four subintervals to approximate $\int_0^\pi \sqrt{\sin x}\, dx$.

Solution $\Delta x = \pi/4$, $x_k = 0 + k\pi/4$, and $f(x_k) = \sqrt{\sin (k\pi/4)}$. By Simpson's Rule

$$\begin{aligned}
\int_0^\pi \sqrt{\sin x}\, dx &\approx \frac{\frac{\pi}{4}}{3}\left[\sqrt{\sin 0} + 4\sqrt{\sin \frac{\pi}{4}}\right. \\
&\quad \left. + 2\sqrt{\sin \frac{2\pi}{4}} + 4\sqrt{\sin \frac{3\pi}{4}} + \sqrt{\sin \pi}\right] \\
&\approx \frac{\pi}{12}\left[0 + 4\sqrt{\frac{1}{\sqrt{2}}} + 2\sqrt{1} + 4\sqrt{\frac{1}{\sqrt{2}}} + \sqrt{0}\right] \\
&\approx \frac{3.14159}{12}[0 + 4(0.8409) + 2 + 4(0.8409) + 0] \\
&\approx 2.285.
\end{aligned}$$

Most of the continuous functions that we can construct, such as $\sqrt[3]{\ln x}$, $e^{\sqrt{x}}$, $e^{\sin x}$, e^{x^2}, and so on, have antiderivatives that cannot be expressed in terms of the "elementary" functions that we have studied in this book. The antiderivatives exist, but they are, essentially, defined only by being antiderivatives of these functions. Most of them have no intrinsic worth, not being essential to the solution of any problem, and have never been studied. The antiderivative of $\sqrt[3]{\ln x}$ is such a function. By using Simpson's Rule the values of this antiderivative could be calculated (for any $x > 0$), but it is not worth the trouble to do so. The only fact of interest about this function is that it is an antiderivative of $\sqrt[3]{\ln x}$. It is simply not worth studying in its own right.

Occasionally the solution of a problem will require the integration of one of these nonstandard functions. In most cases we must then use Simpson's Rule, the Trapezoidal Rule, or some other method to approximate the solution.

EXERCISES 7.6

1. Use Simpson's Rule to approximate the following integrals:

• (a) $\int_0^1 e^{-x^2}\,dx,\ n = 2m = 10$

(b) $\int_0^1 \frac{dx}{1 + x^2},\ n = 2m = 10$

• (c) $\int_1^4 x \cdot \ln x\,dx,\ n = 2m = 6$

(d) $\int_{-1}^3 \sqrt{4 + x^3}\,dx,\ n = 2m = 4$

• (e) $\int_1^4 \frac{e^x}{x}\,dx,\ n = 2m = 6$

(f) $\int_0^1 \sqrt{1 - x^2}\,dx,\ n = 2m = 4$

CALCULUS IN HIGHER DIMENSIONS

8.1 THREE-DIMENSIONAL SPACE

Thus far in this book we have dealt with functions of a single variable. In this chapter we turn our attention to functions of two or more variables. This is a natural extension of the theory, because many quantities depend on two or more related quantities. For example, the demand function for a commodity depends on the prices of the competitors as well as that of the commodity itself. Thus, the demand is a function of several variables.

Function of two variables

DEFINITION A *function of two variables* is a function f that has a set of ordered pairs of real numbers as its domain and a subset of the real numbers as its range.

If f is a function of two variables, we write $f(x, y)$ for the value of f corresponding to the ordered pair (x, y).

Example 1 The volume of a right circular cylinder is a function of the radius (r) and the height (h). The law of correspondence is

$$V = f(r, h) = \pi r^2 h.$$

The domain of f is the set of all ordered pairs (r, h) with $r > 0$ and $h > 0$. The range is the set of all positive real numbers.

Example 2 The demand for Thin tomato soup is given by the equation

$$d = -x + y + 30 - \frac{xy}{100}, \qquad 15 \leq x \leq 25,\ 15 \leq y \leq 25,$$

where x is the retail price of Thin Soup and y is the price of the major competitor. The prices x and y are expressed in cents per can and the demand (d) is expressed in tens of thousands of cans that can be sold per day.

The demand is a function of two variables, x and y. The law of correspondence is

$$f(x, y) = -x + y + 30 - \frac{xy}{100}.$$

The domain of f is the set of all ordered pairs (x, y), where x and y are integers between 15 and 25, inclusive.

It is a simple matter to extend the concept of a function of two variables to functions of several variables. For example, a function g of four variables has a set of ordered quadruples as its domain. We write g(x, y, z, w) for the value of g corresponding to the ordered quadruple (x, y, z, w).

As we know, functions of one variable can be graphed in the plane (*two-dimensional space*). Functions of two variables can be graphed in a similar way in *three-dimensional space*.

Three-Dimensional Space To represent points in three-dimensional space by ordered triples of real numbers we first construct a *coordinate axis system*. This consists of three mutually perpendicular coordinate axes, the x-*axis*, the y-*axis*, and the z-*axis*, which meet at a common origin. It is customary to draw these axes with the y- and z-axes in the plane of the paper, the y-axis horizontal and the z-axis vertical, and the x-axis drawn as if it were pointing straight out from the paper (Figure 8.1).

Coordinate axes

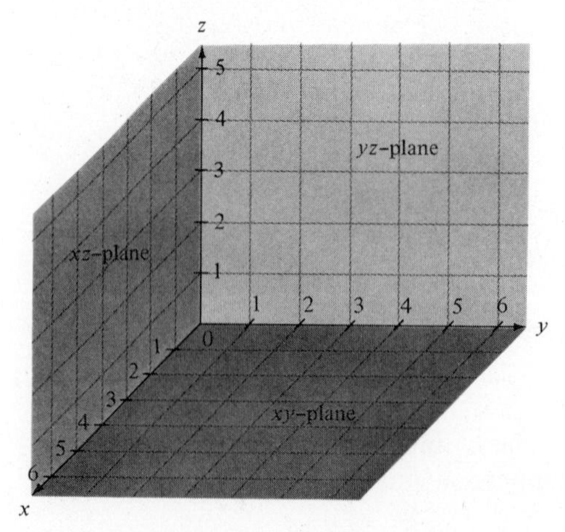

FIGURE 8.1 The coordinate axes and planes.

Coordinate planes

Each pair of coordinate axes determines a *coordinate plane*. Thus we have the xy-*plane*, yz-*plane*, and xz-*plane*. It may help the reader to visualize the coordinate axis system as the corner of a room, the xy-plane being the floor and the xz- and yz-planes the two walls. We now identify the xy-plane with the coordinate plane studied earlier: $\{(x, y) | x \text{ and } y \text{ are real}\}$.

Representation of points by coordinates

The point $P(a, b, c)$ in three-dimensional space is located $|c|$ units above or below the point (a, b) in the xy-plane. (See Figure 8.2.) If $c \geq 0$, the point (a, b, c) is above or on the xy-plane; if $c < 0$, it is below. For example, the point $(1, 5, 7)$ is seven units above $(1, 5)$ and $(-1, 3, -2)$ is two units below $(-1, 3)$. Under this scheme the points in the xy-plane can be written as (x, y) or $(x, y, 0)$.

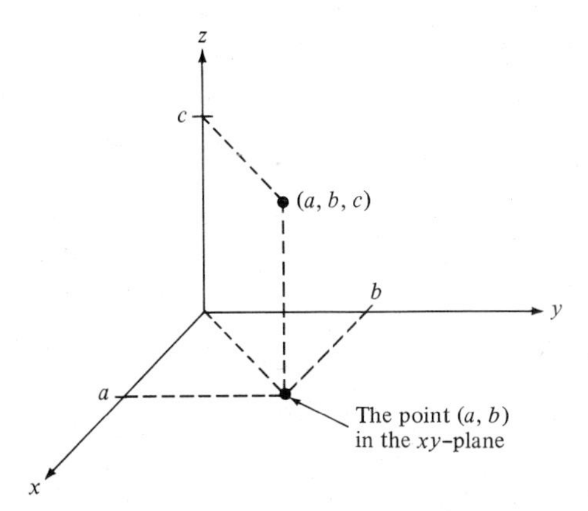

FIGURE 8.2 The point (a, b, c), $c > 0$.

This assignment of coordinates establishes a unique correspondence between points in three-dimensional space and ordered triples of real numbers. That is, two points are different if and only if they have different coordinates. Unfortunately, this uniqueness is not apparent in perspective drawings. When we sketch a figure showing a point, it is usually necessary to draw a few reference lines connecting the point to the coordinate axes. This can be done as in Figure 8.3.

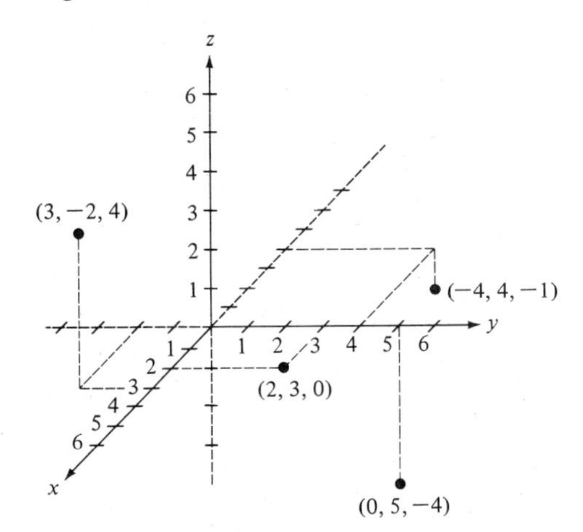

FIGURE 8.3

Octants

The coordinate planes divide three-dimensional space into eight *octants*. *Octant I* is the octant that consists of all points with positive coordinates. There is no standard numbering system for the other seven octants.

The Distance Formula The distance formula can be extended to calculate the distance between two points P_1 and P_2 in three-dimensional space. To establish the new formula we work with the rectangular box which has P_1 and P_2 as opposite corners and sides parallel to the coordinate axes.

The distance formula

Theorem 8.1 The distance between the points $P_1(x_1, y_1, z_1)$ and $P_2(x_2, y_2, z_2)$ is

$$|P_1P_2| = \sqrt{(x_2 - x_1)^2 + (y_2 - y_1)^2 + (z_2 - z_1)^2}.$$

PROOF Let Q be the point $Q(x_2, y_2, z_1)$ (Figure 8.4). Since P_1 and Q are in a plane parallel to the xy-plane, the distance between these two points can be calculated by using the distance formula for points in the xy-plane:

$$|P_1Q| = \sqrt{(x_2 - x_1)^2 + (y_2 - y_1)^2}.$$

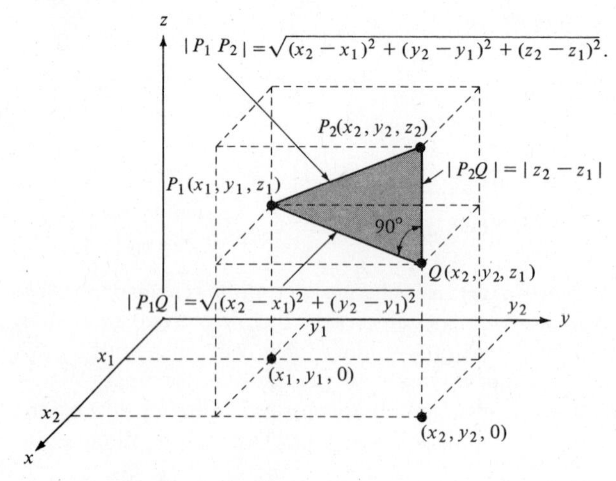

FIGURE 8.4 The distance formula
$|P_1P_2| = \sqrt{(x_2 - x_1)^2 + (y_2 - y_1)^2 + (z_2 - z_1)^2}$.

Since P_2 and Q are on a line parallel to the z-axis, the distance between them is equal to the absolute value of the difference between their z-coordinates:

$$|P_2Q| = |z_2 - z_1|.$$

Observe that the three points P_1, Q, P_2 form a right triangle with sides of length $|P_1Q|$, $|P_2Q|$ and $|P_1P_2|$. It follows from the Pythagorean Theorem that

$$\begin{aligned}|P_1P_2| &= \sqrt{|P_1Q|^2 + |P_2Q|^2} \\ &= \sqrt{\left(\sqrt{(x_2 - x_1)^2 + (y_2 - y_1)^2}\right)^2 + |z_2 - z_1|^2} \\ &= \sqrt{(x_2 - x_1)^2 + (y_2 - y_1)^2 + (z_2 - z_1)^2}.\end{aligned}$$

Example 3 The distance between $P_1(3, 2, -1)$ and $P_2(-1, -1, 2)$ is

$$\begin{aligned}|P_1P_2| &= \sqrt{(-1 - 3)^2 + (-1 - 2)^2 + (2 - (-1))^2} \\ &= \sqrt{4^2 + 3^2 + 3^2} = \sqrt{16 + 9 + 9} = \sqrt{34}.\end{aligned}$$

EXERCISES 8.1

1. Plot the following points:

(a) (0, 0, 1)
(b) (0, 2, 0)
(c) (3, 0, 0)
(d) (1, 1, −1)
(e) (3, 2, 1)
(f) (0, 2, 1)
(g) (2, 1, 0)
(h) (2, 0, 2)
(i) (1, 2, 3)
(j) (−2, −2, −2)
(k) (3, 3, 3)
(l) (1, −2, 1)
(m) (2, −1, 1)
(n) (0, 0, 0)
(o) (1, −1, 1)
(p) (1, 2, 2)

2. Calculate the distance between the points:

(a) (−3, 4, 5), (5, 4, −1)
• (b) (0, 4, −1), (−1, 3, 0)
(c) (−7, 2, 0), (−1, −1, −2)
• (d) (8, 1, −2), (−4, −2, 2)
(e) (1, 2, −3), (3, −2, −1)
(f) (4, 0, −1), (2, 1, 0)
• (g) (0, −3, 2), (1, 0, 0)
(h) (3, 2, −1), (1, 1, 2)

3. Express each of the following functions of two and three variables by an explicit formula involving x, y, z, and so on.

(a) The volume (*V*) of an open-topped box with sides x, y, z.
• (b) The surface area (S) of an open-topped rectangular box with sides x, y, and height z.
(c) The revenue (*R*) obtained from selling *m* items at \$x per item and *n* items at \$*y* per item.
• (d) The cost (C) of producing x items at a cost of \$*d* per item and y item at a cost of \$*e* per item.

8.2 GRAPHS IN THREE-DIMENSIONAL SPACE

Graph of function

DEFINITION Let $f(x, y)$ be a function of x and y. The *graph* of f is the set of all points $P(x, y, z)$ such that $z = f(x, y)$.

Example 1 The graph of the function $f(x, y) = 7$ is

$$\{P(x, y, z) | z = 7\}.$$

If (x, y) is any point in the xy-plane there is a corresponding point (x, y, 7) on the graph seven units above (x, y). The graph is a plane that is parallel to the xy-plane and is seven units above it. (See Figure 8.5.)

In general, the graph of a function $f(x, y)$ is a surface in three-dimensional space. The domain of f is a collection of ordered pairs (x, y) in the xy-plane. The graph is a surface defined over the domain. (See Figure 8.6.)

Graph of equation

The graph of an equation in x, y, and z, such as

$$x^2 + y^2 + z^2 = 4,$$

is defined similarly. It is the set of all points $P(x, y, z)$ with coordinates that satisfy the equation.

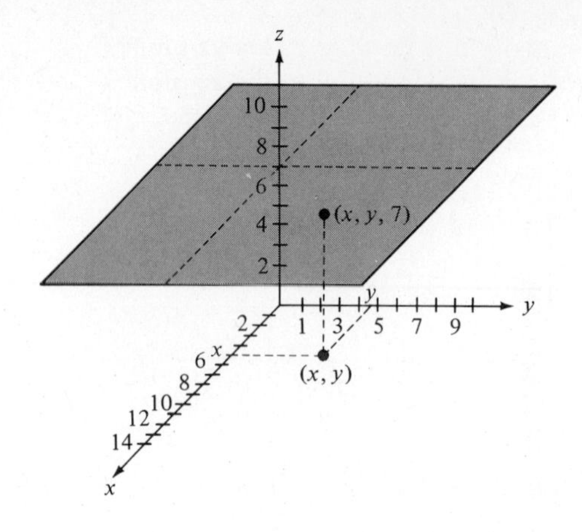

FIGURE 8.5 Example 1. The graph of the constant function $z = f(x, y) = 7$ is a plane parallel to the xy-plane.

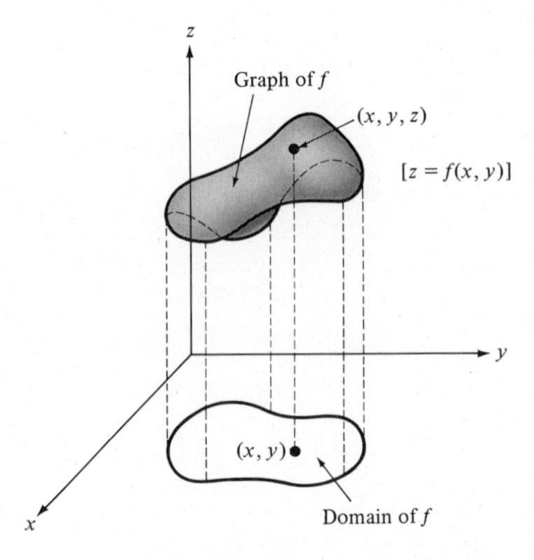

FIGURE 8.6 The graph of $f(x, y)$ is $\{P(x, y, z) | z = f(x, y) \text{ and } (x, y) \text{ is in the domain of } f\}$, a surface in three-dimensional space.

We will denote the general equation in x, y, and z by

$$F(x, y, z) = 0.$$

Then the graph of $F(x, y, z) = 0$ is

$$\{P(x, y, z) | F(x, y, z) = 0\}.$$

Graphs of Linear Equations As we saw in Chapter 1, the graph of a linear equation $ax + by = c$ is a line in the xy-plane. Similarly, the graph of a linear equation

The plane $ax + by + cz = d$

$$ax + by + cz = d \quad (a, b, c \text{ not all zero})$$

is a plane in three-dimensional space.

For example, the graphs of the following equations are planes in three-dimensional space:

$$x - 2y + z = 3,$$
$$2x + 7y = 13,$$
$$x = 2.$$

Linear equations in one variable

Linear equations in one variable are quite easy to graph. It follows as in Example 1 that their graphs are planes parallel to the coordinate planes:

The graph of the equation $x = a$ *is a plane parallel to the yz-plane.*
The graph of the equation $y = b$ *is a plane parallel to the xz-plane.*
The graph of the equation $z = c$ *is a plane parallel to the xy-plane.* (See Figure 8.7.)

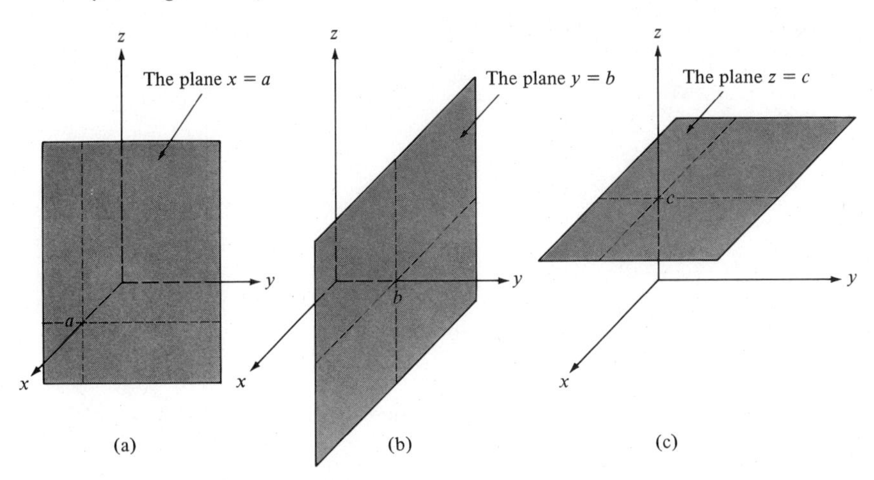

FIGURE 8.7 The planes $x = a$, $y = b$, and $z = c$ are parallel to the coordinate planes.

Traces of Graphs The intersection of a surface with a coordinate plane is called a *trace* of the surface.

Trace of surface

If the surface is the graph of the equation

$$F(x, y, z) = 0,$$

then its trace in the yz-plane can be written as the set of points satisfying the system of equations

$$\begin{cases} F(x, y, z) = 0, \\ x = 0. \end{cases}$$

This system is equivalent to the system

$$\begin{cases} F(0, y, z) = 0, \\ x = 0. \end{cases}$$

Trace in the yz-plane

(See Figure 8.8.) Thus, the trace in the yz-plane can be obtained by substituting $x = 0$ in the defining equation and graphing the resulting equation in the yz-plane. Similarly the traces of the surface in the xy- and xz-planes satisfy

$$\begin{cases} F(x, y, 0) = 0 \\ z = 0 \end{cases}$$

Trace in the xy-plane

and

$$\begin{cases} F(x, 0, z) = 0 \\ y = 0. \end{cases}$$

Trace in the xz-plane

Example 2 Calculate the equations of the traces of the plane

$$x + 2y + 3z = 6.$$

Use the traces to sketch the graph of the plane.

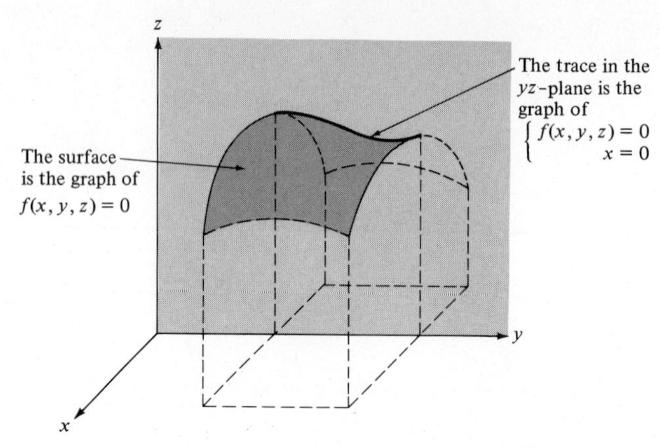

FIGURE 8.8 The trace of a surface in the yz-plane is the curve of intersection of the surface and that plane.

Solution The trace in the xy-plane is obtained by setting $z = 0$. It satisfies the system

$$\begin{cases} x + 2y + 3 \cdot 0 = 6 \\ z = 0 \end{cases} \Leftrightarrow \begin{cases} x + 2y = 6 \\ z = 0. \end{cases}$$

Thus, the trace is the line $x + 2y = 6$ in the xy-plane. The trace in the yz-plane satisfies the system

$$\begin{cases} 0 + 2y + 3z = 6 \\ x = 0 \end{cases} \Leftrightarrow \begin{cases} 2y + 3z = 6 \\ x = 0. \end{cases}$$

The trace is the line $2y + 3z = 6$ in the yz-plane. Similarly, the trace in the xz-plane is the line $x + 3z = 6$. The traces are shown in Figure 8.9a. The plane is drawn by connecting the points on the traces.

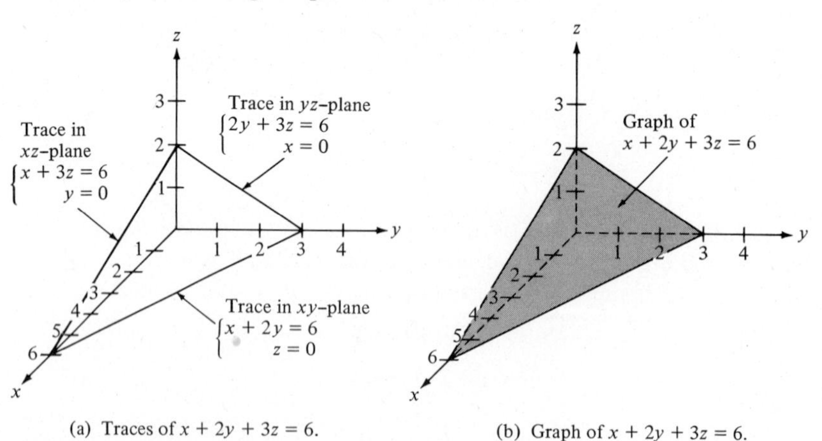

(a) Traces of $x + 2y + 3z = 6$.

(b) Graph of $x + 2y + 3z = 6$.

FIGURE 8.9 Example 2

Example 3 Sketch the graph of the plane $x + 2y = 2$.

Solution The trace in the xy-plane is the line

$$\begin{cases} x + 2y = 2, \\ z = 0. \end{cases}$$

The trace in the xz-plane is the vertical line

$$\begin{cases} x = 2 \\ y = 0. \end{cases}$$

The trace in the yz-plane is the vertical line

$$\begin{cases} y = 1 \\ x = 0. \end{cases}$$

The graph of $x + 2y = 2$ is a plane parallel to the z-axis. It is formed by moving a vertical line along the trace in the xy-plane. (See Figure 8.10.)

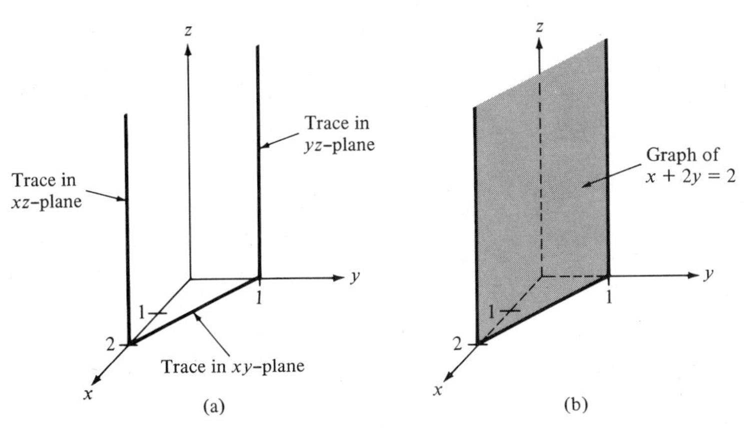

FIGURE 8.10 Example 3. The graph of $x + 2y = 2$.

Cylinders

Graphs of Equations with One Variable Missing If F is a function of x and y then the equation $F(x, y) = 0$ is independent of z. If the point $P(x_0, y_0, 0)$ in the xy-plane is on the graph, then its coordinates satisfy the equation. It follows that $Q(x_0, y_0, z)$ also is on the graph for every value of z. Thus, the graph can be generated by moving a vertical line along the trace in the xy-plane. (See Figure 8.11.) Such a surface is known as a *cylinder*. Similar results hold for equations with x or y missing.

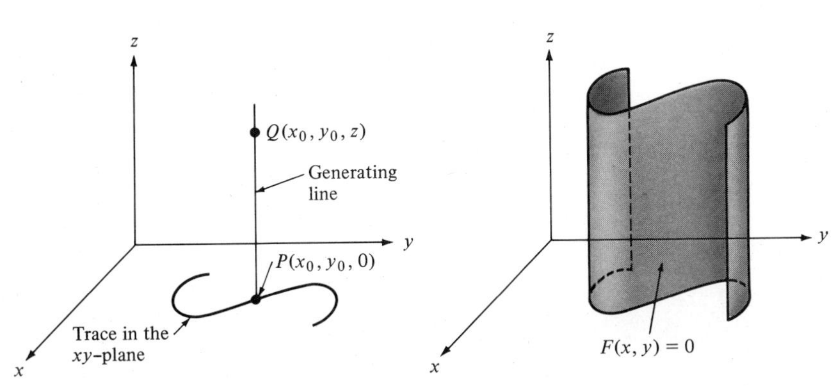

FIGURE 8.11 The graph of the cylinder $F(x, y) = 0$ can be generated by moving a vertical line along the trace in the xy-plane.

Example 4 Sketch the graphs in three-dimensional space of the cylinders (a) $x^2 - y = 0$, (b) $x^2 + z^2 = 1$.

Solution

(a) The variable z is missing from the equation $x^2 - y = 0$. Thus, the graph is obtained by moving a vertical line along the parabola

$$\begin{cases} y = x^2 \\ z = 0 \end{cases}$$

in the xy-plane (Figure 8.12).

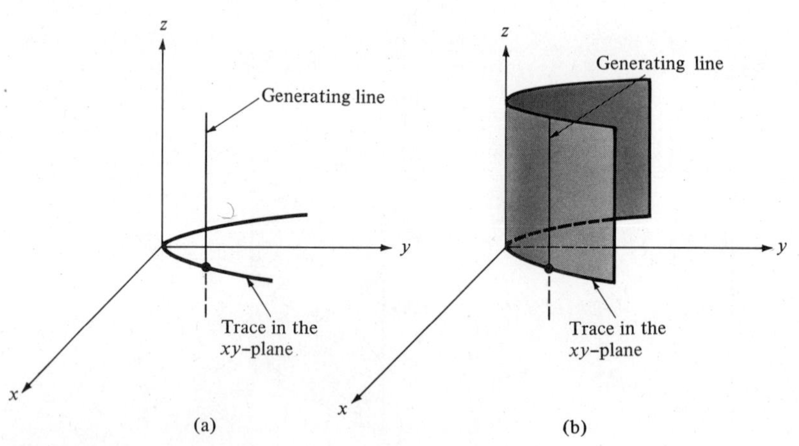

FIGURE 8.12 Example 4a. The cylinder $x^2 - y = 0$.

(b) The variable y is missing from the equation $x^2 + z^2 = 1$. The graph is obtained by moving a line parallel to the y-axis along the unit circle

$$\begin{cases} x^2 + z^2 = 1 \\ y = 0 \end{cases}$$

in the xz-plane. (Figure 8.13).

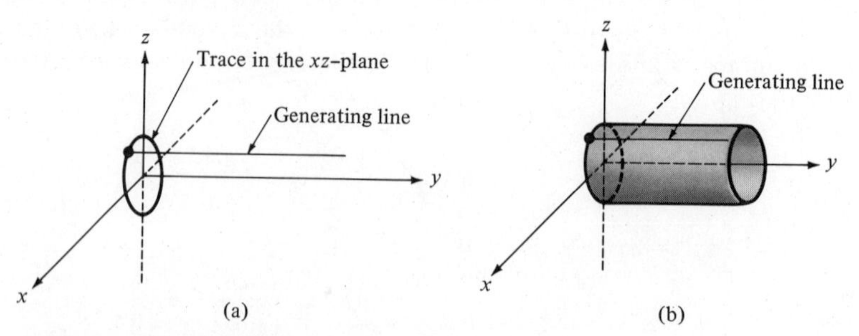

FIGURE 8.13 Example 4b. The cylinder $x^2 + z^2 = 1$.

EXERCISES 8.2

1. Sketch the first octant portion of the planes:

(a) $x = 3$
(b) $y = 2$
(c) $z = 1$
(d) $x + z = 2$
(e) $y + z = 1$
(f) $x + y = 3$
(g) $x + y + z = 2$
(h) $2x + y + z = 4$
(i) $x + 3y + 2z = 6$
(j) $5x + 2y + 2z = 10$

2. Sketch the traces in each of the coordinate planes and sketch the graph.

(a) $z - y + z = 2$
(b) $x + 2y - z = 2$
(c) $x + 2y + 3z = 1$
(d) $2x + 4y + z = 3$
(e) $x - y - z = 2$
(f) $-x + y - z = 4$
(g) $2x - y + 3z = 1$
(h) $x + y = 0$
(i) $x - z = 0$
(j) $2x + 3y - 2z = 6$

3. Sketch the graph of each of the following cylinders in three-dimensional space:

(a) $y^2 + z^2 = 1$
(b) $x = z^2$
(c) $y = 4 - x^2$
(d) $y^2 = x$
(e) $x = 4 - y^2$
(f) $y = z^2$
(g) $x^2 + y^2 = 1$
(h) $(x + 1)^2 + (y + 1)^2 = 1$

*8.3 LEVEL CURVES

In many cases we can make a good perspective drawing of a surface by first sketching the intersections of the surface with planes parallel to the coordinate planes. The traces in the coordinate planes are especially useful as are the intersections with planes $z = c$.

Example 1 Sketch the surface defined by $z = 4x^2 + y^2$.

Solution The trace in the xy-plane is the single point (0, 0, 0). The curve of intersection with the plane $z = c$, $c > 0$, is the ellipse with center at the origin:

$$4x^2 + y^2 = c.$$

The trace in the yz-plane is the parabola

$$z = 4 \cdot 0^2 + y^2,$$
$$z = y^2.$$

The trace in the xz-plane is the parabola $z = 4x^2$. The surface, which is an example of an *elliptic paraboloid*, is shown in Figure 8.14a.

A perspective sketch enables us to visualize the graph, but it cannot be used to obtain quantitative relationships. An alternate method of describing surfaces partially remedies this problem. Unfortunately, we give up the advantages of quick visualization that go with the perspective drawing. This alternate method involves the use of *level curves*.

Level curve C_h

DEFINITION Let h be a real number. The *level curve of height h* of the surface $F(x, y, z) = 0$ is

$$C_h = \{(x, y) | F(x, y, h) = 0\}.$$

Thus, the level curve of height h is the set of all points in the xy-plane that lie directly under the intersection of the surface with the plane $z = h$. We represent the surface by drawing several level curves in the xy-plane, labeling each with the appropriate number h. This method is used by mapmakers to represent height above sea level on contour maps.

Level curves also are called *contour curves*. In economic theory they frequently are called *indifference curves*.

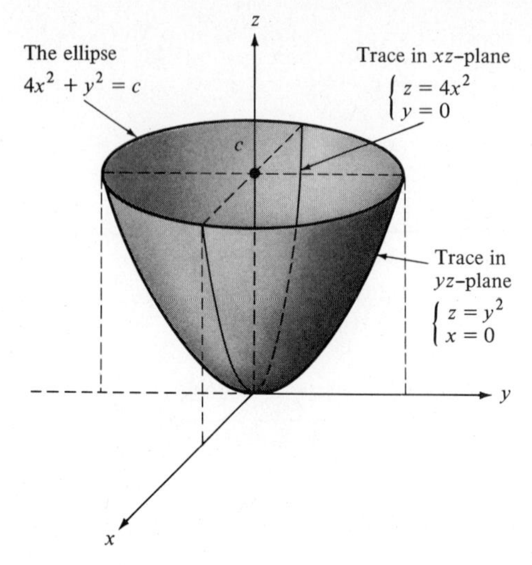

(a) Graph of $z = 4x^2 + y^2$ (perspective drawing)

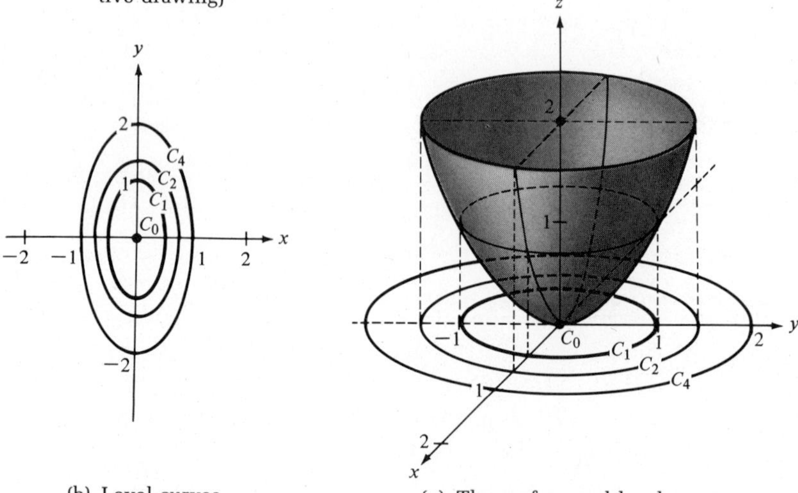

(b) Level curves

(c) The surface and level curves

FIGURE 8.14 Examples 1 and 2. The elliptic paraboloid $z = 4x^2 + y^2$. The trace in the xz-plane is the parabola $z = 4x^2$, the trace in the yz-plane is the parabola $z = y^2$, the intersection with the plane $z = c$ $(c > 0)$ is the ellipse $4x^2 + y^2 = c$.

Example 2 Several level curves for the graph of $z = 4x^2 + y^2$ (see Example 1) are shown in Figure 8.14b. The intersections of the surface with the corresponding planes are shown for comparison in Figure 8.14c.

Example 3 Let $z = y^2 - x^2$.
(a) Make a sketch showing several of the level curves.
(b) Make a perspective drawing.

Solution

(a) The level curve C_h satisfies the equation

$$y^2 - x^2 = h.$$

If $h = 0$, then C_h consists of the two intersecting lines $y = x$, $y = -x$. If $h < 0$, the level curve C_h is an equilateral hyperbola that opens along the x-axis. If $h > 0$, the level curve C_h is an equilateral hyperbola that opens along the y-axis.

(b) To make a perspective sketch we also calculate the traces in the xz- and yz-planes. The trace in the xz-plane is the graph of

$$x^2 - 0^2 = -z,$$
$$z = -x^2,$$

a parabola that opens downward. Similarly, the trace in the yz-plane is $z = y^2$, a parabola that opens upward. The sketch is shown in Figure 8.15b.

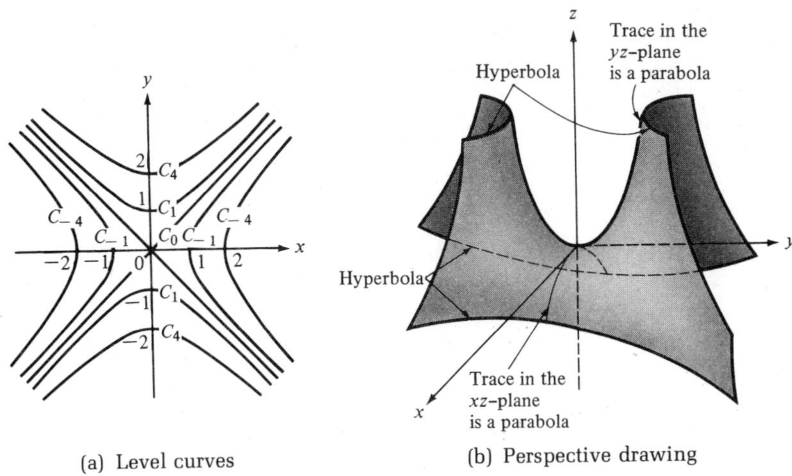

(a) Level curves (b) Perspective drawing

FIGURE 8.15 Example 3. Graph of $z = y^2 - x^2$. The origin is a saddle point.

Saddle point

The surface shown in Figure 8.15 is known as a *saddle surface* (technical name *hyperbolic paraboloid*). The origin is an example of a *saddle point*. Observe that the trace in the xz-plane has a local maximum at the saddle point, while the trace in the yz-plane has a local minimum there.

Applications to Economics Level curves are used in economics to represent graphically the various levels of constant production, constant profit, constant cost, and so forth.

Example 4 The Marvel Marble Company has two machines that can be used for trimming marble blocks. Machine A can trim 30 blocks per hour and machine B can trim 20 per hour. If x and y are the total number of hours that machines A and B, respectively, are operated for this purpose during the week, then the total number of blocks trimmed is

$$z = 30x + 20y, \qquad 0 \leq x \leq 40, 0 \leq y \leq 40.$$

The level "curves" C_h represent the levels of constant production. (See Figure 8.16.) For example,

$$C_{600} = \{(x, y) | 30x + 20y = 600, 0 \leq x \leq 40, 0 \leq y \leq 40\}.$$

Thus, if any point on the line $30x + 20y = 600$, subject to the restrictions above,

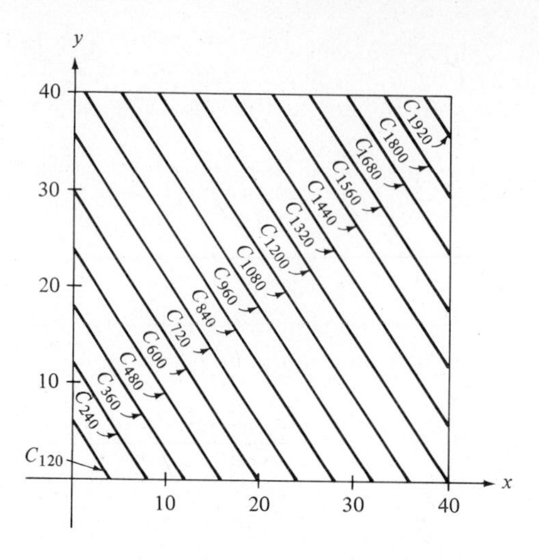

FIGURE 8.16 Level "curves" for Example 4.

is used to assign the times to the two machines, then exactly 600 blocks will be trimmed

Example 5 Charles, a local playboy, has two expensive hobbies which give him great satisfaction—owning and operating sports cars and entertaining women. He can spend up to \$300 per month on these hobbies. Obviously his total satisfaction depends on the way that he allocates his money between them. In order to maximize his satisfaction he consulted with a mathematical economist who made the following analysis.

Let $x =$ the amount of money spent on cars

and

$y =$ the amount spent entertaining women.

The equation

$$S = xy \qquad (x > 0, y > 0)$$

represents the total satisfaction obtained by Charles. For constant values of S the curves $xy = S$ are hyperbolas in quadrant I, as pictured in Figure 8.17a. The further a level curve is from the origin the greater the satisfaction obtained by Charles. Observe that he gets the same satisfaction from any points (x, y) on the same hyperbola, hence the name "indifference curve" as used in economics.

The maximum amount that Charles can spend each month on his two hobbies is \$300. In order to maximize his satisfaction he wishes to find the largest value of S such that the level curve $xy = S$ intersects the triangle defined by

$$x + y = 300, \quad x > 0, \quad y > 0,$$

the total amount of money allocated for cars and entertaining.

The graph in Figure 8.17b pictures the solution of the problem. Because of the symmetry the level curve furthest from the origin that intersects the triangle does it at the point where $x = y$. Since $x + y = 300$, then the maximum satisfaction is obtained by spending \$150 each on cars and on entertaining.

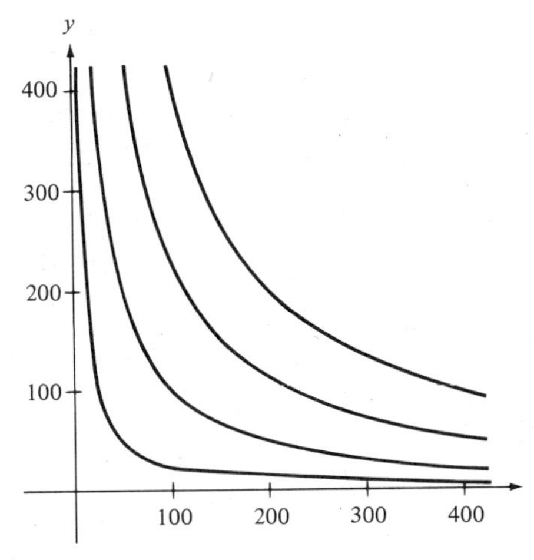

(a) Level curves for $S = xy$ $(x > 0, y > 0)$

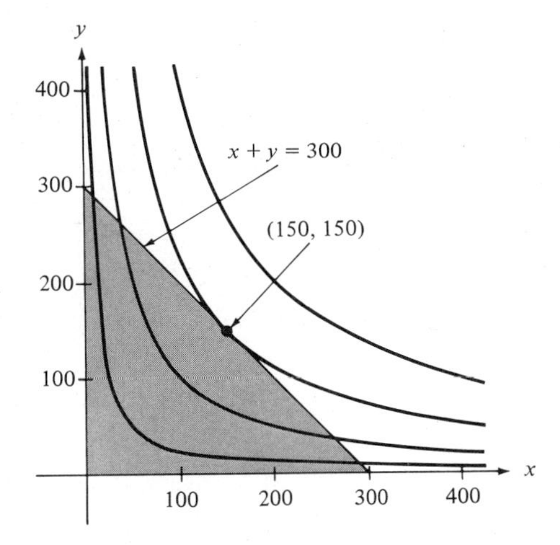

(b) The triangle $x + y = 300$, $x > 0$, $y > 0$, intersects the level curve $S = xy$ for the largest value of S at the point (150,150).

FIGURE 8.17. Example 5.

EXERCISES 8.3

1. Sketch at least three level curves for each of the following functions:

(a) $z = 2x^2 + y^2$
(b) $z = x^2 + y^2$
(c) $z = 2x^2 - y^2$
(d) $z = x^2 + 2y^2$
(e) $z = x - y$
(f) $z = x + y$
(g) $z^2 = 4 - x^2 - y^2$
(h) $z^2 = 4 - x^2 - 2y^2$

2. Make both a perspective drawing and a sketch showing level curves for each of the following equations:

(a) $x^2 + y^2 = z^2$
(b) $x^2 + y^2 = z$
(c) $x^2 + y^2 = z^2 + 1$
(d) $x^2 + 4y^2 = z$
(e) $4x^2 + 9y^2 = 36z^2$
(f) $x^2 + y^2 = z^2 - 1$
(g) $9x^2 + 4y^2 = 36z$
(h) $x^2 + y^2 = 4z$
(i) $y^2 - x^2 = z$
(j) $4x^2 + 9y^2 = 36 - z^2$

•3. Work Example 5 on the assumption that the satisfaction S is given by

$$S = x^2 + y^2 \qquad (x \geq 0,\ y \geq 0).$$

What values of x and y yield the greatest satisfaction if $x + y = 300$?

4. Use level curves to determine the maximum and minimum values of $f(x, y) = x^2 + y^2$ on the rectangle with vertices $(-2, 2)$, $(1, 2)$, $(1, -1)$, and $(-2, -1)$.

•5. Use level curves to determine the maximum and minimum values of $f(x, y) = x^2 + y^2$ on the triangle whose vertices are $(3, -1)$, $(1, 1)$, and $(-1, -1)$.

6. Use level curves to determine the maximum and minimum values of $f(x, y) = xy$ on the circle $x^2 + y^2 = 2$.

•7. Use level curves to determine the maximum and minimum values of $f(x, y) = x + y$ on the rectangle with vertices (1, 1), (1, −1), (−1, 1), and (−1, −1).

8.4 LIMITS AND CONTINUITY

Although we will not pursue the subjects of limits and continuity in detail, we must introduce the terms. The reader should try to develop an intuitive grasp of these and other topics mentioned in this section. He should study the figures carefully and try to draw diagrams of his own that illustrate the concepts.

Speaking loosely,

$$\lim_{(x,y)\to(a,b)} f(x, y) = L$$

provided $f(x, y)$ is close to L whenever (x, y) is close to (a, b).

For example, let $f(x, y) = 3x + 5y$. If (x, y) is close to (1, 2), then $3x$ is close to 3, $5y$ is close to 10, and the sum $3x + 5y$ is close to 13. Therefore,

$$\lim_{(x,y)\to(1,2)} (3x + 5y) = 13.$$

The above idea of "closeness" is not precise enough to enable mathematicians to establish properties of limits. To do this they make a more exact definition. They say that L is the limit of $f(x, y)$ as (x, y) approaches (a, b) provided $|f(x, y) - L|$ can be made less than any preassigned positive number (ϵ) by keeping the distance between (x, y) and (a, b) less than some other number (δ). The formal definition is as follows:

Limit

DEFINITION Let f be a function of x and y. We say that

$$\lim_{(x,y)\to(a,b)} f(x, y) = L$$

provided: If ϵ is any given positive number, there exists a positive number δ (where δ depends on ϵ) such that

$$\text{if} \quad 0 < \sqrt{(x-a)^2 + (y-b)^2} < \delta, \quad \text{then} \quad |f(x, y) - L| < \epsilon.$$

δ-neighborhood

The definition above can be simplified somewhat by the notion of a δ-neighborhood of a point. The *δ-neighborhood of* (a, b) is the set of all points (x, y) inside the circular disk with center (a, b) and radius δ. (See Figure 8.18.) That is, it is

$$\{(x, y) \mid \sqrt{(x-a)^2 + (y-b)^2} < \delta\}.$$

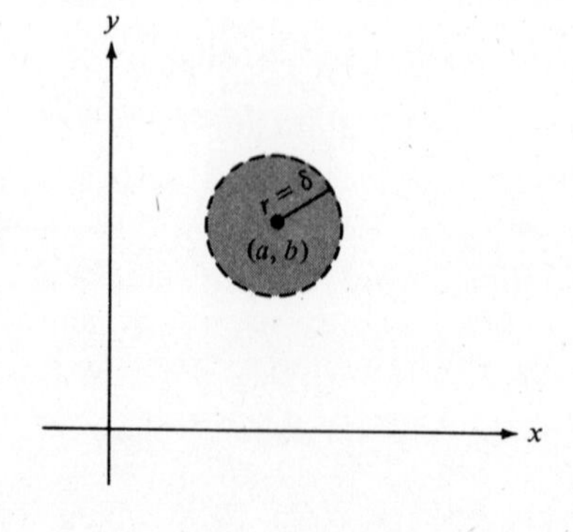

FIGURE 8.18 The δ-neighborhood of $P(a,b)$ consists of all points inside the circular disk with center (a,b) and radius δ.

Using the concept of a δ-neighborhood, we can restate the definition of limit as follows:

$$\lim_{(x,y)\to(a,b)} f(x, y) = L$$

provided: If $\epsilon > 0$, there exists a δ-neighborhood of (a, b) (where δ depends on ϵ) such that if (x, y) is any point of the neighborhood different from (a, b), then $|f(x, y) - L| < \epsilon$.

Observe that this definition is a direct extension of the definition of limit for a function of one variable.

The definition of limit can be used to prove the following extension of the limit theorem.

The limit theorem

Theorem 8.2 (See Theorem 2.1.) Suppose that

$$\lim_{(x,y)\to(a,b)} f(x, y) = L \quad \text{and} \quad \lim_{(x,y)\to(a,b)} g(x, y) = M.$$

Then

(a) $$\lim_{(x,y)\to(a,b)} [f(x,y) \pm g(x,y)] = L \pm M = \lim_{(x,y)\to(a,b)} f(x, y) \pm \lim_{(x,y)\to(a,b)} g(x, y).$$

The limit of the sum (or difference) is the sum (or difference) of the limits.

(b) $$\lim_{(x,y)\to(a,b)} [f(x,y) \cdot g(x,y)] = LM = \lim_{(x,y)\to(a,b)} f(x, y) \cdot \lim_{(x,y)\to(a,b)} g(x, y).$$

The limit of the product is the product of the limits.

(c) $$\lim_{(x,y)\to(a,b)} \frac{f(x, y)}{g(x, y)} = \frac{L}{M} = \frac{\lim_{(x,y)\to(a,b)} f(x, y)}{\lim_{(x,y)\to(a,b)} g(x, y)} \quad \text{provided } M \neq 0.$$

The limit of the quotient is the quotient of the limits, provided the limit of the denominator is not zero.

(d) $$\lim_{(x,y)\to(a,b)} \sqrt[n]{f(x, y)} = \sqrt[n]{L} = \sqrt[n]{\lim_{(x,y)\to(a,b)} f(x, y)}$$

provided $L > 0$ when n is an even integer.

The limit of the nth root is the nth root of the limit, provided the limit is positive when n is even.

Using the results of this theorem, we can evaluate simple limits as in Sec. 2.6.

Example 1 Calculate $\displaystyle\lim_{(x,y)\to(2,1)} \frac{3x^2 + 2xy}{x + y^2}$.

Solution

$$\lim_{(x,y)\to(2,1)} \frac{3x^2+2xy}{x+y^2} = \frac{\lim_{(x,y)\to(2,1)} (3x^2+2xy)}{\lim_{(x,y)\to(2,1)} (x+y^2)}$$

$$= \frac{\lim_{(x,y)\to(2,1)} 3x^2 + \lim_{(x,y)\to(2,1)} 2xy}{\lim_{(x,y)\to(2,1)} x + \lim_{(x,y)\to(2,1)} y^2}$$

$$= \frac{3\cdot 2^2 + 2\cdot 2\cdot 1}{2+1^2} = \frac{16}{3}.$$

Continuity The definition of continuity is similar to the definition for a function of one variable.

Continuity

The function $f(x, y)$ is said to be *continuous at* (a, b) provided

$$\lim_{(x,y)\to(a,b)} f(x, y) = f(a, b).$$

A function is said to be *continuous on the set* S if it is continuous at each point of S.

For example, the function considered in Example 1 is continuous at (2, 1), since

$$\lim_{(x,y)\to(2,1)} f(x, y) = \frac{16}{3}$$

and

$$f(2, 1) = \frac{3\cdot 2^2 + 2\cdot 2\cdot 1}{2+1^2} = \frac{16}{3}.$$

Roughly speaking, a function f is continuous at a point if the surface defined as its graph is unbroken there. Frequently this observation can be used to decide whether or not a function is continuous at a point.

Example 2 Let

$$f(x, y) = \begin{cases} 2 & \text{if } x^2+y^2 \leq 1, \\ 1 & \text{if } x^2+y^2 > 1. \end{cases}$$

The function f is discontinuous at each point of the unit circle (its graph is broken there). It is continuous everywhere else (Figure 8.19).

Limits and continuity are more difficult to determine for functions of two or more variables than for functions of one variable. A function $g(x)$ of one variable has limit L as x approaches a, provided $g(x)$ approaches L as x approaches a from either the left or the right. In order for a function $f(x, y)$ to have L for a limit as (x, y) approaches (a, b), we must have $f(x, y)$ approaching L as (x, y) approaches (a, b) along any possible path. This path may be a line, it may spiral in toward (a, b), or it may be quite irregular.

In order to show that a function has no limit as $(x, y) \to (a, b)$ it is only

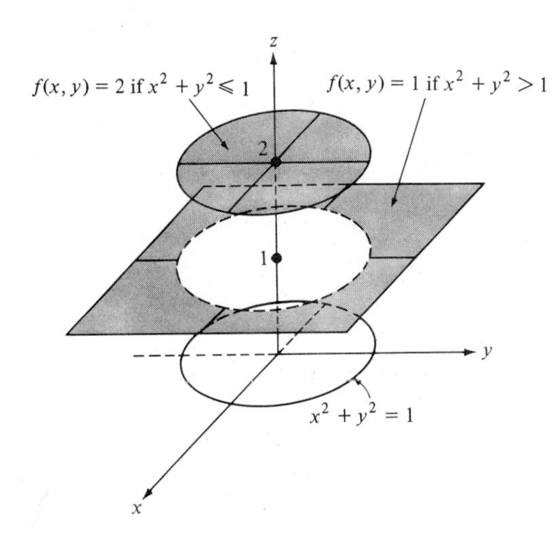

FIGURE 8.19 The graph of the function f defined by

$$f(x,y) = \begin{cases} 2 & \text{if } x^2 + y^2 \leq 1 \\ 1 & \text{if } x^2 + y^2 > 1 \end{cases}$$

The function is discontinuous at points on the circle $x^2 + y^2 = 1$.

necessary to find two different paths which contain (a, b) such that $f(x, y)$ has different limits along these paths or has no limit along one of them.

Example 3 Let $f(x, y) = \dfrac{x^2 + y^2}{x^2 - y^2}$. Show that $f(x, y)$ has no limit as $(x, y) \to (0, 0)$.

Solution We first consider the path defined by $x = 0$ (the y-axis). Along this path

$$f(x, y) = \frac{0 + y^2}{0 - y^2} = -1.$$

Thus,

$$\lim_{\substack{(x,y)\to(0,0) \\ \text{along } y\text{-axis}}} f(x, y) = \lim_{y\to 0} (-1) = -1.$$

We next consider the path defined by $y = 0$ (the x-axis). Along this path

$$f(x, y) = \frac{x^2 + 0}{x^2 - 0} = 1.$$

Thus,

$$\lim_{\substack{(x,y)\to(0,0) \\ \text{along } x\text{-axis}}} f(x, y) = \lim_{x\to 0} 1 = 1.$$

Since $f(x, y)$ has different limits along the two paths, then there is no limit as $(x, y) \to (0, 0)$.

Boundary and Interior Points In order to develop the theory of maxima and minima of functions of two variables we must consider points in the plane that are analogous to interior points and endpoints of intervals. These are *interior points* and *boundary points* of sets in the plane. These points can be defined most easily by using δ-neighborhoods.

Interior point

Let S be a set of points in the xy-plane. The point $P \in S$ is called an *interior point* of S if there exists a δ-neighborhood of P that lies completely in S.

Let S be a set of points in the xy-plane. The point B in the xy-plane is called a *boundary point of* S if every δ-neighborhood of B contains at least one point in S and one point not in S. The collection of all boundary points is called the *boundary of* S.

Example 4 Let S be the set of all points (x, y) such that $1 \leq x < 2$ and $1 \leq y < 2$ (Figure 8.20).

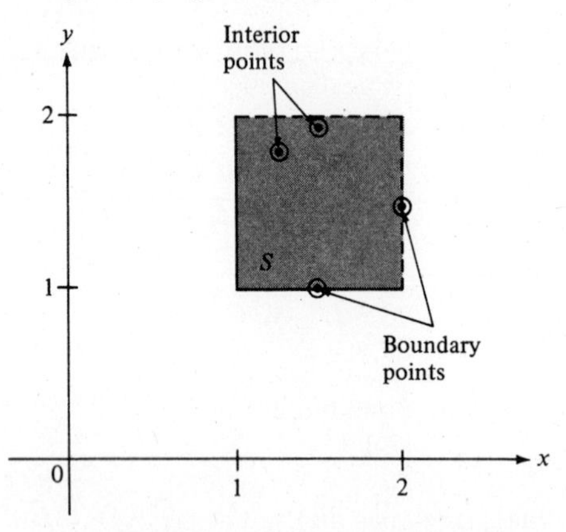

FIGURE 8.20 Example 4. Interior and boundary points of the square $S = \{(x,y) | 1 \leq x < 2, 1 \leq y < 2\}$.

(a) The boundary points of S are those points on the four line segments joining in succession the points (1, 1), (1, 2), (2, 2), and (2, 1). Observe that any circular disk about one of these points will contain points in S and points not in S. This is the case no matter how small the radius of the disk.

(b) The interior points of S are those points located inside the square shown in the figure. That is, the interior points satisfy $1 < x < 2$ and $1 < y < 2$. If (a, b) is any one of these points, then a circular disk can be drawn about (a, b) that lies completely within S. If the point (a, b) is close to the boundary, then the circular disk must be very small; if the point is near the center of the square, the circular disk can be fairly large. In either case such a disk can be drawn.

As we mentioned above the concepts of interior and boundary points in the plane are quite similar to those of *interior points* and *endpoints* of intervals on the real line. The essential difference is that properties true for open intervals have been replaced by properties involving δ-neighborhoods. (The reader should verify that the following properties hold for points on the line.) (1) A point P on the line is an interior point of an interval I if and only if there exists an open interval with P as center that is contained within the interval. A point P in the plane is an interior point of a set if and only if there exists a δ-neighborhood of P that is contained in the set S. (2) A point B on the line is an endpoint of an interval I if and only if each open interval with B as center contains a point in I and a point not in I. A point B in the plane is a boundary point of the set S if and only if each δ-neighborhood about B contains a point in S and a point not in S.

Observe that an interior point of S must belong to S. This is not the case with boundary points. Some of the boundary points in Figure 8.20 are in S and some are not. Observe also that any point of S is either an interior point or a boundary point. That is, either there exists a δ-neighborhood of the point that lies completely in S or every δ-neighborhood contains a point in S (namely the point itself) and a point not in S.

EXERCISES 8.4

1. Describe and sketch the δ-neighborhood of each of the following points:

   (a) $(0, 0)$, $\delta = 1/2$
   • (b) $(1, 1)$, $\delta = 1$
   (c) $(1, -1)$, $\delta = 1/4$
   • (d) $(-1, 2)$, $\delta = 1/2$

2. Use Theorem 8.2 to calculate the limits. Justify all steps.

   (a) $\lim_{(x,y)\to(1,2)} \frac{x^2 - y^2}{xy}$
   • (b) $\lim_{(x,y)\to(2,1)} \frac{x^2 - y^2}{x^2 + y^2}$
   (c) $\lim_{(x,y)\to(0,1)} \frac{x(x - y)}{x + y}$
   • (d) $\lim_{(x,y)\to(1,-1)} \frac{x^2 - y^2}{xy}$
   (e) $\lim_{(x,y)\to(-2,-2)} \frac{2xy}{x + y}$
   • (f) $\lim_{(x,y)\to(2,2)} \frac{x^2y^2}{x^2 + y^2}$
   (g) $\lim_{(x,y)\to(1,-1)} \frac{xy(x^2 - y^2)}{x + y}$
   (h) $\lim_{(x,y)\to(2,2)} \frac{x^2 - y^2}{x - y}$

3. Show that the following functions have no limits as (x, y) approaches $(0, 0)$ by letting (x, y) approach $(0, 0)$ along different paths. (*Hint:* Let $(x, y) \to (0, 0)$ along paths $y = mx$ for different values of m.)

   (a) $f(x, y) = \frac{x^2 - y^2}{x^2 + 2y^2}$
   • (b) $f(x, y) = \frac{xy}{x^2 + y^2}$
   (c) $f(x, y) = \frac{x^2 - y^2}{xy}$

4. Give an example of (a) a set S that has no interior points, (b) a set S that contains all of its boundary points, (c) a set S that has interior points and that contains some but not all of its boundary points.

5. Is it possible to have a nonempty set that has no interior points and no boundary points?

6. Show that if S is a δ-neighborhood of (a, b), then every point of S is an interior point of S. What are the boundary points of S?

7. Sketch each of the following sets and indicate which points are interior points and which are boundary points:

   • (a) $\{(x, y) | x^2 + y^2 < 1\}$
   (b) $\{(x, y) | x^2 + y^2 \leq 1\}$
   • (c) $\{(x, y) | x^2 + y^2 \geq 1\}$
   (d) $\{(x, y) | x^2 + y^2 > 1\}$
   (e) $\{(x, y) | 0 \leq x \leq 1, 0 \leq y \leq 1\}$
   (f) $\{(x, y) | 0 \leq x \leq 1, 0 \leq y < 1\}$
   • (g) $\{(x, y) | 0 \leq x < 1, 0 \leq y < 1\}$
   (h) $\{(x, y) | 0 < x < 1, 0 < y < 1\}$

8.5 PARTIAL DERIVATIVES

Let $z = f(x, y)$ be a function of two variables. We define the *partial derivatives* of z with respect to x and with respect to y by

$$\frac{\partial z}{\partial x} = \frac{\partial f}{\partial x} = \lim_{\Delta x \to 0} \frac{f(x + \Delta x, y) - f(x, y)}{\Delta x}$$

$$\frac{\partial z}{\partial y} = \frac{\partial f}{\partial y} = \lim_{\Delta y \to 0} \frac{f(x, y + \Delta y) - f(x, y)}{\Delta y},$$

provided the limits exist.

Thus, the partial derivative of f with respect to x is calculated by treating y as if it were a constant and f as a function of x alone.

Example 1 Let $z = f(x, y) = x^2 e^{3y}$. To calculate $\partial z/\partial x$ we treat y like a constant:

$$\frac{\partial z}{\partial x} = 2xe^{3y}.$$

To calculate $\partial z/\partial y$ we treat x like a constant:

$$\frac{\partial z}{\partial y} = x^2 e^{3y} \frac{\partial}{\partial y}(3y) = 3x^2 e^{3y}.$$

We write $(\partial x/\partial y)_{(x_0, y_0)}$ to indicate the value of the partial derivative at the point (x_0, y_0). At the point (2, 1) we have the following partial derivatives for the function of Example 1:

$$\left(\frac{\partial z}{\partial x}\right)_{(2,1)} = (2xe^{3y})_{(2,1)} = 2 \cdot 2e^{3 \cdot 1} = 4e^3,$$

$$\left(\frac{\partial z}{\partial y}\right)_{(2,1)} = (3x^2 e^{3y})_{(2,1)} = 3 \cdot 2^2 \cdot e^{3 \cdot 1} = 12e^3.$$

In certain cases it is convenient to use functional notation for the partial derivative. We write

Alternate notation f_x, f_y

$$f_x = \frac{\partial f}{\partial x} \quad \text{and} \quad f_y = \frac{\partial f}{\partial y}.$$

Then $f_x(x_0, y_0)$ and $f_y(x_0, y_0)$ are the values of the partial derivatives at (x_0, y_0).

Example 2 Let $f(x, y) = 3xy + 5x + 7y^{-1}$. Then

$$f_x(x, y) = \frac{\partial f}{\partial x} = 3y + 5 + 0 = 3y + 5,$$

$$f_y(x, y) = \frac{\partial f}{\partial y} = 3x + 0 + 7(-1)y^{-2} = 3x - 7y^{-2}.$$

At the point (0, 1) we have

$$f_x(0, 1) = \left(\frac{\partial f}{\partial x}\right)_{(0,1)} = (3y + 5)_{(0,1)} = 3 \cdot 1 + 5 = 8,$$

$$f_y(0, 1) = \left(\frac{\partial f}{\partial y}\right)_{(0,1)} = (3x - 7y^{-2})_{(0,1)} = 3 \cdot 0 - 7 \cdot 1^{-2} = -7.$$

Second partial derivatives

Second Partial Derivatives The partial derivatives are themselves functions of x and y. Thus we can calculate their partial derivatives with respect to x and y, obtaining the four *second partial derivatives:*

$$\frac{\partial^2 f}{\partial x^2} = \frac{\partial}{\partial x}\left(\frac{\partial f}{\partial x}\right) = f_{xx} \quad \text{(second partial with respect to x),}$$

$$\left.\begin{aligned} \frac{\partial^2 f}{\partial x\,\partial y} &= \frac{\partial}{\partial x}\left(\frac{\partial f}{\partial y}\right) = f_{yx} \\ \frac{\partial^2 f}{\partial y\,\partial x} &= \frac{\partial}{\partial y}\left(\frac{\partial f}{\partial x}\right) = f_{xy} \end{aligned}\right\} \quad \text{(mixed partial derivatives),}$$

$$\frac{\partial^2 f}{\partial y^2} = \frac{\partial}{\partial y}\left(\frac{\partial f}{\partial y}\right) = f_{yy} \quad \text{(second partial with respect to y).}$$

Observe that the order of the symbols is reversed in the two notations for the mixed partial derivatives. That is,

$$\frac{\partial^2 f}{\partial x\,\partial y} = f_{yx}.$$

The first notation stands for $\frac{\partial}{\partial x}\left(\frac{\partial f}{\partial y}\right)$, indicating that the partial derivative is first calculated with respect to y, then with respect to x. The second notation, f_{yx}, stands for $(f_y)_x$, indicating the same order of calculating derivatives. As we shall see, the notation for the mixed partials will cause us no trouble, because in most cases the two mixed partials are equal to each other.

Example 3 Calculate all of the first and second partial derivatives of

$$z = f(x, y) = \frac{x}{y} + ye^x.$$

Solution

$$\frac{\partial z}{\partial x} = \frac{1}{y} + ye^x,$$

$$\frac{\partial z}{\partial y} = x(-1)y^{-2} + e^x = -\frac{x}{y^2} + e^x.$$

The second partial derivatives are:

$$\frac{\partial^2 z}{\partial x^2} = \frac{\partial}{\partial x}\left(\frac{\partial z}{\partial x}\right) = \frac{\partial}{\partial x}\left(\frac{1}{y} + ye^x\right) = 0 + y \cdot \frac{\partial}{\partial x}(e^x) = ye^x,$$

$$\frac{\partial^2 z}{\partial y\,\partial x} = \frac{\partial}{\partial y}\left(\frac{\partial z}{\partial x}\right) = \frac{\partial}{\partial y}\left(\frac{1}{y} + ye^x\right) = -y^{-2} + 1 \cdot e^x = -\frac{1}{y^2} + e^x,$$

$$\frac{\partial^2 z}{\partial x\,\partial y} = \frac{\partial}{\partial x}\left(\frac{\partial z}{\partial y}\right) = \frac{\partial}{\partial x}\left(-\frac{x}{y^2} + e^x\right) = -\frac{1}{y^2} + e^x,$$

$$\frac{\partial^2 z}{\partial y^2} = \frac{\partial}{\partial y}\left(\frac{\partial z}{\partial y}\right) = \frac{\partial}{\partial y}\left(-\frac{x}{y^2} + e^x\right) = -x(-2)y^{-3} + 0 = \frac{2x}{y^3}.$$

Observe that the two mixed partials are equal.

Higher partial derivatives are defined in an analogous manner. For example (considering again the function of Example 3),

$$\frac{\partial^4 z}{\partial x\,\partial y^2\,\partial x} = \frac{\partial}{\partial x}\left(\frac{\partial}{\partial y}\left(\frac{\partial}{\partial y}\left(\frac{\partial z}{\partial x}\right)\right)\right) = \frac{\partial}{\partial x}\left(\frac{\partial}{\partial y}\left(\frac{\partial}{\partial y}\left(\frac{1}{y} + ye^x\right)\right)\right)$$

$$= \frac{\partial}{\partial x}\left(\frac{\partial}{\partial y}\left(-\frac{1}{y^2} + e^x\right)\right) = \frac{\partial}{\partial x}\left(\frac{2}{y^3} + 0\right) = 0.$$

There is a natural extension of these definitions to functions of three or more variables. For example, if

$$z = f(u, v, x, y) = uv + \frac{x}{y} + 2xu,$$

then

$$\frac{\partial z}{\partial u} = v + 0 + 2x = v + 2x = f_u,$$

$$\frac{\partial z}{\partial v} = u + 0 + 0 = u = f_v,$$

$$\frac{\partial z}{\partial x} = 0 + \frac{1}{y} + 2u = \frac{1}{y} + 2u = f_x,$$

$$\frac{\partial z}{\partial y} = 0 + x(-1)y^{-2} + 0 = -\frac{x}{y^2} = f_y,$$

$$\frac{\partial^2 z}{\partial u\, \partial v} = \frac{\partial}{\partial u}\left(\frac{\partial z}{\partial v}\right) = \frac{\partial}{\partial u}(u) = 1 = f_{vu},$$

and so on.

As we mentioned above, the mixed partials $\partial^2 f/\partial x\, \partial y$ and $\partial^2 f/\partial y\, \partial x$ are equal in most cases of interest. The result is stated in the following theorem:

Equality of mixed partial derivatives

Theorem 8.3 Let f, $\partial f/\partial x$, $\partial f/\partial y$, $\partial^2 f/\partial x\, \partial y$ and $\partial^2 f/\partial y\, \partial x$ all be continuous on a δ-neighborhood of (x_0, y_0). Then

$$\left(\frac{\partial^2 f}{\partial x\, \partial y}\right)_{(x_0, y_0)} = \left(\frac{\partial^2 f}{\partial y\, \partial x}\right)_{(x_0, y_0)}.$$

In this book we are concerned primarily with functions that have continuous partial derivatives everywhere in their domains. For such functions we have the following result.

COROLLARY Let f, $\partial f/\partial x$, $\partial f/\partial y$, $\partial^2 f/\partial x\, \partial y$, and $\partial^2 f/\partial y\, \partial x$ be continuous on the interior points of the domain of f. Then

$$\frac{\partial^2 f}{\partial x\, \partial y} = \frac{\partial^2 f}{\partial y\, \partial x}$$

on the interior points of the domain of f.

Geometrical interpretation of partial derivative

Geometrical Interpretation of the Partial Derivatives The partial derivative of f with respect to y, evaluated at (a, b), is obtained by holding x fixed at a and calculating the derivative of the resulting function of y at the point $y = b$. That is,

$$\left(\frac{\partial f}{\partial y}\right)_{(a,b)} = \lim_{\Delta y \to 0} \frac{f(a, b + \Delta y) - f(a, b)}{\Delta y}.$$

To show the true nature of the difference quotient let $g(y) = f(a, y)$. Then

$$\left(\frac{\partial f}{\partial y}\right)_{(a,b)} = \lim_{\Delta y \to 0} \frac{f(a, b + \Delta y) - f(a, b)}{\Delta y} = \lim_{\Delta y \to 0} \frac{g(b + \Delta y) - g(b)}{\Delta y}$$

$$= g'(b).$$

Now $z = g(y)$ is the curve obtained by intersecting the surface $x = f(x, y)$ with the vertical plane $x = a$. (See Figure 8.21.) Therefore $(\partial f/\partial y)_{(a,b)} = g'(b)$ is the

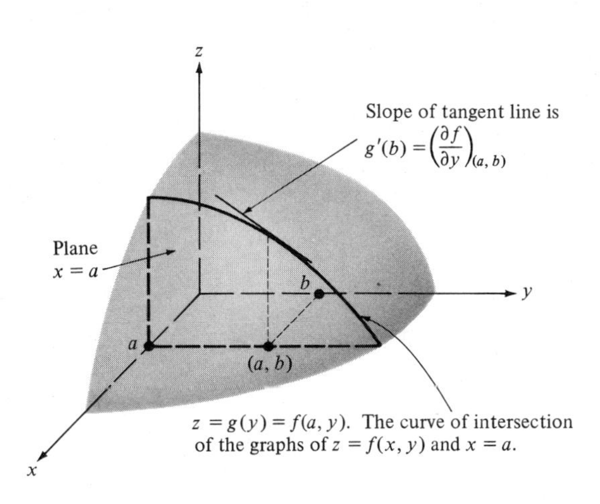

FIGURE 8.21 $\left(\frac{\partial f}{\partial y}\right)_{(a,b)}$ is the slope of the tangent line at $y = b$ to the curve of intersection of the graphs of $z = f\ x,y)$ and $x = a$.

slope of the tangent line to this curve at (a, b). Similarly, if $z = h(x) = f(x, b)$, then $(\partial f/\partial x)_{(a,b)} = h'\ (a)$ is the slope of the tangent line at (a, b) to the curve obtained by intersecting the graph of $f(x, y)$ with the plane $y = b$.

Marginal Analysis In economics partial derivatives are used for marginal analysis. Suppose, for example, that the productivity of a company is considered to be a function of x and y, where x measures the size of the labor force and y the amount of capital invested. We can write $P = f(x, y)$. Then $\partial P/\partial x$ is the *marginal productivity of labor*. For fixed values of x and y this partial derivative is approximately equal to the increase in productivity obtained by adding one man to the labor force. Similarly, $\partial P/\partial y$ is the *marginal productivity of capital,* which is approximately equal to the increase in productivity obtained by increasing the capital by one dollar.

The tangent plane

The Tangent Plane (*Optional*) Recall that the derivative of a function of one variable determines the tangent line at each point where it exists. Similarly, if f is a function of two variables the partial derivatives $\partial f/\partial x$ and $\partial f/\partial y$ determine lines tangent to the surface $z = f(x, y)$ at each point where they both exist. At a given point $P_0(x_0, y_0, z_0)$ these two tangent lines determine a plane in three-dimensional space. If the surface is sufficiently smooth in the vicinity of P_0, then this plane is tangent to the surface and is called the *tangent plane to the surface at* $P_0(x_0, y_0, z_0)$. (See Figure 8.22.)

It is proved in advanced courses that if f, $\partial f/\partial x$, and $\partial f/\partial y$ are continuous on a δ-neighborhood of P_0, then f has a tangent plane at P_0. (This is roughly equivalent to the requirement that the graph of f be smooth in the vicinity of P_0.)

It is a fairly simple matter to determine the equation of the tangent plane at the point $P_0(x_0, y_0, z_0)$. Let

$$z = g(x, y) = ax + by + c$$

be the equation of the tangent plane. If we intersect the surface $z = f(x, y)$ with the vertical plane $y = y_0$ the curve of intersection has a tangent line that lies in the tangent plane as does the line of intersection with the tangent plane itself. Thus,

$$\left(\frac{\partial f}{\partial x}\right)_{(x_0,\,y_0)} = \left(\frac{\partial g}{\partial x}\right)_{(x_0,\,y_0)} = a$$

Similarly,

$$\left(\frac{\partial f}{\partial y}\right)_{(x_0, y_0)} = \left(\frac{\partial g}{\partial y}\right)_{(x_0, y_0)} = b.$$

We now calculate the value of c. Since the point $P_0(x_0, y_0, z_0)$ is on the tangent plane, then

$$z_0 = ax_0 + by_0 + c$$

so that

$$c = z_0 - ax_0 - by_0.$$

Thus, the equation of the tangent plane can be written as

$$z = ax + by + (z_0 - ax_0 - by_0)$$
$$z - z_0 = a(x - x_0) + b(y - y_0)$$

Equation of the tangent plane

$$\boxed{z - z_0 = \left(\frac{\partial f}{\partial x}\right)_{(x_0, y_0)} (x - x_0) + \left(\frac{\partial f}{\partial y}\right)_{(x_0, y_0)} (y - y_0).}$$

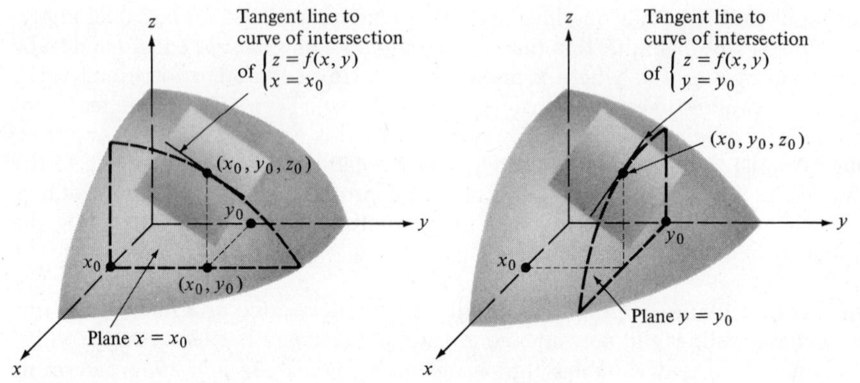

FIGURE 8.22 The tangent plane to the graph of $z = f(x,y)$ at (x_0, y_0, z_0) contains the tangent lines to the curves of intersection with the vertical planes $x = x_0$ and $y = y_0$.

Example 4 Calculate the equation of the tangent plane to the hemisphere

$$f(x, y) = \sqrt{13^2 - x^2 - y^2}$$

at the point (4, 3, 12). (See Figure 8.23.)

Solution $f(x, y) = (13^2 - x^2 - y^2)^{1/2}$.

$$\frac{\partial f}{\partial x} = \frac{1}{2}(13^2 - x^2 - y^2)^{-1/2} \cdot \frac{\partial}{\partial x}(13^2 - x^2 - y^2)$$
$$= \frac{1}{2}(13^2 - x^2 - y^2)^{-1/2}(-2x)$$
$$= \frac{-x}{\sqrt{13^2 - x^2 - y^2}}.$$

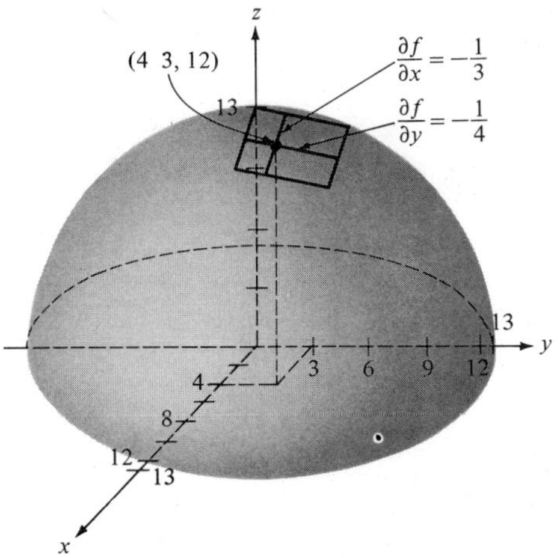

FIGURE 8.23 Example 4.

Similarly,

$$\frac{\partial f}{\partial y} = \frac{-y}{\sqrt{13^2 - x^2 - y^2}}.$$

When $x = 4$ and $y = 3$,

$$\frac{\partial f}{\partial x} = \frac{-4}{\sqrt{13^2 - 4^2 - 3^2}} = -\frac{4}{12} = -\frac{1}{3}$$

and

$$\frac{\partial f}{\partial y} = \frac{-3}{\sqrt{13^2 - 4^2 - 3^2}} = -\frac{3}{12} = -\frac{1}{4}.$$

The equation of the tangent plane is

$$z - z_0 = \frac{\partial f}{\partial x} \cdot (x - x_0) + \frac{\partial f}{\partial y} \cdot (y - y_0),$$

$$z - 12 = -\frac{1}{3}(x - 4) - \frac{1}{4}(y - 3),$$

$$12(z - 12) = -4(x - 4) - 3(y - 3),$$

$$12z - 144 = -4x + 16 - 3y + 9,$$

$$4x + 3y + 12z = 144 + 16 + 9,$$

$$4x + 3y + 12z = 169.$$

EXERCISES 8.5

1. Calculate $\partial f/\partial x$ and $\partial f/\partial y$:

   (a) $f(x, y) = xy^2 - x^2y$

   • (b) $f(x, y) = (xy)^3$

   (c) $f(x, y) = \dfrac{x}{y} - \dfrac{y}{x}$

   • (d) $f(x, y) = \dfrac{1}{x - y}$

   (e) $f(x, y) = (x + y^2)(x^2 - y)$

   (f) $f(x, y) = \dfrac{x - y}{x^2 + y^2}$

   • (g) $f(x, y) = e^{-xy}$

   (h) $f(x, y) = e^{xy}(x^2 - y^2)$

(i) $f(x, y) = \dfrac{e^{xy}}{x^2 - y^2}$

• (j) $f(x, y) = \dfrac{x^3 + y^2}{2xy}$

2. Calculate f_{xx}, f_{xy}, f_{yx}, and f_{yy}:

(a) $f(x, y) = x^2y + xy^2$

• (b) $f(x, y) = \left(\dfrac{x}{y}\right)^3$

(c) $f(x, y) = \dfrac{x^2 - y^2}{xy}$

(d) $f(x, y) = e^{-xy}$

• (e) $f(x, y) = x^2 - y^2$

(f) $f(x, y) = xe^{xy}$

(g) $f(x, y) = \dfrac{1}{x - y}$

• (h) $f(x, y) = (y^2 + e^x)^2$

(i) $f(x, y) = \dfrac{x - y}{xy}$

(j) $f(x, y) = \dfrac{1}{x^2 + y^2}$

3. If $z = f(u, v, x, y) = uvx^2y^3$, calculate

• (a) $\dfrac{\partial z}{\partial v}$

(b) $\dfrac{\partial^2 z}{\partial v^2}$

• (c) $\dfrac{\partial^3 z}{\partial x\, \partial v\, \partial y}$

(d) $\dfrac{\partial^4 z}{\partial x\, \partial u\, \partial y\, \partial v}$

• (e) $\dfrac{\partial^3 z}{\partial x^2\, \partial v}$

(f) $\dfrac{\partial^2 z}{\partial y\, \partial v}$

4. Calculate the equation of the tangent plane to the graph at the given point:

• (a) $z = xy$; $(-1, -1, 1,)$
(b) $z = x^2 + y^2$; $(-1, -1, 2)$
• (c) $z = x \ln y + 1$; $(2, 1, 1)$
(d) $x^2 + y^2 + z^2 = 3$; $(1, 1, 1)$

8.6 LOCAL MAXIMA AND MINIMA

Recall the following concepts from Section 8.4 related to a set S in the xy-plane: (1) the point P is an *interior point* of S if there is a δ-neighborhood of P that lies completely within S; (2) the point B is a *boundary point* of S if every δ-neighborhood of B contains a point in S and a point not in S. Observe that an interior point of S must belong to S, while a boundary point need not.

We can now extend the work in Chapter 3 on maxima-minima problems to functions of two variables.

Maximum on set S

DEFINITION Let f be defined on the set S. We say that M is the *maximum of f on* S provided

1. $f(x, y) \leqslant M$ for every point $(x, y) \in S$.
2. $f(x, y) = M$ for at least one point $(x, y) \in S$.

Minimum on set S

The *minimum* of f on S is defined similarly. If m is the minimum, then

1. $f(x, y) \geqslant m$ for every point $(x, y) \in S$.
2. $f(x, y) = m$ for at least one point $(x, y) \in S$.

There is no guarantee that a function f will have a maximum or a minimum on an arbitrary set. The situation is analogous to that for a function of a single variable. Recall, however, that a continuous function of one variable must have maximum and minimum values over certain types of sets (in particular, the closed intervals). The following theorem states a similar property for a function of two variables. It can be extended to functions of three or more variables.

Existence of extrema

Theorem 8.4 Let S be a set in the xy-plane that is bounded in extent (that is, is contained within some rectangle in the xy-plane) and that contains all of its boundary points. If f is a function of two variables that is continuous on S, then f has a maximum and a minimum on S.

This theorem is the analog of Theorem 3.2 for functions of two variables. The conditions on the set S are similar to the conditions that define closed intervals. The requirement that S be bounded in extent is similar to the requirement that a closed interval have finite length, while the requirement that S contain all of its boundary points is essentially the same as the requirement that a closed interval contain its endpoints.

Observe that an extreme value of f must occur at an interior point of the set S or at a boundary point. (S contains no other types of points.) If an extreme value occurs at an interior point, it is called a *local (relative) maximum* or *minimum*. The precise definition is as follows:

Local Extrema

DEFINITION Let f be a function of two variables. We say that f has a *local* (or *relative*) *maximum* at the point $P(a, b)$ if there exists a δ-neighborhood of P on which f is defined such that the maximum of f on the neighborhood is at P. (See Figure 8.24.)

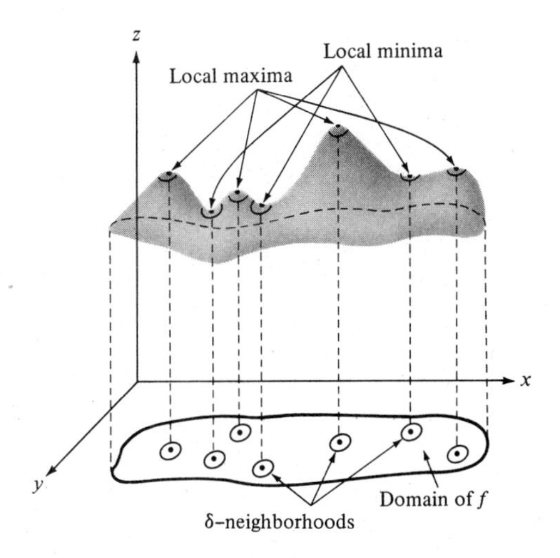

FIGURE 8.24 Local extrema. The extreme values over the δ-neighborhoods occur at the centers of the neighborhoods.

A *local minimum* is defined similarly: There is a *local minimum* at P if there is a δ-neighborhood of P on which f is defined such that the minimum value of f on the neighborhood is at P.

Again, the situation is somewhat similar to that for a function of a single variable. Recall that if g is a function of one variable, then g has a local maximum at the point P if there is an open interval containing P as an interior point such that the maximum of g on the interval is at P. (This is discussed in Sections 3.1

and 3.2.) When we extend the concept to a function of two variables, we replace the open interval with the δ-neighborhood. Recall also that if the function of one variable has a local extremum at a point and is differentiable there, then the value of that derivative is zero. The following theorem generalizes this result. With suitable modifications the theorem can be extended to functions of three or more variables.

Values of partial derivatives at local extremum

Theorem 8.5 (See Theorem 3.1.) Let f be a function of x and y that has a local extremum at the point $P(a, b)$. If $\partial f/\partial x$ and $\partial f/\partial y$ both exist at (a, b), then these partial derivatives are both equal to zero.

PROOF Let f have a local maximum at P. Then there exists a δ-neighborhood of P on which f is defined, having its maximum value at P. For the rest of this discussion we restrict ourselves to points in this neighborhood. If we intersect the graph of f with the vertical plane $x = a$, then the curve of intersection has a local extremum at the point (a, b). Since $\partial f/\partial y$ exists at (a, b), then this number is equal to the slope of the line tangent to the curve at (a, b). But the tangent line to a curve at a local extremum must be horizontal. Therefore,

$$\left(\frac{\partial f}{\partial y}\right)_{(a,b)} = 0.$$

(See Figure 8.25.) Similarly, if we intersect the surface with the plane $y = b$, the curve of intersection has a horizontal tangent line at (a, b), which shows that

$$\left(\frac{\partial f}{\partial x}\right)_{(a,b)} = 0.$$

A similar argument establishes the proof if f has a local minimum at P.

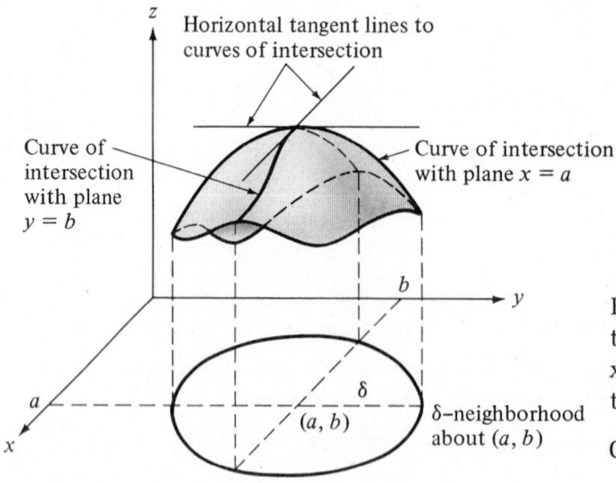

FIGURE 8.25 The curve of intersection of the graphs of $z = f(x,y)$ and $x = a$ has a horizontal tangent line at the local maximum $(a, b, f(a,b))$. Consequently, $\left(\frac{\partial f}{\partial y}\right)_{(a,b)} = 0$.

Example 1 Show that $f(x, y) = x^2 - 6x + y^2 + 4y + 5$ has a local minimum at $(3, -2)$ and no local maximum.

Solution At a local extremum we must have both partial derivatives equal to zero if they exist. We calculate

$$\frac{\partial f}{\partial x} = 2x - 6, \qquad \frac{\partial f}{\partial y} = 2y + 4.$$

The only point that satisfies both $\partial f/\partial x = 0$ and $\partial f/\partial y = 0$ is $x = 3, y = -2$. Thus this is the only possible location of a local extremum.

To show that this is a local minimum we observe that

$$f(x, y) = (x - 3)^2 + (y + 2)^2 - 8.$$

Thus, if $x \neq 3$ or $y \neq -2$, the value of $f(x, y)$ is greater than the value $f(3, -2)$. Consequently, f has a local minimum at $(3, -2)$. Since this is the only possible location for an extremum, there is no local maximum.

The Second Derivative Test In general it can be quite difficult to decide whether a function has a local maximum or a local minimum or neither at a point where both partial derivatives are zero. In the example above we were fortunate in having a rather simple function. The following theorem states a second-derivative test that often can be used to decide which occurs. The proof is omitted.

The second derivative test

Theorem 8.6 (See Theorem 3.4.) Let f have partial derivatives at all points of a δ-neighborhood about (a, b). Suppose that

$$\frac{\partial f}{\partial x} = 0 \quad \text{and} \quad \frac{\partial f}{\partial y} = 0 \qquad \text{at } (a, b).$$

Let

$$\Delta = \frac{\partial^2 f}{\partial x^2} \cdot \frac{\partial^2 f}{\partial y^2} - \left(\frac{\partial^2 f}{\partial x\, \partial y}\right)^2.$$

Then

(1) if

$$\Delta > 0 \quad \text{and} \quad \frac{\partial^2 f}{\partial x^2} < 0,$$

then f *has a local maximum at* (a, b);

(2) if

$$\Delta > 0 \quad \text{and} \quad \frac{\partial^2 f}{\partial x^2} > 0,$$

then f *has a local minimum at* (a, b);

(3) if

$$\Delta < 0,$$

then f *has a saddle point at* (a, b).

(All of these partial derivatives are evaluated at (a, b).)

Example 2 (*Continuation of Example* 1) If we apply the second derivative test to $f(x, y) = x^2 - 6x + y^2 + 4y + 5$ at $(3, -2)$, we find that

$$\frac{\partial f}{\partial x} = 2x - 6, \qquad \frac{\partial f}{\partial y} = 2y + 4,$$

$$\frac{\partial^2 f}{\partial x^2} = 2, \qquad \frac{\partial^2 f}{\partial y^2} = 2, \qquad \frac{\partial^2 f}{\partial x\, \partial y} = 0.$$

Therefore

$$\Delta = \frac{\partial^2 f}{\partial x^2} \cdot \frac{\partial^2 f}{\partial y^2} - \left(\frac{\partial^2 f}{\partial x\, \partial y}\right)^2 = 2 \cdot 2 - 0 = 4 > 0.$$

Since $\partial^2 f/\partial x^2 = 2 > 0$, then part (2) of Theorem 8.6 applies. Thus, f has a local minimum at $(3, -2)$.

Example 3 Show that $f(x, y) = x^2 - y^2$ has a saddle point at the origin and no local extrema.

Solution $\partial f/\partial x = 2x$, $\partial f/\partial y = -2y$. The only point with both partial derivatives equal to zero is $(0, 0)$. We use the second-derivative test:

$$\frac{\partial^2 f}{\partial x^2} = 2, \qquad \frac{\partial^2 f}{\partial y^2} = -2, \qquad \frac{\partial^2 f}{\partial x\, \partial y} = 0.$$

Therefore

$$\Delta = \frac{\partial^2 f}{\partial x^2} \cdot \frac{\partial^2 f}{\partial y^2} - \left(\frac{\partial^2 f}{\partial x\, \partial y}\right)^2 = 2(-2) - 0 = -4 < 0.$$

It follows from part (3) of Theorem 8.6 that f has a saddle point at the point $(0, 0)$. Since this was the only possible location for an extremum, then no extremum exists.

The graph of $f(x, y)$ is similar to the saddle surface shown in Figure 8.15.

Example 4 Find all local maxima, minima, and saddle points for

$$f(x, y) = 2x^2 + 2xy + 14x - 2y^2 + 22y - 8.$$

Solution

$$\frac{\partial f}{\partial x} = 4x + 2y + 14,$$

$$\frac{\partial f}{\partial y} = 2x - 4y + 22.$$

If we set the partial derivatives equal to zero, we obtain the system of equations

$$\begin{cases} 4x + 2y = -14, \\ 2x - 4y = -22. \end{cases}$$

The only solution is $x = -5$, $y = 3$. The only possible location for a local extremum or saddle point is $(-5, 3)$. We use the second-derivative test:

$$\frac{\partial^2 f}{\partial x^2} = 4, \qquad \frac{\partial^2 f}{\partial y^2} = -4, \qquad \frac{\partial^2 f}{\partial x\, \partial y} = 2,$$

$$\Delta = \frac{\partial^2 f}{\partial x^2} \cdot \frac{\partial^2 f}{\partial y^2} - \left(\frac{\partial^2 f}{\partial x\, \partial y}\right)^2 = 4(-4) - 4 = -20 < 0.$$

Therefore f has a saddle point at $(-5, 3)$ and no local extremum.

Remark 1 When we say that $f(x, y)$ has a saddle point at (a, b) we do not necessarily mean that the graph has the classical saddle shape shown in Figure 8.15. We mean that there is at least one path along which f has a local maximum at (a, b) and at least one other path along which f has a local minimum at (a, b). These paths may be straight lines or curves. (See Example 5.)

Remark 2 If

$$\Delta = \frac{\partial^2 f}{\partial x^2} \cdot \frac{\partial^2 f}{\partial y^2} - \left(\frac{\partial^2 f}{\partial x\, \partial y}\right)^2 = 0$$

at a critical point (a, b), then the second derivative test cannot be used. The function f may have a local maximum, a local minimum, a saddle point, or its graph may simply level off at the critical point. (See Exercise 2.)

Example 5 Show that the function

$$f(x, y) = (y - x^2)(y - 2x^2)$$

has a saddle point at the origin and has no local extrema.

Solution

$$\frac{\partial f}{\partial x} = -6xy + 8x^3$$

$$\frac{\partial f}{\partial y} = 2y - 3x^2.$$

To find the critical points we must solve the system of equations

$$\begin{cases} -6xy + 8x^3 = 0 \\ 2y - 3x^2 = 0 \end{cases}$$

If we substitute the value $y = \frac{3}{2}x^2$ from the second equation into the first, it reduces to

$$\begin{aligned} -6x(\tfrac{3}{2}x^2) + 8x^3 &= 0 \\ -9x^3 + 8x^3 &= 0 \\ -x^3 &= 0 \\ x &= 0. \end{aligned}$$

Then $y = \frac{3}{2}x^2 = \frac{3}{2} \cdot 0 = 0$. The only critical point is the origin. Thus, no local extrema exist at any other points.

If we attempt to apply the second derivative test at $(0, 0)$, we find that

$$\Delta = \frac{\partial^2 f}{\partial x^2} \cdot \frac{\partial^2 f}{\partial y^2} - \left(\frac{\partial^2 f}{\partial x\, \partial y}\right)^2 = 0$$

so that a different method must be used.

Observe that $f(0, 0) = 0$. To get an idea of how the function behaves near the origin we graph the level curve defined by $f(x, y) = 0$. If

$$f(x, y) = 0,$$

then

$$(y - x^2)(y - 2x^2) = 0$$

so that

$$y = x^2 \quad \text{or} \quad y = 2x^2.$$

The level curve consists of the two parabolas in Figure 8.26a. These parabolas separate the xy-plane into four regions. We consider the behavior of f in each of these regions.

The region below the two parabolas is defined by

$$y < x^2 \quad \text{and} \quad y < 2x^2.$$

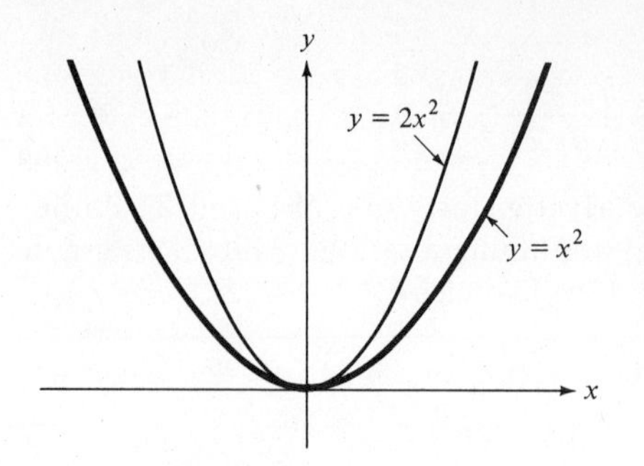

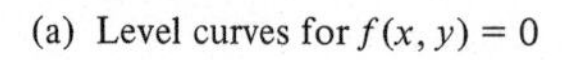
(a) Level curves for $f(x, y) = 0$

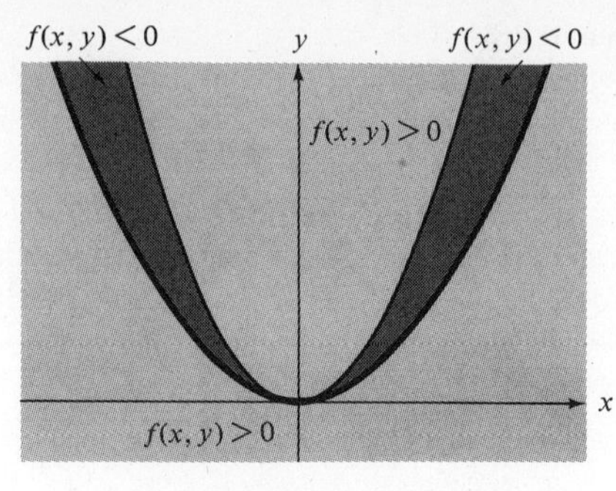

(b) Regions on which $f(x, y) > 0$ and $f(x, y) > 0$

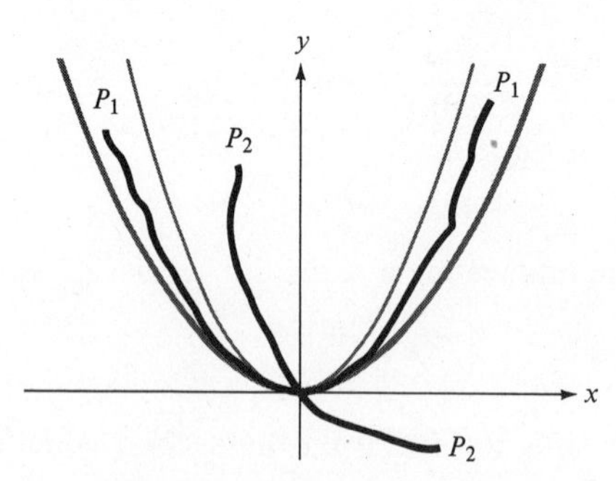

(c) Along path P_1 the function has a local maximum at the origin. Along path P_2 it has a local minimum at the origin.

FIGURE 8.26 Example 5. The function $f(x, y) = (y - x^2)(y - 2x^2)$ has a saddle point at the origin.

On this region we have

$$y - x^2 < 0 \quad \text{and} \quad y - 2x^2 < 0$$

so that

$$f(x, y) = (y - x^2)(y - 2x^2) > 0.$$

(Figure 8.26b.)

The regions between the two parabolas are defined by

$$y < 2x^2 \quad \text{and} \quad y > x^2.$$

On these regions we have

$$f(x, y) = (y - x^2)(y - 2x^2) < 0.$$

The region above the two parabolas is defined by

$$y > 2x^2 \quad \text{and} \quad y > x^2.$$

On this region we have

$$f(x, y) = (y - x^2)(y - 2x^2) > 0.$$

If we now choose a path P_1 through the origin that lies in the regions between the two parabolas we see that $f(x, y)$ has a local maximum at $(0, 0)$ (Figure 8.26c). This follows from the fact that $f(0, 0) = 0$ and $f(x, y) < 0$ at every other point of the path. On the other hand, if we choose a path P_2 that crosses through the origin directly from the region below the parabolas to the region above them, then $f(x, y)$ has a local minimum at (0, 0). Since f has a local minimum at (0, 0) along one path and a local maximum at (0, 0) along another then f has a saddle point at (0, 0). (See Figure 8.26c.)

EXERCISES 8.6

1. Find all local extrema and all saddle points. Use the second-derivative test whenever possible.

   (a) $f(x, y) = x^2 + xy + y^2$
   • (b) $f(x, y) = 4x^2 - y^2$
   (c) $f(x, y) = y^3 - 12y + x^2$
   • (d) $f(x, y) = x^3 + y^3 + 3xy$
   (e) $f(x, y) = y^2 - x^2$
   • (f) $f(x, y) = x^2 - y^3 - 3y$
   (g) $f(x, y) = x^3 - 6xy + y^2$
   • (h) $f(x, y) = x^2 - xy + y^2 + 6y$
   (i) $f(x, y) = x^4 + y^5 - 32x - 5y$
   • (j) $f(x, y) = x^2 - x^3 + y^3 - y^2$

2. Each of the following functions has a critical point at the origin to which the second derivative test cannot be applied. Examine the behavior of $f(x, y)$ near the critical point along the paths defined by $y = mx$ for various values of m. Decide if f has a local maximum, a local minimum, a saddle point, or if the graph levels off at the origin. (*Hint:* Let $g(x) = f(x, mx)$.)

   (a) $f(x, y) = x^3 + 2$
   • (b) $f(x, y) = x^2 + 2xy + y^2$

3. Modify the argument of Example 5 to show that the function

$$f(x, y) = (y - x^2)(y + x^2)$$

   has a saddle point at the origin.

•4. What are the dimensions of the rectangular box with a volume of 343 cubic inches that has the minimum surface area?

5. Show that a tangent plane to the graph of $z = f(x, y)$ at a local extremum must be parallel to the xy-plane.

•6. Find the minimum distance from the point (0, 2, 2) to the plane $x + y + 4z = 4$.

8.7 MAXIMA AND MINIMA

Theorem 8.5 leads us to the following method for finding the extreme values of a continuous function f defined on a set S that contains all of its boundary points and is bounded in extent.

Steps for locating extreme values

1. Locate all points in the interior of S where

$$\frac{\partial f}{\partial x} = 0 \quad \text{or fails to exist}$$

and where

$$\frac{\partial f}{\partial y} = 0 \quad \text{or fails to exist.}$$

2. Locate all points on the boundary of S where f may take extreme values.

3. The maximum (minimum) of f on S is equal to the maximum (minimum) of f on the points located in the above steps.

Critical points

The points located in step 1 are called *interior critical points.* Those located in step 2 are *boundary critical points.* The most difficult part of many maxima-minima problems is locating the boundary critical points. The following example illustrates how these critical points can be found.

Example 1 Calculate the maximum of $f(x, y) = x^2 - 2x + y^2 - 2y + 5$ on the set

$$S = \{(x, y)|x^2 + y^2 \leqslant 4\}.$$

Solution The interior of S is the set of all points inside the circle $x^2 + y^2 = 4$. The boundary is the circle. We calculate the partial derivatives

$$\frac{\partial f}{\partial x} = 2x - 2, \qquad \frac{\partial f}{\partial y} = 2y - 2.$$

Thus the only interior critical point is (1, 1).

The problem of locating the boundary critical points is more complicated. The circle $x^2 + y^2 = 4$ can be broken into the two semicircles

$$y = \sqrt{4 - x^2} \quad \text{and} \quad y = -\sqrt{4 - x^2}, \quad -2 \leqslant x \leqslant 2.$$

(See Figure 8.27.) On the first semicircle the value of f is given by

$$f(x, y) = x^2 - 2x + y^2 - 2y + 5$$

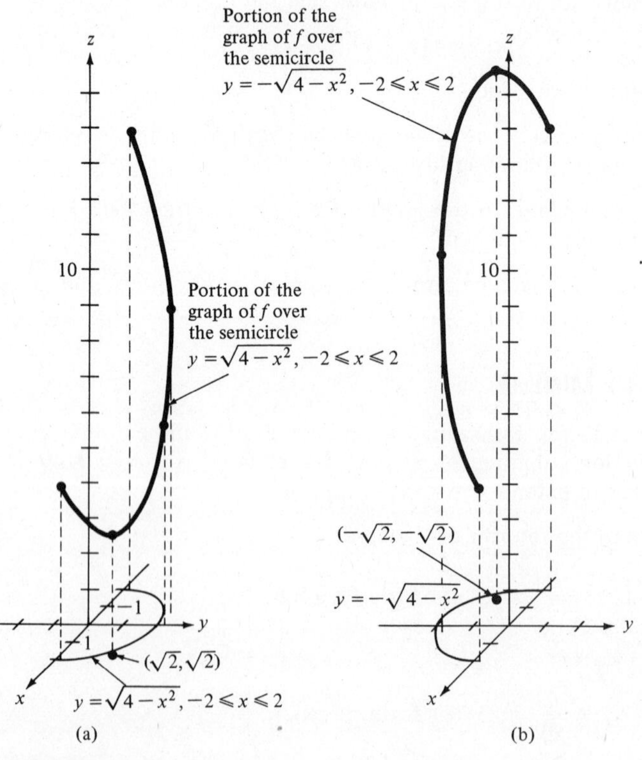

FIGURE 8.27 Example 1.

where $y = \sqrt{4 - x^2}$, $-2 \leq x \leq 2$. Thus, on the semicircle $y = \sqrt{4 - x^2}$

$$\begin{aligned} f(x, y) = f(x, \sqrt{4 - x^2}) &= x^2 - 2x + (\sqrt{4 - x^2})^2 - 2(\sqrt{4 - x^2}) + 5 \\ &= x^2 - 2x + 4 - x^2 - 2\sqrt{4 - x^2} + 5 \\ &= -2x - 2\sqrt{4 - x^2} + 9, \qquad -2 \leq x \leq 2. \end{aligned}$$

Since the value of f on the semicircle depends only on x, we can consider f to be a function of x—say

$$g_1(x) = f(x, \sqrt{4 - x^2}) = -2x - 2\sqrt{4 - x^2} + 9, \qquad -2 \leq x \leq 2.$$

Consequently, the problem of finding the boundary critical points on the semicircle $y - \sqrt{4 - x^2}$ reduces to that of finding the critical points of $g_1(x)$ over the interval $-2 \leq x \leq 2$.

We proceed as in Chapter 3:

$$\begin{aligned} g_1'(x) &= -2 - 2(\tfrac{1}{2})(4 - x^2)^{-1/2}(-2x) \\ &= -2 + \frac{2x}{\sqrt{4 - x^2}}. \end{aligned}$$

If $g_1'(x) = 0$, then

$$2 = \frac{2x}{\sqrt{4 - x^2}}, \qquad \sqrt{4 - x^2} = x, \qquad 4 - x^2 = x^2,$$
$$2x^2 = 4, \qquad x = \pm\sqrt{2}.$$

The only one of these numbers that makes $g_1'(x) = 0$ is $x = \sqrt{2}$.

Since $g_1(x)$ may have its extreme values at the endpoints of the interval, we also must consider the two points $x = -2$, $x = 2$. Thus, we obtain the following *boundary critical points for f on the semicircle* $y = \sqrt{4 - x^2}$, $-2 \leq x \leq 2$:

$$(\sqrt{2}, \sqrt{2}), \quad (-2, 0), \quad (2, 0).$$

We now turn our attention to the other semicircle, $y = -\sqrt{4 - x^2}$. For points on this semicircle we have

$$\begin{aligned} g_2(x) = f(x, -\sqrt{4 - x^2}) &= x^2 - 2x + (-\sqrt{4 - x^2})^2 - 2(-\sqrt{4 - x^2}) + 5 \\ &= x^2 - 2x + 4 - x^2 + 2\sqrt{4 - x^2} + 5 \\ &= -2x + 2\sqrt{4 - x^2} + 9, \qquad -2 \leq x \leq 2. \end{aligned}$$

Then $g_2'(x) = -2 - 2x/\sqrt{4 - x^2}$, so $g_2'(x) = 0$ if and only if $x = -\sqrt{2}$. We also consider the boundary points $x = -2$, $x = 2$ at the end of the interval. Therefore, we have the following *boundary critical points for f on the semicircle* $y = -\sqrt{4 - x^2}$, $-2 \leq x \leq 2$:

$$(-\sqrt{2}, -\sqrt{2}), \quad (-2, 0), \quad (2, 0).$$

Observe that the last two points are two of the boundary critical points calculated on the other semicircle.

Therefore we have one interior critical point: $(1, 1)$; and four boundary critical points:

$$(\sqrt{2}, \sqrt{2}), \quad (-2, 0) \quad (2, 0), \quad (-\sqrt{2}, -\sqrt{2}).$$

The extreme values of f are obtained at these points. We calculate:

$$f(1, 1) = 1 - 2 + 1 - 2 + 5 = 3,$$
$$f(\sqrt{2}, \sqrt{2}) = 2 - 2\sqrt{2} + 2 - 2\sqrt{2} + 5 = 9 - 4\sqrt{2},$$
$$f(-2, 0) = 4 - 2(-2) + 0 - 0 + 5 = 13,$$
$$f(2, 0) = 4 - 2(2) + 0 - 0 + 5 = 5,$$
$$f(-\sqrt{2}, -\sqrt{2}) = 2 - 2(-\sqrt{2}) + 2 - 2(-\sqrt{2}) + 5 = 9 + 4\sqrt{2}.$$

The minimum of f on S is 3 taken at the interior critical point. The maximum is $9 + 4\sqrt{2}$ taken at the point $(-\sqrt{2}, -\sqrt{2})$ on the boundary.

Boundary critical points over a curve $y = h(x)$

The technique illustrated in Example 1 for the determination of boundary critical points can be applied whenever the boundary is the union of a number of simple graphs. Suppose, for example, the portion of the boundary between (a_1, b_1) and (a_2, b_2) is defined by $y = h(x)$, $a_1 \leq x \leq a_2$. Then the portion of the graph of f over this curve is defined by

$$g(x) = f(x, y) = f(x, h(x)), \qquad a_1 \leq x \leq a_2.$$

The critical points of f over this part of the boundary are found among those points between a_1 and a_2 where $g'(x) = 0$, or $g'(x)$ does not exist, along with the endpoints of the interval a_1 and a_2. (See Figure 8.28.)

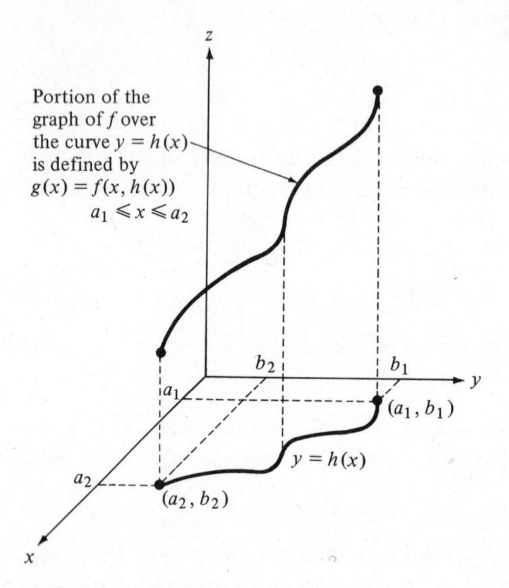

FIGURE 8.28 The portion of the graph of f over the curve

$$y = h(x),\ a_1 \leq x \leq a_2$$

is defined by

$$g(x) = f(x, h(x)),\ a_1 \leq x \leq a_2.$$

The critical points on this curve are those where $g'(x) = 0$ or where $g'(x)$ fails to exist, along with the endpoints $x = a_1$ and $x = a_2$.

Example 2 Let $f(x, y) = 6 - x - y$. Calculate the maximum and minimum of f on the closed triangle with vertices (1, 0), (0, 1), and (2, 3).

Solution The partial derivatives are

$$\frac{\partial z}{\partial x} = -1, \qquad \frac{\partial z}{\partial y} = -1.$$

Since there are no points where both partial derivatives are equal to zero, there are no interior critical points.

The boundary of the triangle consists of the three line segments

$$\begin{array}{lll} y = 1 - x, & 0 \leq x \leq 1 & \text{[segment connecting (0, 1) and (1, 0)]}, \\ y = x + 1, & 0 \leq x \leq 2 & \text{[segment connecting (0, 1) and (2, 3)]}, \\ y = 3x - 3, & 1 \leq x \leq 2 & \text{[segment connecting (1, 0) and (2, 3)]}. \end{array}$$

(See Figure 8.29.)

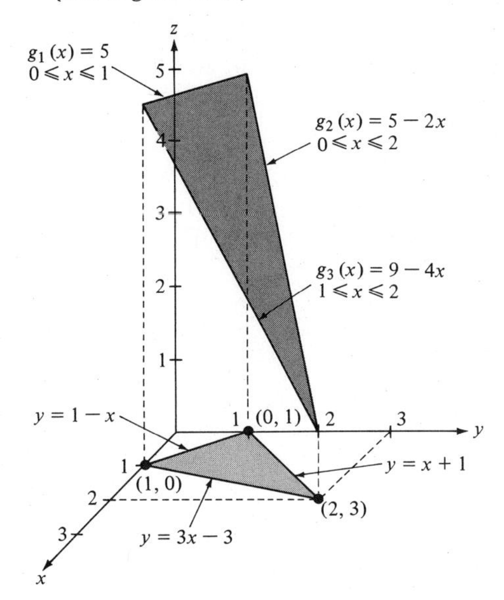

FIGURE 8.29 Example 2.

On the line $y = 1 - x$ the values of f are determined by

$$\begin{cases} f(x, y) = 6 - x - y \\ \qquad y = 1 - x, \qquad 0 \leq x \leq 1. \end{cases}$$

On substituting the second equation into the first we express $f(x, y)$ as a function of x:

$$g_1(x) = f(x, y) = 6 - x - y = 6 - x - (1 - x), \qquad 0 \leq x \leq 1,$$
$$g_1(x) = 5, \qquad 0 \leq x \leq 1.$$

Thus, the function f is constantly equal to 5 above each point of the line segment $y = 1 - x$, $0 \leq x \leq 1$.

On the line $y = x + 1$ the values of f are determined by

$$\begin{cases} f(x, y) = 6 - x - y \\ \qquad y = x + 1, \qquad 0 \leq x \leq 2. \end{cases}$$

On substituting the second equation into the first we express f as a function of x:

$$g_2(x) = f(x, y) = 6 - x - y = 6 - x - (x + 1) = 5 - 2x, \qquad 0 \leq x \leq 2.$$

The minimum value is $g_2(x) = 1$, taken when $x = 2$, the maximum value is $g_2(x) = 5$ taken at $x = 0$.

On the line $y = 3x - 3$ the values of f are determined by

$$\begin{cases} f(x, y) = 6 - x - y \\ \qquad y = 3x - 3, \qquad 1 \leq x \leq 2. \end{cases}$$

This reduces to

$$g_3(x) = f(x, y) = 6 - x - (3x - 3) = 9 - 4x, \qquad 1 \leq x \leq 2.$$

The minimum value is $f(x, y) = 1$ taken at $x = 2$; the maximum is 5 taken at $x = 1$.

The minimum value of $f(x, y)$ over the triangle is 1 taken at the vertex (2, 3). The maximum value is 5 taken at the vertices (1, 0) and (0, 1) (and at each point on the line segment connecting these two vertices).

The procedure used in Example 2 can be generalized to show that a function of form $f(x, y) = ax + by + c$ defined on a closed polygon in the xy-plane takes its extreme values at the vertices of the polygon (Figure 8.30).

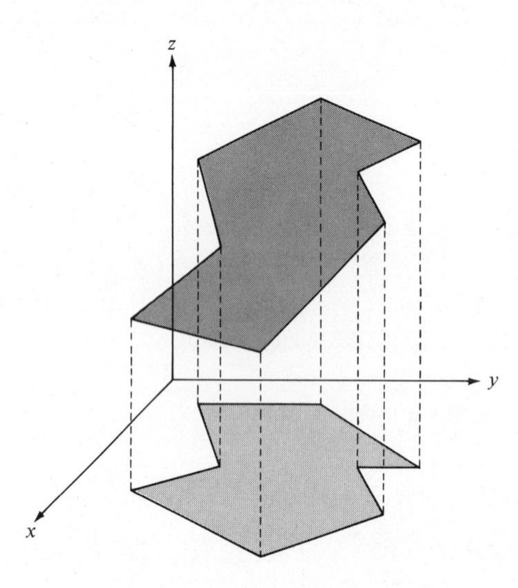

FIGURE 8.30 The maximum and minimum values of a linear function defined over a closed polygon in the xy-plane occur at the vertices (Theorem 8.7).

Extreme values of linear functions on polygons

Theorem 8.7 Let $f(x, y) = ax + by + c$ be defined on the closed polygon P with vertices $(a_1, b_1), (a_2, b_2), \ldots, (a_n, b_n)$. The extreme values of f are taken at the vertices; that is, the maximum of f on P is the largest and the minimum of f on P is the smallest of the numbers $f(a_1, b_1), f(a_2, b_2), \ldots, f(a_n, b_n)$.

Theorem 8.7 and its generalization to linear functions defined on "polygons" in "n-dimensional space" form the basis for many of the elementary methods of linear programming.

Example 3 Calculate the maximum and minimum of

$$f(x, y) = 3x - 2y + 7$$

on the polygon that consists of line segments connecting successive pairs of the points (0, 3), (1, 1), (2, 1), (2, −1), (1, −3), (−1, −2), (0, 3).

Solution Since the function is linear, the extreme values are taken at the vertices. We calculate

$$\begin{aligned}
f(0, 3) &= 0 - 6 + 7 = 1, \\
f(1, 1) &= 3 - 2 + 7 = 8, \\
f(2, 1) &= 6 - 2 + 7 = 11, \\
f(2, -1) &= 6 + 2 + 7 = 15, \\
f(1, -3) &= 3 + 6 + 7 = 16, \\
f(-1, -2) &= -3 + 6 + 7 = 10.
\end{aligned}$$

The maximum of f on the set is 16, taken at $(1, -3)$; the minimum is 1, taken at $(0, 3)$.

EXERCISES 8.7

•1. Let $f(x, y) = 2 - x^2 - y^2$. Find the maximum and minimum of f on the circular disk $\{(x, y)|x^2 + y^2 \leq 1\}$.

2. Let $f(x, y) = x^2 + 2xy + 2y^2 - 3x$. Find the maximum and minimum of f on the rectangle, $\{(x, y)|0 \leq x \leq 3, -2 \leq y \leq 1\}$.

3. Let $f(x, y) = x^2 - 8xy + 8y^2$. Find the maximum and minimum of f on the triangle with vertices $(0, 0)$, $(4, 0)$, $(4, 4)$.

•4. Let $f(x, y) = 8y^2 + 4x^2 + 4x$. Find the maximum and minimum of f on the circular disk $\{(x, y)|x^2 + y^2 \leq 1\}$.

•5. Let $f(x, y) = 3x + 2y + 4$. Find the maximum and minimum of f on the polygon with vertices $(-1, -1)$, $(0, 1)$, $(1, 0)$, $(0, -2)$. (Use Theorem 8.7.)

6. Let $f(x, y) = 2x - y - 3$. Find the maximum and minimum of f on the closed triangle with vertices $(1, 2)$, $(-1, 6)$, $(\frac{1}{2}, -1)$. (Use Theorem 8.7.)

7. Extend the *optimal-lot-size* formula of Section 3.6 to cover the situation where two items are bought and stored. Let x_1 and x_2 be the number of units of the two items ordered in each of the shipments. Suppose the two items are ordered separately and that the assumptions of Section 3.6 are valid for both items. Show that the cost of stocking the two items for a year is

$$C(x) = a_1 N_1 + \frac{b_1 N_1}{x_1} + \frac{k_1 x_1}{2} + a_2 N_2 + \frac{b_2 N_2}{x_2} + \frac{k_2 x_2}{2},$$

where the constants have the same meaning as in Section 3.6. Find the values of x_1 and x_2 that make C a minimum.

8.8 EXTREMAL PROBLEMS WITH CONSTRAINTS. LAGRANGE MULTIPLIERS

Many problems reduce to finding extreme values of a function f (of two or more variables) subject to certain restrictions. Such restrictions arise, for example, when we calculate extreme values on the boundary of a set.

For example, we may wish to find the maximum value of $z = f(x, y) = y - x^2$ subject to the restriction that (x, y) is on the unit circle $x^2 + y^2 = 1$.

As a second example, suppose that the production of a commodity is a function of x (the money spent for labor) and y (the money spent for machinery), say

$$p = ax^{2/3}y^{1/3}.$$

If C is the total amount of money available for labor and machinery we wish to maximize

$$p = ax^{2/3}y^{1/3}$$

subject to the restriction

$$x + y = C.$$

In this section we consider general methods of calculating extreme values of f subject to restrictions that can be expressed as equations relating the variables. That is, we wish to find the extreme values of f, subject to restrictions

$$\begin{aligned} g_1(x, y, z) &= 0 \\ g_2(x, y, z) &= 0 \\ &\vdots \\ g_n(x, y, z) &= 0. \end{aligned}$$

Constraints

The restrictions, $g_1(x, y, z) = 0$, $g_2(x, y, z) = 0$, . . . , are called *constraints* or *side conditions*. In general we must assume that all of the functions involved have partial derivatives at all points considered and that the graphs intersect in some type of curve or surface. We will consider functions of two or three variables in most of the work. The methods can be generalized to functions of more variables.

Substitution method

The most natural method for solving extremal problems with constraints is to solve the equations $g_1(x, y, z) = 0$, $g_2(x, y, z) = 0$, . . . , for certain of the variables and substitute these values into the expression for f, thereby reducing the number of variables. This method is illustrated in the following example.

Example 1 Use substitution to find the maximum value of

$$z = f(x, y) = y - x^2$$

subject to the side condition

$$x^2 + y^2 = 1.$$

Solution We solve the equation $x^2 + y^2 = 1$ for x^2 and substitute in the other equation:

$$z = y - x^2 = y - (1 - y^2) = y^2 + y - 1.$$

We must find the maximum value of

$$z = y^2 + y - 1,$$

where $-1 \leq y \leq 1$.

We calculate the derivative

$$\frac{dz}{dy} = 2y + 1,$$

set it equal to zero, and find the interior critical point $y = -\frac{1}{2}$. The maximum must occur at this critical point or at one of the endpoints $y = -1$, $y = 1$. We calculate the values of z at these points:

$$\begin{aligned} y = -1&: \quad z = y^2 + y - 1 = (-1)^2 + (-1) - 1 = -1 \\ y = -\tfrac{1}{2}&: \quad z = (-\tfrac{1}{2})^2 + (-\tfrac{1}{2}) - 1 = \tfrac{1}{4} - \tfrac{1}{2} - 1 = -1\tfrac{1}{4} \\ y = 1&: \quad z = 1^2 + 1 - 1 = 1. \end{aligned}$$

The maximum is $z = 1$ which occurs at $y = 1$. At this point we must have

$$x = 1 - y^2 = 0.$$

Thus the maximum occurs at the point (0, 1) on the unit circle.

The above method works provided the constraints $g_1(x, y, z) = 0$, $g_2(x, y, z) = 0, \ldots,$ can be solved for one or more variables in terms of the others. In many cases this cannot be done. For example, if

$$g_1(x, y, z) = x^3 + 7x^2y + 13x^2z + 21xz^2 + 17y^3 + 13z^2 + z^3,$$

it would be very difficult to solve the equation $g_1(x, y, z) = 0$ for any one of the variables.

There is an alternate method of solution due to *J. L. Lagrange* (1736–1813). We introduce the method for a function of two variables with a single constraint.

Lagrange's method

Theorem 8.8 (*Lagrange's Method*) Let f and g be differentiable functions of x and y. The critical points for f, subject to the constraint $g(x, y) = 0$, are found among the ordered pairs (x, y) for which there exists a real number λ such that

$$\begin{cases} \dfrac{\partial f}{\partial x} - \lambda \dfrac{\partial g}{\partial x} = 0 \\ \dfrac{\partial f}{\partial y} - \lambda \dfrac{\partial g}{\partial y} = 0 \\ g(x, y) = 0. \end{cases}$$

The number λ is called the *Lagrange multiplier* for the problem.

PROOF We define a function F of three independent variables x, y, and λ by

$$F(x, y, \lambda) = f(x, y) - \lambda\, g(x, y).$$

Observe that if we have an extreme value of $F(x, y, \lambda)$ subject to the constraint $g(x, y) = 0$ then we also have an extreme value of $f(x, y)$ subject to the same constraint, and vice versa.

We will find the extreme values of $F(x, y, \lambda)$, temporarily ignoring the constraint. Because of the differentiability of f and g the extreme values of $F(x, y, \lambda)$ must occur at points where

$$\frac{\partial F}{\partial x} = 0, \quad \frac{\partial F}{\partial y} = 0, \quad \text{and} \quad \frac{\partial F}{\partial \lambda} = 0.$$

Thus, at an extreme value of F we must have

$$\begin{aligned} \frac{\partial f}{\partial x} - \lambda \frac{\partial g}{\partial x} &= 0 \\ \frac{\partial f}{\partial y} - \lambda \frac{\partial g}{\partial y} &= 0 \\ 0 - 1 \cdot g(x, y) &= 0. \end{aligned}$$

But the last equation means that the constraint $g(x, y) = 0$ must hold at an extreme value of $F(x, y, \lambda)$. Thus, the extreme values of $f(x, y)$, subject to the constraint $g(x, y) = 0$, must occur at the extreme values of $F(x, y, \lambda)$. It follows that the extreme values of f, subject to the constraint, occur at points where

$$\frac{\partial f}{\partial x} - \lambda \frac{\partial g}{\partial x} = 0$$
$$\frac{\partial f}{\partial y} - \lambda \frac{\partial g}{\partial y} = 0$$
$$g(x, y) = 0.$$

for some λ.

Example 2 Calculate the maximum and minimum values of

$$z = x^2 - y^2$$

on the unit circle

$$x^2 + y^2 = 1.$$

Solution The problem can be restated as finding the extreme values of $f(x, y) = x^2 - y^2$ subject to the constraint $g(x, y) = x^2 + y^2 - 1 = 0$. The critical points are found among the points (x, y) for which there exists a λ such that

$$\begin{cases} \dfrac{\partial f}{\partial x} - \lambda \dfrac{\partial g}{\partial x} = 0 \\ \dfrac{\partial f}{\partial y} - \lambda \dfrac{\partial g}{\partial y} = 0 \\ g(x, y) = 0. \end{cases}$$

Therefore we must solve the system of equations

$$\begin{cases} (2x - 0) - \lambda\,(2x + 0) = 0 \\ (0 - 2y) - \lambda\,(0 + 2y) = 0 \\ x^2 + y^2 - 1 = 0. \end{cases}$$

This is equivalent to the system

$$\begin{cases} x(1 - \lambda) = 0 \\ y(1 + \lambda) = 0 \\ x^2 + y^2 = 1. \end{cases}$$

Because of the last equation, x and y both cannot be zero. If $x \neq 0$, then, from the first equation, $\lambda = 1$ and so $y = 0$. From the last equation

$$x^2 + 0^2 = 1, \qquad x^2 = 1, \qquad x = \pm 1.$$

Thus, if $x \neq 0$ we have the two critical points $(-1, 0)$ and $(1, 0)$ corresponding to $\lambda = 1$. Similarly, if $y \neq 0$ we obtain the two critical points $(0, -1)$ and $(0, 1)$ corresponding to $\lambda = -1$.

It follows from the theorem that the extreme values of f must be found at these critical points. We calculate the values of f:

$$f(-1, 0) = (-1)^2 - 0^2 = 1,$$
$$f(1, 0) = 1^2 - 0^2 = 1,$$
$$f(0, -1) = 0^2 - (-1)^2 = -1,$$
$$f(0, 1) = 0^2 - 1^2 = -1.$$

The maximum value of f on the circle is 1, taken at $(-1, 0)$ and $(1, 0)$. The minimum value is -1, taken at $(0, -1)$ and $(0, 1)$. The contour map shows the situation rather clearly (Figure 8.31).

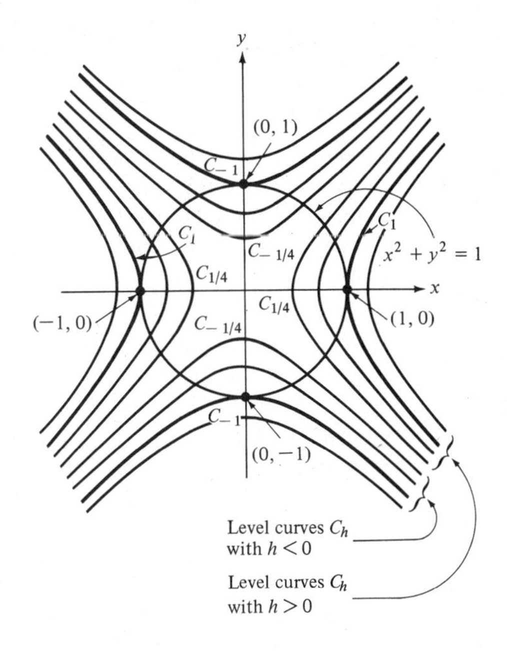

FIGURE 8.31 Example 2. Level curves for $f(x, y) = x^2 - y^2$. The extreme values of f on the circle $x^2 + y^2 = 1$ are taken at the points $(\pm 1, 0)$ and $(0, \pm 1)$.

It should be obvious from the example above that Lagrange's method does not always lead to a reduction of labor. It is an alternate method that may work when direct substitution is not practical.

Example 3 In a certain company the production function is

$$p = ax^{2/3}y^{1/3} \qquad (a > 0),$$

where

$$x = \text{money spent for labor.}$$
$$y = \text{money spent for machinery.}$$

Find the values of x and y that maximize p subject to the constraint

$$x + y - C = 0,$$

where C is a constant.

Solution We use Lagrange's method. The problem reduces to that of solving the system of equations

$$\begin{cases} \frac{2}{3}ax^{-1/3}y^{1/3} - \lambda = 0 \\ \frac{1}{3}ax^{2/3}y^{-2/3} - \lambda = 0 \\ x + y = C. \end{cases}$$

Observe that

$$\lambda = \tfrac{2}{3}ax^{-1/3}y^{1/3} = \tfrac{1}{3}ax^{2/3}y^{-2/3},$$

so that

$$2y^{1/3}y^{2/3} = x^{2/3}x^{1/3}$$
$$2y = x.$$

Since

$$x + y = C,$$

$$x = \frac{2C}{3}, \qquad y = \frac{C}{3}.$$

To show that these values maximize p we note that if either x or y has the value C, then the other must be zero so that $p = 0$. It would appear from the nature of the problem that there should be values of x and y between 0 and C for which p has a maximum. Since only one critical point exists, then p has its maximum value at this critical point: $x = 2C/3$, $y = C/3$.

Lagrange's method generalizes to functions of several variables with several constraints. Essentially, we need a Lagrange multiplier for each constraint and an equation involving partial derivatives for each variable.

The following theorem states the conditions for a function $f(x, y, z)$ of three variables subject to two constraints $g_1(x, y, z) = 0$ and $g_2(x, y, z) = 0$.

Lagrange's method for functions of 3 variables

Theorem 8.9 Let $f(x, y, z)$, $g_1(x, y, z)$, $g_2(x, y, z)$ be differentiable functions. The critical points for f, subject to the constraints

$$g_1(x, y, z) = 0, \qquad g_2(x, y, z) = 0,$$

are among the points (x, y, z) for which there exist constants λ_1 and λ_2 such that

$$\begin{cases} \dfrac{\partial f}{\partial x} - \lambda_1 \dfrac{\partial g_1}{\partial x} - \lambda_2 \dfrac{\partial g_2}{\partial x} = 0 \\ \dfrac{\partial f}{\partial y} - \lambda_1 \dfrac{\partial g_1}{\partial y} - \lambda_2 \dfrac{\partial g_2}{\partial y} = 0 \\ \dfrac{\partial f}{\partial z} - \lambda_1 \dfrac{\partial g_1}{\partial z} - \lambda_2 \dfrac{\partial g_2}{\partial z} = 0 \\ g_1(x, y, z) = 0 \\ g_2(x, y, z) = 0. \end{cases}$$

Example 4 Find the minimum distance from the origin to the line of intersection of the two planes

$$x + y + z = 8$$
$$2x - y + 3z = 28.$$

Solution Let (x, y, z) be the point on the line which is at a minimum distance from the origin. (It is obvious that such a point must exist.) The distance from (x, y, z) to the origin is $\sqrt{x^2 + y^2 + z^2}$. We will find it easier to work with the square of the distance, $x^2 + y^2 + z^2$. Thus we must minimize

$$f(x, y, z) = x^2 + y^2 + z^2,$$

subject to the side conditions

$$x + y + z - 8 = 0$$
$$2x - y + 3z - 28 = 0.$$

We apply Lagrange's method. At the critical point there must exist constants λ_1, λ_2 such that

$$\begin{cases} 2x - \lambda_1 - 2\lambda_2 = 0 & \textit{(partial derivative with respect to x)}, \\ 2y - \lambda_1 + \lambda_2 = 0 & \textit{(partial derivative with respect to y)}, \\ 2z - \lambda_1 - 3\lambda_2 = 0 & \textit{(partial derivative with respect to z)}, \\ \left.\begin{array}{r} x + y + z = 8 \\ 2x - y + 3z = 28 \end{array}\right\} & \text{(constraints)}. \end{cases}$$

The solution of the system is

$$x = 4, \quad y = -2, \quad z = 6, \quad \lambda_1 = 0, \quad \lambda_2 = 4.$$

The critical point is $(4, -2, 6)$. The minimum distance is

$$\sqrt{4^2 + (-2)^2 + 6^2} = \sqrt{56}.$$

EXERCISES 8.8

Use Lagrange's method in all problems.

1. Find the maximum and minimum values of $f(x, y) = x^2 - y^2$ on the circle $x^2 + y^2 = 4$.

•2. Find the extreme values of $f(x, y) = xy$, subject to the constraint $x + y = 2$.

3. Find the extreme values of $f(x, y) = xy$ on the circle $x^2 + y^2 = 2$.

4. Find the point on the line of intersection of the two planes that is closest to the origin.

   • (a) $\begin{array}{r} 2x - 3y + z = 4 \\ x + y - z = -6. \end{array}$ (b) $\begin{array}{r} 2x - y - z = -28 \\ x + 3y + z = 61. \end{array}$

5. Find the points on the parabola $2y = x^2 - 3$ that are nearest to the origin.

•6. Find the local extrema of $f(x, y, z) = xyz$, subject to the constraints $2x + y + 2z - 6 = 0$, $y + z = 3$.

7. Find the maximum of $f(x, y, z) = (xyz)^2$, subject to the constraint $x^2 + y^2 + z^2 = 1$.

•8. The average yield of the Green Thumb Mellon Ranch is

$$5x\sqrt{\frac{y}{100}}\,\sqrt{4z},$$

where x is the number of acres, y the number of man-hours, and z the number of equipment hours. Find the maximum average yield subject to the price constraint

$$400x + 5y + 20z = \$32{,}000.$$

8.9 THE CHAIN RULE

If $z = f(x, y)$ where x and y are functions of t—say $x = g(t)$, $y = h(t)$—then z is itself a function of t:

$$z = f(x, y) = f(g(t), h(t)) = F(t).$$

Example 1 If $z = f(x, y) = x + y + xy$ and $x = 3t^2 - 1$, $y = 2t + 5$, then

$$\begin{aligned} z = f(3t^2 - 1, 2t + 5) &= (3t^2 - 1) + (2t + 5) + (3t^2 - 1)(2t + 5) \\ &= 3t^2 - 1 + 2t + 5 + 6t^3 + 15t^2 - 2t - 5 \\ &= 6t^3 + 18t^2 - 1 = F(t). \end{aligned}$$

We would like to extend the Chain Rule to cover situations of this type. That is, we wish to express dz/dt in terms of $\partial z/\partial x$, $\partial z/\partial y$, dx/dt, and dy/dt. The statement of the extended Chain Rule is given in the following theorem. The proof is omitted.

Theorem 8.10 (*The Chain Rule*) Let $z = f(x, y)$ have continuous partial derivatives. Let x and y be differentiable functions of t. Then

The Chain Rule

$$\boxed{\frac{dz}{dt} = \frac{\partial z}{\partial x} \cdot \frac{dx}{dt} + \frac{\partial z}{\partial y} \cdot \frac{dy}{dt}.}$$

Example 2 Let $z = x + y + xy$, $x = 3t^2 - 1$, $y = 2t + 5$. Calculate dz/dt by the Chain Rule. Compare the answer with that obtained by first expressing z as a function of t and then differentiating.

Solution By the Chain Rule

$$\begin{aligned} \frac{dz}{dt} &= \frac{\partial z}{\partial x} \cdot \frac{dx}{dt} + \frac{\partial z}{\partial y} \cdot \frac{dy}{dt} \\ &= (1 + 0 + y) \cdot (6t) + (0 + 1 + x)(2) \\ &= (1 + y) \cdot 6t + (1 + x) \cdot 2 \\ &= (1 + 2t + 5)6t + (1 + 3t^2 - 1) \cdot 2 \\ &= 36t + 12t^2 + 6t^2 = 18t^2 + 36t. \end{aligned}$$

If we substitute the values of x and y into the expression for z, we have

$$z = x + y + xy = 6t^3 + 18t^2 - 1.$$

Then

$$\frac{dz}{dt} = 18t^2 + 36t.$$

Thus, the two answers agree.

The Chain Rule can be extended to cover functions of two or more variables, each of which is a function of one or more variables. For example, if f is a function of x, y, and z, each of which is a function of t, then (assuming proper conditions of differentiability and continuity)

$$\frac{df}{dt} = \frac{\partial f}{\partial x}\frac{dx}{dt} + \frac{\partial f}{\partial y}\frac{dy}{dt} + \frac{\partial f}{\partial z}\frac{dz}{dt}.$$

If f is a function of u and v, each of which is a function of x and y, then f is a function of x and y and the Chain Rule becomes

The Chain Rule for partial derivatives

$$\boxed{\begin{aligned} \frac{\partial f}{\partial x} &= \frac{\partial f}{\partial u} \cdot \frac{\partial u}{\partial x} + \frac{\partial f}{\partial v} \cdot \frac{\partial v}{\partial x}, \\ \frac{\partial f}{\partial y} &= \frac{\partial f}{\partial u} \cdot \frac{\partial u}{\partial y} + \frac{\partial f}{\partial v} \cdot \frac{\partial v}{\partial y}. \end{aligned}}$$

Similar formulas hold for the Chain Rule if a different number of variables are involved.

Example 3 Let $z = u \ln u + \ln v$, $u = e^{xy} + 1$, $v = x + y^2$. Calculate $\partial z/\partial x$ and $\partial z/\partial y$.

Solution $\dfrac{\partial z}{\partial u} = u \dfrac{\partial}{\partial u} (\ln u) + \ln u \cdot 1 + 0 = u \cdot \dfrac{1}{u} + \ln u = 1 + \ln u,$

$$\frac{\partial z}{\partial v} = 0 + \frac{\partial}{\partial v} (\ln v) = \frac{1}{v},$$

$$\frac{\partial u}{\partial x} = ye^{xy}, \qquad \frac{\partial u}{\partial y} = xe^{xy},$$

$$\frac{\partial v}{\partial x} = 1, \qquad \frac{\partial v}{\partial y} = 2y.$$

Then

$$\begin{aligned}\frac{\partial z}{\partial x} &= \frac{\partial z}{\partial u} \cdot \frac{\partial u}{\partial x} + \frac{\partial z}{\partial v} \cdot \frac{\partial v}{\partial x} \\ &= (1 + \ln u) \cdot ye^{xy} + \frac{1}{v} \cdot 1 \\ &= ye^{xy} (1 + \ln (e^{xy} + 1)) + \frac{1}{x + y^2}. \\ \frac{\partial z}{\partial y} &= \frac{\partial z}{\partial u} \cdot \frac{\partial u}{\partial y} + \frac{\partial z}{\partial v} \cdot \frac{\partial v}{\partial y} \\ &= (1 + \ln u) \cdot xe^{xy} + \frac{1}{v} \cdot 2y \\ &= xe^{xy}(1 + \ln (e^{xy} + 1)) + \frac{2y}{x + y^2}.\end{aligned}$$

Homogeneous functions

Homogeneous Functions Certain functions $f(x, y)$ have the property that

$$f(kx, ky) = k^n f(x, y).$$

These functions are said to be *homogeneous*.

For example, if

$$f(x, y) = 2x - 4\sqrt{xy} + \frac{y^2}{x}$$

and $k > 0$, then

$$\begin{aligned}f(kx, ky) &= 2kx - 4\sqrt{kx \cdot ky} + \frac{(ky)^2}{kx} \\ &= 2kx - 4k\sqrt{xy} + \frac{ky^2}{x} \\ &= kf(x, y).\end{aligned}$$

Therefore, f is homogeneous (with $n = 1$).

Degree of homogeneity

The number n is called the *degree of homogeneity*. The function is said to be *linear homogeneous* if $n = 1$. A large number of the functions in economic theory are homogeneous.

Homogeneous functions appear naturally in many applications. For example, suppose that production (p) is a function of labor (x) and capital (y), say $p = f(x, y)$. We would expect that a change in scale of labor and capital by a factor of k would result in a change in production by a factor of k. If $p_0 = f(x_0, y_0)$ describes the situation before the change in scale then

$$f(kx_0, ky_0) = k\, p_0 = k\, f(x_0, y_0).$$

Thus, production is a linear homogeneous function of labor and capital.

In the next theorem we establish a basic property of homogeneous functions:

$$n \cdot f(x, y) = \frac{\partial f}{\partial x}\, x + \frac{\partial f}{\partial y}\, y.$$

Before proving the theorem we verify this relationship for the linear homogeneous function mentioned above.

Example 4 Let $f(x, y) = 2x - 4\sqrt{xy} + y^2/x$.
Verify that $f(x, y) = \dfrac{\partial f}{\partial x}\, x + \dfrac{\partial f}{\partial y}\, y$.

Solution

$$\begin{aligned}\frac{\partial f}{\partial x} &= 2 - 4 \cdot \frac{1}{2}\, x^{-1/2}\, \sqrt{y} + y^2(-1)\, x^{-2}\\ &= 2 - 2\sqrt{\frac{y}{x}} - \frac{y^2}{x^2}.\\ \frac{\partial f}{\partial y} &= 0 - 4\sqrt{x} \cdot \frac{1}{2}\, y^{-1/2} + 2y/x\\ &= -2\sqrt{\frac{x}{y}} + 2\frac{y}{x}.\end{aligned}$$

Then

$$\begin{aligned}\frac{\partial f}{\partial x}\, x + \frac{\partial f}{\partial y}\, y &= \left(2 - 2\sqrt{\frac{y}{x}} - \frac{y^2}{x^2}\right) x + \left(-2\sqrt{\frac{x}{y}} + 2\frac{y}{x}\right) y\\ &= 2x - 2\sqrt{xy} - \frac{y^2}{x} - 2\sqrt{xy} + 2\frac{y^2}{x}\\ &= 2x - 4\sqrt{xy} + y^2/x = f(x, y).\end{aligned}$$

Theorem 8.11 If f is a homogeneous differentiable function of x and y of degree n then

$$n \cdot f(x, y) = \frac{\partial f}{\partial x}\, x + \frac{\partial f}{\partial y}\, y.$$

PROOF Let $z = f(x, y)$. It can be shown that there exists a function g of a single variable such that

$$z = x^n g\left(\frac{y}{x}\right) = x^n g(u) \qquad \text{where } u = \frac{y}{x}.$$

(See Exercise 4.) We calculate the partial derivatives:

$$\frac{\partial z}{\partial x} = nx^{n-1}\, g(u) + x^n\, g'(u)\, \frac{\partial u}{\partial x}$$

$$= nx^{n-1}\, g(u) + x^n\, g'(u) \left(-\frac{y}{x^2}\right)$$

and

$$\frac{\partial z}{\partial y} = x^n\, g'(u)\, \frac{\partial u}{\partial y} = x^n\, g'(u) \left(\frac{1}{x}\right).$$

Then

$$\frac{\partial z}{\partial x}\, x + \frac{\partial z}{\partial y}\, y = nx^n\, g(u) + x^{n+1}\, g'(u) \left(-\frac{y}{x^2}\right) + x^n\, g'(u) \left(\frac{y}{x}\right)$$

$$= nx^n\, g(u) = nx^n\, g\left(\frac{y}{x}\right) = n \cdot f(x, y).$$

This result has a direct application in economic theory. Suppose that production (p) is a linear homogeneous function of labor (x) and capital (y), say

$$p = f(x, y).$$

By Theorem 8.11

$$p = \frac{\partial p}{\partial x}\, x + \frac{\partial p}{\partial y}\, y.$$

The component parts of this equation are:

1. Marginal productivity of labor ($\partial p/\partial x$).
2. Size of the labor force (x).
3. Marginal productivity of capital ($\partial p/\partial y$).
4. Size of the capital (y).

Thus the contribution of labor to production is given as the product of the marginal productivity of labor and the size of the labor force. A similar interpretation holds for the contribution of capital to productivity.

EXERCISES 8.9

1. In each of the following exercises z is a function of x and y, where x and y are functions of t. Use the Chain Rule to calculate dz/dt.

   - • (a) $z = x^2 + xy + y^2$; $x = 2t + 3$, $y = t^2$
   - (b) $z = (xy)^2$; $x = t^3$, $y = t^2 + t$
   - (c) $z = x^3 - y^3$; $x = t^2 - t + 1$, $y = t^3 + t^2 + 1$
   - • (d) $z = e^{xy}$; $x = e^t$, $y = t^2 - t + 1$
   - (e) $z = xy^2 + x^2y$; $x = te^{-t}$, $y = -te^t$
   - • (f) $z = x^3y^2 - x^2y^3 + xy$; $x = t - t^2$, $y = t^2 + t - 1$

2. Use the Chain Rule to calculate $\partial z/\partial x$ and $\partial z/\partial y$.

   - • (a) $z = e^{u+v}$, $u = x + y + 1$, $v = 2x + y$
   - (b) $z = (uv)^n$, $u = e^{xy}$, $v = \cos x$
   - (c) $z = u^2 + v^2$, $u = \sin x$, $v = x - y$
   - • (d) $z = \dfrac{uv}{u^2 + v^2}$, $u = 2x - y$, $v = x - 2y$

3. Verify that the following functions are homogeneous and that

$$n \cdot f(x, y) = \frac{\partial f}{\partial x} \cdot x + \frac{\partial f}{\partial y} \cdot y,$$

where n is the degree of homogeneity.

• (a) $f(x, y) = x + \frac{x^2}{y} - 2y^{3/2} x^{-1/2}$

(b) $f(x, y) = \frac{3y^3}{x} + \frac{x^3}{y} + x\sqrt{xy}$

4. Let z be a linear homogeneous function of x and y.

(a) Find two examples which illustrate that

$$\frac{z}{x} = g\left(\frac{y}{x}\right)$$

for some function g of a single variable.

(b) Prove that this result holds in general. (*Hint:* Let $z = f(x, y)$. Let g be defined for each number u by $g(u) = f(l, u)$.)

(c) Generalize the result of part (b). Show that if z is a homogeneous function of x and y of degree n, then there exists a function g such that

$$z = x^n g\left(\frac{y}{x}\right).$$

DIFFERENTIAL EQUATIONS

9.1 INTRODUCTION

A *differential equation* is an equation relating a function and one or more of its derivatives. Examples are

Differential equation

$$y' - 3y + x = 0,$$
$$y'' + yy' = 0,$$
$$y''' = x^2 + 4x - 1.$$

The order of the highest derivative that appears is called the *order* of the differential equation. Thus, the first of the equations above has order 1, the second has order 2, while the third has order 3. In this chapter we will concern ourselves mainly with first-order equations.

Order of differential equation

By a *solution* of a differential equation we mean any function $f(x)$, which when substituted for y, makes the equation an identity. In other words, if we substitute $f(x)$ for y in the differential equation, the resulting equation is satisfied by all x in the common part of the domains of f and f'.

Solution of differential equation

Example 1 Show that $f(x) = 5e^{3x} + \frac{1}{3}x + \frac{1}{9}$ is a solution of the differential equation $y' - 3y + x = 0$.

Solution If we substitute $f(x)$ for y in the left-hand side of the differential equation, we obtain

$$\begin{aligned} f'(x) - 3f(x) + x &= (15e^{3x} + \tfrac{1}{3}) - 3(5e^{3x} + \tfrac{1}{3}x + \tfrac{1}{9}) + x \\ &= 15e^{3x} + \tfrac{1}{3} - 15e^{3x} - x - \tfrac{1}{3} + x \\ &= 0 \qquad \text{for all } x. \end{aligned}$$

Since the resulting equation

$$f'(x) - 3f(x) + x = 0$$

is satisfied by all x, the function f is a solution.

In most cases the process of solving a first-order differential equation leads to one integration, that of solving a second-order equation leads to two integrations, and so on. Each of these integrations involves an arbitrary constant. Thus we expect to have one arbitrary constant in the solution of a first-order differential equation, two arbitrary constants in a solution of a second-order equation, and so on.

General solution

Particular solution

A solution of an nth-order differential equation that involves n arbitrary constants is called the *general solution* of the differential equation. A solution that does not contain any arbitrary constants is called a *particular solution*.

Example 2 It follows by direct computation that $g(x) = Ce^{3x} + \frac{1}{3}x + \frac{1}{9}$ is a solution of the first-order differential equation

$$y' - 3y + x = 0$$

of Example 1. Since this solution involves the one arbitrary constant C, it is the general solution. The solution $f(x) = 5e^{3x} + \frac{1}{3}x + \frac{1}{9}$ considered in Example 1 is a particular solution obtained by giving C the particular value 5.

Singular solution

In most cases every particular solution of a differential equation can be obtained by assigning values to the constants in the general solution. There are a few exceptions to this rule. That is, occasionally there exist solutions that are not obtainable from the general solution. These are known as *singular solutions.* We will have examples of singular solutions in the next section.

Actually, we have been solving differential equations since Chapter 4. When we integrate $\int g(x)\,dx$, we are finding a solution of the differential equation $y' = g(x)$. This is the case because $y' = g(x)$ if and only if

$$y = \int g(x)\,dx + C.$$

This observation gives us an easy method for solving differential equations of form

$$y' = g(x), \quad y'' = g(x), \quad y''' = g(x), \quad \ldots .$$

We can solve $y' = g(x)$ by one integration, $y'' = g(x)$ by two integrations, and so on.

Example 3 Solve $y' = e^x + 3x + 1$.

Solution $y = f(x)$ is a solution if and only if

$$y = \int y'\, dx = \int (e^x + 3x + 1)\, dx$$
$$= e^x + \frac{3x^2}{2} + x + C.$$

Since this solution contains an arbitrary constant, it is the general solution.

Example 4 Find the general solution of

$$y''' = e^{2x} + 3x.$$

Solution After one integration we have

$$y'' = \int y'''\, dx = \int (e^{2x} + 3x)\, dx = \frac{1}{2} e^{2x} + \frac{3x^2}{2} + C_1.$$

We now integrate the function y'', obtaining

$$y' = \int y''\, dx = \int \left(\frac{1}{2} e^{2x} + \frac{3x^2}{2} + C_1\right) dx$$
$$= \frac{1}{4} e^{2x} + \frac{3x^3}{2 \cdot 3} + C_1x + C_2$$
$$= \frac{1}{4} e^{2x} + \frac{x^3}{2} + C_1x + C_2.$$

The last step is to integrate y':

$$y = \int y'\, dx = \int \left(\frac{1}{4} e^{2x} + \frac{x^3}{2} + C_1x + C_2\right) dx$$
$$= \frac{1}{8} e^{2x} + \frac{x^4}{2 \cdot 4} + \frac{C_1x^2}{2} + C_2x + C_3$$
$$= \frac{1}{8} e^{2x} + \frac{1}{8} x^4 + \frac{C_1}{2} x^2 + C_2x + C_3.$$

Since the constants C_1, C_2, C_3 are arbitrary, we can modify their form if we wish. Thus, if we let $C_1 = 2A$, $C_2 = B$, $C_3 = C$ the general solution can be written as

$$y = f(x) = \frac{e^{2x}}{8} + \frac{x^4}{8} + Ax^2 + Bx + C.$$

It is standard practice to modify the form of the arbitrary constants as in Example 4 so that they will appear in as simple a final form as possible.

In many practical problems we wish to find a particular solution of a differential equation that satisfies certain conditions. The usual method is to calculate the general solution and then use the stated conditions to evaluate the constants. This technique is illustrated in the following example.

Example 5 Find the particular solution $y = f(x)$ of $y'' = x + 1$ that satisfies the conditions $f(0) = 1$, $f'(0) = 2$.

Solution Let f be a solution. Then

$$f''(x) = x + 1,$$
$$f'(x) = \int f''(x)\,dx + C_1 = \int (x+1)\,dx + C_1 = \frac{x^2}{2} + x + C_1,$$
$$f(x) = \int f'(x)\,dx + C_2 = \int \left(\frac{x^2}{2} + x + C_1\right) dx + C_2$$
$$= \frac{x^3}{6} + \frac{x^2}{2} + C_1x + C_2.$$

This is the general solution. We wish to find the particular solution such that

$$f(0) = \frac{0^3}{6} + \frac{0^2}{2} + C_1 \cdot 0 + C_2 = 1$$

and

$$f'(0) = \frac{0^2}{2} + 0 + C_1 = 2.$$

In order to have these relationships hold we must have $C_1 = 2, C_2 = 1$. The particular solution is

$$f(x) = \frac{x^3}{6} + \frac{x^2}{2} + 2x + 1.$$

Example 6 Find the particular solution $y = f(x)$ of $y'' = x + 1$ such that $f(1) = 3$ and $f(4) = 0$.

Solution We established in Example 5 that the general solution is

$$f(x) = \frac{x^3}{6} + \frac{x^2}{2} + C_1x + C_2.$$

We need

$$f(1) = \tfrac{1}{6} + \tfrac{1}{2} + C_1 + C_2 = 3,$$
$$f(4) = \tfrac{64}{6} + \tfrac{16}{2} + 4C_1 + C_2 = 0.$$

These conditions reduce the problem to solving the system of equations

$$\begin{cases} C_1 + C_2 = 3 - \frac{1}{6} - \frac{1}{2} = \frac{14}{6} = \frac{7}{3}, \\ 4C_1 + C_2 = -\frac{64}{6} - \frac{16}{2} = -\frac{56}{3}. \end{cases}$$

The solution of this system is $C_1 = -7$, $C_2 = 28/3$. Therefore, the particular solution is

$$f(x) = \frac{x^3}{6} + \frac{x^2}{2} - 7x + \frac{28}{3}.$$

Boundary condition

Initial condition

Example 6 is a *boundary-condition* problem. We wish to find a solution that has specified values at one or more points in its domain. Example 5 is an *initial-condition* problem. We wish to find a solution such that it and one or more of its derivatives have specified values at a single point $x = x_0$ in its domain. Both types of problem are important in the theory of differential equations.

Existence Theorems (*Optional*) Frequently we need to know that a differential equation has a solution satisfying certain conditions. For this reason much of the advanced theory of differential equations is concerned with existence theory. A typical theorem will state that any differential equation of a certain form has a solution satisfying certain types of boundary or initial conditions. The following two existence theorems illustrate the types of theorems encountered in advanced courses. They guarantee the existence of solutions for most of the differential equations that we will consider.

Existence theorems

Theorem 9.1 Let $a_1(x)$ and $a_2(x)$ be continuous for all x. Let x_0 and y_0 be real numbers. The differential equation

$$y' + a_1(x)y = a_2(x)$$

has a unique solution $y = f(x)$ such that $f(x_0) = y_0$.

In other words, if we choose any point (x_0, y_0) in the plane, there is a solution of the differential equation with a graph that passes through the point. Furthermore, there is just one such solution. (This is the "uniqueness" part of the theorem.) (See Figure 9.1.)

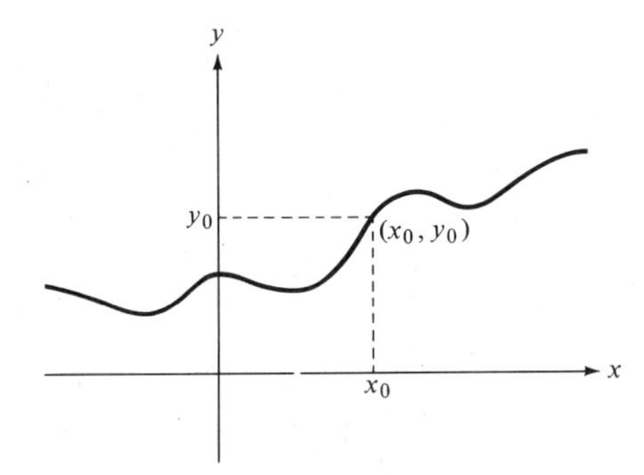

FIGURE 9.1 The differential equation $y' + a_1(x)y = a_2(x)$ has a unique solution $y = f(x)$ such that $f(x_0) = y_0$ (Theorem 9.1).

Theorem 9.2 Let $a_1(x)$, $a_2(x)$, and $a_3(x)$ be continuous for all x. Let x_0, y_0, and m be real numbers. The differential equation

$$y'' + a_1(x)y' + a_2(x)y = a_3(x)$$

has a unique solution $y = f(x)$ such that

$$f(x_0) = y_0, \qquad f'(x_0) = m.$$

In other words, if we choose any point (x_0, y_0) in the plane and any nonvertical line through the point (m is the slope of the line), there is a particular solution $y = f(x)$ with a graph that passes through the point and is tangent to the given line at the point. Furthermore, there is only one such solution. (See Figure 9.2.)

The existence theorems above are not stated in their full generality. The results hold on any open interval on which $a_1(x)$, $a_2(x)$, $a_3(x)$ are continuous. Of course, the number x_0 must be chosen on the open interval.

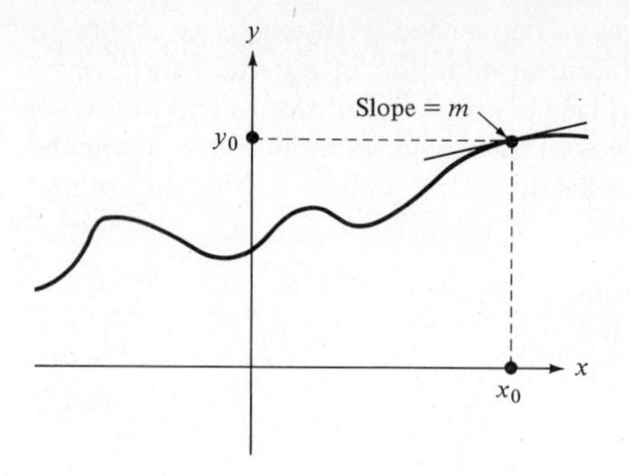

FIGURE 9.2 The differential equation $y'' + a_1(x)y' + a_2(x)y = a_3(x)$ has a unique solution $y = f(x)$ such that $f(x_0) = y_0$ and $f'(x_0) = m$ (Theorem 9.2).

EXERCISES 9.1

1. Show that the given function is a solution of the stated differential equation.
   (a) $f(x) = e^x - 2,\ y'' - y' = 0$
   (b) $f(x) = e^x - e^{-x},\ y'' - y = 0$
   (c) $f(x) = 3e^x + e^{-2x},\ y'' + y' - 2y = 0$
   (d) $f(x) = e^{-x/2}\left(\cos\frac{\sqrt{3}\,x}{2} + \sin\frac{\sqrt{3}\,x}{2}\right),\ y'' + y' + y = 0$
   (e) $f(x) = x^2 + 1,\ y' - 2xy + 2x^3 = 0$
2. Find the general solution of each of the following differential equations:
   • (a) $y' = e^{3x} + x$
   (b) $y' = 3e^{2x} + e^{-x} - 6x^2$
   • (c) $y'' = e^{2x} + e^{-2x} + 2$
   (d) $y'' = 4e^{-2x} - xe^x$
   (e) $y' = 4x^3 - 2x + xe^{x^2}$
   • (f) $y''' = e^x - x + 3$
   (g) $y''' = e^x + e^{-x} + x^3 + 1$
3. Find the particular solution satisfying the stated conditions.
   • (a) $y' = e^x - x^2;\ y = 4$ when $x = 0$
   (b) $y' = 5x^2 - e^{-x};\ y = e^{-1}$ when $x = 1$
   • (c) $y'' = e^x - e^{-x};\ y' = 0$ and $y = -1$ when $x = 0$
   (d) $y'' = x^2 + e^x;\ y' = 2$ and $y = 0$ when $x = 0$
   (e) $y' = xe^x;\ y = 3$ when $x = 0$
   • (f) $y'' = 2x + e^x;\ y = 0$ when $x = 0$
   $y = 0$ when $x = 1$
   (g) $y'' = e^x + e^{-x};\ y = 0$ when $x = 0$
   $y = e + e^{-1}$ when $x = 1$
   (h) $y'' = e^x - e^{-x};\ y = 3$ when $x = 0$
   $y = e - e^{-1}$ when $x = 1$

9.2 SEPARATION OF VARIABLES. SINGULAR SOLUTIONS

A differential equation of form

Separation of variables

$$m(y)\,dy = n(x)\,dx$$

can be solved by integration. The solution is defined implicitly by the equation

$$\int m(y)\,dy = \int n(x)\,dx + C.$$

If a differential equation has been reduced to the above form we say that the variables have been separated.

Differential equations with separable variables were considered briefly in Section 5.5. In this section we give a few additional examples and show how such equations can have singular solutions.

Example 1 Solve $2x(1-y)\dfrac{dy}{dx} = (x+1)(1-y^2)$.

Solution We separate the variables, obtaining

$$\begin{aligned}
\frac{2(1-y)}{1-y^2}\,dy &= \frac{x+1}{x}\,dx,\\
\frac{2}{1+y}\,dy &= \left(1+\frac{1}{x}\right)dx,\\
2\int \frac{1}{1+y}\,dy &= \int \left(1+\frac{1}{x}\right)dx,\\
2\ln|1+y| &= x + \ln|x| + C.
\end{aligned}$$

To solve for y we raise e to the powers indicated:

$$\begin{aligned}
e^{2\ln|1+y|} &= e^{x+\ln|x|+C},\\
(e^{\ln|1+y|})^2 &= e^C e^x e^{\ln|x|},\\
|1+y|^2 &= e^C e^x |x|,\\
1+y &= \pm\sqrt{e^C e^x |x|} = \pm\sqrt{e^C}\cdot e^{x/2}\sqrt{|x|}.
\end{aligned}$$

Since C is an arbitrary constant, so is $\pm\sqrt{e^C}$, which we can replace by k, obtaining

$$\begin{aligned}
1+y &= ke^{x/2}\sqrt{|x|},\\
y &= ke^{x/2}\sqrt{|x|} - 1.
\end{aligned}$$

Example 2 Solve $(x^2-1)\,dy = (y^2-1)\,dx$.

Solution We separate the variables and integrate:

$$\begin{aligned}
\frac{dy}{y^2-1} &= \frac{dx}{x^2-1},\\
\int \frac{dy}{y^2-1} &= \int \frac{dx}{x^2-1} + C.
\end{aligned}$$

We can split the fraction $1/(x^2-1)$ into partial fraction, obtaining

$$\frac{1}{x^2-1} = \frac{1}{2(x-1)} - \frac{1}{2(x+1)}.$$

The fraction $1/(y^2-1)$ can be handled similarly. If we substitute these values into the integrals above we obtain

$$\int \frac{dy}{2(y-1)} - \int \frac{dy}{2(y+1)} = \int \frac{dx}{2(x-1)} - \int \frac{dx}{2(x+1)} + C,$$

$$\int \frac{dy}{y-1} - \int \frac{dy}{y+1} = \int \frac{dx}{x-1} - \int \frac{dx}{x+1} + 2C,$$

$$\ln|y-1| - \ln|y+1| = \ln|x-1| - \ln|x+1| + 2C,$$

$$\ln\left|\frac{y-1}{y+1}\right| = \ln\left|\frac{x-1}{x+1}\right| + 2C.$$

To get rid of the logarithms in the solution we raise e to the powers indicated:

$$e^{\ln|(y-1)/(y+1)|} = e^{\ln|(x-1)/(x+1)|}e^{2C},$$

$$\left|\frac{y-1}{y+1}\right| = \left|\frac{x-1}{x+1}\right| \cdot e^{2C}.$$

If we drop the absolute value signs, we have

$$\frac{y-1}{y+1} = \pm \frac{x-1}{x+1} e^{2C}$$

or

$$\frac{y-1}{y+1} = k \cdot \frac{x-1}{x+1}, \qquad \text{where } k = \pm e^{2C}.$$

We now solve for y:

$$(y-1)(x+1) = k \cdot (x-1)(y+1),$$

$$xy - x + y - 1 = kxy - ky + kx - k,$$

$$xy + y - kxy + ky = x + 1 + kx - k,$$

$$y[(x+1) - k(x-1)] = (x+1) + k(x-1),$$

$$y = \frac{(x+1) + k(x-1)}{(x+1) - k(x-1)}.$$

Example 3 Solve $(1+x)\,dy = x(1+y)\,dx$.

Solution

$$\frac{dy}{1+y} = \frac{x}{1+x}\,dx = \left(1 - \frac{1}{1+x}\right)dx,$$

$$\int \frac{dy}{1+y} = \int \left(1 - \frac{1}{1+x}\right)dx + C,$$

$$\ln|1+y| = x - \ln|1+x| + C,$$

$$\ln|1+y| + \ln|1+x| = x + C.$$

We now raise e to the powers indicated:

$$e^{\ln|1+y| + \ln|1+x|} = e^{x+C},$$

$$e^{\ln|1+y|}e^{\ln|1+x|} = e^C e^x,$$

$$|1+y||1+x| = e^C e^x,$$

$$(1+y)(1+x) = \pm e^C e^x = ke^x \qquad (k = \pm e^C),$$

$$1 + y = \frac{ke^x}{1+x},$$

$$y = \frac{ke^x}{1+x} - 1.$$

Singular Solutions When we carry out the procedure illustrated above, we actually change the original differential equation to a new one and then find the general solution of this new equation. Consequently, it is possible that the original differential equation has solutions that cannot be obtained from the general solution of the new equation. These are singular solutions of the original equation.

Recall that the basic procedure involves the following modifications of the original differential equation:

$$m_1(y)n_1(x)\frac{dy}{dx} = m_2(y)n_2(x) \qquad \textit{(original differential equation)},$$

$$\frac{m_1(y)}{m_2(y)}\,dy = \frac{n_2(x)}{n_1(x)}\,dx \qquad \textit{(new differential equation)},$$

$$\int \frac{m_1(y)}{m_2(y)}\,dy = \int \frac{n_2(x)}{n_1(x)}\,dx + C \quad \textit{(solution of new differential equation)}.$$

All functions that satisfy the new differential equation also satisfy the original one so that

$$\int \frac{m_1(y)}{m_2(y)}\,dy = \int \frac{n_2(x)}{n_1(x)}\,dx + C$$

is the general solution of the original differential equation. The singular solutions (if any) are those solutions lost by the division by $m_2(y)$. In general, if y is any function of x that satisfies both

$$m_2(y) = 0 \quad \text{and} \quad m_1(y)\frac{dy}{dx} = 0,$$

then y is a solution of the original differential equation even though it may not solve the new one. Thus, to find the singular solutions we determine all functions y such that $m_2(y) = 0$ and then check to see which of these functions satisfy the original differential equation. If y is such a function and y cannot be obtained from the general solution, then y is a singular solution.

For example, in Example 1 we had

$$2x(1-y)\frac{dy}{dx} = (x+1)(1-y^2) \qquad \textit{(original differential equation)},$$

$$\frac{2(1-y)}{1-y^2}\,dy = \frac{x+1}{x}\,dx \qquad \textit{(new differential equation)},$$

which we solved, obtaining the general solution

$$y = ke^{x/2}\sqrt{|x|} - 1.$$

In obtaining the new differential equation we divided by $m_2(y) = 1 - y^2$. The only functions $y = f(x)$ that make $1 - y^2$ equal to zero for all x are constant functions

$$y = +1 \quad \text{and} \quad y = -1.$$

Each of these is a solution of the original differential equation. The solution $y = -1$ is a particular solution obtaining by choosing $k = 0$ in the general solution. The solution $y = +1$ is a singular solution that cannot be obtained from the general solution for any choice of k.

EXERCISES 9.2

1. Calculate the general solution of each of the following differential equations:

(a) $\frac{dy}{dx} = \frac{y}{x}$

• (f) $dy = e^{x-y}\,dx$

• (b) $\frac{dy}{dx} = \frac{x-1}{y}$

(g) $2xy\,dx = (x^2 - 1)\,dy$

• (c) $\frac{dy}{dx} = \frac{y}{x^2 - 1}$

• (h) $dy = (xy + 2x)\,dx$

(d) $(x + 1)y\,dy = (x + 1)^2 y^2\,dx$

(i) $x^4\,dy = y^4\,dx$

(e) $\frac{2dy}{3x^2} = (y^2 - 1)\,dx$

• (j) $y\,dx = \frac{x}{2}(x\,dy - y\,dx)$

2. (a to j). Find all of the singular solutions of the differential equations of Exercise 1.

9.3 APPLICATIONS

Logistic Functions The exponential law of growth

$$P = Ce^{kt}$$

considered in Chapter 5 was derived using the assumption that the rate of growth is proportional to the size of the population. In most cases there are inhibiting factors that modify the growth rate when the population is large. Let us assume that because of the inhibiting factors there is a maximum possible population of M. If $P = P(t)$ is the size of the population at time t, then the number

$$M - P,$$

in some sense, measures the capacity of the population to increase. The larger the value of P, the smaller the capacity.

Scientists have determined experimentally that in certain cases the rate of change is proportional to the product of P and $M - P$. That is,

Logistic function

$$\frac{dP}{dt} = kP(M - P),$$

where k is a positive constant. The solutions of this differential equation are known as *logistic functions*. Their graphs are *logistic curves*.

Theorem 9.3 The general solution of the differential equation

$$\frac{dP}{dt} = kP(M - P) \qquad \text{where } 0 < P < M$$

is

$$P = \frac{Me^{at}}{e^{at} + c} = \frac{M}{1 + ce^{-at}} \qquad (a = Mk),$$

where c is an arbitrary positive constant.

PROOF We separate the variables and use the identity

$$\frac{1}{P(M-P)}=\frac{1}{M}\left(\frac{1}{P}+\frac{1}{M-P}\right),$$

obtaining

$$\begin{aligned}
\frac{1}{P(M-P)}\,dP &= k\,dt,\\
\frac{1}{M}\left(\frac{1}{P}+\frac{1}{M-P}\right)dP &= k\,dt,\\
\frac{1}{M}\int\frac{1}{P}\,dP+\frac{1}{M}\int\frac{1}{M-P}\,dP &= \int k\,dt + C,\\
\frac{1}{M}\ln|P|-\frac{1}{M}\ln|M-P| &= kt + C,\\
\ln|P|-\ln|M-P| &= Mkt+MC.
\end{aligned}$$

Since $0 < P < M$, then $|P| = P$ and $|M-P| = M-P$. Therefore we can drop the absolute-value signs and solve for P:

$$\begin{aligned}
\ln P - \ln(M-P) &= Mkt + MC,\\
\ln\frac{P}{M-P} &= Mkt+MC,\\
\frac{P}{M-P} &= e^{\ln[P/(M-P)]} = e^{Mkt}\,e^{MC} = e^{MC}\,e^{Mkt},\\
P &= (e^{MC}\,e^{Mkt})(M-P) = Me^{MC}\,e^{Mkt} - Pe^{MC}\,e^{Mkt},\\
P(e^{MC}\,e^{Mkt}+1) &= Me^{MC}\,e^{Mkt},\\
P &= \frac{Me^{MC}\,e^{Mkt}}{e^{MC}\,e^{Mkt}+1} = \frac{Me^{Mkt}}{e^{Mkt}+\dfrac{1}{e^{MC}}}
\end{aligned}$$

If we let $a = Mk$ and $c = 1/e^{MC}$, we obtain

$$P=\frac{Me^{at}}{e^{at}+c}=\frac{M}{1+ce^{-at}}.$$

It is customary to make the above substitution $a = Mk$ to simplify the logistic function because k, the constant of proportionality, has no meaningful interpretation in the final form.

The logistic curve is shown in Figure 9.3. The curve is symmetric about the point of inflection $((\ln c)/a, M/2)$. The maximum growth rate occurs at the point of inflection. That is, dP/dt has its maximum value when $t = (\ln c)/a$, $M = M/2$. The lines $P = 0$ and $P = M$ are horizontal asymptotes to the curve.

Example 1 Two years ago a small lake was stocked with 1000 fish. At present the number has increased to 3000. The maximum possible population is 10,000. Assuming that the conditions for growth do not change, determine the formula for the population at the end of t years.

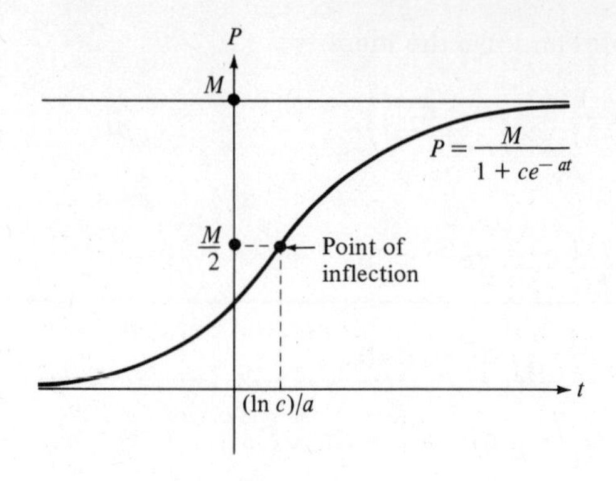

FIGURE 9.3 The logistic curve.

Solution We measure the population in thousands. The formula is

$$P = \frac{10e^{at}}{e^{at} + c},$$

where a and c are constants. When $t = 0$, $P = 1$:

$$1 = \frac{10e^0}{e^0 + c} = \frac{10}{1 + c},$$

$$1 + c = 10,$$
$$c = 9.$$

When $t = 2$, $P = 3$:

$$3 = \frac{10e^{2a}}{e^{2a} + 9},$$

$$3e^{2a} + 27 = 10e^{2a},$$
$$27 = 7e^{2a},$$
$$e^{2a} = \frac{27}{7},$$
$$2a = \ln\left(\frac{27}{7}\right) = \ln 27 - \ln 7,$$
$$a = \frac{\ln 27 - \ln 7}{2}.$$

The formula for the population after t years is

$$P = P(t) = \frac{10e^{(\ln 27 - \ln 7)(t/2)}}{e^{(\ln 27 - \ln 7)(t/2)} + 9}$$
$$= \frac{10e^{[\ln(27/7)](t/2)}}{(e^{\ln(27/7)})^{t/2} + 9} = \frac{10(\frac{27}{7})^{t/2}}{(\frac{27}{7})^{t/2} + 9}.$$

A great deal of information can be obtained from the logistic function. For example, the maximum growth rate is maintained when the population is kept at $M/2$. For the example above this means that the greatest annual yield of fish can be obtained if the population is stabilized at 5000 fish. This maximum annual yield is approximately equal to the value of the derivative

$$\frac{dP}{dt} = kP(M - P) = \frac{a}{M}P(M - P)$$

evaluated at $P = 5$ (thousand). Thus, the maximum yield that can be maintained year after year is approximately

$$\frac{\frac{\ln 27 - \ln 7}{2}}{10} \cdot 5 \cdot (10 - 5) = \frac{\ln 27 - \ln 7}{20} \cdot 25$$

$$\approx (3.2958 - 1.9459)\frac{5}{4} \approx 1.687 \quad \text{(thousand).}$$

In round numbers the maximum annual yield, obtained when the population is kept at 5000, is approximately 1700 fish.

Chemical Reactions The following example illustrates a rate-of-change problem typical of many types of chemical reaction.

Example 2 In a certain chemical process one molecule of substance Z is formed by combining one molecule of X and one of Y. The rate of formation of Z at time t is proportional to the product of the amount of X and the amount of Y present at time t. At the beginning of the process there are 15 units of X, 20 units of Y, and no units of Z. After 10 minutes there are 12 units of Z. Derive the formula for the amount of Z present at time t.

Solution Let $z =$ the number of units of Z present at time t. Then $15 - z =$ the number of units of X present at time t and $20 - z =$ the number of units of Y present at time t. The conditions of the rate of formation can be expressed as the differential equation

$$\frac{dz}{dt} = k(15 - z)(20 - z),$$

$$\frac{dz}{(15 - z)(20 - z)} = k\,dt.$$

By partial fractions

$$\frac{1}{(15 - z)(20 - z)} = \frac{1}{5(15 - z)} - \frac{1}{5(20 - z)},$$

so that

$$\int \frac{dz}{(15 - z)(20 - z)} = k \int dt + C,$$

$$\frac{1}{5}\int \frac{dz}{15 - z} - \frac{1}{5}\int \frac{dz}{20 - z} = kt + C,$$

$$\frac{1}{5} \ln |15 - z| + \frac{1}{5} \ln |20 - z| = kt + C.$$

Since $0 \leqslant z \leqslant 15$ (by the original conditions), we can drop the absolute-value sign. If we multiply by 5 and simplify, we have

$$-\ln (15 - z) + \ln (20 - z) = 5kt + 5C.$$

$$\ln \frac{20 - z}{15 - z} = 5C + 5kt,$$

$$\frac{20 - z}{15 - z} = e^{5C}\, e^{5kt} = Ke^{5kt} \qquad \text{where } K = e^{5C}.$$

We now use the initial conditions: When $t = 0$, $z = 0$.

$$\frac{20-0}{15-0} = Ke^0, \qquad K = \frac{4}{3}.$$

When $t = 10$, $z = 12$.

$$\frac{20-12}{15-12} = \frac{4}{3}e^{50k}, \qquad \frac{8}{3} = \frac{4}{3}e^{50k},$$

$$e^{50k} = 2, \qquad 50k = \ln 2, \qquad k = \frac{\ln 2}{50}.$$

Therefore

$$\frac{20-z}{15-z} = e^{5kt} = \frac{4}{3}e^{5[(\ln 2)/50]\cdot t} = \frac{4}{3}e^{\ln 2\cdot(t/10)},$$

$$\frac{20-z}{15-z} = \frac{4}{3}\cdot 2^{t/10}.$$

We now solve for z:

$$3(20-z) = 4\cdot 2^{t/10}(15-z),$$
$$60-3z = 4\cdot 2^{t/10}\cdot 15 - 4\cdot 2^{t/10}z,$$
$$(4\cdot 2^{t/10}-3)z = 60\cdot 2^{t/10} - 60,$$
$$z = \frac{60(2^{t/10}-1)}{4\cdot 2^{t/10}-3}.$$

Theoretically, the above reaction never ends. From a practical standpoint, however, it eventually slows to the point where no further reaction is measurable.

Predator-Prey Relationships Consider an ecological system that contains two species of animals—one species (the *prey*) which feeds on vegetable matter and one (the *predator*) which feeds on the prey.

Let $x = x(t)$ = the number of prey at time t and $y = y(t)$ = the number of predators at time t. It was observed early in the twentieth century that x and y are periodic functions of t, that growth periods of x are followed by growth periods of y (more food is available), which are followed by periods of decline of x, which, in turn, are followed by periods of decline of y (less food is available), and so on. These observations led to a great deal of experimentation and mathematical analysis.

Observe that for each species

rate of change of population = (*birth rate*) − (*death rate*).

For the prey population the birth rate is proportional to the population. The death rate, however, depends on the size of both populations. (For large x and small y many of the prey die of natural causes; for small x and large y most of the prey are eaten by the predators.) It seems reasonable to assume that the death rate of the prey is proportional to the product xy. Our assumptions lead to the differential equation

$$\frac{dx}{dt} = ax - bxy,$$

where a and b are positive constants.

The situation is reversed for the predator population. If x is large, then y increases rapidly. If x is small, then starvation acts as an inhibiting factor for the population and cuts down the birth rate. Thus the birth rate of the predators depends on both x and y. We assume that it is proportional to the product xy. The death rate, however, depends primarily on the size of y. Thus we obtain the second differential equation

$$\frac{dy}{dt} = mxy - ny,$$

where m and n are positive.

We must consider these two differential equations jointly, as a system of differential equations

$$\begin{cases} \dfrac{dx}{dt} = ax - bxy \\ \dfrac{dy}{dt} = mxy - ny. \end{cases}$$

Thus we wish to find functions x and y which simultaneously solve both differential equations.

Unfortunately, there is no general way to solve the system. If, however, we have specific values of the constants and specific initial conditions we can use advanced numerical methods to graph x and y as functions of t. Typical graphs are shown in Figure 9.4.

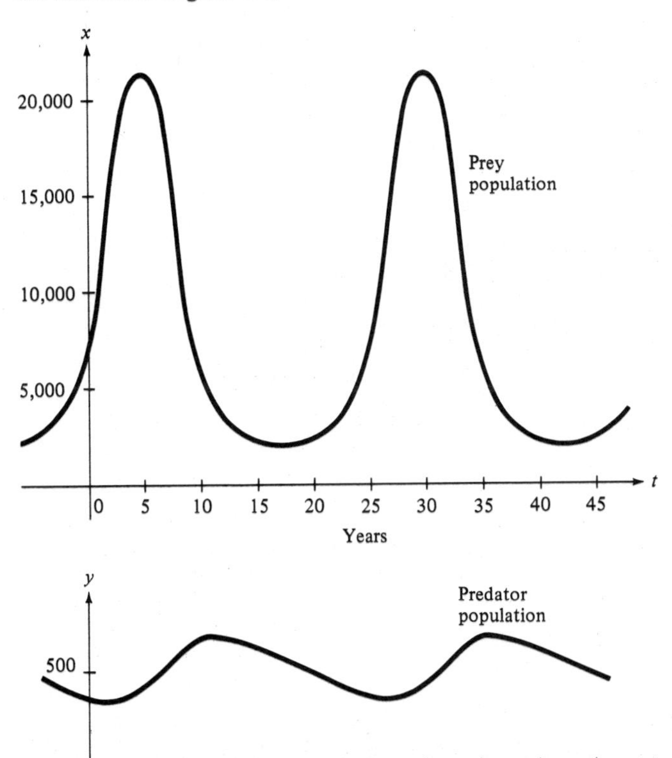

FIGURE 9.4 The predator-prey relationship graphed with respect to time.

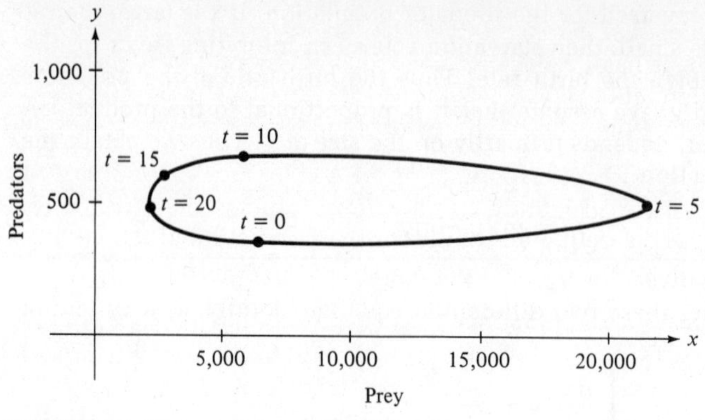

FIGURE 9.5 The predator-prey relationship.

There is a way to obtain a single differential equation which relates x and y. Although the resulting equation does not give information about how x and y vary with time, it does show how x and y relate to each other.

We multiply each of the differential equations by dt and divide the second by the first, obtaining

$$\frac{dy}{dx} = \frac{(mxy - ny)dt}{(ax - bxy)dt} = \frac{y(mx - n)}{x(a - by)}.$$

We now separate the variables and integrate

$$\int \frac{a - by}{y}\, dy = \int \frac{mx - n}{x}\, dx.$$

$$a \int \frac{dy}{y} - b \int dy = m \int dx - n \int \frac{dx}{x},$$

$$a \ln y - by = mx - n \ln x + C,$$

where C is a constant that depends on the initial conditions.

This equation relates x and y, but cannot be solved explicitly for y as a function of x. By using advanced numerical methods, however, it can be graphed for particular values of a, b, m, n, C. In each case the graph is a closed curve in the xy-plane similar to the one in Figure 9.5. As we trace along the curve the coordinates of the points give the values of x and y at various times t.

EXERCISES 9.3

•1. Three years ago a lake was stocked with 5000 fish. Currently the population is 8000. If the maximum number of fish the lake can support is 10,000, determine the formula for the population at the end of t years.

•2. The population of a county in 1960 was 10,000. By 1970 the population had increased to 20,000. It has been determined that the county can sustain a maximum population of 40,000. Assuming that the law of growth is a logistic function, determine a formula for the population.

3. A culture is observed to have a population of 10,000 after one month of growth, and 15,000 after two months. If the culture can sustain a maximum population of 60,000, determine a formula for the population of the culture after t months. What was the original population of the culture?

4. Newton's law of cooling asserts that the rate of change of the temperature (T) of a body at time t is directly proportional to the difference of the temperature of the body and the temperature of the surrounding medium (M). Show that Newton's law of cooling can be stated by the differential equation $dT/dt = k(T - M)$ and that the solutions of this equation satisfy $\ln(T - M) = kt + C$ or $T = M + C'e^{kt}$, where C, C', and k are constants.

•5. An object whose temperature is 100° is placed on a porch where the temperature is 20°. After one hour the temperature of the object is 60°. If the air remains at 20°, what will be the temperature of the object after five hours? (Use Exercise 4.)

6. An object is placed in a room where the temperature is constantly 30°. After one hour the temperature of the object is observed to be 100°, and after two hours, 80°. At what temperature was the object when it was placed in the room? When will the temperature be equal to 66°? (Use Exercise 4.)

•7. In a process for making compound W one unit of compound A is combined with one unit of compound B to yield one unit of compound W. The rate of formation of W at time t is directly proportional to the product of the amount of compound A and the amount of compound B present. If 20 units of A and 40 units of B and zero units of W are present when the process is started, and 10 units of W are present after five minutes, derive a formula for the amount of W present at time t.

8. A certain process involves the combination of equal parts of compounds A and B to produce a single unit of compound C. The rate of formation of compound C is directly proportional to the product of the amount of A and the amount of B present at time t. If 20 units of both A and B and zero units of C are present when the process is started, and 10 units of C are formed in five minutes, derive a formula for the amount of C present at time t. How long will it take to produce 16 units of compound C?

9.4 FIRST-ORDER LINEAR DIFFERENTIAL EQUATIONS

In this section we derive a formula for the solution of the differential equation

$$\frac{dy}{dx} + p(x)y = q(x).$$

The formula holds on any open interval where both $p(x)$ and $q(x)$ are continuous. The differential equation is called a *linear differential equation*.

Linear differential equation

Theorem 9.4 The general solution of the linear differential equation

$$\frac{dy}{dx} + p(x)y = q(x)$$

is

General solution of linear differential equation

$$\boxed{y = e^{-\int p(x)dx}\left\{\int e^{\int p(x)dx}\, q(x)\, dx + C\right\}}$$

where C is a constant. This formula holds on any open interval where both p and q are continuous.

PROOF We multiply both sides of the differential equation by $e^{\int p(x)dx}$, obtaining

$$e^{\int p(x)dx}\frac{dy}{dx} + e^{\int p(x)dx}p(x)y = e^{\int p(x)dx}q(x).$$

We now make the substitution $u = e^{\int p(x)dx}y$. It follows from D6 and D10 that

$$\frac{du}{dx} = e^{\int p(x)dx}\frac{dy}{dx} + y \cdot \frac{d}{dx}\left(e^{\int p(x)dx}\right)$$

$$= e^{\int p(x)dx}\frac{dy}{dx} + y \cdot e^{\int p(x)dx} \cdot p(x)$$

which is equal to the left-hand side of the above differential equation. Therefore, the differential equation reduces to

$$\frac{du}{dx} = e^{\int p(x)dx}q(x).$$

We now separate the variables and integrate, obtaining

$$du = e^{\int p(x)dx}q(x)\, dx$$

$$u = \int e^{\int p(x)dx}q(x)\, dx + C.$$

Since $u = e^{\int p(x)dx}y$, then

$$e^{\int p(x)dx}y = \int e^{\int p(x)dx}q(x)\, dx + C$$

$$y = e^{-\int p(x)dx}\left\{\int e^{\int p(x)dx}\, q(x)\, dx + C\right\}.$$

Example 1 Solve $\frac{dy}{dx} + \frac{1}{x}y = 3x^2 + 5$.

Solution $p(x) = 1/x$, $q(x) = 3x^2 + 5$. $\int p(x)\, dx = \int (1/x)\, dx = \ln|x|$. (We drop the constant of integration—it will cancel out later if we keep it.) Then

$$e^{-\int p(x)dx} = e^{-\ln|x|} = e^{\ln(1/|x|)} = \frac{1}{|x|}$$

and

$$e^{\int p(x)dx} = e^{\ln|x|} = |x|.$$

The general solution is

$$y = e^{-\int p(x)dx}\left\{\int e^{\int p(x)dx}q(x)\,dx + C\right\}$$
$$= \frac{1}{|x|}\left\{\int |x|\,(3x^2+5)\,dx + C\right\}.$$

If $x > 0$, then $|x| = x$. The solution becomes

$$y = \frac{1}{x}\left\{\int x(3x^2+5)\,dx + C\right\} = \frac{1}{x}\left\{\int (3x^3+5x)\,dx + C\right\}$$
$$= \frac{1}{x}\left\{\frac{3x^4}{4} + \frac{5x^2}{2} + C\right\} = \frac{3x^3}{4} + \frac{5x}{2} + \frac{C}{x}.$$

We leave it to the reader to show that the general solution for $x < 0$ has the same form.

Example 2 Solve $y' = \cos x \cdot (y+1)$.

Solution We rewrite the differential equation in the form

$$\frac{dy}{dx} - \cos x \cdot y = \cos x.$$

Then $p(x) = -\cos x$, $q(x) = \cos x$,

$$\int p(x)\,dx = \int (-\cos x)\,dx = -\sin x.$$

The solution of the differential equation is

$$y = e^{-\int p(x)dx}\left\{\int e^{\int p(x)dx}q(x)\,dx + C\right\}$$
$$= e^{\sin x}\left\{\int e^{-\sin x}\cos x\,dx + C\right\}.$$

To integrate the function we let $u = \sin x$, $du = \cos x\,dx$. Then

$$y = e^u\left\{\int e^{-u}\,du + C\right\}$$
$$= -e^u\left\{\int e^{-u}(-du) - C\right\}$$
$$= -e^u\{e^{-u} - C\} = -1 + Ce^u$$
$$= -1 + Ce^{\sin x}.$$

Application. Learning Curves Psychologists have made a number of theoretical studies of the learning process. For simplicity we consider the case where a simple skill is learned by a completely unskilled person. When he is first exposed to the process there is a period of rapid learning. As his proficiency increases, the rate of learning slows until eventually there is no learning at all.

Learning curve

Assume that the learner practices the skill continuously so that his proficiency is a function of time, say $y = p(t)$. Then $y' = p'(t)$ is the rate at which the learning occurs at time t. If m is the maximum level of proficiency of which the learner is capable, then

$$m - y$$

is a measure of his capacity to learn at time t. It is reasonable to assume that the rate of learning is proportional to this capacity to learn. That is,

$$y' = k(m - y),$$

where k is the constant of proportionality.

The above differential equation can be rewritten as the linear differential equation

$$y' + ky = km.$$

To solve this differential equation we let

$$p(t) = k \quad \text{and} \quad q(t) = km,$$

so that

$$\int p(t)\, dt = kt.$$

The solution is

$$\begin{aligned} y &= e^{-\int p(t)dt} \left\{ \int e^{\int p(t)dt} q(t)\, dt + C \right\} \\ &= e^{-kt} \left\{ \int e^{kt} km\, dt + C \right\} \\ &= e^{-kt} \{ me^{kt} + C \} \\ &= m + Ce^{-kt}. \end{aligned}$$

To evaluate the constant C recall that $y = 0$ when $t = 0$. (The learner was completely unskilled at the beginning of the learning process.) Then

$$\begin{aligned} y &= m + Ce^{-kt} \\ 0 &= m + Ce^{0} = m + C \qquad \text{(Substituting } t = 0\text{)} \\ C &= -m. \end{aligned}$$

The solution of the differential equation is

$$y = m(1 - e^{-kt}),$$

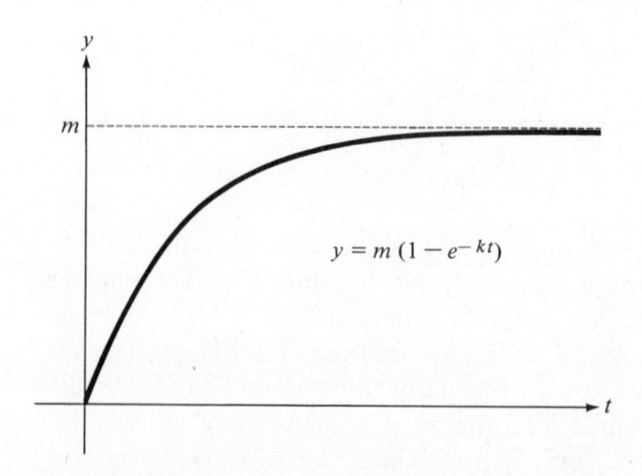

FIGURE 9.6 Typical learning curve.

where m is the maximum level of proficiency and k is a measure of the rate of learning.

The graph of a typical learning curve is shown in Figure 9.6.

EXERCISES 9.4

1. Solve the following differential equations:

(a) $\dfrac{dy}{dx} + y = e^{3x}$

• (b) $y' + 2xy = 3x^2e^{-x^2}$

(c) $(2y - 15x^3)\,dx + x\,dy = 0$

• (d) $xy' + 3y = 5x^2$

(e) $xy' - y = x^2e^{-x}$

(f) $\dfrac{dy}{dx} - y = 2xe^{x}$

• (g) $x\dfrac{dy}{dx} = 4xe^{3x} + y(3x - 1)$

(h) $\dfrac{dy}{dx} + y - 4x = 6$

• (i) $\dfrac{dy}{dx} = x(1 + 2y)$

(j) $(x - 1)y' + 2y = e^{x}(x - 1)^{-1}$

2. Thomas Foreman must train workers in a new skill on an assembly line. All workers are completely untrained at the beginning so that the equation

$$y = m(1 - e^{-kt})$$

describes the level of proficiency at time t. Thomas has measured the number of times per hour each worker can perform the necessary operations at the end of one day and at the end of two days of training. Calculate m, the maximum number of times each worker can perform the operations when fully trained.

• (a) Worker A: 50 times at end of one day, 75 times at the end of two days;
(b) Worker B: 40 times at the end of one day, 75 times at the end of two days;
(c) Worker C: 60 times at the end of one day, 70 times at the end of two days.

9.5 APPLICATIONS. DIFFUSION EQUATIONS

Diffusion equation

Diffusion equations are mathematical statements that describe how characteristics or properties are diffused in a population. They can be used to explain the spread of epidemics, the dissemination of information, the effect of advertising on sales, and so on. In this section we show how to derive certain differential equations which describe diffusion processes. We begin with a classic example which explains how certain types of infectious disease are diffused in a population during an epidemic.

Example 1 In a fixed population of size P, $y = y(t)$ persons are infected with a mild communicable disease which cannot be cured. These persons circulate freely in the population infecting other members. On the average, each infected person contacts m persons per day. Each contact with an uninfected person results in a new case of the disease. Derive a formula for y in terms of t.

Solution On the tth day each of the $y(t)$ persons contacts m persons, resulting in a total of $m \cdot y(t)$ persons being contacted. The number of new cases is equal to the product of $m \cdot y(t)$ and $\frac{P - y(t)}{P}$, the proportion of uninfected persons in the population. Over an "average" period of time from t to $t + \Delta t$ we would expect the number of new cases to be

$$\Delta y = m \cdot y \cdot \left(\frac{P - y}{P}\right) \cdot \Delta t.$$

To obtain a differential equation we divide by Δt and take the limit as $\Delta t \to 0$:

$$\frac{dy}{dt} = \lim_{\Delta t \to 0} \frac{\Delta y}{\Delta t} = \lim_{\Delta t \to 0} my \left(\frac{P - y}{P}\right) = my \left(\frac{P - y}{P}\right)$$

$$\frac{dy}{dt} = \frac{m}{P} \cdot y \cdot (P - y).$$

Except for notation, this differential equation is the same as the one we considered in Section 9.3 for the logistic function. Thus, the solution is

$$y = \frac{P}{1 + ce^{-mt}}$$

where c is a constant that depends on the number of infected persons at time $t = 0$. Observe that $y \to P$ as $t \to \infty$. Thus, over a long period of time the entire population will be infected.

The assumptions made in Example 1 are realistic for mild, highly contagious diseases. Thus, the formula could be used with reasonable accuracy to predict the spread of a disease such as athlete's foot. The assumptions do not hold if the disease is severe. In that case the disease itself affects the size of the population or inhibits infected persons from circulating freely, or both. Thus, the formula obtained in Example 1 is not valid for a disease such as smallpox. Even then, however, it may be reasonably good in predicting the spread of a disease during the early stages of an epidemic while the disease is incubating in most of the infected individuals.

It was noticed quickly that although the assumptions stated in Example 1 are not always valid for epidemics they do hold for the dissemination of information by word-of-mouth. That is, the persons holding the special information circulate freely in the population (which obviously does not change in size due to the spread of the information), and each person knowing the information contacts a certain number of individuals during a day. If the information is of sufficient importance each contact with a new individual results in it being mentioned. Thus the number of individuals with the special knowledge satisfies the differential equation

$$\frac{dy}{dt} = my \frac{P - y}{P}$$

and the equation

$$y = \frac{P}{1 + ce^{-mt}},$$

where P is the size of the population, m is the average number of contacts made during a day by each person and c is a constant that depends on the number of persons knowing the information at time $t = 0$.

Use of advertising is one way to speed the dissemination of information. The diffusion process described above can be modified to describe the effect of advertising on sales.

Example 2 There is a stable market for hair tonic in the United States. The Hansum Hair Company plans to spend a fixed amount each month on continuous advertising. From past experience it knows how much new business each dollar will bring in from the "average" potential customer. Furthermore, due to advertising by its competitors, Hansum Hair loses a certain proportion of its established market each month. Derive a formula to measure the sales effect of the advertising campaign.

Solution

Let n = the number of persons in the market.

$y = y(t)$ = the number of customers in the tth month.

a = the amount Hansum Hair plans to spend on advertising each month.

b = the number of persons per month per dollar spent for advertising who buy Hansum Hair Tonic.

c = the proportion of established customers lost each month to competitors.

Then $\dfrac{n - y(t)}{n}$ = the proportion of the market controlled by Hansum Hair's competitors during the tth month.

$ab\left(\dfrac{n - y(t)}{n}\right)$ = the number of new customers in the tth month due to the advertising campaign.

$cy(t)$ = the number of established customers lost to competitors during the tth month.

It follows that

$$ab\left(\frac{n - y(t)}{n}\right) - cy(t)$$

is the net change in the number of customers during the tth month. Over an "average" period of time from t to $t + \Delta t$ the net change in the number of customers is

$$\Delta y = \left[ab\left(\frac{n - y(t)}{n}\right) - cy(t)\right]\Delta t.$$

We divide by Δt and take the limit as $\Delta t \to 0$:

$$\frac{dy}{dt} = \lim_{\Delta t \to 0} \frac{\Delta y}{\Delta t} = \lim_{\Delta t \to 0}\left[ab\left(\frac{n - y(t)}{n}\right) - cy(t)\right]$$

$$= ab\left(\frac{n - y(t)}{n}\right) - cy(t).$$

Thus y satisfies the differential equation

$$\frac{dy}{dt} = \frac{ab}{n}(n - y) - cy.$$

To solve this differential equation we rewrite it as a linear differential equation and apply the formula of Section 9.4.

$$\frac{dy}{dt} + \left(\frac{ab}{n} + c\right) y = ab.$$

Here

$$p(t) = \frac{ab}{n} + c; \; q(t) = ab,$$

$$\int p(t)\,dt = \left(\frac{ab}{n} + c\right) t,$$

and

$$y = e^{-\int p(t)dt}\left[\int e^{\int p(t)dt} q(t)\,dt + C\right]$$

$$= e^{-(ab/n+c)t}\left[\int e^{(ab/n+c)t} ab\,dt + C\right]$$

$$= e^{-(ab/n+c)t}\left[\frac{ab}{\frac{ab}{n} + c} e^{(ab/n+c)t} + C\right]$$

$$y = \frac{ab}{\frac{ab}{n} + c} + Ce^{-(ab/n+c)t}.$$

If we assume that the advertising campaign starts at time $t = 0$ and that y_0 is the number of customers at that time then

$$y_0 = \frac{ab}{\frac{ab}{n} + c} + Ce^0 = \frac{ab}{\frac{ab}{n} + c} + C$$

so that

$$C = y_0 - \frac{ab}{\frac{ab}{n} + c}.$$

Thus, the formula is

$$y = \frac{ab}{\frac{ab}{n} + c} + \left(y_0 - \frac{ab}{\frac{ab}{n} + c}\right) e^{-(ab/n+c)t}.$$

A number of results can be obtained from the differential equation of Example 2 and from the explicit equation for y in terms of t. For example, since

$$\frac{dy}{dt} = \frac{ab}{n}(n - y) - cy$$

then sales *increase* (the derivative is positive) if

$$a > \frac{ncy}{b(n-y)}$$

and sales *decrease* if

$$a < \frac{ncy}{b(n-y)}.$$

Observe also that since

$$y = \frac{ab}{\frac{ab}{n}+c} + \left(y_0 - \frac{ab}{\frac{ab}{n}+c}\right) e^{-(ab/n+c)t}$$

then

$$\lim_{n\to\infty} y(t) = \frac{ab}{\frac{ab}{n}+c} + 0 = \frac{ab}{\frac{ab}{n}+c}.$$

Thus over a long period of time the number of customers will tend to stabilize at

$$\frac{ab}{\frac{ab}{n}+c}.$$

In this chapter we have done little more than touch on the theory of differential equations. We have restricted ourselves to first-order differential equations—most of which were linear—ignoring higher-order equations, equations involving partial derivatives, and other difficult types.

There is no general formula for solving all differential equations. Rather, there is a large number of special formulas and special techniques that can be used to solve various special types of differential equations. For this reason much of the elementary work is concerned with the classification of types of equations (separable variables, linear, and so on) and the development of special techniques for solving equations of given types.

In actual practice it may not be possible to solve a given differential equation by elementary methods. Then the mathematician attempts to find an *approximate solution*. This is a computed function $f(x)$ which approximates the desired particular solution at each point of an interval. This is somewhat analogous to our work on approximate integration in Chapter 7 (Trapezoidal Rule and Simpson's Rule).

The applications considered in this chapter indicate the power of differential equations. Whenever a problem involves a rate of change, rate of growth, velocity, or acceleration, there is a good chance that it can be formulated in terms of a differential equation. The immediate practical problem then reduces to finding a particular solution of the equation that satisfies certain initial or boundary conditions. This approach has been utilized extensively for many years in physics, engineering, and statistics. It is now becoming standard in the social sciences as well.

APPENDIX

SUMMARY OF ELEMENTARY ALGEBRA

A.1 SETS

The word "*set*" is undefined in mathematics. Basically, a set consists of objects called its *elements*. We write $a \in A$ to indicate that a is an element of the set A. Examples of sets are

Set

1. The set of integers.
2. The set of vowels in the English language.
3. The empty set, $\varnothing$.

This last set, the empty set, is the set that contains no elements.

Empty set $\varnothing$

Sets can be specified by listing the elements, such as

$$\{a, e, i, o, u\}.$$

They also can be specified by stating a condition satisfied by the elements of the set and no other objects, such as

The set of all vowels in the English language.

Set-Building Notation The notation

Set-building notation

$$\{x | \text{"statement about x"}\}$$

means "the set of all objects for which the statement is true." For example, the set of real numbers between 0 and 1 (inclusive) can be described as

$$\{x | 0 \leq x \leq 1\}.$$

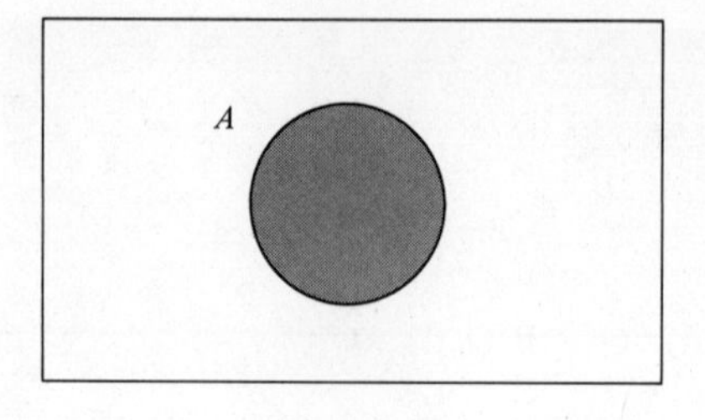

FIGURE A-1 A Venn diagram. The points in the circle represent the elements of the set A. The points that are in the rectangle, but outside of the circle, represent the elements under consideration that are not in A.

Venn diagram

Venn Diagrams Sets can be represented schematically by *Venn diagrams*, boxes in which shaded figures represent sets. Figure A.1 is a Venn diagram representing a set A.

Intersections and Unions The *intersection* of sets A and B is the set of all elements common to both A and B. It is denoted by the symbol $A \cap B$. Thus,

Intersection of sets

$$A \cap B = \{x|x \in A \text{ and } x \in B\} \qquad \text{(Figure A.2a)}.$$

If A and B have no common elements, then $A \cap B = \varnothing$.

The *union* of A and B, denoted by the symbol $A \cup B$, is the set of all elements that are in A or in B.[1] Thus

Union of sets

$$A \cup B = \{x|x \in A \text{ or } x \in B\} \qquad \text{(Figure A.2b)}.$$

Example 1 If $A = \{1,2,3,4,5\}$ and $B = \{0,2,4,6,8\}$ then

$$A \cup B = \{0,1,2,3,4,5,6,8\}$$

and

$$A \cap B = \{2,4\}.$$

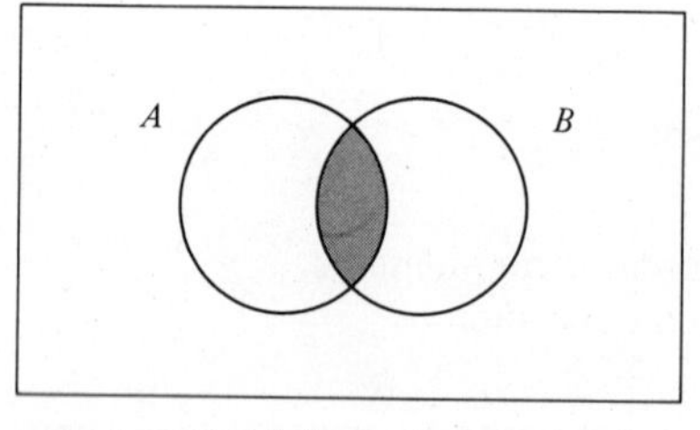

(a) $A \cap B = \{x|x \in A \text{ and } x \in B\}$.

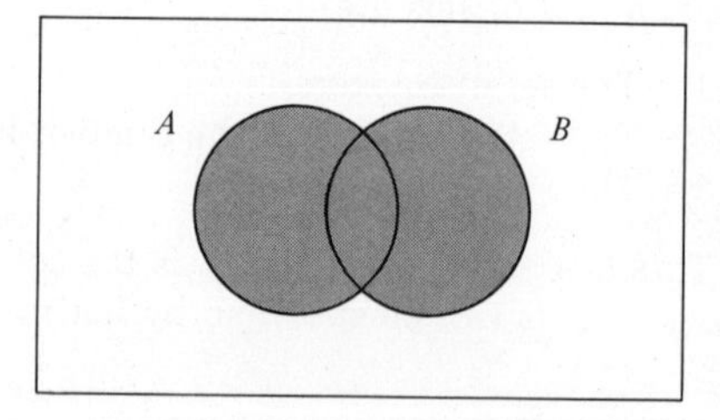

(b) $A \cup B = \{x|x \in A \text{ or } x \in B\}$.

FIGURE A-2 Intersection and union.

EXERCISES A.1

1. Describe the following sets by using set-building notation and by listing the elements:

[1] In mathematics we use the "inclusive" *either-or*. Thus, $A \cup B$ contains the elements common to A and B as well as the elements that are in one set, but not the other.

- (a) The set of letters in the word *Mississippi*.
  (b) The set of positive integers less than 6.
- (c) The set of consonants in the English language.

2. Let $A = \{1, 2, 4, 5, 7, 8\}$, $B = \{2, 3, 5, 6, 8, 9\}$, $C = \{1, 2, 3, 6, 7, 8\}$. List the elements in each of the following sets:

- (a) $A \cap B$
  (b) $B \cup C$
  (c) $A \cup B$
- (d) $A \cap (B \cup C)$
  (e) $A \cup (B \cap C)$
- (f) $(A \cap B) \cap C$

3. Let A, B, C be arbitrary sets. Draw Venn diagrams to represent the following sets:

(a) $A \cap B$
(b) $B \cup C$
(c) $A \cup B$
(d) $A \cap (B \cup C)$
(e) $A \cup (B \cap C)$
(f) $(A \cap B) \cap C$

A.2 THE REAL NUMBER LINE

The real number line is pictured in Figure A.3. Every point on the line represents a real number and every real number can be represented as a point on the line. It is customary to write the positive real numbers to the right of zero and the negative numbers to the left.

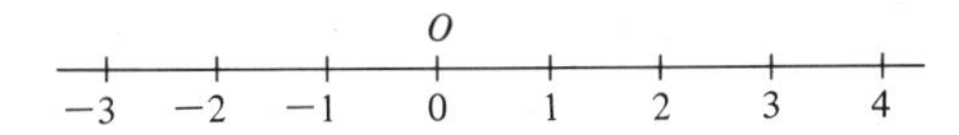

The number scale

FIGURE A-3

A.3 ORDER

If a is to the left of b on the number line we write

Less than

$$a < b \quad (a \text{ is less than } b)$$

and

Greater than

$$b > a \quad (b \text{ is greater than } a) \qquad \text{(see Figure A.4).}$$

For example, $5 > 2$, $7 > -1$, $-5 < 0$, and $-2 < -1$.

Observe that the positive numbers p satisfy the inequality $p > 0$ and the negative numbers n satisfy $n < 0$.

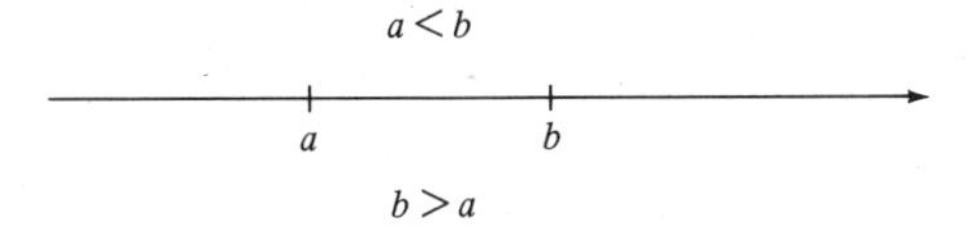

FIGURE A-4 Inequality. We say $a < b$ provided $b - a$ is positive. This is the case if, and only if, a is to the left of b on the number line.

Order properties

Properties of Order The following properties can be established from the elementary properties of positive and negative numbers.

O1 If $a < b$ and $b < c$, then $a < c$ — *the transitive property.*

O2 If a and b are real numbers, then $a < b$, or $a = b$ or $a > b$ — *the law of trichotomy.*

O3 If $a < b$, then $a + c < b + c$. — An inequality is preserved if a constant is added to both sides.

O4 If $a < b$ and $c > 0$ then $ac < bc$. — An inequality is preserved if both sides are multiplied by a positive constant.

O5 If $a < b$ and $c < 0$, then $ac > bc$. — An inequality is reversed if both sides are multiplied by a negative constant.

Less-than-or-equal

Greater-than-or-equal

We say that a *is less-than-or-equal* to b ($a \leqslant b$) if a is less than b or is equal to b. Under the same conditions we say that b is *greater-than-or-equal* to a. Properties similar to O1 to O5 hold for these relationships as well.

A.4 ABSOLUTE VALUE

Absolute value

The *absolute value* of a real number a, written $|a|$, is its distance from the number zero on the real line. Thus

$|3| = 3$ since 3 is three units from zero,
$|-4| = 4$ since -4 is four units from zero, and
$|0| = 0$ since 0 is zero units from zero (see Figure A.5).

$$|a| = a \qquad (a \text{ nonnegative}),$$

while if $a < 0$, then

$$|a| = -a \qquad (a \text{ negative}).$$

For example,

$$|3| = 3 \text{ since 3 is nonnegative,}$$

while

$$|-4| = -(-4) = 4 \text{ since } -4 \text{ is negative.}$$

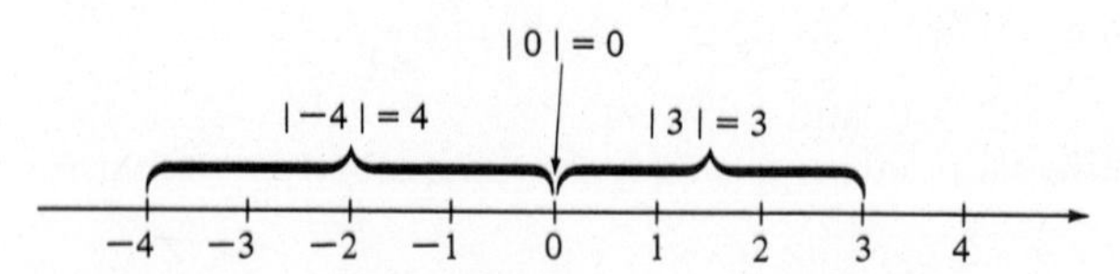

FIGURE A-5 Absolute value.

$$|a| = \begin{cases} a & \text{if } a \geqslant 0 \\ -a & \text{if } a < 0. \end{cases}$$

The absolute value of a number is equal to its distance from the origin on the number line.

Properties of Absolute Value

Properties of absolute value

$|ab| = |a| \cdot |b|$. — *The absolute value of a product is equal to the product of the absolute values.*

$\left|\dfrac{a}{b}\right| = \dfrac{|a|}{|b|}$. — *The absolute value of a quotient is equal to the quotient of the absolute values.*

$|a + b| \leq |a| + |b|$. *The Triangle Inequality. The absolute value of a sum is less than or equal to the sum of the absolute values.*

These properties can be established by using the relationship

$$|a| = \begin{cases} a & \text{if } a \geq 0 \\ -a & \text{if } a < 0. \end{cases}$$

Example 1 Show that the rule $|ab| = |a||b|$ holds for all a and b.

Solution We must consider the following cases:

Case 1: $a \geq 0$, $b \geq 0$. Then $ab \geq 0$ and $|ab| = ab = |a| \cdot |b|$, since $|a| = a$ and $|b| = b$.

Case 2: $a \geq 0$, $b < 0$. Then $ab \leq 0$ and $|ab| = -ab = a(-b) = |a| \cdot |b|$, since $|a| = a$ and $|b| = -b$.

Case 3: $a < 0$, $b \geq 0$. Similar to Case 2.

Case 4: $a < 0$, $b < 0$. Then $ab > 0$ and $|ab| = ab = (-a)(-b) = |a| \cdot |b|$, since $|a| = -a$ and $|b| = -b$.

Thus, the relationship $|ab| = |a| \cdot |b|$ holds in all cases.

EXERCISES A.4

1. Show that $-|a| \leq a \leq |a|$ for every real number a.
2. Show that if $a < 0$ then $-a > 0$.
3. Show that if $|a| = b$ then $a = b$ or $a = -b$.
4. Prove the triangle inequality.
5. Use the order properties to prove that if

$$a < b \quad \text{and} \quad c < d \quad \text{then} \quad a + c < b + d.$$

A.5 POWERS AND ROOTS

Powers of a

Positive integral powers of the number a are defined by

$$\begin{aligned} a^1 &= a \\ a^2 &= a \cdot a^1 = a \cdot a \\ a^3 &= a \cdot a^2 = a \cdot a \cdot a \\ &\cdots\cdots\cdots \\ a^n &= a \cdot a^{n-1} = a \cdot a \cdot a \cdots a \qquad (n \text{ factors}) \end{aligned}$$

If $a \neq 0$ we define

$$a^0 = 1 \quad \text{and} \quad a^{-n} = \frac{1}{a^n}.$$

The following rules of exponents hold (provided only nonzero numbers are raised to zero or negative powers):

Rules of exponents

E1 $a^m \cdot a^n = a^{m+n}$ — {*We add exponents when multiplying two powers of the same number.*

E2 $\dfrac{a^m}{a^n} = a^{m-n}$ — {*We subtract exponents when dividing two powers of the same number.*

E3 $(a^m)^n = a^{mn}$. — {*We multiply exponents when raising a power to a power.*

E4 $\begin{cases} (ab)^n = a^n b^n \\ \left(\dfrac{a}{b}\right)^n = \dfrac{a^n}{b^n} \end{cases}$ — {*We raise each term to the power when raising a product or a quotient to a power.*

Example 1

(a)
$$(3^{-2})^{-3} = 3^{(-2)(-3)} = 3^6.$$

(b)
$$\frac{xy}{x^3y^{-1}} \cdot \frac{1}{x^{-1}} = x^{1-3}y^{1-(-1)}x^{-(-1)} = x^{-2}y^2x^1$$
$$= x^{-2+1}y^2 = x^{-1}y^2 = \frac{y^2}{x}.$$

Roots If n is a positive integer and $a > 0$, then there exists a unique number $r > 0$ such that

$$r^n = a.$$

Principal root of positive number

This number r, which is denoted by the symbol $r = \sqrt[n]{a}$, is called the *principal nth root of a*. (In the special case $n = 2$ the principal square root is denoted by the symbol $\sqrt{a}$.)

Example 2

(a) Since $2^4 = 16$, then $\sqrt[4]{16} = 2$.
(b) Since $(2.75892)^3 \approx 21$, then[2]

$$\sqrt[3]{21} \approx 2.75892.$$

If a is a negative number, then a has no real square root, fourth root, sixth root, and so on. The number a does, however, have exactly one negative cube root, one negative fifth root, and so on. These roots are also called principal roots and are denoted by the symbol $\sqrt[n]{a}$.

Principal root of negative number

Example 3

(a) $\sqrt[5]{-32} = -2$
(b) $\sqrt[3]{-1} = -1$
(c) There is no real square root of -16.

Remark Most numbers have several nth roots. Only the positive nth root of a positive number or the negative nth root of a negative number is denoted by the symbol $\sqrt[n]{a}$. For example, both 2 and -2 are fourth roots of 16, but $\sqrt[4]{16} = 2$.

The roots of most numbers are irrational. Table I at the end of the book contains approximations to square roots and cube roots of selected numbers.

---

[2] The symbol "≈" is used to indicate that two numbers are approximately equal.

Fractional Exponents The principal nth root of a can be denoted by the symbol $a^{1/n}$. For example,

$$16^{1/4} = \sqrt[4]{16} = 2,$$
$$(-27)^{1/3} = \sqrt[3]{-27} = -3,$$

and

$$2^{1/2} = \sqrt[2]{2} \approx 1.4142.$$

Fractional exponents

We define $a^{m/n}$ (for positive n) by

$$a^{m/n} = (a^{1/n})^m = (\sqrt[n]{a})^m.$$

For example,

$$16^{3/4} = (16^{1/4})^3 = 2^3 = 8$$

and

$$2^{-3/2} = \frac{1}{2^{3/2}} = \frac{1}{(2^{1/2})^3} = \frac{1}{(2^{1/2} \cdot 2^{1/2}) \cdot 2^{1/2}}$$
$$= \frac{1}{2 \cdot 2^{1/2}} \approx \frac{1}{2(1.4142)} \approx \frac{1}{2.8284}.$$

The rules of exponents E1 to E4 all hold for rational exponents provided all of the roots are real and only nonzero numbers are raised to zero or negative powers.

Example 4

(a) $(3^{1/2}7^{2/3})^6 = 3^{(1/2)\cdot 6}7^{(2/3)\cdot 6} = 3^3 7^4 = 27 \cdot 2401 = 64{,}827.$

(b) $$\frac{a^{1/2} + b^{1/2}}{a^{-1/2}b^{-1/2}} = \frac{a^{1/2} + b^{1/2}}{\dfrac{1}{a^{1/2}b^{1/2}}} = (a^{1/2} + b^{1/2})a^{1/2}b^{1/2}$$
$$= a^{1/2+1/2}b^{1/2} + a^{1/2}b^{1/2+1/2} = ab^{1/2} + a^{1/2}b.$$

EXERCISES A.5

1. Calculate the roots. Use Table I if necessary.

- • (a) $\sqrt[3]{27}$
- • (b) $\sqrt{24}$
- (c) $\sqrt[3]{-64}$
- • (d) $(-36)^{1/3}$
- (e) $(300)^{1/2}$
- • (f) $(-87)^{1/3}$

2. Simplify to a form free of zero and negative exponents.

- • (a) $\dfrac{3^2 \cdot 3^{-4}}{3^3 \cdot 3^4}$
- (b) $\left(\dfrac{a^2b^3}{a^3}\right)^4$
- • (c) $(x^2y)^2x^{-1}y^{-3}$
- (d) $\left(\dfrac{a^{1/3}b^{2/3}}{a}\right)^{3/2}$
- • (e) $\left(\dfrac{x^2y^{1/2}}{x^{1/2}y}\right)^2$
- • (f) $(x^2yz^{1/2})(x^{1/2}y^{1/3}z^2)$
- • (g) $(x^{1/2} + x^{3/2})^2$
- (h) $(x^{-1} + x)^{-1}$

A.6 FUNCTIONS

Function
Domain
Range
Variable

A *function* is a law of correspondence between two sets that assigns to each element of the first set (known as the *domain*) a unique element of the second set (the *range*). Elements of the domain are called *variables*.

An example is the function that assigns to the number $s > 0$ the volume of the cube with sides of length s. The law of correspondence is

$$V = s^3.$$

The domain of the function is the set of all possible values of $s > 0$, the range is the set of all possible volumes V.

f(x) notation

If f is a function and x is an element of its domain, then there is a unique element y assigned to x by the function. It is customary to denote this number by the symbol $y = f(x)$.[3]

Example 1 If f is the function described above that assigns the number s^3 to every positive number s, then

$$f(s) = s^3 \text{ (for } s > 0),$$
$$f(x) = x^3 \text{ (for } x > 0),$$
$$f(2) = 2^3 = 8,$$
$$f(7) = 7^3 = 343,$$

while

$$f(-1) \text{ is not defined.}$$

A function can be represented schematically by the diagram in Figure A.6. The elements of the domain are represented by the points in the region on the left, those of the range by the points in the one on the right. The arrow indicates the assignment of $f(x)$ to the variable x of the domain.

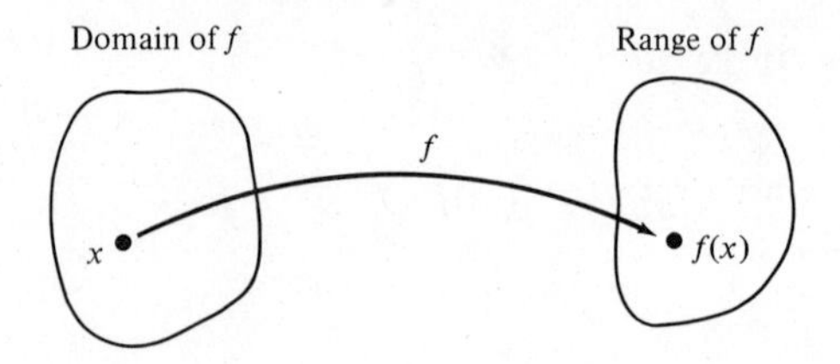

FIGURE A-6 Schematic representation of a function. The number $f(x)$ in the range is assigned to the number x of the domain by the function f.

Unspecified domain

Unspecified Domains If a function is defined by an algebraic equation and the domain is not specified, then, by convention, it consists of every real number to which the law of correspondence assigns a real number in the range.

Example 2

(a) The domain of

$$f(x) = \frac{1}{x - 1}$$

is the set of all real numbers x except $x = 1$.

(b) The domain of

$$g(x) = \sqrt{x}$$

---

[3] "$f(x)$" is a single symbol (read "f of x"). It should not be confused with the product of f and x.

is the set of all nonnegative real numbers.

(c) The domain of

$$h(x) = 3x^2 + 2$$

is the set of all real numbers.

Functions of several variables

Functions of Several Variables Functions of two or more variables are common in mathematics. For example, the volume of a right circular cylinder is a function of the radius, r, and the height, h:

$$V = f(r, h) = \pi r^2 h.$$

The domain of this function is the set of all ordered pairs (r, h) such that $r > 0$ and $h > 0$. For each positive r and h there is a corresponding value of V. For example,

$$f(1, 1) = \pi \cdot 1^2 \cdot 1 = \pi,$$
$$f(3, 7) = \pi \cdot 3^2 \cdot 7 = 63\pi,$$

while

$$f(-1, 0) \text{ is not defined.}$$

Algebraic expression

Algebraic Expressions Functions are sometimes represented by algebraic expressions. For example, the expression

$$3x^2 - 2xy + 5y^2 - 7$$

represents the function of two variables

$$f(x, y) = 3x^2 - 2xy + 5y^2 - 7.$$

Such algebraic expressions can be added, subtracted, multiplied, and divided by using the standard rules of elementary algebra.

Example 3

(a) $(2x^2 + 3)(x - 2) = 2x^2(x - 2) + 3(x - 2)$
$\quad = 2x^3 - 4x^2 + 3x - 6.$

(b) $(x + y^2) + (x - y) = 2x + y^2 - y.$

(c) $\dfrac{5x^2 + 2x + 3}{x - 1} = 5x + 7 + \dfrac{10}{x - 1}.$

The result of Example 3(c) was obtained by first dividing $5x^2 + 2x + 3$ by $x - 1$, using the familiar long division algorithm. Since

$$5x^2 + 2x + 3 = (5x + 7)(x - 1) + 10,$$

then

$$\frac{5x^2 + 2x + 3}{x - 1} = 5x + 7 + \frac{10}{x - 1}.$$

EXERCISES A.6

1. Determine the domain of each function. Calculate $f(1)$ and $f(2)$ if these numbers are defined.

(a) $f(x) = 3x + 1$

• (b) $f(x) = \dfrac{x+1}{x-1}$

(c) $f(x) = \dfrac{2}{x}$

• (d) $f(x) = \dfrac{x-1}{2x+1}$

• (e) $f(x) = \dfrac{x+3}{x^2-4}$

(f) $f(x) = \dfrac{1}{x-2}$

2. Simplify the following algebraic expressions:

• (a) $(5x^2 + 3) - (2x^2 - 3x + 5)$

(b) $(2x - 1)(x + 3)$

• (c) $(x^2 - x + 5)(3x^3 - 4x + 5)$

(d) $(3x^2 - 2x + 1) + (3x - 2)$

• (e) $\dfrac{x^3 - 2x^2 + x - 1}{x + 4}$

• (f) $\dfrac{x^4 - 3x^3 + 2x^2 - x - 5}{x^2 + x - 2}$

3. Simplify to a form free of negative and zero exponents:

(a) $\left(\dfrac{x^{-1} + x^{-2}}{x^{-1} - x^{-2}}\right)^{-1}$

• (b) $(x^{1/2} + x^{-1/2})^2$

(c) $(x^{-1} + y^{-1})(x^{-1} - y^{-1})^{-1}$

(d) $(\sqrt{x} + \sqrt{y})^{-2}$

4. Let $f(x, y) = x^2 + y$. Calculate:

(a) $f(1, 2)$

• (b) $f(-1, -1)$

(c) $f(a, b)$

• (d) $f(2x, 3y)$

(e) $f(\sqrt{x}, y^2)$

(f) $f(x + 1, y - 1)$

A.7 EQUATIONS

Equation

An *equation* is formed by setting two functions equal to each other. Examples of equations are

$$2x + 7y = 5y - 4,$$
$$3x - 9 = 0,$$

and

$$(x - 2)^2 = 9.$$

Solution

A *solution* of an equation is a set of numbers that make the equation a true statement when substituted for the variables x, y, and so on. The central problem of elementary algebra is that of solving equations.

Example 1

(a) The equation $3x - 9 = 0$ has the unique solution $x = 3$.

(b) The equation $2x + 7y = 5y - 4$ has an infinite number of solutions. Two of them are $x = 1$, $y = -3$ and $x = 7$, $y = -9$. These two solutions can be represented by the ordered pairs $(1, -3)$ and $(7, -9)$.

Equivalent equations

Two equations are *equivalent* provided they have the same solution set. For example, the equations

$$3x - 9 = 0 \quad \text{and} \quad 3x = 9$$

are equivalent. The equations have the same solution, $x = 3$.

Operations for Transforming Equations The elementary methods of solving equations involve two operations that are used to change equations into simpler equivalent equations. We discuss the operations for equations in a single variable, x.

Operations for obtaining equivalent equations

Operation 1 If a constant or a multiple of x is *added* (or *subtracted*) to both sides of an equation then the new equation is equivalent to the original one.

Operation 2 If both sides of an equation are *multiplied* (or *divided*) by a nonzero constant, then the new equation is equivalent to the original one.

Example 2
(a) The equation

$$4x + 1 = 3x + 3$$

is equivalent to the equation

$$x = 2.$$

(Subtract $3x + 1$ from both sides of the original equation.)
(b) The equations

$$3x = 9$$

and

$$x = 3$$

are equivalent. (Divide both sides of the original equation by 3.)
(c) The equations

$$3x^2 - 2x = 0$$

and

$$3x - 2 = 0$$

are not equivalent. (The first equation was divided by x, which is not a constant.)

Example 3 Solve $5x - 7 = 2x + 3$.

Solution

$$\begin{aligned} 5x - 7 &= 2x + 3 \\ 3x - 7 &= 3 && \text{(subtracting } 2x\text{)} \\ 3x &= 10 && \text{(adding 7)} \\ x &= \frac{10}{3} && \text{(dividing by 3).} \end{aligned}$$

The only solution of the final equation is $x = 10/3$. Since all of the equations are equivalent then $x = 10/3$ is the only solution of the original equation.

EXERCISES A.7

1. Which of the following operations will change an equation in x into an equivalent equation?

   (a) Adding 3 to both sides of the equation.

- • (b) Multiplying both sides of the equation by $x^2 - 1$.
- (c) Multiplying both sides of the equation by 3.
- • (d) Subtracting $5x^2 + 7x - 1$ from both sides of the equation.
- (e) Multiplying both sides of the equation by 0.
- (f) Dividing both sides of the equation by x.

2. Solve each of the following equations by constructing a sequence of simpler equivalent equations:

- • (a) $7x = -8$
- • (b) $15x + 2 = 3x - 4$
- (c) $2x + 3 = 3x + 4 - x$
- (d) $-4x + 1 = 2x + 7$
- • (e) $3x + 5 = 3x + 5$
- (f) $2x + 5 = 7x + 5$

A.8 LINEAR EQUATIONS IN x

Linear equation
$ax = b$

Linear equations in x are those that can be written in the form

$$ax = b \qquad (a \neq 0).$$

The solution of this linear equation is obtained by dividing both sides of the equation by a. The unique solution is

$$x = \frac{b}{a}.$$

Example 1

(a) The solution of the equation

$$7x = 3$$

is

$$x = \frac{3}{7}$$

(b) The solution of the equation

$$5x - 2 = 7x + 3$$

is found in two steps:

$$5x - 2 = 7x + 3$$
$$-2x = 5$$
$$x = -\frac{5}{2}.$$

EXERCISES A.8

1. Solve the equations

- (a) $5x = 15$
- • (b) $-3x = 18$
- • (c) $x - 2 = 3x + 7$
- (d) $4x - 3 + x = 2x - 1$
- • (e) $3x - 2 = 5x + 5$
- (f) $2x + 1 = 7x + 1$
- • (g) $2x - x + 3 = 4 + x - 1$
- (h) $x + 3 = x + 5$

A.9 LINEAR INEQUALITIES IN x

The solutions of the linear inequalities

$$ax < b, \quad ax > b, \quad ax \leq b, \quad \text{and} \quad ax \geq b$$

can be obtained by a method similar to the one used to solve linear equations: *we divide by a*. The situation is more complicated, however, than with equations:

(a) If $a > 0$ we preserve the inequality when we divide by a. (This follows from O4 and the fact that dividing by a is equivalent to multiplying by $1/a$.)
(b) If $a < 0$ we reverse the inequality when we divide by a. (This follows from O5.)

The basic operations used in solving inequalities are similar to the ones used in solving equations:

Operations for obtaining equivalent inequalities

1. An equivalent inequality is obtained if the constant or a multiple of x is added to or subtracted from both sides.
2. An equivalent inequality is obtained if both sides are divided by a nonzero constant.

Example 1 Solve the linear inequality

$$-2x < 7.$$

Solution When we divide by -2, we reverse the inequality:

$$-2x < 7$$
$$x > \frac{7}{-2} = -\frac{7}{2}.$$

The solution set is $\{x|x > -\frac{7}{2}\}$. (See Figure A.7a.)

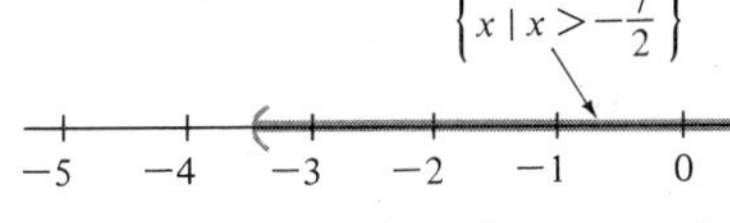

(a) Example 1. The solution set of $-2x < 7$ is $\{x|x > -\frac{7}{2}\}$.

$\left\{x \mid x < \frac{3}{4}\right\}$

−2 −1 0 1 2 3

(b) Example 2. The solution set of $3x - 2 > 7x - 5$ is $\{x|x < \frac{3}{4}\}$.

FIGURE A-7

Example 2 Solve $3x - 2 > 7x - 5$.

Solution

$$3x - 2 > 7x - 5$$
$$-4x - 2 > -5 \qquad \text{(subtracting } 7x\text{)}$$
$$-4x > -3 \qquad \text{(adding 2)}$$
$$x < \frac{-3}{-4} = \frac{3}{4} \qquad \text{(dividing by } -4\text{)}$$
$$x < \frac{3}{4}.$$

The solution set is $\{x|x < \frac{3}{4}\}$. (See Figure A.7b.)

EXERCISES A.9

1. Solve the inequalities

(a) $5x > 15$

• (b) $-3x \leqslant 18$

(c) $x - 2 < 3x + 7$
• (d) $4x - 3 + x < 2x - 1$
(e) $3x - 2 > 5x + 5$
(f) $2x + 1 \geqslant 7x + 1$
• (g) $2x - x + 3 > 4 + x - 1$
• (h) $x + 3 < x + 5$

A.10 INEQUALITIES INVOLVING ABSOLUTE VALUES

The inequality $|x - a| < b$

The following result is needed in Section 2.5: *The solution of the inequality*

$$|x - a| < b \qquad \text{(where } b > 0\text{)}$$

is

$$\{x | a - b < x < a + b\}.$$

(See Figure A.8.)

To establish this result we consider two cases.

Case I: $x - a \geqslant 0$. Then $|x - a| = x - a$. The inequality becomes

$$\begin{aligned} |x - a| &< b, \\ x - a &< b, \\ x &< a + b. \end{aligned}$$

Since $x \geqslant a$ throughout this case the solution set for Case I is

$$S_1 = \{x | a \leqslant x < a + b\}.$$

Case II: $x - a < 0$. Then $|x - a| = -(x - a) = -x + a$. The inequality becomes

$$\begin{aligned} |x - a| &< b, \\ -x + a &< b, \\ -x &< b - a, \\ x &> a - b. \end{aligned}$$

Since $x < a$ throughout this case the solution set for Case II is

$$S_2 = \{x | a - b < x < a\}.$$

Since either Case I or Case II can hold, then the solution set S for the original inequality is the union of S_1 and S_2, the two solutions sets for the individual cases:

$$\begin{aligned} S = S_1 \cup S_2 &= \{x | a \leqslant x < a + b\} \cup \{x | a - b < x < a\} \\ &= \{x | a - b < x < a + b\}. \end{aligned}$$

The inequality

$$|x - a| < b$$

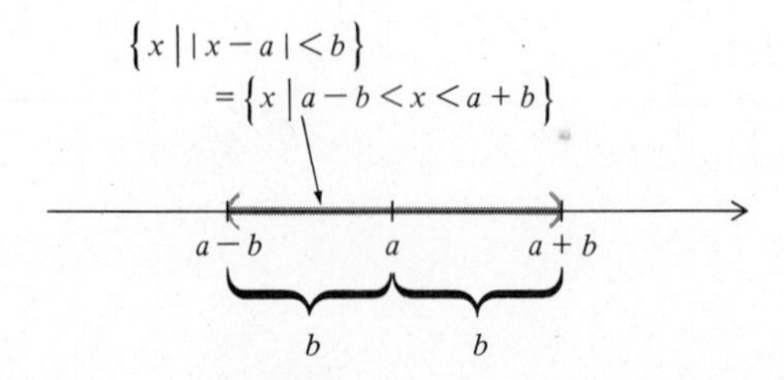

FIGURE A-8 The solution set of $|x - a| < b$ is $\{x | a - b < x < a + b\}$, all numbers x within b units of a.

can be interpreted geometrically as follows: Since

$$|x - a| = \text{the distance from } x \text{ to } a,$$

then the solution set of the inequality is

$$\begin{aligned} S &= \{x \mid \text{the distance from } x \text{ to } a \text{ is less than } b\} \\ &= \{x \mid x \text{ is between } a - b \text{ and } a + b\}. \end{aligned}$$

(See Figure A.8.)

EXERCISES A.10

1. Solve the following inequalities:

   • (a) $|x - 2| < 3$
   (b) $|x - 3| < 7$
   (c) $|x + 1| < 3$
   • (d) $|2x + 5| < 7$

2. Modify the argument in the text to solve the following inequalities:

   • (a) $|x + 2| > 1$
   (b) $|3x + 7| > 5$
   • (c) $|2x - 3| > 0$
   (d) $|4x - 1| < -1$

A.11 SYSTEMS OF LINEAR EQUATIONS

Equations such as

$$3x + 2y = 7$$

and

$$5x - 2z + 3t + w = z - 5y + 1$$

are called *linear equations*. The variables occur only to the first power and no products of variables occur.

System of equations

A system of linear equations is a set of two or more linear equations in one, two, or more variables. We group the equations in braces to indicate a system. For example,

$$\begin{cases} 2x - 3y = 7 \\ 5x + 2y = 20 \end{cases} \quad \text{and} \quad \begin{cases} x + y = 2 \\ x - y = 0 \end{cases}$$

are systems of linear equations.

Solution of a system of equations

By a *solution* of a system of equations we mean any common solution of all of the equations in the system.

Example 1 Consider the system

$$\begin{cases} x + y = 2 \\ y - z = 0. \end{cases}$$

(a) The numbers $x = 1$, $y = 1$, $z = 1$ constitute a solution of the system. These numbers form a solution of both equations in the system.

(b) The numbers $x = 2$, $y = 0$, $z = 3$ form a solution of the first equation of the system, but not of the second equation. Thus, $x = 2$, $y = 0$, $z = 3$ is not a solution of the system.

Equivalent Systems Two systems are *equivalent* if they have the same solution set. There are three operations that can be used to transform a system of equations into an equivalent system:

Operations for obtaining equivalent systems

Operation 1 Multiply an equation by a nonzero constant.
Operation 2 Interchange any two equations.
Operation 3 Add a nonzero multiple of one equation to a nonzero multiple of a different equation.

In applying operations 1 and 3 we assume that all equations remain unchanged except for the one equation we specifically change.

$\Leftrightarrow$

We will use the symbol "$\Leftrightarrow$" to indicate equivalent systems.

Example 2

(a)
$$\begin{cases} 2x - y = 3 \\ x + y = 4 \end{cases} \Leftrightarrow \begin{cases} 2x - y = 3 \\ 3x = 7. \end{cases}$$

(The first equation was added to the second.)

(b)
$$\begin{cases} 2x - y = 3 \\ 3x = 7 \end{cases} \Leftrightarrow \begin{cases} 2x - y = 3 \\ x = \dfrac{7}{3} \end{cases}.$$

(The second equation was divided by 3.)

Echelon Form A system of equations written in the "triangular" form

Echelon form

$$\begin{cases} a_1x + b_1y + c_1z = d_1 \\ a_2x + b_2y = d_2 \\ a_3x = d_3 \end{cases}$$

is said to be in *echelon form*. Such a system can be solved by first solving the third equation for x, then the second equation for y, then the first equation for z. Similar echelon forms can be written for systems in any number of variables.

Example 3 The solution of the system

$$\begin{cases} x - 2y + z = 3 \\ x + y = 5 \\ x = 1 \end{cases}$$

can be found in three steps:

$x = 1$ (from the third equation).
$y = 5 - x = 5 - 1 = 4$ (from the second equation).
$z = 3 - x + 2y = 3 - 1 + 8 = 10$ (from the first equation).

The solution is $x = 1$, $y = 4$, $z = 10$.

Reduction to Echelon Form A system of linear equations can be reduced to an equivalent system in echelon form by the three operations stated above. The following example illustrates the procedure.

Example 4 Solve the system

$$\begin{cases} x - 2y + 2z = -3 \\ 3x - y - z = 20 \\ 2x - y + 5z = 1 \end{cases}$$

by reduction to echelon form.

Example of reduction to echelon form

Solution

1. The first major step is to get an equivalent system in which only the first equation has a nonzero z-coefficient. To accomplish this we first need to get an equivalent system in which the first equation has a z-coefficient of ± 1. This can be done in several ways (by dividing the first equation by 2, for example.) To avoid fractions we interchange the first two equations.

$$\begin{cases} x - 2y + 2z = -3 \\ 3x - y - z = 20 \\ 2x - y + 5z = 1 \end{cases} \Leftrightarrow \begin{cases} 3x - y - z = 20 \\ x - 2y + 2z = -3 \\ 2x - y + 5z = 1. \end{cases} \quad \text{(Interchange first two equations.)}$$

We now add the appropriate multiples of the first equation to the other two equations that will yield equations with the z-coefficients equal to zero

$$\begin{cases} 3x - y - z = 20 \\ x - 2y + 2z = -3 \\ 2x - y + 5z = 1 \end{cases} \Leftrightarrow \begin{cases} 3x - y - z = 20 \\ 7x - 4y = 37 \\ 2x - y + 5z = 1. \end{cases} \quad \text{(Add twice the first equation to the second equation.)}$$

$$\Leftrightarrow \begin{cases} 3x - y - z = 20 \\ 7x - 4y = 37 \\ 17x - 6y = 101. \end{cases} \quad \text{(Add five times the first equation to the third equation.)}$$

2. We now wish to get rid of the y-coefficient in the third equation. A simple trick can help us to avoid fractions. Observe that if the coefficient of y in the third equation were a multiple of 4, then we could subtract a multiple of the second equation from it and obtain a system in echelon form. In order to make this y-coefficient a multiple of 4, we first multiply the third equation by 2, then reduce to echelon form:

$$\begin{cases} 3x - y - z = 20 \\ 7x - 4y = 37 \\ 17x - 6y = 101 \end{cases} \Leftrightarrow \begin{cases} 3x - y - z = 20 \\ 7x - 4y = 37 \\ 34x - 12y = 202. \end{cases} \quad \text{(Multiply third equation by 2.)}$$

$$\Leftrightarrow \begin{cases} 3x - y - z = 20 \\ 7x - 4y = 37 \\ 13x = 91. \end{cases} \quad \text{(Subtract three times the second equation from the third equation.)}$$

The system has been reduced to echelon form. The solution ($x = 7$, $y = 3$, $z = -2$) can now be found by repeated substitution as in Example 3.

Inconsistent system

Inconsistent Systems When we reduce a system to echelon form we may obtain an "impossible equation" such as

$$0 = 3.$$

If this occurs, then the new system has no solution. Since the original system is equivalent to the reduced system, then it also has no solution. A system of equations which has no solution is said to be *inconsistent*.

Example 5 The systems

$$\begin{cases} 2x + y = 5 \\ -2x - y = 4 \end{cases} \quad \text{and} \quad \begin{cases} 2x + y = 5 \\ 0 = 9 \end{cases}$$

are equivalent. The second system is inconsistent. Therefore the first system is inconsistent.

Dependent system

Parameter

Dependent Systems If, when reduced to echelon form, a consistent system of equations contains more variables than nontrivial equations, then it has an infinity of solutions. Such a system is said to be *dependent*. The solutions can be obtained as functions of auxiliary variables called *parameters*. The number of parameters is equal to the difference in the number of variables and the number of nontrivial equations.

The procedure for solving dependent systems is illustrated in the following example.

Example 6 Solve the system

$$\begin{cases} x - y + 3z - t = 7 \\ 2x + y + z \qquad = 5 \\ 0 = 0 \\ 0 = 0. \end{cases}$$

Solution The system has two nontrivial equations. We solve those equations for z and t in terms of x and y:

$$\begin{cases} 3z - t = 7 - x + y \\ z \qquad = 5 - 2x - y. \end{cases}$$

Let p and q be auxiliary variables (parameters). Let

$$x = p, \qquad y = q.$$

Then

$$z = 5 - 2p - q \qquad \text{(from the second equation).}$$

If we substitute this value of z into the first equation, we obtain

$$\begin{aligned} -t &= 7 - x + y - 3z \\ &= 7 - p + q - 3 \cdot (5 - 2p - q) \\ &= 7 - p + q - 15 + 6p + 3q \\ &= -8 + 5p + 4q. \end{aligned}$$

so that

$$t = 8 - 5p - 4q$$

The solutions are given by

$$x = p, \quad y = q, \quad z = 5 - 2p - q, \quad t = 8 - 5p - 4q.$$

The parameters p and q can be any real numbers. Several solutions corresponding to specific values of p and q are shown in Table A.1.

TABLE A-1

p	q	x	y	z	t
0	0	0	0	5	8
1	2	1	2	1	-5
2	5	2	5	-4	-22
-3	π	-3	π	$11-\pi$	$23-4\pi$

EXERCISES A.11

1. Solve the following systems of equations by reducing them to echelon form. If a system has an infinite number of solutions, determine a general form for the solutions in terms of a parameter k.

• (a) $\begin{cases} x-y+z=0 \\ 2x-y+2z=1 \\ 3x+y+z=2 \end{cases}$

(b) $\begin{cases} x+2y-z=1 \\ 3x+2y-3z=-1 \\ 2x+y-2z=-1 \end{cases}$

• (c) $\begin{cases} x+y+z=6 \\ 2x+2y-3z=-3 \\ x+y-2z=-3 \end{cases}$

(d) $\begin{cases} 2x-y+z=-5 \\ x-2y+2z=-5 \\ x-3y+2z=-7 \end{cases}$

(e) $\begin{cases} x-y+z=0 \\ x+y+2z=7 \\ x+y-z=4 \\ 2x-y+z=2 \end{cases}$

• (f) $\begin{cases} 3x+2z-w=0 \\ 2x-y+z-2w=0 \\ 5x+y+z+2w=0 \end{cases}$

A.12 COMPLEX NUMBERS

Complex number

Imaginary unit i

Complex numbers are numbers that can be written in the form $a+bi$, where a and b are real numbers and i is the *imaginary unit* with the property that $i^2=-1$.

Example 1

(a) $3+2i$ is a complex number.
(b) $i(3-2i)=3i-2i^2=3i-2(-1)=2+3i$ is a complex number.
(c) $-5i=0+(-5)i$ is a complex number.
(d) $2=2+0\cdot i$ is a complex number.

Complex numbers obey the same arithmetical rules as do real numbers. In calculations involving products we replace i^2 with -1 whenever it occurs.

Example 2

$$\begin{aligned}(2+i)(4+5i) &= 2(4+5i)+i(4+5i)\\ &= 8+10i+4i+5i^2\\ &= 8+10i+4i-5\\ &= 3+14i.\end{aligned}$$

Principal square root of negative number

Square Roots If $b \geqslant 0$ the principal square root of $-b^2$ is defined to be bi. Thus, for example,

$$\sqrt{-25} = 5i,$$
$$\sqrt{-17} = i\sqrt{17},$$

and so on.

Conjugate

Conjugates The *conjugate* of the complex number $a + bi$ (where a and b are real) is the number $a - bi$. For example, the conjugate of $2 + 7i$ is $2 - 7i$, the conjugate of $-5i$ is $5i$, the conjugate of 3 is 3. Observe that the product of a nonzero complex number $a + bi$ and its conjugate is $a^2 + b^2$, which is a positive real number.

To simplify a fraction, such as $\dfrac{2+5i}{3-2i}$, we multiply numerator and denominator by the conjugate of the denominator and simplify the resulting products.

Example 3

$$\frac{2+5i}{3-2i} = \frac{2+5i}{3-2i} \cdot \frac{3+2i}{3+2i} = \frac{(2+5i)(3+2i)}{(3-2i)(3+2i)} = \frac{-4+19i}{9+4}$$
$$= -\frac{4}{13} + \frac{19}{13}i$$

EXERCISES A.12

1. Simplify the following complex numbers:

(a) $(8 - i)(3 + 2i)$

(b) $(3 - \sqrt{-1})(3 + \sqrt{-4})$

• (c) $\dfrac{5 + \sqrt{-4}}{6 + \sqrt{-49}}$

(d) $(8 + i)/\sqrt{-1}$

• (e) $\dfrac{4 - 3i}{4 + 3i}$

(f) $\dfrac{12 + i}{4 - i}$

A.13 QUADRATIC EQUATIONS. FACTORIZATIONS

Quadratic equation
$ax^2 + bx + c = 0$

A *quadratic equation* is one that can be written in the form

$$ax^2 + bx + c = 0 \qquad (\text{where } a \neq 0).$$

If the left-hand side of the equation can be factored as a product of linear terms, then the solutions can be obtained immediately. The method is discussed in the following example.

Example 1 Solve the quadratic equation

$$2x^2 - 3x - 2 = 0.$$

Solution The left-hand side of the equation can be factored:

$$2x^2 - 3x - 2 = 0$$
$$(2x + 1)(x - 2) = 0.$$

One of the basic properties of numbers, real and complex, is that a product can be zero only in case at least one of the factors is zero. Thus, if x is a solution of the original equation, then

$$2x + 1 = 0 \quad \text{or} \quad x - 2 = 0$$
$$x = -\frac{1}{2} \quad \text{or} \quad x = 2.$$

If we substitute these values into the original equation we find that they are indeed solutions. Thus the solutions are

$$x = -\frac{1}{2} \quad \text{and} \quad x = 2.$$

Difference of squares

Differences of Squares One important quadratic equation can be factored and solved without difficulty. The equation

$$x^2 = a^2$$

can be solved as follows:

$$\begin{aligned} x^2 - a^2 &= 0, \\ (x - a)(x + a) &= 0, \\ x - a = 0 \quad \text{or} \quad x + a &= 0, \\ x = a \quad \text{or} \quad x &= -a. \end{aligned}$$

The solutions are $x = -a$ and $x = +a$. We frequently combine these two solutions into one form, written as

$$x = \pm a$$

(read "plus-or-minus a").

Example 2

(a) The solutions of

$$x^2 - 25 = 0 \quad \text{are} \quad x = \pm 5.$$

(b) The solutions of

$$x^2 = 17 \quad \text{are} \quad x = \pm\sqrt{17}.$$

(c) The solutions of

$$x^2 = -16 \quad \text{are} \quad x = \pm 4i.$$

(Write -16 as $(4i)^2$.)

(d) The solutions of $(2x - 1)^2 = 7$ can be obtained after making the substitution $y = 2x - 1$.

$$\begin{aligned} (2x - 1)^2 &= 7, \\ y^2 &= 7, \\ y &= \pm\sqrt{7}. \end{aligned}$$

Therefore

$$\begin{aligned} 2x - 1 &= \pm\sqrt{7}, \\ 2x &= 1 \pm \sqrt{7}, \\ x &= \frac{1 \pm \sqrt{7}}{2}. \end{aligned}$$

Factorizations The standard techniques of factorization involve a great amount of trial-and-error work. The basic problem is to factor

$$ax^2 + bx + c$$

as a product of linear terms:

$$ax^2 + bx + c = (dx + e)(fx + g).$$

If we multiply the terms on the right we obtain

$$ax^2 + bx + c = dfx^2 + (dg + ef)x + eg.$$

Thus we must have

$$a = df,$$
$$b = dg + ef,$$

and

$$c = eg.$$

Observe that d and f are factors of a, while e and g are factors of c. The usual methods of factorization involve finding the integer factors of a and c, writing down the possible combinations of factors that could conceivably yield $ax^2 + bx + c$ as a product, then checking to see which, if any, of the products are actually equal to $ax^2 + bx + c$.

Example 3 Solve $2x^2 + x - 28 = 0$.

Solution If $2x^2 + x - 28$ can be factored as

$$2x^2 + x - 28 = (dx + e)(fx + g),$$

then $df = 2$ and $eg = -28$. Thus d and f can be ± 1 or ± 2, while e and g can be $\pm 1, \pm 2, \pm 4, \pm 7, \pm 14$, or ± 28. A few of the possible factorizations are

$$(x - 1)(2x + 28), \qquad (x + 1)(2x - 28),$$
$$(x - 2)(2x + 14), \qquad (x + 2)(2x - 14),$$
$$(x - 4)(2x + 7), \qquad (x + 4)(2x - 7),$$

and so on. When we multiply these pairs of possible factors together we find that

$$(x + 4)(2x - 7) = 2x^2 + x - 28.$$

Returning now to the original equation we have

$$2x^2 + x - 28 = 0,$$
$$(x + 4)(2x - 7) = 0,$$
$$x = -4 \quad \text{or} \quad x = \tfrac{7}{2}.$$

The solutions of the equation are $x = -4$ and $x = \frac{7}{2}$.

The method illustrated in Example 3 can be used to find rational solutions of quadratic equations which have integer coefficients. If such a quadratic equation cannot be factored by this method, then its solutions are irrational or complex numbers.

EXERCISES A.13

1. Solve the quadratic equations by factoring. If an equation has no rational solutions, state that fact.

(a) $x^2 - 4x + 3 = 0$
• (b) $x^2 + 7x + 12 = 0$
(c) $x^2 - x - 12 = 0$
• (d) $x^2 - 4x - 10 = 0$
(e) $y^2 + 2y = 13$
• (f) $z^2 - 28 = 3z$
(g) $4x^2 = 4x + 3$
• (h) $7x^2 + 48x = 7$

2. Solve by rewriting the equation as $x^2 = a^2$ or an equivalent form. (*Hint:* In (e) to (h) make a substitution.)

(a) $x^2 - 16 = 0$
• (b) $4x^2 = 25$
(c) $y^2 - 13 = 0$
(d) $z^2 = 27$
• (e) $(x - 1)^2 - 9 = 0$
(f) $\left(\frac{y + 2}{5}\right)^2 = 3$
• (g) $(2x + 3)^2 = 11$
(h) $(5z - 8)^2 - 15 = 0$

A.14 "COMPLETING THE SQUARE." THE QUADRATIC FORMULA

Completing the square

The method of "completing the square" rests on the fact that

$$x^2 + 2ax + a^2 = (x + a)^2.$$

Thus, if we are given

$$x^2 + 2ax = m,$$

we can make the left-hand side a perfect square by adding a^2 to both sides of the equation:

$$\begin{aligned} x^2 + 2ax \quad &= m \\ x^2 + 2ax + a^2 &= m + a^2 \\ (x + a)^2 &= m + a^2. \end{aligned}$$

Observe that we add the square of one half the coefficient of the x-term to both sides of the equation.

Example 1 Solve $x^2 + 8x = 2$.

Solution We complete the square by adding $(\frac{8}{2})^2 = 16$ to both sides:

$$\begin{aligned} x^2 + 8x \quad &= 2 \\ x^2 + 8x + 16 &= 2 + 16 = 18 \\ (x + 4)^2 &= (\sqrt{18})^2 \\ x + 4 &= \pm\sqrt{18} \\ x &= -4 \pm \sqrt{18}. \end{aligned}$$

The solutions are $x = -4 + \sqrt{18}$ and $x = -4 - \sqrt{18}$.

The method of "completing the square" requires that the coefficient of x^2 be unity. If this is not the case we first divide by that coefficient.

Example 2 Solve $2x^2 + 6x + 7 = 0$.

Solution

$$2x^2 + 6x + 7 = 0$$

$$
\begin{aligned}
x^2 + 3x \qquad &= -\frac{7}{2} \\
x^2 + 3x + \frac{9}{4} &= -\frac{7}{2} + \frac{9}{4} = \frac{-14+9}{4} = -\frac{5}{4} \\
\left(x + \frac{3}{2}\right)^2 &= \left(\frac{i\sqrt{5}}{2}\right)^2 \\
x + \frac{3}{2} &= \frac{\pm i\sqrt{5}}{2} \\
x &= \frac{-3 \pm i\sqrt{5}}{2}.
\end{aligned}
$$

The Quadratic Formula If we divide the equation

$$ax^2 + bx + c = 0 \qquad (a \neq 0)$$

by a, complete the square and solve for x, we obtain:

$$
\begin{aligned}
x^2 + \frac{b}{a}x \qquad &= -\frac{c}{a} \\
x^2 + \frac{b}{a}x + \frac{b^2}{4a^2} &= \frac{b^2}{4a^2} - \frac{c}{a} = \frac{b^2 - 4ac}{4a^2} \\
\left(x + \frac{b}{2a}\right)^2 &= \left(\frac{\sqrt{b^2 - 4ac}}{2a}\right)^2 \\
x + \frac{b}{2a} &= \pm \frac{\sqrt{b^2 - 4ac}}{2a}.
\end{aligned}
$$

The quadratic formula

$$x = \frac{-b \pm \sqrt{b^2 - 4ac}}{2a} \qquad \textit{(the quadratic formula).}$$

Example 3 Solve $2x^2 - 3x + 7 = 0$ by the quadratic formula.

Solution $a = 2$, $b = -3$, $c = 7$. The solutions are

$$x = \frac{-b \pm \sqrt{b^2 - 4ac}}{2a} = \frac{3 \pm \sqrt{9 - 56}}{4} = \frac{3 \pm i\sqrt{47}}{4}.$$

EXERCISES A.14

1. Solve the following equations by "completing the square":

   (a) $x^2 + 4x + 1 = 0$
   • (b) $x^2 + 2x = -3$
   (c) $x^2 - 3x = 1$
   • (d) $5x^2 + 2x = 3$
   (e) $3x^2 + 7x + 3 = 0$
   • (f) $3x^2 - 4x + 7 = -2$

2. Solve the following equations by use of the quadratic formula:

   (a) $3x^2 - 4x - 2 = 0$
   • (b) $2x^2 - 3x + 4 = 0$
   (c) $x^2 = 4(x - 1)$
   • (d) $3x^2 + 2 = 5x$
   (e) $x(x + 1) = -1$
   • (f) $2x^2 + x = 2$
   (g) $4x^2 + 5x + 2 = 0$
   • (h) $2x^2 + 3x + 3 = 0$

A.15 THE COORDINATE PLANE

Axes

We can associate ordered pairs of real numbers with points in a plane by the following process. We construct a horizontal line and a vertical line crossing at a point O. We mark number scales on the two lines, locating the positive numbers to the right of O on the horizontal line and above O on the vertical line. The horizontal line is called the *x-axis*. The vertical line is the *y-axis*. As in the case of a single line, the point O is called the *origin*.

Let P be a point in the plane. A vertical line through P will intersect the *x-axis* at a point that represents a real number, say a. Similarly a horizontal line through P intersects the *y-axis* at a point b. Observe that P uniquely determines the two real numbers a and b, and, conversely, the two numbers uniquely determine P. Thus, we may label the point P with the ordered pair of real numbers (a, b). The number a is called the *x-coordinate* (or *abscissa*) of P and b is called the *y-coordinate* (or *ordinate*). We write $P(a, b)$ [Figure A.9a].

Coordinates

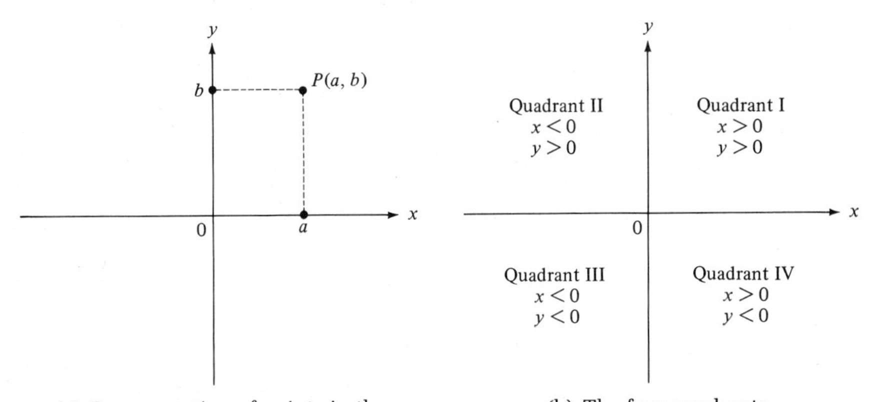

(a) Representation of points in the xy-plane by coordinates

(b) The four quadrants

FIGURE A-9 The coordinate plane.

Observe that the point labeled a on the x-axis should now be relabeled $(a, 0)$ and the point b on the y-axis should be relabeled $(0, b)$. We shall follow this scheme in subsequent discussion unless we obviously are referring to a point located on a particular axis.

The coordinate axes divide the plane into four quadrants. The portion containing the points with positive x- and y-coordinates is called *quadrant I*. If we move around the plane in a counterclockwise direction, we pass through *quadrant I, quadrant II, quadrant III,* and *quadrant IV* in order (Figure A.9b).

Quadrants

A.16 GRAPHS OF FUNCTIONS AND EQUATIONS

Graph of function

The *graph of a function* f is the set of all points $P(x, y)$ such that $y = f(x)$:

$$\begin{aligned} \text{Graph of } f &= \{P(x, y) | y = f(x)\} \\ &= \{P(x, f(x)) | x \text{ is in the domain of } f\}. \end{aligned}$$

In most cases a graph consists of one or more curves or lines.

Example 1

(a) The graph of $f(x) = 2x - 1$ is

$$\{P(x, y) | y = 2x - 1\},$$

the line shown in Figure A.10a.

(b) The graph of $g(x) = x^2$ is

$$\{P(x, y) | y = x^2\},$$

the curve (an example of a *parabola*) shown in Figure A.10b.

(c) The graph of the function with domain $\{x | -1 \leq x \leq 1\}$ and law of correspondence

$$h(x) = 1$$

is the horizontal line segment shown in Figure A.10c.

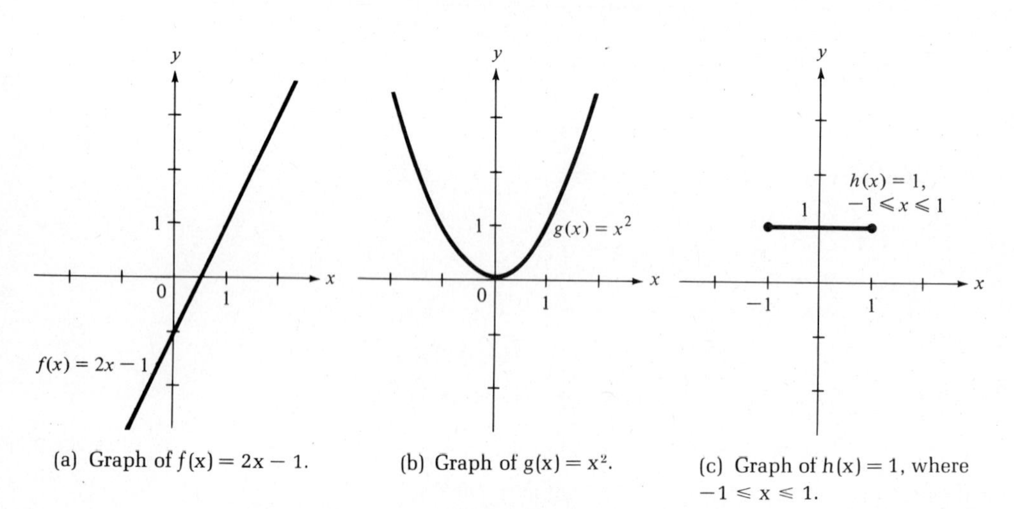

(a) Graph of $f(x) = 2x - 1$.

(b) Graph of $g(x) = x^2$.

(c) Graph of $h(x) = 1$, where $-1 \leq x \leq 1$.

FIGURE A-10 Example 1.

Graph of equation

The *graph of an equation* relating x and y is defined similarly to the graph of a function. The graph is the set of all points $P(x, y)$ such that the pair (x, y) satisfy the equation.

Example 2 The graph of the equation

$$x^2 - y^2 = 1$$

is

$$\{P(x, y) | x^2 - y^2 = 1\}.$$

This graph, shown in Figure A.11, is an example of a *hyperbola*. It consists of two distinct branches.

A crude approximation to the graph of a function or equation can be constructed by plotting a large number of points on the graph and connecting them with one or more smooth curves. More sophisticated methods of graphing are based on the use of the calculus. These are discussed in Section 3.4.

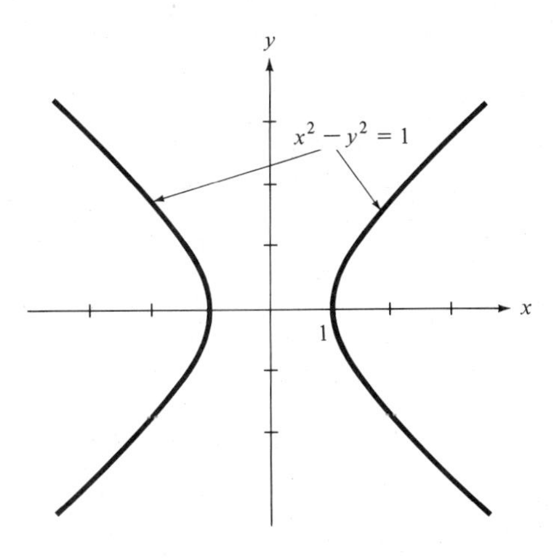

FIGURE A-11 Graph of the equation $x^2 - y^2 = 1$.

EXERCISES A.16

1. Plot several points on the graphs of the following functions and equations. Sketch the graphs.

(a) $f(x) = 3x + 1$
(b) $y = 2x + 3$
(c) $y = -2x^2$ (parabola)
(d) $g(x) = 1/x$ (hyperbola)
(e) $x^2 + y^2 = 1$ (circle)
(f) $-x + 2y = 3$
(g) $h(x) = x^2 + 1$
(h) $g(x) = -x + 2,\ 1 \leq x \leq 3$.

A.17 GRAPHS OF POLYNOMIALS. POWER AND ROOT FUNCTIONS

Functions of form

(1) $$f(x) = a_n x^n + a_{n-1} x^{n-1} + \cdot \cdot \cdot + a_2 x^2 + a_1 x + a_0,$$

where the exponents are nonnegative integers, are called *polynomial functions in x*.

Polynomial function

The *degree* of the polynomial (1) is the largest integer n such that the coefficient a_n is not equal to zero. A first-degree polynomial is said to be *linear*, a second-degree polynomial is *quadratic*, a third degree is *cubic*, a fourth degree is *quartic*, and so on. If all of the coefficients are zero, the polynomial does not have a degree.

Degree

Example 1

(a) $f(x) = 5x^2 - 2x + 7$ is quadratic (degree 2).
(b) $g(x) = 3x^3 - 7x^2 + 2x$ is cubic (degree 3).
(c) $h(x) = -x + \sqrt{17}$ is linear (degree 1).
(d) $j(x) = -2x^2 + 3x^5 - 4x + 1$ is quintic (degree 5).
(e) $k(x) = 5$ is a constant polynomial (degree 0).
(f) $l(x) = 0$ has no degree.

The graph of a polynomial is a smooth curve which may have several "turning points." (See Figure A.12.) The number of "turning points," if any, is less than the degree of the polynomial. (One of the problems that we consider in Chapter 3 is the location of the "turning points" of a polynomial.)

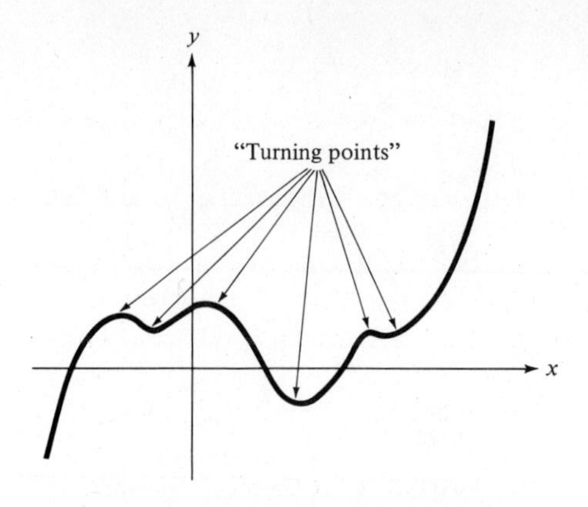

FIGURE A-12 Graph of a typical polynomial.

Power function

Power Functions The polynomials $f(x) = x^2, x^3, x^4, \ldots$, called *power functions*, are comparatively easy to graph. (See Figure A.13.) Observe that the graphs of power functions of even degree are similar in form to the graph of $f(x) = x^2$, whereas the graphs of power functions of odd degree are similar to the graph of $f(x) = x^3$.

Root function

Root Functions The graphs of $y = x^{1/2}, x^{1/3}, x^{1/4}, x^{1/5}, x^{1/6}$, and $x^{1/7}$ are shown in Figure A.14. Observe that the graph of $y = x^{1/3}$ is identical to the graph of $y = x^3$ except that it is "flipped" over the 45° line. This is illustrated in Figure A.15a. This result follows from the fact that

$$y = x^{1/3} \qquad \text{if and only if} \qquad x = y^3.$$

Thus the power function $x = y^3$ has a graph identical to the graph of $y = x^3$, except that the roles of x and y have been reversed. It can be shown that this is equivalent to "flipping" the graph over the 45° line.

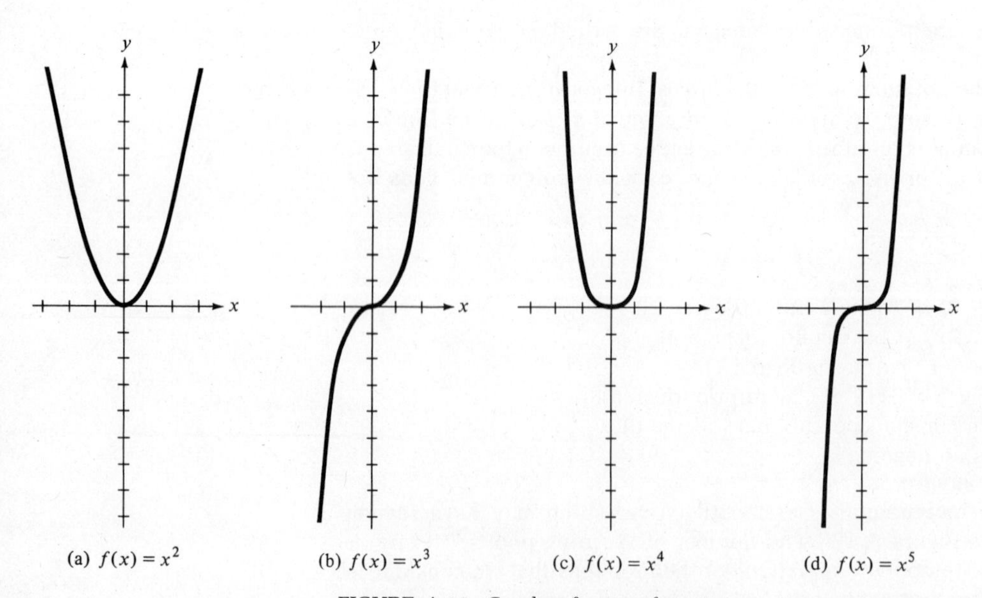

(a) $f(x) = x^2$ (b) $f(x) = x^3$ (c) $f(x) = x^4$ (d) $f(x) = x^5$

FIGURE A-13 Graphs of power functions.

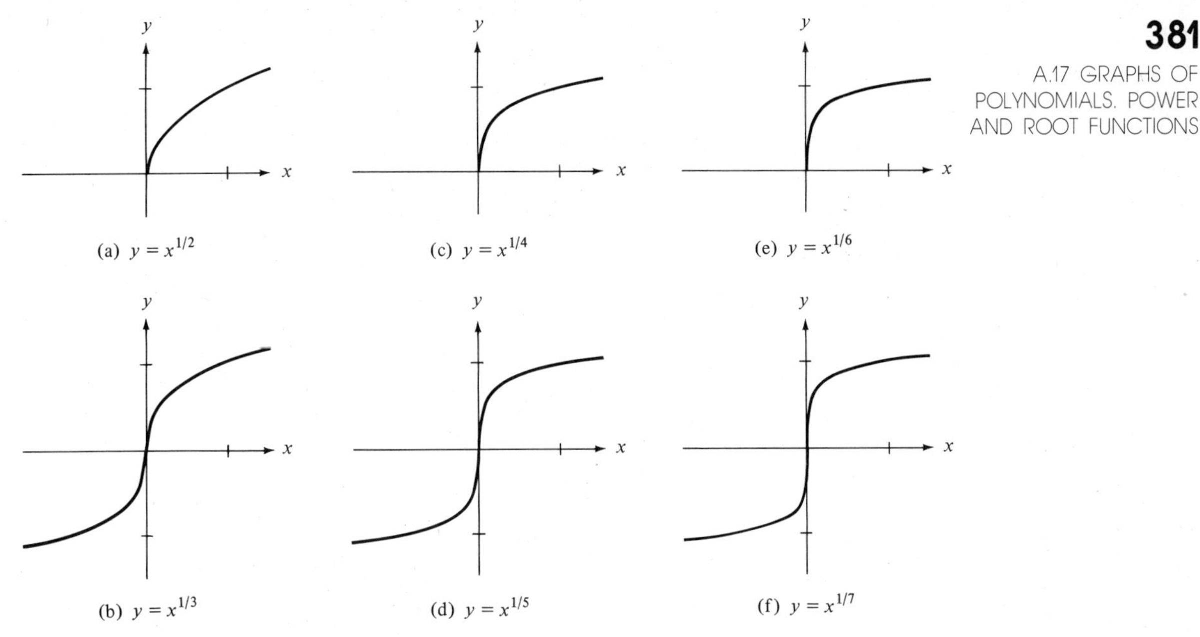

FIGURE A-14 Graphs of root functions.

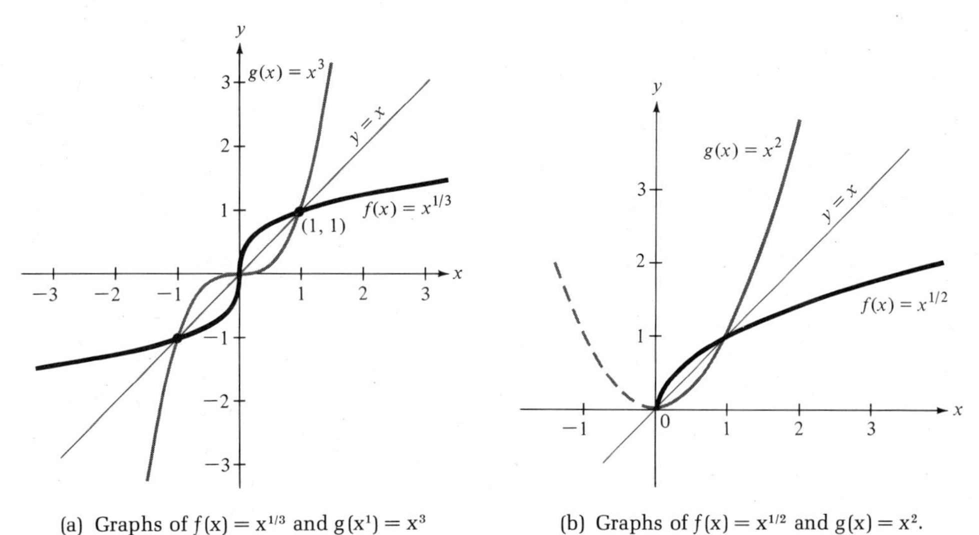

(a) Graphs of $f(x) = x^{1/3}$ and $g(x^1) = x^3$

(b) Graphs of $f(x) = x^{1/2}$ and $g(x) = x^2$.

FIGURE A-15 Root and power functions.

Similarly, the graphs of $y = x^{1/5}$, $y = x^{1/7}$, and so on, can be obtained by "flipping" the corresponding graphs of $y = x^5$, $y = x^7$, and so on, over the 45° line.

The graph of $y = x^{1/2}$ is slightly more difficult to describe. Recall that $x^{1/2} = \sqrt{x}$ is only defined as a real number when x and y are both nonnegative.

Thus

$$y = x^{1/2} \quad \text{if and only if} \quad x = y^2, \quad \text{where } y \geqslant 0.$$

Consequently, graphing $y = x^{1/2}$ is equivalent to graphing $x = y^2$, for $y \geqslant 0$. This graph is identical to the graph of $y = x^2$, for $x \geqslant 0$, except that the roles of x and y have been interchanged. It follows that the graph of $y = x^{1/2}$ can be obtained by "flipping" the quadrant I portion of the graph of $y = x^2$ over the 45° line. (See Figure A.15b.)

Similarly, the graphs of $y = x^{1/4}$, $y = x^{1/6}$, and so on, can be obtained by "flipping" the quadrant I portions of the graphs of $y = x^4$, $y = x^6$, and so on, over the 45° line.

EXERCISES A.17

1. Identify each of the following polynomials by type (linear, quadratic, cubic, and so on). What is the degree of each polynomial?

   (a) $2x^2 - 5x + 7$
   (b) $2 - 3x + x^3$
   (c) $3x^3 - 5x^2 + 2x - 3x^3$
   (d) $x^5 - x^2 + 1$

2. Make cardful drawings of each of the following power and root functions.

   (a) $f(x) = x^2$
   (b) $f(x) = x^3$
   (c) $f(x) = x^4$
   (d) $f(x) = x^5$
   (e) $g(x) = x^{1/2}$
   (f) $g(x) = x^{1/3}$
   (g) $g(x) = x^{1/4}$
   (h) $g(x) = x^{1/5}$

3. Plot several points on the graphs of the following polynomials. Sketch the graphs.

   (a) $f(x) = 3x^2 - x + 2$
   (b) $f(x) = x^3 - x$
   (c) $f(x) = x^3 - 3x^2 + x - 1$
   (d) $f(x) = x^4 - x^2 + 1$

TABLES

TABLE I Powers, roots, reciprocals 1–50

N	N^2	$\sqrt{N}$	$\sqrt{10N}$	N^3	$\sqrt[3]{N}$	$\sqrt[3]{10N}$	$\sqrt[3]{100N}$	$1000/N$
1	1	1.00 000	3.16 228	1	1.00 000	2.15 443	4.64 159	1000.00
2	4	1.41 421	4.47 214	8	1.25 992	2.71 442	5.84 804	500.00 0
3	9	1.73 205	5.47 723	27	1.44 225	3.10 723	6.69 433	333.33 3
4	16	2.00 000	6.32 456	64	1.58 740	3.41 995	7.36 806	250.00 0
5	25	2.23 607	7.07 107	125	1.70 998	3.68 403	7.93 701	200.00 0
6	36	2.44 949	7.74 597	216	1.81 712	3.91 487	8.43 433	166.66 7
7	49	2.64 575	8.36 660	343	1.91 293	4.12 129	8.87 904	142.85 7
8	64	2.82 843	8.94 427	512	2.00 000	4.30 887	9.28 318	125.00 0
9	81	3.00 000	9.48 683	729	2.08 008	4.48 140	9.65 489	111.11 1
10	100	3.16 228	10.00 00	1 000	2.15 443	4.64 159	10.00 00	100.00 0
11	121	3.31 662	10.48 81	1 331	2.22 398	4.79 142	10.32 28	90.90 91
12	144	3.46 410	10.95 45	1 728	2.28 943	4.93 242	10.62 66	83.33 33
13	169	3.60 555	11.40 18	2 197	2.35 133	5.06 580	10.91 39	76.92 31
14	196	3.74 166	11.83 22	2 744	2.41 014	5.19 249	11.18 69	71.42 86
15	225	3.87 298	12.24 74	3 375	2.46 621	5.31 329	11.44 71	66.66 67
16	256	4.00 000	12.64 91	4 096	2.51 984	5.42 884	11.69 61	62.50 00
17	289	4.12 311	13.03 84	4 913	2.57 128	5.53 966	11.93 48	58.82 35
18	324	4.24 264	13.41 64	5 832	2.62 074	5.64 622	12.16 44	55.55 56
19	361	4.35 890	13.78 40	6 859	2.66 840	5.74 890	12.38 56	52.63 16
20	400	4.47 214	14.14 21	8 000	2.71 442	5.84 804	12.59 92	50.00 00
21	441	4.58 258	14.49 14	9 261	2.75 892	5.94 392	12.80 58	47.61 90
22	484	4.69 042	14.83 24	10 648	2.80 204	6.03 681	13.00 59	45.45 45
23	529	4.79 583	15.16 58	12 167	2.84 387	6.12 693	13.20 01	43.47 83
24	576	4.89 898	15.49 19	13 824	2.88 450	6.21 446	13.38 87	41.66 67
25	625	5.00 000	15.81 14	15 625	2.92 402	6.29 961	13.57 21	40.00 00
26	676	5.09 902	16.12 45	17 576	2.96 250	6.38 250	13.75 07	38.46 15
27	729	5.19 615	16.43 17	19 683	3.00 000	6.46 330	13.92 48	37.03 70
28	784	5.29 150	16.73 32	21 952	3.03 659	6.54 213	14.09 46	35.71 43
29	841	5.38 516	17.02 94	24 389	3.07 232	6.61 911	14.26 04	34.48 28
30	900	5.47 723	17.32 05	27 000	3.10 723	6.69 433	14.42 25	33.33 33
31	961	5.56 776	17.60 68	29 791	3.14 138	6.76 790	14.58 10	32.25 81
32	1 024	5.65 685	17.88 85	32 768	3.17 480	6.83 990	14.73 61	31.25 00
33	1 089	5.74 456	18.16 59	35 937	3.20 753	6.91 042	14.88 81	30.30 30
34	1 156	5.83 095	18.43 91	39 304	3.23 961	6.97 953	15.03 69	29.41 18
35	1 225	5.91 608	18.70 83	42 875	3.27 107	7.04 730	15.18 29	28.57 14
36	1 296	6.00 000	18.97 37	46 656	3.30 193	7.11 379	15.32 62	27.77 78
37	1 369	6.08 276	19.23 54	50 653	3.33 222	7.17 905	15.46 68	27.02 70
38	1 444	6.16 441	19.49 36	54 872	3.36 198	7.24 316	15.60 49	26.31 58
39	1 521	6.24 500	19.74 84	59 319	3.39 121	7.30 614	15.74 06	25.64 10
40	1 600	6.32 456	20.00 00	64 000	3.41 995	7.36 806	15.87 40	25.00 00
41	1 681	6.40 312	20.24 85	68 921	3.44 822	7.42 896	16.00 52	24.39 02
42	1 764	6.48 074	20.49 39	74 088	3.47 603	7.48 887	16.13 43	23.80 95
43	1 849	6.55 744	20.73 64	79 507	3.50 340	7.54 784	16.26 13	23.25 58
44	1 936	6.63 325	20.97 62	85 184	3.53 035	7.60 590	16.38 64	22.72 73
45	2 025	6.70 820	21.21 32	91 125	3.55 689	7.66 309	16.50 96	22.22 22
46	2 116	6.78 233	21.44 76	97 336	3.58 305	7.71 944	16.63 10	21.73 91
47	2 209	6.85 565	21.67 95	103 823	3.60 883	7.77 498	16.75 07	21.27 66
48	2 304	6.92 820	21.90 89	110 592	3.63 424	7.82 974	16.86 87	20.83 33
49	2 401	7.00 000	22.13 59	117 649	3.65 931	7.88 374	16.98 50	20.40 82
50	2 500	7.07 107	22.36 07	125 000	3.68 403	7.93 701	17.09 98	20.00 00

These tables have been adapted from the *Rinehart Mathematical Tables, Formulas and Curves*, H. D. Larsen, Ed. Holt, Rinehart and Winston, 1953.

TABLE I Powers, roots, reciprocals 51–100

N	N^2	$\sqrt{N}$	$\sqrt{10N}$	N^3	$\sqrt[3]{N}$	$\sqrt[3]{10N}$	$\sqrt[3]{100N}$	$1000/N$
51	2 601	7.14 143	22.58 32	132 651	3.70 843	7.98 957	17.21 30	19.60 78
52	2 704	7.21 110	22.80 35	140 608	3.73 251	8.04 145	17.32 48	19.23 08
53	2 809	7.28 011	23.02 17	148 877	3.75 629	8.09 267	17.43 51	18.86 79
54	2 916	7.34 847	23.23 79	157 464	3.77 976	8.14 325	17.54 41	18.51 85
55	3 025	7.41 620	23.45 21	166 375	3.80 295	8.19 321	17.65 17	18.18 18
56	3 136	7.48 331	23.66 43	175 616	3.82 586	8.24 257	17.75 81	17.85 71
57	3 249	7.54 983	23.87 47	185 193	3.84 850	8.29 134	17.86 32	17.54 39
58	3 364	7.61 577	24.08 32	195 112	3.87 088	8.33 955	17.96 70	17.24 14
59	3 481	7.68 115	24.28 99	205 379	3.89 300	8.38 721	18.06 97	16.94 92
60	3 600	7.74 597	24.49 49	216 000	3.91 487	8.43 433	18.17 12	16.66 67
61	3 721	7.81 025	24.69 82	226 981	3.93 650	8.48 093	18.27 16	16.39 34
62	3 844	7.87 401	24.89 98	238 328	3.95 789	8.52 702	18.37 09	16.12 90
63	3 969	7.93 725	25.09 98	250 047	3.97 906	8.57 262	18.46 91	15.87 30
64	4 096	8.00 000	25.29 82	262 144	4.00 000	8.61 774	18.56 64	15.62 50
65	4 225	8.06 226	25.49 51	274 625	4.02 073	8.66 239	18.66 26	15.38 46
66	4 356	8.12 404	25.69 05	287 496	4.04 124	8.70 659	18.75 78	15.15 15
67	4 489	8.18 535	25.88 44	300 763	4.06 155	8.75 034	18.85 20	14.92 54
68	4 624	8.24 621	26.07 68	314 432	4.08 166	8.79 366	18.94 54	14.70 59
69	4 761	8.30 662	26.26 79	328 509	4.10 157	8.83 656	19.03 78	14.49 28
70	4 900	8.36 660	26.45 75	343 000	4.12 129	8.87 904	19.12 93	14.28 57
71	5 041	8.42 615	26.64 58	357 911	4.14 082	8.92 112	19.22 00	14.08 45
72	5 184	8.48 528	26.83 28	373 248	4.16 017	8.96 281	19.30 98	13.88 89
73	5 329	8.54 400	27.01 85	389 017	4.17 934	9.00 411	19.39 88	13.69 86
74	5 476	8.60 233	27.20 29	405 224	4.19 834	9.04 504	19.48 70	13.51 35
75	5 625	8.66 025	27.38 61	421 875	4.21 716	9.08 560	19.57 43	13.33 33
76	5 776	8.71 780	27.56 81	438 976	4.23 582	9.12 581	19.66 10	13.15 79
77	5 929	8.77 496	27.74 89	456 533	4.25 432	9.16 566	19.74 68	12.98 70
78	6 084	8.83 176	27.92 85	474 552	4.27 266	9.20 516	19.83 19	12.82 05
79	6 241	8.88 819	28.10 69	493 039	4.29 084	9.24 434	19.91 63	12.65 82
80	6 400	8.94 427	28.28 43	512 000	4.30 887	9.28 318	20.00 00	12.50 00
81	6 561	9.00 000	28.46 05	531 441	4.32 675	9.32 170	20.08 30	12.34 57
82	6 724	9.05 539	28.63 56	551 368	4.34 448	9.35 990	20.16 53	12.19 51
83	6 889	9.11 043	28.80 97	571 787	4.36 207	9.39 780	20.24 69	12.04 82
84	7 056	9.16 515	28.98 28	592 704	4.37 952	9.43 539	20.32 79	11.90 48
85	7 225	9.21 954	29.15 48	614 125	4.39 683	9.47 268	20.40 83	11.76 47
86	7 396	9.27 362	29.32 58	636 056	4.41 400	9.50 969	20.48 80	11.62 79
87	7 569	9.32 738	29.49 58	658 503	4.43 105	9.54 640	20.56 71	11.49 43
88	7 744	9.38 083	29.66 48	681 472	4.44 796	9.58 284	20.64 56	11.36 36
89	7 921	9.43 398	29.83 29	704 969	4.46 475	9.61 900	20.72 35	11.23 60
90	8 100	9.48 683	30.00 00	729 000	4.48 140	9.65 489	20.80 08	11.11 11
91	8 281	9.53 939	30.16 62	753 571	4.49 794	9.69 052	20.87 76	10.98 90
92	8 464	9.59 166	30.33 15	778 688	4.51 436	9.72 589	20.95 38	10.86 96
93	8 649	9.64 365	30.49 59	804 357	4.53 065	9.76 100	21.02 94	10.75 27
94	8.836	9.69 536	30.65 94	830 684	4.54 684	9.79 586	21.10 45	10.63 83
95	9 025	9.74 679	30.82 21	857 375	4.56 290	9.83 048	21.17 91	10.52 63
96	9 216	9.79 796	30.98 39	884 736	4.57 886	9.86 485	21.25 32	10.41 67
97	9 409	9.84 886	31.14 48	912 673	4.59 470	9.89 898	21.32 67	10.30 93
98	9 604	9.89 949	31.30 50	941 192	4.61 044	9.93 288	21.39 97	10.20 41
99	9 801	9.94 987	31.46 43	970 299	4.62 607	9.96 655	21.47 23	10.10 10
100	10 000	10.00 000	31.62 28	1 000 000	4.64 159	10.00 000	21.54 43	10.00 00

TABLE II Natural values of the trigonometric functions for angles in radians

Rad.	Sin	Tan	Cot	Cos
0.00	.00000	.00000	–	1.00000
.01	.01000	.01000	99.997	0.99995
.02	.02000	.02000	49.993	.99980
.03	.03000	.03001	33.323	.99955
.04	.03999	.04002	24.987	.99920
.05	.04998	.05004	19.983	.99875
.06	.05996	.06007	16.647	.99820
.07	.06994	.07011	14.262	.99755
.08	.07991	.08017	12.473	.99680
.09	.08988	.09024	11.081	.99595
0.10	.09983	.10033	9.9666	.99500
.11	.10978	.11045	9.0542	.99396
.12	.11971	.12058	8.2933	.99281
.13	.12963	.13074	7.6489	.99156
.14	.13954	.14092	7.0961	.99022
.15	.14944	.15114	6.6166	.98877
.16	.15932	.16138	6.1966	.98723
.17	.16918	.17166	5.8256	.98558
.18	.17903	.18197	5.4954	.98384
.19	.18886	.19232	5.1997	.98200
0.20	.19867	.20271	4.9332	.98007
.21	.20846	.21314	4.6917	.97803
.22	.21823	.22362	4.4719	.97590
.23	.22798	.23414	4.2709	.97367
.24	.23770	.24472	4.0864	.97134
.25	.24740	.25534	3.9163	.96891
.26	.25708	.26602	3.7591	.96639
.27	.26673	.27676	3.6133	.96377
.28	.27636	.28755	3.4776	.96106
.29	.28595	.29841	3.3511	.95824
0.30	.29552	.30934	3.2327	.95534
.31	.30506	.32033	3.1218	.95233
.32	.31457	.33139	3.0176	.94924
.33	.32404	.34252	2.9195	.94604
.34	.33349	.35374	2.8270	.94275
.35	.34290	.36503	2.7395	.93937
.36	.35227	.37640	2.6567	.93590
.37	.36162	.38786	2.5782	.93233
.38	.37092	.39941	2.5037	.92866
.39	.38019	.41105	2.4328	.92491
0.40	.38942	.42279	2.3652	.92106
.41	.39861	.43463	2.3008	.91712
.42	.40776	.44657	2.2393	.91309
.43	.41687	.45862	2.1804	.90897
.44	.42594	.47078	2.1241	.90475
.45	.43497	.48306	2.0702	.90045
.46	.44395	.49545	2.0184	.89605
.47	.45289	.50797	1.9686	.89157
.48	.46178	.52061	1.9208	.88699
.49	.47063	.53339	1.8748	.88233
0.50	.47943	.54630	1.8305	.87758

Rad.	Sin	Tan	Cot	Cos
0.50	.47943	.54630	1.8305	.87758
.51	.48818	.55936	1.7878	.87274
.52	.49688	.57256	1.7465	.86782
.53	.50553	.58592	1.7067	.86281
.54	.51414	.59943	1.6683	.85771
.55	.52269	.61311	1.6310	.85252
.56	.53119	.62695	1.5950	.84726
.57	.53963	.64097	1.5601	.84190
.58	.54802	.65517	1.5263	.83646
.59	.55636	.66956	1.4935	.83094
0.60	.56464	.68414	1.4617	.82534
.61	.57287	.69892	1.4308	.81965
.62	.58104	.71391	1.4007	.81388
.63	.58914	.72911	1.3715	.80803
.64	.59720	.74454	1.3431	.80210
.65	.60519	.76020	1.3154	.79608
.66	.61312	.77610	1.2885	.78999
.67	.62099	.79225	1.2622	.78382
.68	.62879	.80866	1.2366	.77757
.69	.63654	.82534	1.2116	.77125
0.70	.64422	.84229	1.1872	.76484
.71	.65183	.85953	1.1634	.75836
.72	.65938	.87707	1.1402	.75181
.73	.66687	.89492	1.1174	.74517
.74	.67429	.91309	1.0952	.73847
.75	.68164	.93160	1.0734	.73169
.76	.68892	.95045	1.0521	.72484
.77	.69614	.96967	1.0313	.71791
.78	.70328	.98926	1.0109	.71091
.79	.71035	1.0092	.99084	.70385
0.80	.71736	1.0296	.97121	.69671
.81	.72429	1.0505	.95197	.68950
.82	.73115	1.0717	.93309	.68222
.83	.73793	1.0934	.91455	.67488
.84	.74464	1.1156	.89635	.66746
.85	.75128	1.1383	.87848	.65998
.86	.75784	1.1616	.86091	.65244
.87	.76433	1.1853	.84365	.64483
.88	.77074	1.2097	.82668	.63715
.89	.77707	1.2346	.80998	.62941
0.90	.78333	1.2602	.79355	.62161
.91	.78950	1.2864	.77738	.61375
.92	.79560	1.3133	.76146	.60582
.93	.80162	1.3409	.74578	.59783
.94	.80756	1.3692	.73034	.58979
.95	.81342	1.3984	.71511	.58168
.96	.81919	1.4284	.70010	.57352
.97	.82489	1.4592	.68531	.56530
.98	.83050	1.4910	.67071	.55702
.99	.83603	1.5237	.65631	.54869
1.00	.84147	1.5574	.64209	.54030

Rad.	Sin	Tan	Cot	Cos
1.00	.84147	1.5574	.64209	.54030
1.01	.84683	1.5922	.62806	.53186
1.02	.85211	1.6281	.61420	.52337
1.03	.85730	1.6652	.60051	.51482
1.04	.86240	1.7036	.58699	.50622
1.05	.86742	1.7433	.57362	.49757
1.06	.87236	1.7844	.56040	.48887
1.07	.87720	1.8270	.54734	.48012
1.08	.88196	1.8712	.53441	.47133
1.09	.88663	1.9171	.52162	.46249
1.10	.89121	1.9648	.50897	.45360
1.11	.89570	2.0143	.49644	.44466
1.12	.90010	2.0660	.48404	.43568
1.13	.90441	2.1198	.47175	.42666
1.14	.90863	2.1759	.45959	.41759
1.15	.91276	2.2345	.44753	.40849
1.16	.91680	2.2958	.43558	.39934
1.17	.92075	2.3600	.42373	.39015
1.18	.92461	2.4273	.41199	.38092
1.19	.92837	2.4979	.40034	.37166
1.20	.93204	2.5722	.38878	.36236
1.21	.93562	2.6503	.37731	.35302
1.22	.93910	2.7328	.36593	.34365
1.23	.94249	2.8198	.35463	.33424
1.24	.94578	2.9119	.34341	.32480
1.25	.94898	3.0096	.33227	.31532
1.26	.95209	3.1133	.32121	.30582
1.27	.95510	3.2236	.31021	.29628
1.28	.95802	3.3413	.29928	.28672
1.29	.96084	3.4672	.28842	.27712
1.30	.96356	3.6021	.27762	.26750

Rad.	Sin	Tan	Cot	Cos
1.30	.96356	3.6021	.27762	.26750
1.31	.96618	3.7471	.26687	.25785
1.32	.96872	3.9033	.25619	.24818
1.33	.97115	4.0723	.24556	.23848
1.34	.97438	4.2556	.23498	.22875
1.35	.97572	4.4552	.22446	.21901
1.36	.97786	4.6734	.21398	.20924
1.37	.97991	4.9131	.20354	.19945
1.38	.98185	5.1774	.19315	.18964
1.39	.98370	5.4707	.18279	.17981
1.40	.98545	5.7979	.17248	.16997
1.41	.98710	6.1654	.16220	.16010
1.42	.98865	6.5811	.15195	.15023
1.43	.99010	7.0555	.14173	.14033
1.44	.99146	7.6018	.13155	.13042
1.45	.99271	8.2381	.12139	.12050
1.46	.99387	8.9886	.11125	.11057
1.47	.99492	9.8874	.10114	.10063
1.48	.99588	10.983	.09105	.09067
1.49	.99674	12.350	.08097	.08071
1.50	.99749	14.101	.07091	.07074
1.51	.99815	16.428	.06087	.06076
1.52	.99871	19.670	.05084	.05077
1.53	.99917	24.498	.04082	.04079
1.54	.99953	32.461	.03081	.03079
1.55	.99978	48.078	.02080	.02079
1.56	.99994	92.621	.01080	.01080
1.57	1.00000	1255.8	.00080	.00080
1.58	.99996	−108.65	−.00920	−.00920
1.59	.99982	−52.067	−.01921	−.01920
1.60	.99957	−34.233	−.02921	−.02920

TABLE III Radians to degrees, minutes, and seconds

	Radians	Tenths	Hundredths	Thousandths	Ten-thousandths
1	57° 17′ 44.″8	5° 43′ 46.″5	0° 34′ 22.″6	0° 3′ 26.″3	0° 0′ 20.″6
2	114° 35′ 29.″6	11° 27′ 33.″0	1° 8′ 45.″3	0° 6′ 52.″5	0° 0′ 41.″3
3	171° 53′ 14.″4	17° 11′ 19.″4	1° 43′ 07.″9	0° 10′ 18.″8	0° 1′ 01.″9
4	229° 10′ 59.″2	22° 55′ 05.″9	2° 17′ 30.″6	0° 13′ 45.″1	0° 1′ 22.″5
5	286° 28′ 44.″0	28° 38′ 52.″4	2° 51′ 53.″2	0° 17′ 11.″3	0° 1′ 43.″1
6	343° 46′ 28.″8	34° 22′ 38.″9	3° 26′ 15.″9	0° 20′ 37.″6	0° 2′ 03.″8
7	401° 4′ 13.″6	40° 6′ 25.″4	4° 0′ 38.″5	0° 24′ 03.″9	0° 2′ 24.″4
8	458° 21′ 58.″4	45° 50′ 11.″8	4° 35′ 01.″2	0° 27′ 30.″1	0° 2′ 45.″0
9	515° 39′ 43.″3	51° 33′ 58.″3	5° 9′ 23.″8	0° 30′ 56.″4	0° 3′ 05.″6

TABLE IV Four-place natural logarithms 1.00–5.49 ($\ln x = \log_e x$)

N	.00	.01	.02	.03	.04	.05	.06	.07	.08	.09
1.0	0.0000	0.0100	0.0198	0.0296	0.0392	0.0488	0.0583	0.0677	0.0770	0.0862
1.1	0.0953	0.1044	0.1133	0.1222	0.1310	0.1398	0.1484	0.1570	0.1655	0.1740
1.2	0.1823	0.1906	0.1989	0.2070	0.2151	0.2231	0.2311	0.2390	0.2469	0.2546
1.3	0.2624	0.2700	0.2776	0.2852	0.2927	0.3001	0.3075	0.3148	0.3221	0.3293
1.4	0.3365	0.3436	0.3507	0.3577	0.3646	0.3716	0.3784	0.3853	0.3920	0.3988
1.5	0.4055	0.4121	0.4187	0.4253	0.4318	0.4383	0.4447	0.4511	0.4574	0.4637
1.6	0.4700	0.4762	0.4824	0.4886	0.4947	0.5008	0.5068	0.5128	0.5188	0.5247
1.7	0.5306	0.5365	0.5423	0.5481	0.5539	0.5596	0.5653	0.5710	0.5766	0.5822
1.8	0.5878	0.5933	0.5988	0.6043	0.6098	0.6152	0.6206	0.6259	0.6313	0.6366
1.9	0.6419	0.6471	0.6523	0.6575	0.6627	0.6678	0.6729	0.6780	0.6831	0.6881
2.0	0.6931	0.6981	0.7031	0.7080	0.7129	0.7178	0.7227	0.7275	0.7324	0.7372
2.1	0.7419	0.7467	0.7514	0.7561	0.7608	0.7655	0.7701	0.7747	0.7793	0.7839
2.2	0.7885	0.7930	0.7975	0.8020	0.8065	0.8109	0.8154	0.8198	0.8242	0.8286
2.3	0.8329	0.8372	0.8416	0.8459	0.8502	0.8544	0.8587	0.8629	0.8671	0.8713
2.4	0.8755	0.8796	0.8838	0.8879	0.8920	0.8961	0.9002	0.9042	0.9083	0.9123
2.5	0.9163	0.9203	0.9243	0.9282	0.9322	0.9361	0.9400	0.9439	0.9478	0.9517
2.6	0.9555	0.9594	0.9632	0.9670	0.9708	0.9746	0.9783	0.9821	0.9858	0.9895
2.7	0.9933	0.9969	1.0006	1.0043	1.0080	1.0116	1.0152	1.0188	1.0225	1.0260
2.8	1.0296	1.0332	1.0367	1.0403	1.0438	1.0473	1.0508	1.0543	1.0578	1.0613
2.9	1.0647	1.0682	1.0716	1.0750	1.0784	1.0818	1.0852	1.0886	1.0919	1.0953
3.0	1.0986	1.1019	1.1053	1.1086	1.1119	1.1151	1.1184	1.1217	1.1249	1.1282
3.1	1.1314	1.1346	1.1378	1.1410	1.1442	1.1474	1.1506	1.1537	1.1569	1.1600
3.2	1.1632	1.1663	1.1694	1.1725	1.1756	1.1787	1.1817	1.1848	1.1878	1.1909
3.3	1.1939	1.1969	1.2000	1.2030	1.2060	1.2090	1.2119	1.2149	1.2179	1.2208
3.4	1.2238	1.2267	1.2296	1.2326	1.2355	1.2384	1.2413	1.2442	1.2470	1.2499
3.5	1.2528	1.2556	1.2585	1.2613	1.2641	1.2669	1.2698	1.2726	1.2754	1.2782
3.6	1.2809	1.2837	1.2865	1.2892	1.2920	1.2947	1.2975	1.3002	1.3029	1.3056
3.7	1.3083	1.3110	1.3137	1.3164	1.3191	1.3218	1.3244	1.3271	1.3297	1.3324
3.8	1.3350	1.3376	1.3403	1.3429	1.3455	1.3481	1.3507	1.3533	1.3558	1.3584
3.9	1.3610	1.3635	1.3661	1.3686	1.3712	1.3737	1.3762	1.3788	1.3813	1.3838
4.0	1.3863	1.3888	1.3913	1.3938	1.3962	1.3987	1.4012	1.4036	1.4061	1.4085
4.1	1.4110	1.4134	1.4159	1.4183	1.4207	1.4231	1.4255	1.4279	1.4303	1.4327
4.2	1.4351	1.4375	1.4398	1.4422	1.4446	1.4469	1.4493	1.4516	1.4540	1.4563
4.3	1.4586	1.4609	1.4633	1.4656	1.4679	1.4702	1.4725	1.4748	1.4770	1.4793
4.4	1.4816	1.4839	1.4861	1.4884	1.4907	1.4929	1.4951	1.4974	1.4996	1.5019
4.5	1.5041	1.5063	1.5085	1.5107	1.5129	1.5151	1.5173	1.5195	1.5217	1.5239
4.6	1.5261	1.5282	1.5304	1.5326	1.5347	1.5369	1.5390	1.5412	1.5433	1.5454
4.7	1.5476	1.5497	1.5518	1.5539	1.5560	1.5581	1.5602	1.5623	1.5644	1.5665
4.8	1.5686	1.5707	1.5728	1.5748	1.5769	1.5790	1.5810	1.5831	1.5851	1.5872
4.9	1.5892	1.5913	1.5933	1.5953	1.5974	1.5994	1.6014	1.6034	1.6054	1.6074
5.0	1.6094	1.6114	1.6134	1.6154	1.6174	1.6194	1.6214	1.6233	1.6253	1.6273
5.1	1.6292	1.6312	1.6332	1.6351	1.6371	1.6390	1.6409	1.6429	1.6448	1.6467
5.2	1.6487	1.6506	1.6525	1.6544	1.6563	1.6582	1.6601	1.6620	1.6639	1.6658
5.3	1.6677	1.6696	1.6715	1.6734	1.6752	1.6771	1.6790	1.6808	1.6827	1.6845
5.4	1.6864	1.6882	1.6901	1.6919	1.6938	1.6956	1.6974	1.6993	1.7011	1.7029

$\ln .1 = .6974–3$ $\ln .01 = .3948–5$ $\ln .001 = .0922–7$

TABLE IV (continued) Four-place natural logarithms 5.50–9.99

N	.00	.01	.02	.03	.04	.05	.06	.07	.08	.09
5.5	1.7047	1.7066	1.7084	1.7102	1.7120	1.7138	1.7156	1.7174	1.7192	1.7210
5.6	1.7228	1.7246	1.7263	1.7281	1.7299	1.7317	1.7334	1.7352	1.7370	1.7387
5.7	1.7405	1.7422	1.7440	1.7457	1.7475	1.7492	1.7509	1.7527	1.7544	1.7561
5.8	1.7579	1.7596	1.7613	1.7630	1.7647	1.7664	1.7681	1.7699	1.7716	1.7733
5.9	1.7750	1.7766	1.7783	1.7800	1.7817	1.7834	1.7851	1.7867	1.7884	1.7901
6.0	1.7918	1.7934	1.7951	1.7967	1.7984	1.8001	1.8017	1.8034	1.8050	1.8066
6.1	1.8083	1.8099	1.8116	1.8132	1.8148	1.8165	1.8181	1.8197	1.8213	1.8229
6.2	1.8245	1.8262	1.8278	1.8294	1.8310	1.8326	1.8342	1.8358	1.8374	1.8390
6.3	1.8405	1.8421	1.8437	1.8453	1.8469	1.8485	1.8500	1.8516	1.8532	1.8547
6.4	1.8563	1.8579	1.8594	1.8610	1.8625	1.8641	1.8656	1.8672	1.8687	1.8703
6.5	1.8718	1.8733	1.8749	1.8764	1.8779	1.8795	1.8810	1.8825	1.8840	1.8856
6.6	1.8871	1.8886	1.8901	1.8916	1.8931	1.8946	1.8961	1.8976	1.8991	1.9006
6.7	1.9021	1.9036	1.9051	1.9066	1.9081	1.9095	1.9110	1.9125	1.9140	1.9155
6.8	1.9169	1.9184	1.9199	1.9213	1.9228	1.9242	1.9257	1.9272	1.9286	1.9301
6.9	1.9315	1.9330	1.9344	1.9359	1.9373	1.9387	1.9402	1.9416	1.9430	1.9445
7.0	1.9459	1.9473	1.9488	1.9502	1.9516	1.9530	1.9544	1.9559	1.9573	1.9587
7.1	1.9601	1.9615	1.9629	1.9643	1.9657	1.9671	1.9685	1.9699	1.9713	1.9727
7.2	1.9741	1.9755	1.9769	1.9782	1.9796	1.9810	1.9824	1.9838	1.9851	1.9865
7.3	1.9879	1.9892	1.9906	1.9920	1.9933	1.9947	1.9961	1.9974	1.9988	2.0001
7.4	2.0015	2.0028	2.0042	2.0055	2.0069	2.0082	2.0096	2.0109	2.0122	2.0136
7.5	2.0149	2.0162	2.0176	2.0189	2.0202	2.0215	2.0229	2.0242	2.0255	2.0268
7.6	2.0281	2.0295	2.0308	2.0321	2.0334	2.0347	2.0360	2.0373	2.0386	2.0399
7.7	2.0412	2.0425	2.0438	2.0451	2.0464	2.0477	2.0490	2.0503	2.0516	2.0528
7.8	2.0541	2.0554	2.0567	2.0580	2.0592	2.0605	2.0618	2.0631	2.0643	2.0656
7.9	2.0669	2.0681	2.0694	2.0707	2.0719	2.0732	2.0744	2.0757	2.0769	2.0782
8.0	2.0794	2.0807	2.0819	2.0832	2.0844	2.0857	2.0869	2.0882	2.0894	2.0906
8.1	2.0919	2.0931	2.0943	2.0956	2.0968	2.0980	2.0992	2.1005	2.1017	2.1029
8.2	2.1041	2.1054	2.1066	2.1078	2.1090	2.1102	2.1114	2.1126	2.1138	2.1150
8.3	2.1163	2.1175	2.1187	2.1199	2.1211	2.1223	2.1235	2.1247	2.1258	2.1270
8.4	2.1282	2.1294	2.1306	2.1318	2.1330	2.1342	2.1353	2.1365	2.1377	2.1389
8.5	2.1401	2.1412	2.1424	2.1436	2.1448	2.1459	2.1471	2.1483	2.1494	2.1506
8.6	2.1518	2.1529	2.1541	2.1552	2.1564	2.1576	2.1587	2.1599	2.1610	2.1622
8.7	2.1633	2.1645	2.1656	2.1668	2.1679	2.1691	2.1702	2.1713	2.1725	2.1736
8.8	2.1748	2.1759	2.1770	2.1782	2.1793	2.1804	2.1815	2.1827	2.1838	2.1849
8.9	2.1861	2.1872	2.1883	2.1894	2.1905	2.1917	2.1928	2.1939	2.1950	2.1961
9.0	2.1972	2.1983	2.1994	2.2006	2.2017	2.2028	2.2039	2.2050	2.2061	2.2072
9.1	2.2083	2.2094	2.2105	2.2116	2.2127	2.2138	2.2148	2.2159	2.2170	2.2181
9.2	2.2192	2.2203	2.2214	2.2225	2.2235	2.2246	2.2257	2.2268	2.2279	2.2289
9.3	2.2300	2.2311	2.2322	2.2332	2.2343	2.2354	2.2364	2.2375	2.2386	2.2396
9.4	2.2407	2.2418	2.2428	2.2439	2.2450	2.2460	2.2471	2.2481	2.2492	2.2502
9.5	2.2513	2.2523	2.2534	2.2544	2.2555	2.2565	2.2576	2.2586	2.2597	2.2607
9.6	2.2618	2.2628	2.2638	2.2649	2.2659	2.2670	2.2680	2.2690	2.2701	2.2711
9.7	2.2721	2.2732	2.2742	2.2752	2.2762	2.2773	2.2783	2.2793	2.2803	2.2814
9.8	2.2824	2.2834	2.2844	2.2854	2.2865	2.2875	2.2885	2.2895	2.2905	2.2915
9.9	2.2925	2.2935	2.2946	2.2956	2.2966	2.2976	2.2986	2.2996	2.3006	2.3016

$\ln .0001 = .7897-10$ $\ln .00001 = .4871-12$ $\ln .000\,001 = .1845-14$

TABLE IV (continued) Four-place natural logarithms 10.0–54.9

N	.0	.1	.2	.3	.4	.5	.6	.7	.8	.9
10	2.3026	2.3125	2.3224	2.3321	2.3418	2.3514	2.3609	2.3702	2.3795	2.3888
11	2.3979	2.4069	2.4159	2.4248	2.4336	2.4423	2.4510	2.4596	2.4681	2.4765
12	2.4849	2.4932	2.5014	2.5096	2.5177	2.5257	2.5337	2.5416	2.5494	2.5572
13	2.5649	2.5726	2.5802	2.5878	2.5953	2.6027	2.6101	2.6174	2.6247	2.6319
14	2.6391	2.6462	2.6532	2.6603	2.6672	2.6741	2.6810	2.6878	2.6946	2.7014
15	2.7081	2.7147	2.7213	2.7279	2.7344	2.7408	2.7473	2.7537	2.7600	2.7663
16	2.7726	2.7788	2.7850	2.7912	2.7973	2.8034	2.8094	2.8154	2.8214	2.8273
17	2.8332	2.8391	2.8449	2.8507	2.8565	2.8622	2.8679	2.8736	2.8792	2.8848
18	2.8904	2.8959	2.9014	2.9069	2.9124	2.9178	2.9232	2.9285	2.9339	2.9392
19	2.9444	2.9497	2.9549	2.9601	2.9653	2.9704	2.9755	2.9806	2.9857	2.9907
20	2.9957	3.0007	3.0057	3.0106	3.0155	3.0204	3.0253	3.0301	3.0350	3.0397
21	3.0445	3.0493	3.0540	3.0587	3.0634	3.0681	3.0727	3.0773	3.0819	3.0865
22	3.0910	3.0956	3.1001	3.1046	3.1091	3.1135	3.1179	3.1224	3.1268	3.1311
23	3.1355	3.1398	3.1442	3.1485	3.1527	3.1570	3.1612	3.1655	3.1697	3.1739
24	3.1781	3.1822	3.1864	3.1905	3.1946	3.1987	3.2027	3.2068	3.2108	3.2149
25	3.2189	3.2229	3.2268	3.2308	3.2347	3.2387	3.2426	3.2465	3.2504	3.2542
26	3.2581	3.2619	3.2658	3.2696	3.2734	3.2771	3.2809	3.2847	3.2884	3.2921
27	3.2958	3.2995	3.3032	3.3069	3.3105	3.3142	3.3178	3.3214	3.3250	3.3286
28	3.3322	3.3358	3.3393	3.3429	3.3464	3.3499	3.3534	3.3569	3.3604	3.3638
29	3.3673	3.3707	3.3742	3.3776	3.3810	3.3844	3.3878	3.3911	3.3945	3.3979
30	3.4012	3.4045	3.4078	3.4111	3.4144	3.4177	3.4210	3.4243	3.4275	3.4308
31	3.4340	3.4372	3.4404	3.4436	3.4468	3.4500	3.4532	3.4563	3.4595	3.4626
32	3.4657	3.4689	3.4720	3.4751	3.4782	3.4812	3.4843	3.4874	3.4904	3.4935
33	3.4965	3.4995	3.5025	3.5056	3.5086	3.5115	3.5145	3.5175	3.5205	3.5234
34	3.5264	3.5293	3.5322	3.5351	3.5381	3.5410	3.5439	3.5467	3.5496	3.5525
35	3.5553	3.5582	3.5610	3.5639	3.5667	3.5695	3.5723	3.5752	3.5779	3.5807
36	3.5835	3.5863	3.5891	3.5918	3.5946	3.5973	3.6000	3.6028	3.6055	3.6082
37	3.6109	3.6136	3.6163	3.6190	3.6217	3.6243	3.6270	3.6297	3.6323	3.6350
38	3.6376	3.6402	3.6428	3.6454	3.6481	3.6507	3.6533	3.6558	3.6584	3.6610
39	3.6636	3.6661	3.6687	3.6712	3.6738	3.6763	3.6788	3.6814	3.6839	3.6864
40	3.6889	3.6914	3.6939	3.6964	3.6988	3.7013	3.7038	3.7062	3.7087	3.7111
41	3.7136	3.7160	3.7184	3.7209	3.7233	3.7257	3.7281	3.7305	3.7329	3.7353
42	3.7377	3.7400	3.7424	3.7448	3.7471	3.7495	3.7519	3.7542	3.7565	3.7589
43	3.7612	3.7635	3.7658	3.7682	3.7705	3.7728	3.7751	3.7773	3.7796	3.7819
44	3.7842	3.7865	3.7887	3.7910	3.7932	3.7955	3.7977	3.8000	3.8022	3.8044
45	3.8067	3.8089	3.8111	3.8133	3.8155	3.8177	3.8199	3.8221	3.8243	3.8265
46	3.8286	3.8308	3.8330	3.8351	3.8373	3.8395	3.8416	3.8437	3.8459	3.8480
47	3.8501	3.8523	3.8544	3.8565	3.8586	3.8607	3.8628	3.8649	3.8670	3.8691
48	3.8712	3.8733	3.8754	3.8774	3.8795	3.8816	3.8836	3.8857	3.8877	3.8898
49	3.8918	3.8939	3.8959	3.8979	3.9000	3.9020	3.9040	3.9060	3.9080	3.9100
50	3.9120	3.9140	3.9160	3.9180	3.9200	3.9220	3.9240	3.9259	3.9279	3.9299
51	3.9318	3.9338	3.9357	3.9377	3.9396	3.9416	3.9435	3.9455	3.9474	3.9493
52	3.9512	3.9532	3.9551	3.9570	3.9589	3.9608	3.9627	3.9646	3.9665	3.9684
53	3.9703	3.9722	3.9741	3.9759	3.9778	3.9797	3.9815	3.9834	3.9853	3.9871
54	3.9890	3.9908	3.9927	3.9945	3.9964	3.9982	4.0000	4.0019	4.0037	4.0055

$\ln 100 = 4.6052$ $\ln 1000 = 6.9078$ $\ln 10{,}000 = 9.2103$

TABLE IV (continued) Four-place natural logarithms 55.0–99.9

N	.0	.1	.2	.3	.4	.5	.6	.7	.8	.9
55	4.0073	4.0091	4.0110	4.0128	4.0146	4.0164	4.0182	4.0200	4.0218	4.0236
56	4.0254	4.0271	4.0289	4.0307	4.0325	4.0342	4.0360	4.0378	4.0395	4.0413
57	4.0431	4.0448	4.0466	4.0483	4.0500	4.0518	4.0535	4.0553	4.0570	4.0587
58	4.0604	4.0622	4.0639	4.0656	4.0673	4.0690	4.0707	4.0724	4.0741	4.0758
59	4.0775	4.0792	4.0809	4.0826	4.0843	4.0860	4.0877	4.0893	4.0910	4.0927
60	4.0943	4.0960	4.0977	4.0993	4.1010	4.1026	4.1043	4.1059	4.1076	4.1092
61	4.1109	4.1125	4.1141	4.1158	4.1174	4.1190	4.1207	4.1223	4.1239	4.1255
62	4.1271	4.1287	4.1304	4.1320	4.1336	4.1352	4.1368	4.1384	4.1400	4.1415
63	4.1431	4.1447	4.1463	4.1479	4.1495	4.1510	4.1526	4.1542	4.1558	4.1573
64	4.1589	4.1604	4.1620	4.1636	4.1651	4.1667	4.1682	4.1698	4.1713	4.1728
65	4.1744	4.1759	4.1775	4.1790	4.1805	4.1821	4.1836	4.1851	4.1866	4.1881
66	4.1897	4.1912	4.1927	4.1942	4.1957	4.1972	4.1987	4.2002	4.2017	4.2032
67	4.2047	4.2062	4.2077	4.2092	4.2106	4.2121	4.2136	4.2151	4.2166	4.2180
68	4.2195	4.2210	4.2224	4.2239	4.2254	4.2268	4.2283	4.2297	4.2312	4.2327
69	4.2341	4.2356	4.2370	4.2384	4.2399	4.2413	4.2428	4.2442	4.2456	4.2471
70	4.2485	4.2499	4.2513	4.2528	4.2542	4.2556	4.2570	4.2584	4.2599	4.2613
71	4.2627	4.2641	4.2655	4.2669	4.2683	4.2697	4.2711	4.2725	4.2739	4.2753
72	4.2767	4.2781	4.2794	4.2808	4.2822	4.2836	4.2850	4.2863	4.2877	4.2891
73	4.2905	4.2918	4.2932	4.2946	4.2959	4.2973	4.2986	4.3000	4.3014	4.3027
74	4.3041	4.3054	4.3068	4.3081	4.3095	4.3108	4.3121	4.3135	4.3148	4.3162
75	4.3175	4.3188	4.3202	4.3215	4.3228	4.3241	4.3255	4.3268	4.3281	4.3294
76	4.3307	4.3320	4.3334	4.3347	4.3360	4.3373	4.3386	4.3399	4.3412	4.3425
77	4.3438	4.3451	4.3464	4.3477	4.3490	4.3503	4.3516	4.3529	4.3541	4.3554
78	4.3567	4.3580	4.3593	4.3605	4.3618	4.3631	4.3644	4.3656	4.3669	4.3682
79	4.3694	4.3707	4.3720	4.3732	4.3745	4.3758	4.3770	4.3783	4.3795	4.3808
80	4.3820	4.3833	4.3845	4.3858	4.3870	4.3883	4.3895	4.3907	4.3920	4.3932
81	4.3944	4.3957	4.3969	4.3981	4.3994	4.4006	4.4018	4.4031	4.4043	4.4055
82	4.4067	4.4079	4.4092	4.4104	4.4116	4.4128	4.4140	4.4152	4.4164	4.4176
83	4.4188	4.4200	4.4212	4.4224	4.4236	4.4248	4.4260	4.4272	4.4284	4.4296
84	4.4308	4.4320	4.4332	4.4344	4.4356	4.4368	4.4379	4.4391	4.4403	4.4415
85	4.4427	4.4438	4.4450	4.4462	4.4473	4.4485	4.4497	4.4509	4.4520	4.4532
86	4.4543	4.4555	4.4567	4.4578	4.4590	4.4601	4.4613	4.4625	4.4636	4.4648
87	4.4659	4.4671	4.4682	4.4694	4.4705	4.4716	4.4728	4.4739	4.4751	4.4762
88	4.4773	4.4785	4.4796	4.4807	4.4819	4.4830	4.4841	4.4853	4.4864	4.4875
89	4.4886	4.4898	4.4909	4.4920	4.4931	4.4942	4.4954	4.4965	4.4976	4.4987
90	4.4998	4.5009	4.5020	4.5031	4.5042	4.5053	4.5065	4.5076	4.5087	4.5098
91	4.5109	4.5120	4.5131	4.5142	4.5152	4.5163	4.5174	4.5185	4.5196	4.5207
92	4.5218	4.5229	4.5240	4.5250	4.5261	4.5272	4.5283	4.5294	4.5304	4.5315
93	4.5326	4.5337	4.5347	4.5358	4.5369	4.5380	4.5390	4.5401	4.5412	4.5422
94	4.5433	4.5444	4.5454	4.5465	4.5475	4.5486	4.5497	4.5507	4.5518	4.5528
95	4.5539	4.5549	4.5560	4.5570	4.5581	4.5591	4.5602	4.5612	4.5623	4.5633
96	4.5643	4.5654	4.5664	4.5675	4.5685	4.5695	4.5706	4.5716	4.5726	4.5737
97	4.5747	4.5757	4.5768	4.5778	4.5788	4.5799	4.5809	4.5819	4.5829	4.5839
98	4.5850	4.5860	4.5870	4.5880	4.5890	4.5901	4.5911	4.5921	4.5931	4.5941
99	4.5951	4.5961	4.5971	4.5981	4.5992	4.6002	4.6012	4.6022	4.6032	4.6042

$\ln 100{,}000 = 11.5129$ $\ln 1{,}000{,}000 = 13.8155$ $\ln 10{,}000{,}000 = 16.1181$

TABLE V Values of exponential functions 0.00–1.50

x	e^x	e^{-x}	x	e^x	e^{-x}	x	e^x	e^{-x}
0.00	1.0000	1.00 000	**0.50**	1.6487	.60 653	**1.00**	2.7183	.36 788
0.01	1.0101	0.99 005	0.51	1.6653	.60 050	1.01	2.7456	.36 422
0.02	1.0202	.98 020	0.52	1.6820	.59 452	1.02	2.7732	.36 059
0.03	1.0305	.97 045	0.53	1.6989	.58 860	1.03	2.8011	.35 701
0.04	1.0408	.96 079	0.54	1.7160	.58 275	1.04	2.8292	.35 345
0.05	1.0513	.95 123	**0.55**	1.7333	.57 695	**1.05**	2.8577	.34 994
0.06	1.0618	.94 176	0.56	1.7507	.57 121	1.06	2.8864	.34 646
0.07	1.0725	.93 239	0.57	1.7683	.56 553	1.07	2.9154	.34 301
0.08	1.0833	.92 312	0.58	1.7860	.55 990	1.08	2.9447	.33 960
0.09	1.0942	.91 393	0.59	1.8040	.55 433	1.09	2.9743	.33 622
0.10	1.1052	.90 484	**0.60**	1.8221	.54 881	**1.10**	3.0042	.33 287
0.11	1.1163	.89 583	0.61	1.8404	.54 335	1.11	3.0344	.32 956
0.12	1.1275	.88 692	0.62	1.8589	.53 794	1.12	3.0649	.32 628
0.13	1.1388	.87 810	0.63	1.8776	.53 259	1.13	3.0957	.32 303
0.14	1.1503	.86 936	0.64	1.8965	.52 729	1.14	3.1268	.31 982
0.15	1.1618	.86 071	**0.65**	1.9155	.52 205	**1.15**	3.1582	.31 664
0.16	1.1735	.85 214	0.66	1.9348	.51 685	1.16	3.1899	.31 349
0.17	1.1853	.84 366	0.67	1.9542	.51 171	1.17	3.2220	.31 037
0.18	1.1972	.83 527	0.68	1.9739	.50 662	1.18	3.2544	.30 728
0.19	1.2092	.82 696	0.69	1.9937	.50 158	1.19	3.2871	.30 422
0.20	1.2214	.81 873	**0.70**	2.0138	.49 659	**1.20**	3.3201	.30 119
0.21	1.2337	.81 058	0.71	2.0340	.49 164	1.21	3.3535	.29 820
0.22	1.2461	.80 252	0.72	2.0544	.48 675	1.22	3.3872	.29 523
0.23	1.2586	.79 453	0.73	2.0751	.48 191	1.23	3.4212	.29 229
0.24	1.2712	.78 663	0.74	2.0959	.47 711	1.24	3.4556	.28 938
0.25	1.2840	.77 880	**0.75**	2.1170	.47 237	**1.25**	3.4903	.28 650
0.26	1.2969	.77 105	0.76	2.1383	.46 767	1.26	3.5254	.28 365
0.27	1.3100	.76 338	0.77	2.1598	.46 301	1.27	3.5609	.28 083
0.28	1.3231	.75 578	0.78	2.1815	.45 841	1.28	3.5966	.27 804
0.29	1.3364	.74 826	0.79	2.2034	.45 384	1.29	3.6328	.27 527
0.30	1.3499	.74 082	**0.80**	2.2255	.44 933	**1.30**	3.6693	.27 253
0.31	1.3634	.73 345	0.81	2.2479	.44 486	1.31	3.7062	.26 982
0.32	1.3771	.72 615	0.82	2.2705	.44 043	1.32	3.7434	.26 714
0.33	1.3910	.71 892	0.83	2.2933	.43 605	1.33	3.7810	.26 448
0.34	1.4049	.71 177	0.84	2.3164	.43 171	1.34	3.8190	.26 185
0.35	1.4191	.70 469	**0.85**	2.3396	.42 741	**1.35**	3.8574	.25 924
0.36	1.4333	.69 768	0.86	2.3632	.42 316	1.36	3.8962	.25 666
0.37	1.4477	.69 073	0.87	2.3869	.41 895	1.37	3.9354	.25 411
0.38	1.4623	.68 386	0.88	2.4109	.41 478	1.38	3.9749	.25 158
0.39	1.4770	.67 706	0.89	2.4351	.41 066	1.39	4.0149	.24 908
0.40	1.4918	.67 032	**0.90**	2.4596	.40 657	**1.40**	4.0552	.24 660
0.41	1.5068	.66 365	0.91	2.4843	.40 252	1.41	4.0960	.24 414
0.42	1.5220	.65 705	0.92	2.5093	.39 852	1.42	4.1371	.24 171
0.43	1.5373	.65 051	0.93	2.5345	.39 455	1.43	4.1787	.23 931
0.44	1.5527	.64 404	0.94	2.5600	.39 063	1.44	4.2207	.23 693
0.45	1.5683	.63 763	**0.95**	2.5857	.38 674	**1.45**	4.2631	.23 457
0.46	1.5841	.63 128	0.96	2.6117	.38 289	1.46	4.3060	.23 224
0.47	1.6000	.62 500	0.97	2.6379	.37 908	1.47	4.3492	.22 993
0.48	1.6161	.61 878	0.98	2.6645	.37 531	1.48	4.3929	.22 764
0.49	1.6323	.61 263	0.99	2.6912	.37 158	1.49	4.4371	.22 537
0.50	1.6487	.60 653	**1.00**	2.7183	.36 788	**1.50**	4.4817	.22 313

TABLE V (continued) Values of exponential functions 1.50–3.00

x	e^x	e^{-x}	x	e^x	e^{-x}	x	e^x	e^{-x}
1.50	4.4817	.22 313	**2.00**	7.3891	.13 534	**2.50**	12.182	.082 085
1.51	4.5267	.22 091	2.01	7.4633	.13 399	2.51	12.305	.081 268
1.52	4.5722	.21 871	2.02	7.5383	.13 266	2.52	12.429	.080 460
1.53	4.6182	.21 654	2.03	7.6141	.13 134	2.53	12.554	.079 659
1.54	4.6646	.21 438	2.04	7.6906	.13 003	2.54	12.680	.078 866
1.55	4.7115	.21 225	**2.05**	7.7679	.12 873	**2.55**	12.807	.078 082
1.56	4.7588	.21 014	2.06	7.8460	.12 745	2.56	12.936	.077 305
1.57	4.8066	.20 805	2.07	7.9248	.12 619	2.57	13.066	.076 536
1.58	4.8550	.20 598	2.08	8.0045	.12 493	2.58	13.197	.075 774
1.59	4.9037	.20 393	2.09	8.0849	.12 369	2.59	13.330	.075 020
1.60	4.9530	.20 190	**2.10**	8.1662	.12 246	**2.60**	13.464	.074 274
1.61	5.0028	.19 989	2.11	8.2482	.12 124	2.61	13.599	.073 535
1.62	5.0531	.19 790	2.12	8.3311	.12 003	2.62	13.736	.072 803
1.63	5.1039	.19 593	2.13	8.4149	.11 884	2.63	13.874	.072 078
1.64	5.1552	.19 398	2.14	8.4994	.11 765	2.64	14.013	.071 361
1.65	5.2070	.19 205	**2.15**	8.5849	.11 648	**2.65**	14.154	.070 651
1.66	5.2593	.19 014	2.16	8.6711	.11 533	2.66	14.296	.069 948
1.67	5.3122	.18 825	2.17	8.7583	.11 418	2.67	14.440	.069 252
1.68	5.3656	.18 637	2.18	8.8463	.11 304	2.68	14.585	.068 563
1.69	5.4195	.18 452	2.19	8.9352	.11 192	2.69	14.732	.067 881
1.70	5.4739	.18 268	**2.20**	9.0250	.11 080	**2.70**	14.880	.067 206
1.71	5.5290	.18 087	2.21	9.1157	.10 970	2.71	15.029	.066 537
1.72	5.5845	.17 907	2.22	9.2073	.10 861	2.72	15.180	.065 875
1.73	5.6407	.17 728	2.23	9.2999	.10 753	2.73	15.333	.065 219
1.74	5.6973	.17 552	2.24	9.3933	.10 646	2.74	15.487	.064 570
1.75	5.7546	.17 377	**2.25**	9.4877	.10 540	**2.75**	15.643	.063 928
1.76	5.8124	.17 204	2.26	9.5831	.10 435	2.76	15.800	.063 292
1.77	5.8709	.17 033	2.27	9.6794	.10 331	2.77	15.959	.062 662
1.78	5.9299	.16 864	2.28	9.7767	.10 228	2.78	16.119	.062 039
1.79	5.9895	.16 696	2.29	9.8749	.10 127	2.79	16.281	.061 421
1.80	6.0496	.16 530	**2.30**	9.9742	.10 026	**2.80**	16.445	.060 810
1.81	6.1104	.16 365	2.31	10.074	.09 9261	2.81	16.610	.060 205
1.82	6.1719	.16 203	2.32	10.176	.09 8274	2.82	16.777	.059 606
1.83	6.2339	.16 041	2.33	10.278	.09 7296	2.83	16.945	.059 013
1.84	6.2965	.15 882	2.34	10.381	.09 6328	2.84	17.116	.058 426
1.85	6.3598	.15 724	**2.35**	10.486	.09 5369	**2.85**	17.288	.057 844
1.86	6.4237	.15 567	2.36	10.591	.09 4420	2.86	17.462	.057 269
1.87	6.4883	.15 412	2.37	10.697	.09 3481	2.87	17.637	.056 699
1.88	6.5535	.15 259	2.38	10.805	.09 2551	2.88	17.814	.056 135
1.89	6.6194	.15 107	2.39	10.913	.09 1630	2.89	17.993	.055 576
1.90	6.6859	.14 957	**2.40**	11.023	.09 0718	**2.90**	18.174	.055 023
1.91	6.7531	.14 808	2.41	11.134	.08 9815	2.91	18.357	.054 476
1.92	6.8210	.14 661	2.42	11.246	.08 8922	2.92	18.541	.053 934
1.93	6.8895	.14 515	2.43	11.359	.08 8037	2.93	18.728	.053 397
1.94	6.9588	.14 370	2.44	11.473	.08 7161	2.94	18.916	.052 866
1.95	7.0287	.14 227	**2.45**	11.588	.08 6294	**2.95**	19.106	.052 340
1.96	7.0993	.14 086	2.46	11.705	.08 5435	2.96	19.298	.051 819
1.97	7.1707	.13 946	2.47	11.822	.08 4585	2.97	19.492	.051 303
1.98	7.2427	.13 807	2.48	11.941	.08 3743	2.98	19.688	.050 793
1.99	7.3155	.13 670	2.49	12.061	.08 2910	2.99	19.886	.050 287
2.00	7.3891	.13 534	**2.50**	12.182	.08 2085	**3.00**	20.086	.049 787

TABLE V (continued) Values of exponential functions 3.00–4.50

x	e^x	e^{-x}	x	e^x	e^{-x}	x	e^x	e^{-x}
3.00	20.086	.04 9787	**3.50**	33.115	.030 197	**4.00**	54.598	.01 8316
3.01	20.287	.04 9292	3.51	33.448	.029 897	4.01	55.147	.01 8133
3.02	20.491	.04 8801	3.52	33.784	.029 599	4.02	55.701	.01 7953
3.03	20.697	.04 8316	3.53	34.124	.029 305	4.03	56.261	.01 7774
3.04	20.905	.04 7835	3.54	34.467	.029 013	4.04	56.826	.01 7597
3.05	21.115	.04 7359	**3.55**	34.813	.028 725	**4.05**	57.397	.01 7422
3.06	21.328	.04 6888	3.56	35.163	.028 439	4.06	57.974	.01 7249
3.07	21.542	.04 6421	3.57	35.517	.028 156	4.07	58.557	.01 7077
3.08	21.758	.04 5959	3.58	35.874	.027 876	4.08	59.145	.01 6907
3.09	21.977	.04 5502	3.59	36.234	.027 598	4.09	59.740	.01 6739
3.10	22.198	.04 5049	**3.60**	36.598	.027 324	**4.10**	60.340	.01 6573
3.11	22.421	.04 4601	3.61	36.966	.027 052	4.11	60.947	.01 6408
3.12	22.646	.04 4157	3.62	37.338	.026 783	4.12	61.559	.01 6245
3.13	22.874	.04 3718	3.63	37.713	.026 516	4.13	62.178	.01 6083
3.14	23.104	.04 3283	3.64	38.092	.026 252	4.14	62.803	.01 5923
3.15	23.336	.04 2852	**3.65**	38.475	.025 991	**4.15**	63.434	.01 5764
3.16	23.571	.04 2426	3.66	38.861	.025 733	4.16	64.072	.01 5608
3.17	23.807	.04 2004	3.67	39.252	.025 476	4.17	64.715	.01 5452
3.18	24.047	.04 1586	3.68	39.646	.025 223	4.18	65.366	.01 5299
3.19	24.288	.04 1172	3.69	40.045	.024 972	4.19	66.023	.01 5146
3.20	24.533	.04 0762	**3.70**	40.447	.024 724	**4.20**	66.686	.01 4996
3.21	24.779	.04 0357	3.71	40.854	.024 478	4.21	67.357	.01 4846
3.22	25.028	.03 9955	3.72	41.264	.024 234	4.22	68.033	.01 4699
3.23	25.280	.03 9557	3.73	41.679	.023 993	4.23	68.717	.01 4552
3.24	25.534	.03 9164	3.74	42.098	.023 754	4.24	69.408	.01 4408
3.25	25.790	.03 8774	**3.75**	42.521	.023 518	**4.25**	70.105	.01 4264
3.26	26.050	.03 8388	3.76	42.948	.023 284	4.26	70.810	.01 4122
3.27	26.311	.03 8006	3.77	43.380	.023 052	4.27	71.522	.01 3982
3.28	26.576	.03 7628	3.78	43.816	.022 823	4.28	72.240	.01 3843
3.29	26.843	.03 7254	3.79	44.256	.022 596	4.29	72.966	.01 3705
3.30	27.113	.03 6883	**3.80**	44.701	.022 371	**4.30**	73.700	.01 3569
3.31	27.385	.03 6516	3.81	45.150	.022 148	4.31	74.440	.01 3434
3.32	27.660	.03 6153	3.82	45.604	.021 928	4.32	75.189	.01 3300
3.33	27.938	.03 5793	3.83	46.063	.021 710	4.33	75.944	.01 3168
3.34	28.219	.03 5437	3.84	46.525	.021 494	4.34	76.708	.01 3037
3.35	28.503	.03 5084	**3.85**	46.993	.021 280	**4.35**	77.478	.01 2907
3.36	28.789	.03 4735	3.86	47.465	.021 068	4.36	78.257	.01 2778
3.37	29.079	.03 4390	3.87	47.942	.020 858	4.37	79.044	.01 2651
3.38	29.371	.03 4047	3.88	48.424	.020 651	4.38	79.838	.01 2525
3.39	29.666	.03 3709	3.89	48.911	.020 445	4.39	80.640	.01 2401
3.40	29.964	.03 3373	**3.90**	49.402	.020 242	**4.40**	81.451	.01 2277
3.41	30.265	.03 3041	3.91	49.899	.020 041	4.41	82.269	.01 2155
3.42	30.569	.03 2712	3.92	50.400	.019 841	4.42	83.096	.01 2034
3.43	30.877	.03 2387	3.93	50.907	.019 644	4.43	83.931	.01 1914
3.44	31.187	.03 2065	3.94	51.419	.019 448	4.44	84.775	.01 1796
3.45	31.500	.03 1746	**3.95**	51.935	.019 255	**4.45**	85.627	.01 1679
3.46	31.817	.03 1430	3.96	52.457	.019 063	4.46	86.488	.01 1562
3.47	32.137	.03 1117	3.97	52.985	.018 873	4.47	87.357	.01 1447
3.48	32.460	.03 0807	3.98	53.517	.018 686	4.48	88.235	.01 1333
3.49	32.786	.03 0501	3.99	54.055	.018 500	4.49	89.121	.01 1221
3.50	33.115	.03 0197	**4.00**	54.598	.018 316	**4.50**	90.017	.01 1109

TABLE V (continued) Values of exponential functions 4.50–10.00

x	e^x	e^{-x}	x	e^x	e^{-x}	x	e^x	e^{-x}
4.50	90.017	.011 109	**5.00**	148.41	.00 67379	**7.50**	1 808.0	.000 5531
4.51	90.922	.010 998	5.05	156.02	.00 64093	7.55	1 900.7	.000 5261
4.52	91.836	.010 889	5.10	164.02	.00 60967	7.60	1 998.2	.000 5005
4.53	92.759	.010 781	5.15	172.43	.00 57994	7.65	2 100.6	.000 4760
4.54	93.691	.010 673	5.20	181.27	.00 55166	7.70	2 208.3	.000 4528
4.55	94.632	.010 567	**5.25**	190.57	.00 52475	**7.75**	2 321.6	.000 4307
4.56	95.583	.010 462	5.30	200.34	.00 49916	7.80	2 440.6	.000 4097
4.57	96.544	.010 358	5.35	210.61	.00 47482	7.85	2 565.7	.000 3898
4.58	97.514	.010 255	5.40	221.41	.00 45166	7.90	2 697.3	.000 3707
4.59	98.494	.010 153	5.45	232.76	.00 42963	7.95	2 835.6	.000 3527
4.60	99.484	.010 052	**5.50**	244.69	.00 40868	**8.00**	2 981.0	.000 3355
4.61	100.48	.009 9518	5.55	257.24	.00 38875	8.05	3 133.8	.000 3191
4.62	101.49	.009 8528	5.60	270.43	.00 36979	8.10	3 294.5	.000 3035
4.63	102.51	.009 7548	5.65	284.29	.00 35175	8.15	3 463.4	.000 2887
4.64	103.54	.009 6577	5.70	298.87	.00 33460	8.20	3 641.0	.000 2747
4.65	104.58	.009 5616	**5.75**	314.19	.00 31828	**8.25**	3 827.6	.000 2613
4.66	105.64	.009 4665	5.80	330.30	.00 30276	8.30	4 023.9	.000 2485
4.67	106.70	.009 3723	5.85	347.23	.00 28799	8.35	4 230.2	.000 2364
4.68	107.77	.009 2790	5.90	365.04	.00 27394	8.40	4 447.1	.000 2249
4.69	108.85	.009 1867	5.95	383.75	.00 26058	8.45	4 675.1	.000 2139
4.70	109.95	.009 0953	**6.00**	403.43	.00 24788	**8.50**	4 914.8	.000 2035
4.71	111.05	.009 0048	6.05	424.11	.00 23579	8.55	5 166.8	.000 1935
4.72	112.17	.008 9152	6.10	445.86	.00 22429	8.60	5 431.7	.000 1841
4.73	113.30	.008 8265	6.15	468.72	.00 21335	8.65	5 710.1	.000 1751
4.74	114.43	.008 7386	6.20	492.75	.00 20294	8.70	6 002.9	.000 1666
4.75	115.58	.008 6517	**6.25**	518.01	.00 19305	**8.75**	6 310.7	.000 1585
4.76	116.75	.008 5656	6.30	544.57	.00 18363	8.80	6 634.2	.000 1507
4.77	117.92	.008 4804	6.35	572.49	.00 17467	8.85	6 974.4	.000 1434
4.78	119.10	.008 3960	6.40	601.85	.00 16616	8.90	7 332.0	.000 1364
4.79	120.30	.008 3125	6.45	632.70	.00 15805	8.95	7 707.9	.000 1297
4.80	121.51	.008 2297	**6.50**	665.14	.00 15034	**9.00**	8 103.1	.000 1234
4.81	122.73	.008 1479	6.55	699.24	.00 14301	9.05	8 518.5	.000 1174
4.82	123.97	.008 0668	6.60	735.10	.00 13604	9.10	8 955.3	.000 1117
4.83	125.21	.007 9865	6.65	772.78	.00 12940	9.15	9 414.4	.000 1062
4.84	126.47	.007 9071	6.70	812.41	.00 12309	9.20	9 897.1	.000 1010
4.85	127.74	.007 8284	**6.75**	854.06	.00 11709	**9.25**	10 405	.000 0961
4.86	129.02	.007 7505	6.80	897.85	.00 11138	9.30	10 938	.000 0914
4.87	130.32	.007 6734	6.85	943.88	.00 10595	9.35	11 499	.000 0870
4.88	131.63	.007 5970	6.90	992.27	.00 10078	9.40	12 088	.000 0827
4.89	132.95	.007 5214	6.95	1 043.1	.00 09586	9.45	12 708	.000 0787
4.90	134.29	.007 4466	**7.00**	1 096.6	.00 09119	**9.50**	13 360	.000 0749
4.91	135.64	.007 3725	7.05	1 152.9	.00 08674	9.55	14 045	.000 0712
4.92	137.00	.007 2991	7.10	1 212.0	.00 08251	9.60	14 765	.000 0677
4.93	138.38	.007 2265	7.15	1 274.1	.00 07849	9.65	15 522	.000 0644
4.94	139.77	.007 1546	7.20	1 339.4	.00 07466	9.70	16 318	.000 0613
4.95	141.17	.007 0834	**7.25**	1 408.1	.00 07102	**9.75**	17 154	.000 0583
4.96	142.59	.007 0129	7.30	1 480.3	.00 06755	9.80	18 034	.000 0555
4.97	144.03	.006 9431	7.35	1 556.2	.00 06426	9.85	18 958	.000 0527
4.98	145.47	.006 8741	7.40	1 636.0	.00 06113	9.90	19 930	.000 0502
4.99	146.94	.006 8057	7.45	1 719.9	.00 05814	9.95	20 952	.000 0477
5.00	148.41	.006 7379	**7.50**	1 808.0	.00 05531	**10.00**	22 026	.000 0454

TABLE VI Basic derivative formulas

D1 $\frac{dC}{dx} = 0$

D2 $\frac{dx}{dx} = 1$ *(Special case of D5)*

D3 $\frac{d}{dx}(u \pm v) = \frac{du}{dx} \pm \frac{dv}{dx}$

D4 $\frac{d}{dx}(Cu) = C\frac{du}{dx}$

D5 $\frac{d}{dx}(x^n) = nx^{n-1}$ *(Power Rule)*

$\frac{d}{dx}(u^n) = nu^{n-1}\frac{du}{dx}$ *(General Power Rule)*

D6 $\frac{d}{dx}(uv) = u\frac{dv}{dx} + v\frac{du}{dx}$ *(Product Rule)*

D7 $\frac{d}{dx}\left(\frac{u}{v}\right) = \frac{v\frac{du}{dx} - u\frac{dv}{dx}}{v^2}$ $(v(x) \neq 0)$ *(Quotient Rule)*

D8 $\frac{dy}{dx} = \frac{dy}{du} \cdot \frac{du}{dx}$ *(Chain Rule)*

D9 $\frac{d}{dx}(\ln u) = \frac{1}{u} \cdot \frac{du}{dx}$

D10 $\frac{d}{dx}(e^u) = e^u \cdot \frac{du}{dx}$

D11 $\frac{d}{dx}(\sin u) = \cos u \cdot \frac{du}{dx}$

D12 $\frac{d}{dx}(\cos u) = -\sin u \cdot \frac{du}{dx}$

D13 $\frac{d}{dx}(\tan u) = \sec^2 u \cdot \frac{du}{dx}$

D14 $\frac{d}{dx}(\cot u) = -\csc^2 u\frac{du}{dx}$

D15 $\frac{d}{dx}(\sec u) = \sec u \tan u\frac{du}{dx}$

D16 $\frac{d}{dx}(\csc u) = -\csc u \cot u\frac{du}{dx}$

D17 $\frac{d}{dx}(\text{Sin}^{-1} u) = \frac{1}{\sqrt{1-u^2}} \cdot \frac{du}{dx}$ $(-1 < u(x) < 1)$

D18 $\frac{d}{dx}(\text{Cos}^{-1} u) = \frac{-1}{\sqrt{1-u^2}} \cdot \frac{du}{dx}$ $(-1 < u(x) < 1)$

D19 $\frac{d}{dx}(\text{Tan}^{-1} u) = \frac{1}{1+u^2} \cdot \frac{du}{dx}$

D20 $\frac{d}{dx}(\text{Cot}^{-1} u) = \frac{-1}{1+u^2} \cdot \frac{du}{dx}$

A constant of integration should be added to each of the following formulas. All arguments of the trigonometric functions are in radians and all logarithms are in the natural system.

FUNDAMENTAL FORMS

I1 $\int x^n\,dx = \dfrac{x^{n+1}}{n+1} \qquad (n \neq -1)$

I2 $\int [f(x) \pm g(x)]dx = \int f(x)\,dx \pm \int g(x)\,dx$

I3 $\int kf(x)\,dx = k\int f(x)\,dx$

I4 $\int \dfrac{dx}{x} = \ln|x|, \qquad x \neq 0$

I5 $\int \ln x\,dx = x\ln x - x$

I6 $\int e^x\,dx = e^x$

I7 $\int \cos x\,dx = \sin x$

I8 $\int \sin x\,dx = -\cos x$

I9 $\int \sec^2 x\,dx = \tan x$

I10 $\int \csc^2 x\,dx = -\cot x$

I11 $\int \sec x \tan x\,dx = \sec x$

I12 $\int \csc x \cot x\,dx = -\csc x$

I13 $\int \dfrac{dx}{1+x^2} = \mathrm{Tan}^{-1}\,x$

I14 $\int \dfrac{dx}{\sqrt{1-x^2}} = \mathrm{Sin}^{-1}\,x, \qquad -1 < x < 1$

I15 $\int u\,dv = uv - \int v\,du \qquad$ (integration by parts)

FORMS INVOLVING $ax + b$

1. $$\int (ax+b)^m\,dx = \frac{(ax+b)^{m+1}}{a(m+1)} \qquad (m \neq -1)$$

2. $$\int \frac{dx}{ax+b} = \frac{1}{a}\ln|ax+b|$$

3. $$\int \frac{dx}{(ax+b)^2} = -\frac{1}{a(ax+b)}$$

4. $$\int \frac{dx}{(ax+b)^3} = -\frac{1}{2a(ax+b)^2}$$

5. $$\int x(ax+b)^m\,dx = \frac{(ax+b)^{m+2}}{a^2(m+2)} - \frac{b(ax+b)^{m+1}}{a^2(m+1)}, \qquad (m \neq -1, -2)$$

6. $$\int \frac{x\,dx}{ax+b} = \frac{x}{a} - \frac{b}{a^2}\ln|ax+b|$$

7. $$\int \frac{x\,dx}{(ax+b)^2} = \frac{b}{a^2(ax+b)} + \frac{1}{a^2}\ln|ax+b|$$

8. $$\int \frac{x\,dx}{(ax+b)^3} = \frac{b}{2a^2(ax+b)^2} - \frac{1}{a^2(ax+b)}$$

9. $$\int x^m(ax+b)^n\,dx$$
$$= \frac{1}{a(m+n+1)}\left[x^m(ax+b)^{n+1} - mb\int x^{m-1}(ax+b)^n\,dx\right]$$
$$= \frac{1}{m+n+1}\left[x^{m+1}(ax+b)^n + nb\int x^m(ax+b)^{n-1}\,dx\right] \ (m > 0,\ m+n+1 \neq 0)$$

FORMS INVOLVING $ax + b$ AND $cx + d$ $\quad (bc \neq ad)$

10. $$\int \frac{dx}{(ax+b)(cx+d)} = \frac{1}{bc-ad}\ln\left|\frac{cx+d}{ax+b}\right|$$

11. $$\int \frac{x\,dx}{(ax+b)(cx+d)} = \frac{1}{bc-ad}\left[\frac{b}{a}\ln|ax+b| - \frac{d}{c}\ln|cx+d|\right]$$

12. $$\int \frac{dx}{(ax+b)^2(cx+d)} = \frac{1}{bc-ad}\left[\frac{1}{ax+b} + \frac{c}{bc-ad}\ln\left|\frac{cx+d}{ax+b}\right|\right]$$

13. $$\int \frac{dx}{(ax+b)^m(cx+d)^n} = \frac{-1}{(n-1)(bc-ad)}\left[\frac{1}{(ax+b)^{m-1}(cx+d)^{n-1}} + a(m+n-2)\int \frac{dx}{(ax+b)^m(cx+d)^{n-1}}\right]$$

FORMS INVOLVING $ax^2 + bx + c$

Let $X = ax^2 + bx + c$, $\quad D = b^2 - 4ac$.

14. $$\int X^m\,dx = \frac{1}{2a(2m+1)}\left[(2ax+b)X^m - Dm\int X^{m-1}\,dx\right]$$

15. $\displaystyle\int \frac{dx}{X} = \frac{1}{\sqrt{D}} \ln \left| \frac{2ax + b - \sqrt{D}}{2ax + b + \sqrt{D}} \right| \qquad (D > 0)$

16. $\displaystyle\int \frac{dx}{X} = \frac{2}{\sqrt{-D}} \operatorname{Tan}^{-1} \frac{2ax + b}{\sqrt{-D}} \qquad (D < 0)$

FORMS INVOLVING $\sqrt{ax + b}$, WHERE $ax + b > 0$

17. $\displaystyle\int \sqrt{ax + b}\, dx = \frac{2}{3a} \sqrt{(ax + b)^3}$

18. $\displaystyle\int x \sqrt{ax + b}\, dx = \frac{2(3ax - 2b)}{15a^2} \sqrt{(ax + b)^3}$

19. $\displaystyle\int x^2 \sqrt{ax + b}\, dx = \frac{2(15a^2x^2 - 12abx + 8b^2)}{105a^3} \sqrt{(ax + b)^3}$

20. $\displaystyle\int x^m \sqrt{ax + b}\, dx = \frac{2}{a(2m + 3)} \left[x^m \sqrt{(ax + b)^3} - mb \int x^{m-1} \sqrt{ax + b}\, dx \right]$

FORMS INVOLVING $\sqrt{a^2 + x^2}$

21 $\displaystyle\int \sqrt{a^2 + x^2}\, dx = \frac{x}{2} \sqrt{a^2 + x^2} + \frac{a^2}{2} \ln (x + \sqrt{a^2 + x^2})$

22. $\displaystyle\int x \sqrt{a^2 + x^2}\, dx = \frac{1}{3} \sqrt{(a^2 + x^2)^3}$

23. $\displaystyle\int x^2 \sqrt{a^2 + x^2}\, dx = \frac{x}{4} \sqrt{(a^2 + x^2)^3} - \frac{a^2x}{8} \sqrt{a^2 + x^2} - \frac{a^4}{8} \ln (x + \sqrt{a^2 + x^2})$

24. $\displaystyle\int x^3 \sqrt{a^2 + x^2}\, dx = \left(\frac{1}{5} x^2 - \frac{2}{15} a^2 \right) \sqrt{(a^2 + x^2)^3}$

25. $\displaystyle\int \frac{\sqrt{a^2 + x^2}\, dx}{x} = \sqrt{a^2 + x^2} - a \ln \left| \frac{a + \sqrt{a^2 + x^2}}{x} \right|$

26. $\displaystyle\int \frac{\sqrt{a^2 + x^2}\, dx}{x^2} = -\frac{\sqrt{a^2 + x^2}}{x} + \ln (x + \sqrt{a^2 + x^2})$

27. $\displaystyle\int \frac{\sqrt{a^2 + x^2}\, dx}{x^3} = -\frac{\sqrt{a^2 + x^2}}{2x^2} - \frac{1}{2a} \ln \left| \frac{a + \sqrt{a^2 + x^2}}{x} \right|$

28. $\displaystyle\int \frac{dx}{\sqrt{a^2 + x^2}} = \ln (x + \sqrt{a^2 + x^2})$

29. $\displaystyle\int \frac{x\, dx}{\sqrt{a^2 + x^2}} = \sqrt{a^2 + x^2}$

FORMS INVOLVING $\sqrt{a^2 - x^2}$ WHERE $a^2 > x^2$, $a > 0$

30. $\displaystyle\int \sqrt{a^2 - x^2}\, dx = \frac{1}{2} \left(x \sqrt{a^2 - x^2} + a^2 \operatorname{Sin}^{-1} \frac{x}{a} \right)$

31. $\displaystyle\int x \sqrt{a^2 - x^2}\, dx = -\frac{1}{3} \sqrt{(a^2 - x^2)^3}$

32. $\int x^2\sqrt{a^2-x^2}\,dx = -\frac{x}{4}\sqrt{(a^2-x^2)^3} + \frac{a^2}{8}\left(x\sqrt{a^2-x^2} + a^2\,\mathrm{Sin}^{-1}\frac{x}{a}\right)$

33. $\int x^3\sqrt{a^2-x^2}\,dx = \left(-\frac{1}{5}x^2 - \frac{2}{15}a^2\right)\sqrt{(a^2-x^2)^3}$

34. $\int \frac{\sqrt{a^2-x^2}\,dx}{x} = \sqrt{a^2-x^2} - a\ln\left|\frac{a+\sqrt{a^2-x^2}}{x}\right|$

35. $\int \frac{\sqrt{a^2-x^2}\,dx}{x^2} = -\frac{\sqrt{a^2-x^2}}{x} - \mathrm{Sin}^{-1}\frac{x}{a}$

36. $\int \frac{\sqrt{a^2-x^2}\,dx}{x^3} = -\frac{\sqrt{a^2-x^2}}{2x^2} + \frac{1}{2a}\ln\left|\frac{a+\sqrt{a^2-x^2}}{x}\right|$

37. $\int \frac{dx}{\sqrt{a^2-x^2}} = \mathrm{Sin}^{-1}\frac{x}{a}$

38. $\int \frac{x\,dx}{\sqrt{a^2-x^2}} = -\sqrt{a^2-x^2}$

39. $\int \frac{x^2\,dx}{\sqrt{a^2-x^2}} = -\frac{x}{2}\sqrt{a^2-x^2} + \frac{a^2}{2}\,\mathrm{Sin}^{-1}\frac{x}{a}$

40. $\int \frac{x^3\,dx}{\sqrt{a^2-x^2}} = \frac{1}{3}\sqrt{(a^2-x^2)^3} - a^2\sqrt{a^2-x^2}$

FORMS INVOLVING sin x

41. $\int \sin x\,dx = -\cos x$

42. $\int \sin^2 x\,dx = \frac{x}{2} - \frac{\sin 2x}{4}$

43. $\int \sin^3 x\,dx = \frac{\cos^3 x}{3} - \cos x$

44. $\int \sin^4 x\,dx = \frac{3x}{8} - \frac{\sin 2x}{4} + \frac{\sin 4x}{32}$

45. $\int \sin^{2m} x\,dx = -\frac{\sin^{2m-1} x\cos x}{2m} + \frac{2m-1}{2m}\int \sin^{2(m-1)} x\,dx$

46. $\int \frac{dx}{\sin^m x} = \int \csc^m x\,dx = -\frac{\cos x}{(m-1)\sin^{m-1} x} + \frac{m-2}{m-1}\int \frac{dx}{\sin^{m-2} x}\quad (m \neq 1)$

47. $\int x\sin x\,dx = \sin x - x\cos x$

48. $\int x^2\sin x\,dx = 2x\sin x - (x^2-2)\cos x$

49. $\int x^m\sin x\,dx = -x^m\cos x + mx^{m-1}\sin x - m(m-1)\int x^{m-2}\sin x\,dx$

FORMS INVOLVING cos x

50. $\int \cos x\,dx = \sin x$

51. $\int \cos^2 x \, dx = \frac{x}{2} + \frac{\sin 2x}{4}$

52. $\int \cos^3 x \, dx = \sin x - \frac{\sin^3 x}{3}$

53. $\int \cos^4 x \, dx = \frac{3x}{8} + \frac{\sin 2x}{4} + \frac{\sin 4x}{32}$

54. $\int \cos^{2m} x \, dx = \frac{1}{2m} \cos^{2m-1} x \sin x + \frac{2m-1}{2m} \int \cos^{2(m-1)} x \, dx$

55. $\int \cos^{2m+1} x \, dx = \int (1 - \sin^2 x)^m \cos x \, dx$

56. $\int \frac{dx}{\cos x} = \int \sec x \, dx = \ln |\sec x + \tan x| = \ln \left|\tan\left(\frac{\pi}{4} + \frac{x}{2}\right)\right|$

57. $\int \frac{dx}{\cos^2 x} = \int \sec^2 x \, dx = \tan x$

58. $\int \frac{dx}{\cos^m x} = \int \sec^m x \, dx = \frac{\sin x}{(m-1)\cos^{m-1} x} + \frac{m-2}{m-1} \int \frac{dx}{\cos^{m-2} x} \quad (m \neq 1)$

59. $\int x \cos x \, dx = \cos x + x \sin x$

60. $\int x^2 \cos x \, dx = 2x \cos x + (x^2 - 2) \sin x$

61. $\int x^m \cos x \, dx = x^m \sin x + mx^{m-1} \cos x - m(m-1) \int x^{m-2} \cos x \, dx$

EXPONENTIAL FORMS

For integrals involving a^x, substitute $a^x = e^{(\ln a)x}$ and use the forms that follow.

62. $\int e^{ax} \, dx = \frac{1}{a} e^{ax}$

63. $\int xe^{ax} \, dx = \frac{e^{ax}}{a^2} (ax - 1)$

64. $\int x^m e^{ax} \, dx = \frac{1}{a} x^m e^{ax} - \frac{m}{a} \int x^{m-1} e^{ax} \, dx$

65. $\int x^p e^{ax} \, dx = \frac{e^{ax}}{a^{p+1}} [(ax)^p - p(ax)^{p-1} + p(p-1)(ax)^{p-2} - \cdots + (-1)^p p!] \quad (p = \text{integer})$

66. $\int \frac{e^{ax} \, dx}{x} = \ln |x| + ax + \frac{(ax)^2}{2 \cdot 2!} + \frac{(ax)^3}{3 \cdot 3!} + \cdots \quad (0 < x^2 < \infty)$

67. $\int \frac{e^{ax} \, dx}{x^m} = -\frac{e^{ax}}{(m-1)x^{m-1}} + \frac{a}{m-1} \int \frac{e^{ax} \, dx}{x^{m-1}} \quad (m \neq 1)$

68. $\int \frac{e^{ax} \, dx}{b + ce^{ax}} = \frac{1}{ac} \ln |b + ce^{ax}|$

69. $\int \frac{dx}{b + ce^{ax}} = \frac{1}{ab} \ln \left|\frac{e^{ax}}{b + ce^{ax}}\right|$

70. $\displaystyle\int \frac{dx}{ae^{cx} + be^{-cx}} = \frac{1}{c\sqrt{ab}} \operatorname{Tan}^{-1}\left(e^{cx}\sqrt{\frac{a}{b}}\right)$ $(ab > 0)$

71. $\displaystyle\int e^{ax} \sin bx \, dx = \frac{e^{ax}}{a^2 + b^2}(a \sin bx - b \cos bx)$

LOGARITHMIC FORMS

In these forms, $x > 0$.

72. $\displaystyle\int \ln x \, dx = x(\ln x - 1)$

73. $\displaystyle\int (\ln x)^m \, dx = x(\ln x)^m - m\int (\ln x)^{m-1} \, dx$

74. $\displaystyle\int \frac{dx}{(\ln x)^m} = -\frac{x}{(m-1)(\ln x)^{m-1}} + \frac{1}{m-1}\int \frac{dx}{(\ln x)^{m-1}}$

75. $\displaystyle\int x^m \ln x \, dx = x^{m+1}\left[\frac{\ln x}{m+1} - \frac{1}{(m+1)^2}\right]$ $(m \neq -1)$

76. $\displaystyle\int \frac{(\ln x)^m \, dx}{x} = \frac{(\ln x)^{m+1}}{m+1}$ $(m \neq -1)$

77. $\displaystyle\int \frac{dx}{x \ln x} = \ln |\ln x|$

78. $\displaystyle\int x^m (\ln x)^n \, dx = \frac{x^{m+1}}{m+1}(\ln x)^n - \frac{n}{m+1}\int x^m (\ln x)^{n-1} \, dx$ $(m, n \neq -1)$

79. $\displaystyle\int \frac{x^m \, dx}{(\ln x)^n} = -\frac{x^{m+1}}{(n-1)(\ln x)^{n-1}} + \frac{m+1}{n-1}\int \frac{x^m \, dx}{(\ln x)^{n-1}}$ $(n \neq 1)$

TABLE VII Basic trigonometric identities

T1 $\tan \alpha = \dfrac{\sin \alpha}{\cos \alpha} \qquad (\cos \alpha \neq 0)$

T2 $\cot \alpha = \dfrac{\cos \alpha}{\sin \alpha} \qquad (\sin \alpha \neq 0)$

T3 $\sec \alpha = \dfrac{1}{\cos \alpha} \qquad (\cos \alpha \neq 0)$

T4 $\csc \alpha = \dfrac{1}{\sin \alpha} \qquad (\sin \alpha \neq 0)$

T5 $\cos^2 \alpha + \sin^2 \alpha = 1$

T6 $1 + \tan^2 \alpha = \sec^2 \alpha$

T7 $\cot^2 \alpha + 1 = \csc^2 \alpha$

T8 $\sin (\alpha \pm \beta) = \sin \alpha \cos \beta \pm \cos \alpha \sin \beta$

T9 $\cos (\alpha \pm \beta) = \cos \alpha \cos \beta \mp \sin \alpha \sin \beta$

T10 $\cos (-\alpha) = \cos \alpha, \qquad \sin (-\alpha) = -\sin \alpha$

T11 $\sin \left(\dfrac{\pi}{2} - \alpha\right) = \cos \alpha, \qquad \cos \left(\dfrac{\pi}{2} - \alpha\right) = \sin \alpha$

T12 $\sin 2\alpha = 2 \sin \alpha \cos \alpha$

T13 $\cos 2\alpha = \cos^2 \alpha - \sin^2 \alpha$

T14 $\cos 2\alpha = 2 \cos^2 \alpha - 1$

T15 $\cos 2\alpha = 1 - 2 \sin^2 \alpha$

T16 $\sin \dfrac{\theta}{2} = \pm \sqrt{\dfrac{1 - \cos \theta}{2}}$

T17 $\cos \dfrac{\theta}{2} = \pm \sqrt{\dfrac{1 + \cos \theta}{2}}$

T18 $\tan \dfrac{\theta}{2} = \pm \sqrt{\dfrac{1 - \cos \theta}{1 + \cos \theta}} = \dfrac{1 - \cos \theta}{\sin \theta} = \dfrac{\sin \theta}{1 + \cos \theta}$

In T16, T17, and T18 the sign before the radical is determined by the quadrant.

ANSWERS TO MARKED EXERCISES

Chapter 1

Page 8 **Section 1.2** **1. (b)** 5; **(d)** 5. **2. (b)** Isosceles, right. **3. (b)** $(2, -\frac{1}{2})$; **(c)** $(-1, 1)$. **4. (b)** $x^2 - 2x + y^2 + 2y = 2$; **(f)** $x^2 - 6x + y^2 + 8y + 9 = 0$; **(h)** $x^2 - 8x + y^2 + 2y + 13 = 0$. **5. (a)** Circle of radius 3, center at origin; **(e)** The single point $(0, -2)$; **(h)** Circle of radius 2, center $(3, -3)$; **(k)** $\varnothing$. **6. (b)** Center $(\frac{1}{2}, -\frac{1}{2})$, $r = 1/\sqrt{2}$; **(d)** $\varnothing$; **(f)** Center $(-\frac{1}{2}, \frac{3}{2})$, $r = 1/\sqrt{2}$. **8.** $x + y = 6$.

Page 17 **Section 1.3** **1. (c)** -3; **(e)** $\frac{1}{8}$; **(g)** $\frac{8}{5}$. **2. (b)** $4x + y = 12$; **(e)** $2x - y = 3$. **3. (a)** $5x + y = 6$; **(b)** $x + 2y = 5$; **(h)** $x + y = 5$. **4. (b)** $y = -2x + \frac{5}{2}$; **(e)** $y = -x + 1$; **(i)** $y = -2x/3$.

Page 21 **Section 1.4** **1. (a)** x-axis; **(c)** None; **(g)** Both axes and origin; **(k)** Both axes and origin.

Page 26 **Section 1.5** **1. (a)** $y' = x'^2$, new origin at $(-2, 0)$; **(c)** $y' = x'^2$, new origin at $(1, -4)$; **(e)** $y' = 6x'^2$, new origin at $(-1, -13)$. **2. (b)** $5x' + y' = 0$; **(e)** $x'^2 + y'^2 = \frac{9}{4}$; **(h)** $x'^2 + y'^2 = -2$.

Page 33 **Section 1.6** **4. (b)** $x'y' = 1, x' = x + 2, y' = y - 2$; **(d)** $x'y' = -10, x' = x + 3, y' = y - 4$; **(f)** $y' = 2x'^2, x' = x + 2, y' = y$; **(h)** $9x'^2 + 4y'^2 = 36, x' = x + 1, y' = y + 1$; **(j)** $2x'^2 + 3y'^2 = 6, x' = x + 1, y' = y + 5$.

Chapter 2

Section 2.2 **1.** **(b)** -4; **(d)** $\frac{3}{8}$. **2.** **(a)** $\frac{1}{2}$; **(d)** -2; **(e)** $5x - 3$. **3.** **(b)** 4; **(c)** 5; **(g)** 4. **4.** **(b)** $4x - y = 4$; **(c)** $y = 5x - 7$; **(g)** $4x - y = 2$. Page 41

Section 2.3 **1.** **(c)** 2; **(f)** 3; **(h)** 13; **(j)** 7. **2.** **(b)** $9x^2 + 4x + 6$; **(e)** $6x + 5$; **(g)** $2u$; **(i)** $2t - 1$. **3.** **(b)** $3x - y = 2$; **(e)** $40x - y + 85 = 0$; **(h)** $y = x$; **(i)** $x + 9y = 6$. **4.** **(c)** $x = \frac{1}{2}$; **(e)** $x = -1, \frac{1}{3}$; **(i)** Nowhere. Page 45

Section 2.4 **1.** **(c)** $10x$; **(d)** $14x - 32$. **2.** **(b)** $v(t) = -16t + 16, a(t) = -16$; **(d)** $v(t) = t - 6, a(t) = 1$; **(h)** $v(t) = 2 - t^{-2}, a(t) = 2t^{-3}$; **(j)** $v(t) = 6t + 2$; $a(t) = 6$. **4.** Rises for 3 seconds, maximum height is 144 feet. **6.** $MC(x) = C'(x) = 14x/1000 + 9/100$. Page 49

Section 2.5 **1.** **(a)** If $\epsilon > 0$ there exists $\delta > 0$ such that if $0 < |x - 2| < \delta$, then $|(3x - 1) - 5| < \epsilon$. **2.** **(a)** No limit; **(d)** -1. **3.** **(c)** 3; **(g)** No limit. **4.** **(c)** δ can be any positive number $\leqslant 0.1$. Page 54

Section 2.6 **1.** The values of the limits are given: **(b)** -1; **(d)** 0; **(p)** $\frac{3}{4}$. Page 57

Section 2.7 **1.** **(a)** Continuous; **(b)** Discontinuous. **2.** **(c)** Everywhere except $x = -1$; **(e)** Everywhere except $x = -3$; **(f)** Everywhere except $x = -3, 1$; **(j)** Everywhere; **(m)** Everywhere. Page 61

Section 2.8 **1.** **(d)** 0; **(f)** $10x$; **(k)** $6x^2 + 6x - 5$; **(o)** $9x^8 + 12x - 1$; **(r)** $24x^{11} + 7x^6 + 10x^4 + 4$. **2.** **(c)** $-\frac{1}{3}$; **(f)** $-1, 0$; **(i)** $-\frac{1}{3}$. **3.** $(\frac{4}{3}, 0), (0, -8)$. **6.** $2\pi r$. Page 65

Section 2.9 **1.** **(b)** $8x^3 + 6x^2 + 2x - 1$; **(e)** $45x^8 + 5x^4 + 20x^3$. **2.** **(b)** $-2/x^2$; **(d)** $-4x/(x^2 - 1)^2$; **(f)** $(-4x^2 - 2x + 1)/(x^2 - x)^2$; **(i)** $(x^2 + 4x - 1)/(x + 2)^2$. **3.** **(b)** $-8(3 - 2x)^3$; **(d)** $9(2x - 2)(x^2 - 2x + 3)^8$; **(i)** $\frac{1}{5}(x^2 + x + 17)^{-4/5}(2x + 1)$; **(k)** $2/\sqrt{8 + 4x}$; **(n)** $5x(x^2 + 4)^{3/2}$; **(r)** $\frac{25}{2}(x + 1)(5x^2 + 10x - 7)^{1/4}$. **4.** **(b)** $(x^2 - 1)^4 \cdot (9x^2 + 1)/4x^2$; **(d)** $(9x^2 - 13x)/2(x - 1)^{3/2}$; **(g)** $(3x^3 + x)/\sqrt{x^2 + 1}$. **5.** **(a)** $2x + y = 7$; **(d)** $x - 6y + 9 = 0$. Page 69

Section 2.10 **1.** **(a)** $6x^2 + 20$; **(d)** $135x^2 - 13$. **2.** **(b)** $dy/dx = 9u^2(u^3 + 1)^2\, du/dx = \frac{9}{2}((x - 1)^{3/2} + 1)^2(x - 1)^{1/2}$; **(d)** $12(u + 3)^{11}\, du/dx = 6((x - 1)^{1/2} + 3)^{11}/\sqrt{x - 1}$; **(f)** $2(3u^2 - 2u + 5)(6u - 2)du/dx = (3x + 2 - 2\sqrt{x - 1})(6\sqrt{x - 1} - 2)/\sqrt{x - 1}$. **3.** **(a)** $[2(x^2 + 1) + 5] \cdot (2x)$; **(c)** $\frac{1}{2}[(1/x + x)^{-1/2} - (1/x + x)^{-3/2}](1 - 1/x^2)$; **(e)** $\dfrac{x + 4\sqrt{x} - 7}{(\sqrt{x} + 2)^2} \cdot \dfrac{1}{2\sqrt{x}}$; **4.** **(b)** 0, **(d)** $-1/24$. Page 74

Chapter 3

Section 3.1 **1.** **(b)** Endpoints $x = -2, 1$, neither open nor closed; **(g)** Endpoints $x = -1, 1$, closed. **2.** **(b)** Max value $f(-1) = 0$, local max at $x = -1$; **(d)** Max value $f(0) = -1$, Min value $f(1) = -4$, no local extrema; **(g)** Max value at $x = \pm 2$, local min at $x = 0$. Page 80

Section 3.2 **1.** **(b)** Max at $x = \pm 3$, min at $x = \pm 1$; **(d)** Max at $x = -1$, min at $x = 3$; **(g)** Max at $x = \frac{1}{2}$, min at $x = -1$; **(i)** Max at $x = -2$, min at $x = -1$; **(l)** Max at $x = -1$, min at $x = 2$. **5.** $3 \times 3 \times 3$. **8.** \$55 per set. Page 85

Section 3.3 **1.** **(a)** Local min at $x = -1$; **(d)** Local max at $x = -\sqrt{2}$, local min at $x = \sqrt{2}$; **(h)** Local min at $x = -1, 3$, local max at $x = 1$; **(j)** Local min at $x = \frac{1}{16}$; **(l)** Local max at $x = 1$, minimum of zero at $x = 0$, no local min. **3.** 50. **6.** Height = radius = $\sqrt[3]{V_0/\pi}$. **7.** **(a)** $3 \times 3 \times 3$. Page 94

Page 100 **Section 3.4** **1. (c)** Local max at $x = -\sqrt{\frac{2}{3}}$, local min at $x = \sqrt{\frac{2}{3}}$, point of inflection at $x = 0$; **(e)** No local extrema, point of inflection at $x = 0$; **(h)** Local max at $x = -1$, local min at $x = 1$, no point of inflection; **(j)** Local max at $x = -1$, local min at $x = -5, 3$, points of inflection at $x = (-3 \pm 4\sqrt{3})/3$. **2. (b)** Concave upward if $x < -1$ or $x > 1$, concave downward if $-1 < x < 1$; **(d)** Concave upward if $x < -1$ or $x > 0$, concave downward if $-1 < x < 0$; **(g)** Concave upward if $x < 0$ or $x > 1$, concave downward if $0 < x < 1$; **(j)** Concave upward if $x < -1/\sqrt{3}$ or $x > 1/\sqrt{3}$, concave downward if $-1/\sqrt{3} < x < 1/\sqrt{3}$.

Page 103 **Section 3.5** **1. (b)** Local max at $x = -6$, local min at $x = 0$, inflection point at $x = -3$; **(d)** Local max at $x = \frac{1}{4}$, points of inflection at $x = \frac{1}{2}, 1$; **(g)** Local min at $x = -1, 0$, local max at $x = -\frac{1}{2}$, points of inflection at $x = (-3 \pm \sqrt{3})/6$; **(i)** Local max at $x = -1$, local min at $x = 3$, point of inflection at $x = 2$. **6.** $x = \pm\frac{1}{2}$, line: $x = 4y$.

Page 108 **Section 3.6** **1. (b)** Max profit at $x = 10$. **3.** $2000\sqrt{10} \approx 6300$. **5.** $x = 10$. **7. (a)** 33 shipments; **(b)** About 32 shipments. **8.** $x = 3$. **9. (a)** 4000.

Page 113 **Section 3.7** **1. (b)** $x = \frac{2}{3}$. **4. (b)** No. **5.** One such function is $f(x) = x^3$.

Page 118 **Section 3.8** **1. (a)** $-4x/3y$; **(d)** $\sqrt{y/x}$. **2. (b)** -2; **(d)** -1. **3. (b)** $-\dfrac{1}{4y} - \dfrac{x^2}{16y^3} = -\dfrac{1}{4y^3}$; **(d)** $2y/x^2$.

Page 122 **Section 3.9** **2.** $2/\pi$ ft/min. **3. (b)** $-\frac{9}{4}$ ft/sec. **5. (a)** $dC/dt = 109{,}600$ at $t = 5$; **(c)** $dP/dt = -19/200$ at $t = 5$.

Page 127 **Section 3.10** **1. (a)** $3x^2dx$; **(d)** $6x^2(x^3 + 1)dx$; **(h)** $\dfrac{2 - 2x^2}{(x^2 + 1)^2}\,dx$. **2. (b)** 12; **(d)** 7; **(f)** 7. **3. (b)** $6x(3 + 1/\sqrt{x^2 + 1})$; **(d)** $2(u^2 - 1)(5u^2 + 2u - 1)3x/\sqrt{v}$.

Page 131 **Section 3.11** **1. (a)** 10.005 (approx.); **(c)** 4.0208 (approx.). **2. (b)** 17.7. **3. (b)** $17.7/(-45) \approx -.393$. **4. (a)** $95,000.

Chapter 4

Page 137 **Section 4.1** **1. (a)** $3x^2/2 + C$; **(d)** $4x^{3/2}/3 + C$; **(f)** $x^2/2 - 1/x + C$; **(g)** $14x^{3/2}/3 - 9x^{4/3}/4 + C$; **(j)** $3x^5 + 2x^4 - x^3 + 2x + C$; **(k)** $2x^{3/2}/3 + 3x^{4/3}/4 + C$. **2. (a)** $(x - 1)^4/4 + C$; **(c)** $-(2 - x)^3/3 + C$; **(f)** $\frac{2}{3}x^{3/2} + x + C$; **(i)** $-(x^2 - x - 1)^{-2}/2 + C$; **(l)** $x^4/4 + 2x^3/3 + x^2/2 + C$; **(n)** $-(3x^4 - 3x^2 - 6)^{-1} + C$.

Page 141 **Section 4.2** **1. (a)** $x^3 - x^2 + 7x + 3$; **(c)** $\frac{2}{3}x^{3/2} - \frac{25}{3}$. **2. (a)** $x^2/100 + 2x + 40$; **(c)** $.001x^3/3 + .03x^2/2 + 9x + 40$. **3. (b)** $s(t) = t^4/12 + 4t^2 - 3t + 10$; **(d)** $s(t) = (t + 2)^{-1}/2 + 9t/8 + \frac{7}{4}$. **5.** 4 seconds; $t = 1$, $t = 3$.

Page 146 **Section 4.3** **1. (a)** $A = \frac{10}{3}$; **(c)** $A = \frac{16}{3}$.

Page 149 **Section 4.4** **1. (a)** $\frac{1}{3} + \frac{2}{3} + \frac{3}{3} + \frac{4}{3} + \frac{5}{3} + \frac{6}{3}$; **(b)** $1 + 2 + 3 + 4 + 5$; **(f)** $[4f(4)]^4 + [5f(5)]^5 + [6f(6)]^6 + [7f(7)]^7$.

Page 154 **Section 4.5** **1. (a)** 6; **(c)** -12; **(e)** 63. **2. (b)** No, the integral is zero, but the area is a positive quantity.

Page 157 **Section 4.6** **1. (b)** 0; **(c)** -5; **(f)** -24. **2. (b)** $F(x) = (x^3 + 1)/3$.

Page 160 **Section 4.7** **1. (b)** $-\frac{1}{6}$; **(d)** $\frac{8}{3}$; **(f)** 6; **(h)** $-\frac{15}{4}$; **(j)** 6.

Section 4.8 **1. (a)** $\frac{9}{2}$; **(c)** $\frac{2}{3}$; **(f)** $\frac{1}{3}$; **(h)** $\frac{9}{2}$; **(j)** $\frac{4}{3}$. **2. (b)** 7278. **3. (b)** 727.80. **4. (b)** $s = \frac{119}{6}$, average velocity $= \frac{119}{30}$; **(d)** $s = \frac{290}{3}$, average velocity $= \frac{145}{12}$. **5. (b)** $-\frac{1}{6}$; **(d)** 3. Page 168

Section 4.9 **1. (a)** ∞; **(b)** $\frac{2}{3}$; **(c)** ∞; **(e)** 0; **(g)** 0. **2. (b)** ∞; **(d)** ∞; **(f)** $\frac{7}{4}$. Page 176

Section 4.10 **1. (b)** ∞; **(c)** 0; **(f)** No limit; **(h)** 0. **4. (a)** $\frac{13}{2}$; **(c)** 4; **(e)** 1. Page 181

Section 4.11 **1. (b)** $p = \frac{1}{2}$; **(d)** $p = \frac{3}{4}$. **2. (b)** $p = 14(.6)^{13} - 13(.6)^{14} \approx .008$. **3.** $\frac{5}{9}$. Page 185

Chapter 5

Section 5.1 **1. (b)** $7^{1/6}$; **(e)** 3^{a+b-c}. **2. (b)** $\frac{1}{6}$; **(c)** 4; **(e)** -16.2702. **3. (a)** 28.3; **(b)** 1.92; **(d)** 3.71; **(e)** 13.1. **4. (a)** $7.08 = e^{1.9573}$; **(d)** $\log_{10}\left(\frac{1}{1000}\right) = -3$. Page 191

Section 5.2 **1. (a)** $1/x$; **(d)** $1 + \ln x$; **(f)** $(x^2 + 1 - 2x^2 \ln x)/x(x^2+1)^2$; **(h)** $10[\ln(x-1)]^9/(x-1)$; **(k)** $-3/\sqrt{x^2+1}$; **(l)** $[(x-1)(1 + \ln x) - 6x \ln x]/(x-1)^7$. **2. (a)** Concave upward for all $x > 0$, local min at $x = 1/e$. **3. (b)** $x \log_8 e/(x^2+1)$; **(c)** $3(2x-1)[\log_3 (x^2 - x - 1)]^2 \log_3 e/(x^2 - x - 1)$. **4. (a)** $15x^2/(x^3+1) + 2x/3(x^2+5) - 16x/9(x^2+3)$. **6. (a)** $x - 3y = 6 - 3 \ln 36$. Page 196

Section 5.3 **1. (b)** $(7x^3 - 5 \ln|x|)/3 + C$; **(d)** $\ln|x-3| + C$; **(f)** $(\ln x)^3/3 + C$; **(h)** $\frac{(\ln x)^3}{3} + \frac{3(\ln x)^2}{2} + 7 \ln x + C$; **(i)** $4x + 3 \ln|x| - x^{-1} + C$. **2. (b)** Improper; **(d)** Improper; **(f)** $(\ln 5)/2$. **5.** $2 \ln 6$. Page 199

Section 5.4 **1. (a)** $2e^x$; **(d)** $4e^{2x}/(e^{2x}+1)^2$; **(f)** $2xe^{x^2}$; **(g)** $e^x/(e^x+1)$. **3. (b)** $2^{x+1}3^{x-1} \ln 6$; **(c)** $-\ln 3 \cdot 3^{(x+1)/2x^3}\,[(2x+3)/2x^4]$. **4. (a)** $e^{3x}/3 + C$; **(c)** $e^{x^2-2x+1}/2 + C$; **(d)** $(1 - e^{-2})/2$; **(h)** $-e^{x^{-2}}/2 + C$; **(i)** $\frac{1}{2}(1 - e^{2x}) + C$. **6.** 1. **8. (a)** .3413; **(d)** .0049. **9. (b)** .1587; **(c)** .0227. Page 205

Section 5.5 **1. (c)** $y = 1 + Ce^{x+x^2/2}$; **(f)** $y = (1 + Ce^{-2x^2})/4$; **(h)** $y = Ce^{-7x}$. **2. (b)** $P = -e^{t/5}$; **(d)** $y = 3e^{x+x^2/2}$. **3.** 9.9 million (approx.). **4.** July 21. Page 211

Section 5.6 **1.** 3170 years (approx.). **3.** Half-life = 2.7 years (approx.). **5.** 0.512 mg. **7.** \$4635 (approx.). **9.** \$1,485,000 (approx.) if valued together, \$1,482,000 (approx.) if valued separately. Page 216

Section 5.7 **1. (a)** 9.8 percent (approx.). **2. (b)** $\ln|5+x| - \ln|5-x| = 10t + C$, or $x = \frac{5(Ke^{10t}-1)}{Ke^{10t}+1}$; **(c)** $x = \frac{1 - Ce^{14t}}{1 + Ce^{14t}}$. **3.** 97. **4. (a)** \$2744.05. **5.** \$13,272. Page 220

Chapter 6

Section 6.1 **1. (a)** $\pi/12$; **(e)** $31\pi/18$; **(g)** $\pi/5$. **2. (b)** 15°; **(d)** 240°; **(f)** 72°. **3. (b)** 10π cm²; **(d)** 8π in²; **(f)** $3\pi/4$ mi². Page 224

Section 6.2 **1. (b)** $\pi/4$; **(d)** $\pi/4$; **(f)** 0.47; **(h)** 0.77; **(j)** 0.5. **2.** Coordinates are given for P_α: **(b)** $(-\sqrt{2}/2, -\sqrt{2}/2)$; **(c)** (0, 1); **(f)** (−.89157, −.45289); **(h)** (−.71791, −.69614); **(j)** (−.87758, −.47943). **3. (b)** 1; **(e)** $1/\sqrt{2}$; **(g)** $\sqrt{3}$; **(h)** −.76433; **(j)** −49.993; **(l)** −.15932. Page 231

Section 6.3 **1. (b)** Period $= \pi$, amplitude = 1; **(c)** Period $= \pi$, amplitude = 2; **(f)** Period $= 2\pi/3$, amplitude = 2. Page 238

Section 6.4 **1. (a)** $y = \frac{1}{2} \sin 2x$; **(c)** $y = (\sin x - \cos x)\sqrt{2}$; **(e)** $y = \sin(x + \pi/4)$. Page 240

2. (a) $\dfrac{\sqrt{2}-\sqrt{3}}{2}$; **(d)** $\dfrac{-1-\sqrt{3}}{2\sqrt{2}} \approx -.9657$; **(f)** $\dfrac{1-\sqrt{3}}{1+\sqrt{3}} \approx .2679$. **3.** $\cos\alpha = -\frac{3}{5}$, $\tan\alpha = \frac{4}{3}$, etc. **4. (a)** $-\frac{56}{65}$; **(d)** $\frac{7}{25}$; **(e)** $5/\sqrt{26}$; **(g)** $\frac{7}{24}$.

Page 247

Section 6.5 **1. (b)** $\lim_{x\to 0^+} \dfrac{\cos 2x}{x} = \infty$, $\lim_{x\to 0^-} \dfrac{\cos 2x}{x} = -\infty$; **(c)** $\frac{3}{2}$; **(e)** 0; **(f)** 1. **2. (d)** $2x \cdot \cos(x^2+5)$; **(e)** $-(2x-1)\sin(x^2-x-2)$; **(f)** $6\sin 3x\cos 3x = 3\sin 6x$; **(h)** $-e^x \cdot \csc^2 e^x$; **(j)** $2x \cdot \cos x^2 \cdot e^{\sin x^2}$. **3. (a)** Local max at $x = 0, \pi, 2\pi$, local min at $x = \pi/2, 3\pi/2$; **(c)** Local max at $x = 3\pi/4$, local min at $x = \pi/4$. **6. (a)** Answers for the interval $[0, 2\pi]$: inflection points at $x = \pi/2$ and $3\pi/2$, concave downward for $0 \le x < \pi/2$, concave upward for $\pi/2 < x < 3\pi/2$, concave downward for $3\pi/2 < x \le 2\pi$; **(d)** Answers for the interval $(-\pi/2, \pi/2)$: inflection point at $x = 0$, concave downward if $-\pi/2 < x < 0$, concave upward if $0 < x < \pi/2$.

Page 251

Section 6.6 **1. (b)** $-\dfrac{\cos(3x+3)}{3} + C$; **(c)** $-\dfrac{\cos 2x^2}{4} + C$; **(e)** $-e^{\cos x} + C$; **(h)** $\dfrac{4x - \sin 4x}{32} + C$; **(j)** $-\frac{1}{2}\cot 2x + C$; **(l)** $-\frac{1}{6}\cos^3 2x + C$; **(m)** $-\ln(1+\cos x) + C$; **(o)** $\frac{1}{3}(1+2\sin x)^{3/2} + C$. **2. (b)** $\frac{1}{2}$; **(c)** $\frac{1}{2}\ln 2$. **3.** $2\sqrt{2}$.

Page 254

Section 6.7 **1. (b)** 0.72; **(d)** -1.10; **(f)** 2.81. **2. (b)** $x = \frac{1}{2}\operatorname{Sin}^{-1} y$; **(d)** $x = \frac{1}{3}\cos y$; **(f)** $x = \sin(2y+1) - 2$. **3. (a)** $\pi/4$; **(d)** $\frac{2}{3}\operatorname{Sin}^{-1}(3x) + C$.

Chapter 7

Page 258

Section 7.1 **1. (a)** $2e^{\sqrt{t}} + C$; **(d)** $\frac{1}{3}(\ln 31 - \ln 4)$; **(f)** $\frac{1}{2}(x - \ln x)^2 + C$; **(i)** $2\sqrt{1+e} - 2\sqrt{2}$. **2. (a)** -2; **(b)** $e + \ln(e+1) - 1 - \ln 2$; **(e)** $-(\ln x)^{-1} + C$.

Page 260

Section 7.2 **1. (b)** $\dfrac{x^2}{4}(2\ln x - 1) + C$; **(d)** $\dfrac{x^3}{3}\ln x - \dfrac{x^3}{9} + C$; **(e)** $x\ln x - x + C$; **(h)** $\dfrac{e^x}{2}(\sin x + \cos x) + C$. **5. (a)** $-x^3\cos x + 3x^2\sin x + 6x\cos x - 6\sin x + C$.

Page 264

Section 7.3 **1. (b)** $-5\ln|x-2| + 7\ln|x-3| + C$; **(d)** $-\frac{1}{7}\ln|x| + \frac{22}{7}\ln|x+7| + C$; **(f)** $3\ln|x+2| + 2\ln|x+1| + C$; **(h)** $\dfrac{1}{a-b}(\ln|x-a| - \ln|x-b|) + C$ if $a \ne b$, $-(x-a)^{-1} + C$ if $a = b$.

Page 265

Section 7.4 **1. (a)** $\sqrt{16+x^2} - 4\ln\left|\dfrac{4+\sqrt{16+x^2}}{x}\right| + C$ (by **25**); **(c)** $\dfrac{1}{\sqrt{7}}\operatorname{Tan}^{-1}\left(\dfrac{4x+1}{\sqrt{7}}\right) + C$ (by **16**); **(f)** Use **21; (i)** Use **32; (h)** Use **77; (n)** Use **75; (p)** Use **69.**

Page 269

Section 7.5 **1. (b)** $\dfrac{1}{3}\left[\dfrac{\sqrt{2}}{2} + \dfrac{\sqrt{163}}{9} + \dfrac{\sqrt{178}}{9} + \sqrt{3} + \dfrac{\sqrt{418}}{9} + \dfrac{\sqrt{787}}{9} + \dfrac{\sqrt{18}}{2}\right] \approx 4.28$; **(d)** $\dfrac{1}{5}\left[\dfrac{1}{4} + \dfrac{5}{11} + \dfrac{5}{12} + \dfrac{5}{13} + \dfrac{5}{14} + \dfrac{1}{6}\right] \approx 0.406$; **(f)** $\dfrac{1}{3}\left[\dfrac{1}{2} + \dfrac{\sqrt[3]{28}}{3} + \dfrac{\sqrt[3]{35}}{3} + \sqrt[3]{2} + \dfrac{\sqrt[3]{91}}{3} + \dfrac{\sqrt[3]{152}}{3} + \dfrac{\sqrt[3]{9}}{2}\right] \approx 2.727$; **(h)** $\dfrac{1}{2}\left[\dfrac{\ln 2}{2} + \ln 3 + \ln 4 + \ln 5 + \ln 6 + \ln 7 + \dfrac{\ln 8}{2}\right] \approx 4.609$.

Section 7.6 **1. (a)** $\frac{1}{30}[1 + 4e^{-0.01} + 2e^{-0.04} + 4e^{-0.09} + \cdots + e^{-1}] \approx .747$; **(c)** $\frac{1}{6}[\ln 1 + 6 \ln \frac{3}{2} + 4 \ln 2 + 10 \ln \frac{5}{2} + 6 \ln 3 + 14 \ln \frac{7}{2} + 4 \ln 4] \approx 7.34$;
(e) $\frac{1}{6}\left[e + \frac{8}{3}e^{3/2} + e^2 + \frac{8}{5}e^{5/2} + \frac{2e^3}{3} + \frac{8}{7}e^{7/2} + \frac{e^4}{4}\right] \approx 17.739$. Page 273

Chapter 8

Section 8.1 **2. (b)** $\sqrt{3}$; **(d)** 13; **(g)** $\sqrt{14}$. **3. (b)** $S = xy + 2xz + 2yz$; **(d)** $C = dx + ey$. Page 279

Section 8.3 **3.** $x = 0, y = 300$ or $x = 300, y = 0$. **5.** Max at $(3, -1)$, min at $(0, 0)$. **7.** Max at $(1, 1)$, min at $(-1, -1)$. Page 289

Section 8.4 **1. (b)** Circular disk with center $(1, 1)$ and radius 1, $\{(x, y) | (x - 1)^2 + (y - 1)^2 < 1\}$; **(d)** $\{(x, y) | (x + 1)^2 + (y - 2)^2 < \frac{1}{4}\}$. **2. (b)** $L = \frac{3}{5}$; **(d)** 0; **(f)** 2. **3. (b)** $f(x, y) \to \frac{m}{1 + m^2}$ as $(x, y) \to (0, 0)$ along the path $y = mx$. If $m = 0, f(x, y) \to 0$, if $m = 1, f(x, y) \to \frac{1}{2}$. **7. (a)** All points in the set are interior points; **(c)** The points on the circle $x^2 + y^2 = 1$ are boundary points, those outside the circle are interior points; **(g)** Boundary points: points on the rectangle with vertices $(0, 0)$, $(1, 0)$, $(0, 1)$, and $(1, 1)$. Page 295

Section 8.5 **1. (b)** $\partial f/\partial x = 3x^2y^3$, $\partial f/\partial y = 3x^3y^2$; **(d)** $\partial f/\partial x = -(x - y)^{-2}$; $\partial f/\partial y = (x - y)^{-2}$; **(g)** $\partial f/\partial x = -ye^{-xy}$, $\partial f/\partial y = -xe^{-xy}$; **(j)** $\frac{\partial f}{\partial x} = \frac{2x^3 - y^2}{2x^2y}$, $\frac{\partial f}{\partial y} = \frac{y^2 - x^3}{2xy^2}$
2. (b) $f_{xx} = 6x/y^3$, $f_{xy} = f_{yx} = -9x^2/y^4$, $f_{yy} = 12x^3/y^5$; **(e)** $f_{xx} = 2$, $f_{xy} = f_{yx} = 0$, $f_{yy} = -2$; **(h)** $f_{xx} = 2e^x(y^2 + 2e^x)$, $f_{xy} = f_{yx} = 4ye^x$, $f_{yy} = 12y^2 + 4e^x$. **3. (a)** ux^2y^3; **(c)** $6uxy^3$; **(e)** $2uy^3$. **4. (a)** $x + y + z = -1$; **(c)** $z - 2y = -1$. Page 301

Section 8.6 **1. (b)** Saddle point at $(0, 0)$; **(d)** Saddle point at $(0, 0)$, local max at $(-1, -1)$; **(f)** No real critical points; **(h)** Local min at $(-2, -4)$; **(j)** Local max at $(\frac{2}{3}, 0)$, local min at $(0, \frac{2}{3})$, saddle points at $(0, 0)$ and $(\frac{2}{3}, \frac{2}{3})$. **2. (b)** Local min at $(0, 0)$. **4.** $7 \times 7 \times 7$. **6.** $\sqrt{2}$. Page 309

Section 8.7 **1.** Max of 2 at $(0, 0)$, min of 1 at every point on the boundary. **4.** Min at $(-\frac{1}{2}, 0)$, max at $(\frac{1}{2}, \pm\sqrt{3}/2)$. **5.** Max is 7, min is -1. Page 315

Section 8.8 **2.** Max of 1 at $(1, 1)$. **4. (a)** $(-2, -2, 2)$. **6.** Local max at $(1, 2, 1)$, local min at $(0, 0, 3)$. **8.** $x = 40$, $y = 1600$, $z = 400$. Page 321

Section 8.9 **1. (a)** $2(2x + y) + 2t(x + 2y)$; **(d)** $(ye^{xy})e^t + (xe^{xy})(2t - 1)$; **(f)** $(3x^2y^2 - 2xy^3 + y)(1 - 2t) + (2x^3y - 3x^2y^2 + x)(2t + 1)$. **2. (a)** $\partial z/\partial x = 3e^{u+v}$, $\partial z/\partial y = 2e^{u+v}$; **(d)** $\frac{\partial z}{\partial x} = \frac{u^3 - 2u^2v - uv^2 + 2v^3}{(u^2 + v^2)^2}$, $\frac{\partial z}{\partial y} = \frac{-2u^3 + vu^2 + 2uv^2 - v^3}{(u^2 + v^2)^2}$.
3. (a) $n = 1$. Page 325

Chapter 9

Section 9.1 **2. (a)** $y = e^{3x}/3 + x^2/2 + C$; **(c)** $y = (e^{2x} + e^{-2x})/4 + x^2 + Ax + B$; **(f)** $y = e^x - x^4/24 + x^3/2 + Ax^2 + Bx + C$. **3. (a)** $y = e^x - x^3/3 + 3$; **(c)** $y = e^x - e^{-x} - 2x - 1$; **(f)** $y = x^3/3 + e^x + (\frac{2}{3} - e)x - 1$. Page 332

Page 336 **Section 9.2** **1.** **(b)** $y = \sqrt{x^2 - 2x + C}$; **(c)** $y = \sqrt{C(x-1)/(x+1)}$; **(f)** $y = \ln|e^x + C|$; **(h)** $y = Ce^{x^2/2} - 2$; **(j)** $y = Cx/e^{2x}$. **2.** **(b)**, **(c)**, **(f)**, **(h)**, **(j)** No singular solutions.

Page 342 **Section 9.3** **1.** $P(t) = \dfrac{10 \cdot 4^{t/3}}{4^{t/3} + 1}$. **2.** $P(t) = \dfrac{40 \cdot 3^{t/10}}{3^{t/10} + 3}$. **5.** 22.5° (approx.). **7.** $W = \dfrac{40[(1.5)^{t/5} - 1]}{2(1.5)^{t/5} - 1}$.

Page 347 **Section 9.4** **1.** **(b)** $y = e^{-x^2}(x^3 + C)$; **(d)** $y = x^2 + Cx^{-3}$; **(g)** $y = e^{3x}(2x + Cx^{-1})$; **(i)** $y = Ce^{x^2} - \frac{1}{2}$. **2.** **(a)** $m = 100$.

Appendix

Page 354 **Section A.1** **1.** **(a)** $\{x|x$ is a letter in the word "Mississippi"$\} = \{M, I, S, P\}$; **(c)** $\{x|x$ is a consonant in the English language$\} = \{b, c, d, f, g, h, j, k, l, m, n, p, q, r, s, t, v, w, x, y, z\}$. **2.** **(a)** $\{2, 5, 8\}$; **(d)** $\{1, 2, 5, 7, 8\}$; **(f)** $\{2, 8\}$.

Page 359 **Section A.5** **1.** **(a)** 3; **(b)** $2\sqrt{6} \approx 4.89898$; **(d)** -3.30193 (approx.); **(f)** -4.43105 (approx.). **2.** **(a)** $1/3^9$; **(c)** x^3/y; **(e)** x^3/y; **(f)** $x^{5/2}y^{4/3}z^{5/2}$; **(g)** $x + 2x^2 + x^3$.

Page 361 **Section A.6** **1.** **(b)** Domain $= \{x|x \neq 1\}$, $f(2) = 3$; **(d)** Domain $= \{x|x \neq -\frac{1}{2}\}$, $f(1) = 0$, $f(2) = \frac{1}{5}$; **(e)** Domain $= \{x|x \neq \pm 2\}$, $f(1) = -\frac{4}{3}$. **2.** **(a)** $3x^2 + 3x - 2$; **(c)** $3x^5 - 3x^4 + 11x^3 + 9x^2 - 25x + 25$; **(e)** $x^2 - 6x + 25 - \dfrac{101}{x+4}$; **(f)** $x^2 - 4x + 8 + \dfrac{-17x + 11}{x^2 + x - 2}$. **3.** **(b)** $x + 2 + 1/x$. **4.** **(b)** 0; **(d)** $4x^2 + 3y$.

Page 363 **Section A.7** **1.** **(b)** No; **(d)** Yes. **2.** **(a)** $x = -\frac{8}{7}$; **(b)** $x = -\frac{1}{2}$; **(e)** Any number x is a solution.

Page 364 **Section A.8** **1.** **(b)** $x = -6$; **(c)** $x = -\frac{9}{2}$; **(e)** $x = -\frac{7}{2}$; **(g)** Any number x.

Page 365 **Section A.9** **1.** **(b)** $x \geqslant -6$; **(d)** $x < \frac{2}{3}$; **(g)** No solution; **(h)** Any real number x.

Page 367 **Section A.10** **1.** **(a)** $-1 < x < 5$; **(d)** $-6 < x < 1$. **2.** **(a)** $x < -3$ or $x > -1$; **(c)** $x \neq \frac{3}{2}$.

Page 371 **Section A.11** **1.** **(a)** $x = 0, y = z = 1$; **(c)** $x = 3 - p, y = p, z = 3$; **(f)** $x = -p/4, y = -13p/8, z = 7p/8, w = p$. (The parametric solution can be simplified by letting $p = 8q$. Then $x = -2q, y = -13q, z = 7q, w = 8q$.]

Page 372 **Section A.12** **1.** **(c)** $\frac{44}{85} - \frac{23}{85}i$; **(e)** $\frac{7}{25} - \frac{24}{25}i$.

Page 374 **Section A.13** **1.** **(b)** $-3, -4$; **(d)** No rational solutions; **(f)** $-4, 7$; **(h)** $\frac{1}{7}, -7$. **2.** **(b)** $\pm\frac{5}{2}$; **(e)** $-2, 4$; **(g)** $(-3 \pm \sqrt{11})/2$.

Page 376 **Section A.14** **1.** **(b)** $-1 \pm i\sqrt{2}$; **(d)** $-1, \frac{3}{5}$; **(f)** $(2 \pm i\sqrt{23})/3$. **2.** **(b)** $(3 \pm i\sqrt{23})/4$; **(d)** $1, \frac{2}{3}$; **(f)** $(-1 \pm \sqrt{17})/4$; **(h)** $(-3 \pm i\sqrt{15})/4$.

INDEX

LIMITS

L1 $\lim_{x \to a} [u \pm v] = \lim_{x \to a} u \pm \lim_{x \to a} v$

L2 $\lim_{x \to a} uv = \lim_{x \to a} u \cdot \lim_{x \to a} v$

L3 $\lim_{x \to a} \frac{u}{v} = \frac{\lim_{x \to a} u}{\lim_{x \to a} v}$ provided $\lim_{x \to a} v \neq 0$

L4 $\lim_{x \to a} \sqrt[n]{u} = \sqrt[n]{\lim_{x \to a} u}$ provided $\lim_{x \to a} u > 0$ when n is even

L5 $\lim_{x \to a} C = C$

L6 $\lim_{x \to a} x = a$

L7 $\lim_{x \to a} Cu = C \lim_{x \to a} u$

ORDER PROPERTIES

O1 If $a < b$ and $b < c$ then $a < c$. *(Transitive Property)*

O2 If a and b are real numbers then exactly one of the following is true: $a < B$ or $a = b$ or $a > b$. *(Law of Trichotomy)*

O3 If $a < b$ then $a + c < b + c$.

O4 If $a < b$ and $c > 0$ then $ac < bc$.

O5 If $a < b$ and $c < 0$ then $ac > bc$.